Medicinal Plants and Their Bioactives in Human Diseases

Kamil Kuca • Jiri Patocka
Vikas Kumar • Rajni Dhalaria
Editors

Medicinal Plants and Their Bioactives in Human Diseases

Springer

Editors
Kamil Kuca
Department of Chemistry
University of Hradec Kralove
Hradec Kralove, Czech Republic

Jiri Patocka
Faculty of Science
University of Hradec Kralove
Hradec Kralove, Czech Republic

Vikas Kumar
Department of Biotechnology
Chandigarh University
Mohali, Punjab, India

Rajni Dhalaria
Department of Biosciences
Minerva PG College of Arts, Science
and Commerce
Kangra, Himachal Pradesh, India

ISBN 978-3-032-01355-2 ISBN 978-3-032-01356-9 (eBook)
https://doi.org/10.1007/978-3-032-01356-9

This Springer imprint is published by the registered company Springer Nature Switzerland AG
The registered company address is: Gewerbestrasse 11, 6330 Cham, Switzerland

Preface

Chronic and neurodegenerative diseases constitute a major health problem for the worldwide population today. With the increasing aging population and lifestyle changes, the prevalence of these diseases is rising sharply. Despite advances in modern medicine and medical technologies, there remains an urgent need for safer, cost-effective, and more holistic therapeutic alternatives. Medicinal plants, with their rich diversity of bioactive compounds, offer immense potential in addressing this need. The use of medicinal plants could provide safer, more holistic approaches to managing chronic and neurodegenerative diseases, reducing side effects and improving patient outcomes. This book, *Applications of Medicinal Plants and Their Bioactive Compounds Against Chronic and Neurodegenerative Human Diseases*, aims to investigate and strengthen present-day scientific knowledge and emerging research on the applications of medicinal plants and their bioactive compounds. This book originated from a collective vision of the editors and contributing scholars to compile evidence-based insights regarding the role of phytochemicals in the management of various human diseases, including hepatic and cardiovascular conditions, infectious diseases, cancers, respiratory disorders, and neurodegenerative disorders. This book offers a comprehensive and multidisciplinary opinion on herbal remedies by integrating contributions from renowned experts in pharmacology, ethnobotany, neurology, and clinical medicine.

The chapters are organized thematically to facilitate a structured exploration—starting from the historical relevance and regulatory concerns in the use of medicinal plants, to disease-specific applications and mechanistic insights. Special attention has been given to safety, dosage, synergism with conventional drugs, and translational challenges, making this book both academically enriching and practically relevant.

This book is intended for a broad readership, including researchers, healthcare professionals, industry scientists, and postgraduate students. We hope that it not only serves as a reference for current knowledge but also inspires future innovations in phytotherapy, drug discovery, and integrative medicine. We express our deep gratitude to all contributors for their scholarly efforts and to Springer for supporting this important initiative. We also thank our peer reviewers for their valuable insights that enhanced the quality of this work.

We are confident that this book will be a valuable resource in advancing both scientific understanding and practical applications of medicinal plants in combating chronic and neurodegenerative diseases.

Hradec Kralove, Czech Republic Kamil Kuca
Hradec Kralove, Czech Republic Jiri Patocka
Mohali, Punjab, India Vikas Kumar
Kangra, Himachal Pradesh, India Rajni Dhalaria

Contents

Historical Overview and Current Trends in Use of Medicinal Plants

1

Nsovo N. Mayimele, David R. Katerere, and Manikandan Gurusamy

Abstract

Medicinal plants, which have been used for centuries to cure and manage various illnesses, are used by approximately 80% of the world's population. This widespread use is attributed to the diverse purposes these plants serve, including healing diseases, food preparation, and spiritual practices. Furthermore, local knowledge and cultural traditions guide the use of medicinal plants, often requiring specialized training or education. Medicinal plants have been shown to successfully treat a wide range of ailments, including memory loss, cardiovascular disease, digestive disorders, arthritis, reproductive problems, skin illnesses, neurological disorders, hypertension, cancer, and diabetes. Recent developments in pharmacological and phytochemical research have enabled novel approaches to treating diseases that were previously incurable, building on the therapeutic potential of these plants. The antibacterial, anti-inflammatory, and antioxidant qualities of medicinal plants are attributed to phytochemicals such as flavonoids, alkaloids, and terpenoids. Plants such as *Bacopa* spp., *Catharanthus roseus, Ginkgo biloba*, *Gloriosa superba*, and *Rhodiola litvinovii* have been found to have significant pharmacological properties. Modern bioassays and fractionation techniques have enabled the isolation of therapeutically important compounds. This chapter provides a comprehensive historical overview and examines current trends in the use of medicinal plants, highlighting their potential to produce potent medications and pharmacologically active constituents.

N. N. Mayimele · D. R. Katerere · M. Gurusamy (✉)
Pharmaceutical and Biotech Advancement in Africa, Department of Pharmaceutical Sciences, Tshwane University of Technology, Arcadia Campus, Pretoria, South Africa
e-mail: mayimelenn@tut.ac.za; katereredr@tut.ac.za; gurusamyg@tut.ac.za

K. Kuca et al. (eds.), *Medicinal Plants and Their Bioactives in Human Diseases*,
https://doi.org/10.1007/978-3-032-01356-9_1

Keywords

Plant-based therapeutics · Pharmacognosy · Ethnomedicine · *Catharanthus roseus* bioactive compounds · Disease management

1 Introduction

The concept of "medicinal plants" was formally introduced in 1967, initially focusing on hallucinogenic species (Heinrich et al. 2012). Medicinal plants are defined as plant species whose various parts are utilized directly or incorporated into preparations for therapeutic purposes, aiming to treat or alleviate specific ailments or diseases (Heinrich et al. 2012). Remarkably, out of approximately 391,000 vascular plant species globally, around 35,000 are believed to possess therapeutic properties. Indeed, traditional herbal remedies play a vital role in meeting the medicinal needs of approximately 80% of the global population, who rely on these natural therapies for their health and well-being (Kew et al. 2016; García de Alba García et al. 2012). Moreover, plants have been a rich source of bioactive compounds, contributing to around 25% of the pharmacologically active substances utilized in modern medicines, as highlighted by Kew et al. (2016) and García de Alba García et al. (2012).

Medicinal plants are a treasure trove of phytochemical compounds, which play a pivotal role in the development of therapeutics. For thousands of years, people have harnessed the medicinal properties of plants, often through traditional knowledge and practices. Scientific research has now conclusively established that virtually every part of a plant possesses medicinal properties, offering a vast and largely untapped reservoir of potential therapeutic agents (Wink 2010). With a long history of use those dates back to the prehistoric era, plants have been an essential source of medicine for thousands of years. Plants have been a universal source of medicinal chemicals, utilized by people from various cultures and continents for their therapeutic properties. These substances, which plants make to protect themselves from the flu, have been used for their antimicrobial and therapeutic properties. Plants have therefore helped create several strong and effective medicines. Plants have therapeutic significance because of the chemicals they carry, which may have a pharmacological effect (Packiyalakshmi et al. 2017). Despite their potential, traditional medicines have received limited scientific attention. Infectious diseases remain the leading cause of death worldwide, accounting for approximately one-half of all fatalities globally. Furthermore, 50–75% of hospital deaths are attributed to infectious diseases (Aiyegoro et al. 2008). In response, scientists from diverse fields are re-examining plants for their antimicrobial potential, seeking alternative solutions to existing drugs. The vast array of chemical constituents found in plants presents a promising source of biological properties, offering new avenues for combating infectious diseases (Aiyegoro et al. 2008; Manikandan et al. 2017).

In recent decades, medicinal plants, the cornerstone of traditional medicine, have been the subject of extensive pharmacological research. The realization that these plants constitute a treasure trove of novel chemicals with therapeutic potential has sparked a surge of interest, driving innovation in drug development. Furthermore, medications made from plants are useful models for creating safer and more effective medications, opening the door for advancements in medical research (Manikandan et al. 2019). This chapter provides a comprehensive historical overview and examines current trends in the use of medicinal plants, highlighting their potential to produce potent medications and pharmacologically active constituents.

2 Historical Overview of Medicinal Plants

Ancient times, people have used plants healing qualities (Ojha et al. 2020). Because many contemporary medications are derived from medicinal plants, human medicine and plant biodiversity are thus closely related (Kiguba et al. 2016). The use of therapeutic plants has grown gradually throughout history, especially in emerging nations (Tabuti et al. 2012). The availability, cultural acceptance, perceived safety, and effectiveness of medicinal herbs are major factors contributing to this development (Chinsembu and Hedimbi 2010). According to current estimates, between 50% and 95% of people worldwide utilize medicinal plants, demonstrating their broad use and importance in healthcare (Wanjohi et al. 2020). Due to the lack of contemporary medical facilities, the effectiveness of traditional treatments, and the impact of cultural priorities and individual preferences, traditional medicines offer a cost-effective and easily accessible substitute for primary healthcare in developing nations (Svarstad and Dhillion 2000). Due to the increased interest in complementary and alternative therapies, the use of traditional herbal medicines is expanding quickly in both developed and developing countries. One prominent example is China, where 30–50% of all drug use is attributed to traditional herbal remedies. Herbal remedies are also the first line of treatment for about 60% of children with high malarial fever in several African nations, including Nigeria, Ghana, Zambia, and Mali. Furthermore, because of things like cultural acceptability, cost, and restricted access to contemporary medications, an astounding 80% of Ethiopians still use traditional remedies (Kassaye et al. 2006). This pervasive use of traditional medicines highlights how important they are in meeting the medical needs of many populations around the world.

The earliest recorded evidence of plant-based medicine dates to around 2500 BC, with Emperor Shen Nung's "Pen T'Sao" cataloguing 365 medicinal plants, many of which, such as ginseng, ephedra, and cinnamon bark, remain in use today. Similarly, ancient Indian texts, the Vedas, documented the medicinal properties of native plants, including clove, pepper, and nutmeg, while the Ebers Papyrus (c. 1550 BC) listed 700 plant species, including aloe, garlic, and willow, used for therapeutic purposes, demonstrating the enduring importance of plant-based medicine across cultures and centuries (Wiart 2007).

The significant usage of plants for ritual and medical purposes is documented in ancient texts and traditions. For example, the use of aromatic herbs such as myrtle and incense in various ceremonies that precede therapies is described in the Bible and the Talmud (Dimitrova 1999). In a similar vein, 63 plant species from Minoan,

Mycenaean and Egyptian Assyrian pharmacotherapy are mentioned in Homer's epic poems, The Iliad and The Odyssey (c. 800 BC). It is interesting to note that some of these plants were given names derived from mythological figures. For example, Elecampane (*Inula helenium* L. Asteraceae) went by the name of Elena, a key player in the Trojan War (Sovilj and Spasojević 2001). Named after the Greek word "artemis," which means "healthy," the genus *Artemisia* is thought to maintain health and restore strength. Botany and medicine both benefited greatly from the work of ancient Greek philosophers and intellectuals. Orpheus described garlic and the aromatic hellebore, while Herodotus (c. 500 BC) wrote about the castor oil plant. Pythagoras mentioned mustard, cabbage, and sea onions. Hippocrates (459–370 BC) categorized 300 therapeutic herbs according to their physiological activity in his books. For instance, garlic was used to fight intestinal parasites, and Wormwood and common centaury were used to treat fever. Fancy hellebore and haselwort were employed as emetics, while other plants like opium, henbane, deadly nightshade, and mandrake were utilized as narcotics (Gorunovic and Lukic 2001).

Theophrastus (371–287 BC) laid the foundation for botanical science with his seminal works "De Causis Plantarium" (Plant Etiology) and "De Historia Plantarium" (Plant History). These texts classified over 500 medicinal plants known during that era, including cinnamon, iris rhizome and pomegranate (Pelagic 1970). Theophrastus' descriptions of plant toxicology emphasized the importance of gradual dose escalation to facilitate human adaptation. His trailblazing efforts in classifying and describing medicinal plants earned him the esteemed title of "father of botany" (Nikolovski 1995). Furthermore, the renowned medical writer Celsus, in his seminal work "De re medica" (25 BC–50 AD), meticulously documented approximately 250 medicinal plants, including notable examples such as aloe, henbane, flax, poppy, pepper and cinnamon, demonstrating the significance of botanical knowledge in ancient medical practices (Tucakov 1948).

"The father of pharmacognosy," Dioscorides, penned the seminal treatise "De Materia Medica" in 77 AD. Around 657 of 944 drugs described in this classic text are plant-based. In addition to providing information on their native names and growth habitats, Dioscorides described their appearance, collection, preparation, and therapeutic properties (Thorwald 1991). While mentioning plants with potent effects, like opium and deadly nightshade, he focused on those with mild effects (Milinčević 1980). Dioscorides used common household herbs, such as garlic, chamomile, and willow, to cure a variety of illnesses. Additionally, he issued a warning regarding the potential for drug fraud, citing examples of counterfeit opium and oriental pharmaceuticals (Dimitrova 1999).

The Slavic people used a variety of plants for practical, cosmetic, and medical uses around the seventh century AD. Plants, such as mint (*Mentha viridis*), iris (*Iris germanica*), basil (*Ocimum basilicum*), and rosemary (*Rosmarinus officinalis*), were used to make cosmetics. While yarrow (*Achillea millefolium*), cucumber (*Cucumis sativus*), nettle (*Urtica dioica*), and white hellebore (*Veratrum album*) were used to ward off insects including lice, fleas, moths, mosquitoes, and spiders, garlic (*Allium sativum*) was utilized as a treatment. In hunting, monk's hood (*Aconitum napellus*) was also used as a poison (Bojadzievski 1992).

The knowledge and application of medicinal herbs underwent a dramatic shift in the early nineteenth century. Modern scientific pharmacy was founded on the identification, isolation, and classification of alkaloids from plants, including quinine (1820), ipecacuanha (1817), and poppies (1806). Other active substances found in medicinal plants, such as tannins, saponins, essential oils, vitamins and hormones, were discovered as chemical techniques advanced (Dervendzi 1992). However, by the late nineteenth and early twentieth centuries, the use of medicinal plants for therapeutic purposes was seriously in danger of disappearing. Enzymatic degradation during drying processes, according to critics, limited the therapeutic qualities of plant-based medications. Consequently, isolated alkaloids and glycosides became more well-liked and replaced conventional plant-based treatments. However, it quickly became clear that plant-based medications offered a more thorough and long-lasting therapeutic impact, even though pure alkaloids had a quicker effect. As a result, early twentieth-century researchers created stabilization techniques for fresh medical plants, especially those that contained unstable medicinal ingredients. Studies have concentrated on improving the circumstances for medicinal plant production and cultivation (Kovacevic 2000). Nowadays, plant-based medications with demonstrated therapeutic benefit are recognized by pharmacopoeias around the world, including Ph Eur 6, USP XXXI, and BP 2007. There are distinct herbal pharmacopoeias in nations including Germany, Russia, and the United Kingdom. Evidence-based phytotherapy is becoming more popular, especially in Germany, where standardized plant extracts are produced and validated for safety and efficacy, even though unofficial herbal medicines are still extensively utilized (Pharmacopoeia 2007).

3 Current Trends in Medicinal Plants

3.1 Global Use of Medicinal Plants

Medicinal plants have been a foundation of traditional healing practices worldwide, offering a safe and effective therapeutic approach. The efficacy of medicinal plants is rooted in the collective experience of countless healers across centuries, passed down through generations, shared among practitioners, and refined through personal experience. Despite the advent of modernity and cultural shifts, the wisdom of this natural medical paradigm remains unchanged. As a result, no single modern medical system can claim ownership of this traditional knowledge. Instead, it coexists alongside various well-established systems of treatment, including allopathic, homeopathic, ayurvedic, and Chinese medicine (Gurib-Fakim 2006).

3.2 Uses of Medicinal Plants in Different Culture

3.2.1 Traditional Indian Medicine

Regarded as the "mother of all therapies," traditional Indian medicine is primarily recognized as the world's oldest healthcare system, with a rich history that dates back to the Vedic era about 5000 years ago. Sacred scriptures like the Rig-Veda and Atharva-Veda carefully document the fundamental ideas and methods of Ayurveda (Hsu and Barrett 2008). "Ayurveda" is derived from Sanskrit and literally means "knowledge of life," which reflects the system's all-encompassing approach to health and well-being. Ayurveda addresses the physical, mental, and spiritual aspects of health to improve overall welfare by skillfully fusing physiological and holistic medicine. This approach acknowledges the complex interdependence of the human body (Routh and Bhowmik 1999).

3.2.2 Greek Civilization

In the history of philosophy and science, the ancient Greek civilization was a pivotal period. Particularly in phytopharmaceuticals, Greeks made significant contributions to the pharmaceutical sciences. About 500 rudimentary medications used to treat a variety of medical disorders were mentioned by Aristotle (Chatard 1908). The first scientific medical paradigm was developed by Hippocrates, who is regarded as the founder of allopathic medicine. He postulated that many ailments were caused by disruptions in human physiology. He discovered around 400 plant-based therapeutic compounds (Sykiotis et al. 2006). In his work, Aristotle's pupil Theophrastus listed 500 unrefined substances (Scarborough 1978). Prominent physician Claudius Galen originated the idea of pharmaceutical formulation and created Galenicals, a process for extracting vegetable medications. He made substantial contributions to the science of pharmacy by writing almost 300 volumes on plants (Buerki and Higby 2006).

3.2.3 Traditional Chinese Medicine

With a history that dates back thousands of years, Traditional Chinese Medicine (TCM) is one of the most unique and ancient medical systems. This successful medical system is highly significant in medical history and is well-known throughout the world for its evidence-based practice (Patwardhan et al. 2005). TCM has stayed mostly independent of other influences and is distinctive in its beliefs, treatments, and treatment approaches. It is thought that Fu His (2953 BC) was the first person to use TCM. Later, the development of TCM was greatly aided by the emperors Shen Nung and Hong Ti. There were several treatments for different medical issues in the Chinese pharmacopoeia, Pen Tsao. The Shen Nong Ben Cao Jin (22–250 AD) is considered the pinnacle of textual Chinese medicine (Kopp et al. 2003). The seminal work Zhu Bing Yuan Ji Lun, written by Cao Yuan Fang (550–630), detailed the causes and signs of several illnesses. Chinese medical students continue to use this book as a standard reference (Kopp et al. 2003). With the publication of Waitai Miyao, which detailed over 600 prescriptions, Wang Tao (702–772) made a substantial contribution to TCM. The extensive pharmacopoeia

Ben Ca Gang Mu was written by Li Shizen, a well-known Chinese physician and naturalist, and it was published in 1596. This text, which still serves as a useful reference and manual for research and teaching in China and other communities, includes 1894 prescriptions. Despite being passed down through the generations, traditional Chinese medicine was not formally incorporated under academic educational training until the 1950s (Xu and Yang 2009). TCM still has a significant impact on healthcare today, both in China and around the world.

3.2.4 Arabic Medicine

Following the fall of the Roman Empire, Arab scholars made monumental strides in science and medicine. They meticulously translated ancient Greek and Roman texts, laying the groundwork for groundbreaking discoveries. Arab physicians pioneered the concept of holistic healthcare, emphasizing diet control, exercise, and medication. They also established the foundation of modern pharmacy practices, including the creation of drug stores, defining the roles of physicians and pharmacists, and developing drug extraction and formulation techniques (Azaizeh et al. 2006). The Islamic faith played a pivotal role in shaping the science of medicine in Arabia. The Quran and Sunnah provided guidance on maintaining a hygienic lifestyle, which became the cornerstone of Tibb al-Nabi (Qureshi and Ghufran 2007). Eminent Muslim scientists, such as Ali Ibn Rabban Al Tabri (782–855 AD), made significant contributions to the field. His seminal work, Firdous Al Hikmat, featured a dedicated section on drugs and poisons (Chatard 1908). The legendary Abu Ali Al Hussan Ibn Sina (Avicenna, 980–1037 AD) revolutionized medicine with his iconic book, Canon, which catalogued over 1000 drugs and became a staple textbook in European medical schools (Said 1980). His magnum opus, Kitab Ash-Shifa, remains a revered scientific encyclopedia. Arab scholars also delved into the toxic properties of plants, with Abu Musa Jabir ben Hayyan authoring a comprehensive treatise on plant poisons and antidotes, The Book on Poisons and Antidotes (Saad et al. 2006).

3.3 Integration with Conventional Medicine

Integrative medicine represents a transformative shift in healthcare, extending beyond the mere combination of complementary and alternative medicine (CAM) and conventional treatment. As defined by Maizes et al. (2002), integrative medicine is a holistic approach that prioritizes the patient-physician relationship and seamlessly integrates the most effective elements of complementary and alternative therapies with evidence-based conventional medicine, fostering a comprehensive and patient-centered healing experience. This definition emphasizes the value of a patient-centered approach, which promotes cooperation between patients and healthcare professionals. A similar description is provided by Berndtson (1998), who emphasizes the critical role that evidence-based practice plays in integrated medicine. Integrative medicine is characterized as a patient-centered clinical approach that synergistically blends the best of conventional and alternative therapies. This approach prioritizes treatment options that are evidenced to be safe, effective, and offer a reasonable expectation of benefit to the patient, based on a thorough review of available evidence. While recognizing the importance of complementary therapies, this focus on evidence-based practice guarantees that integrative medicine is based on scientific excellence.

Snyderman and Weil (2002) assert that integrated medicine is a revolutionary trend that has drawn interest from prestigious academic health facilities due to consumer demand. According to Maizes, the fundamental goal of integrative medicine is health and healing, with a focus on the critical role that the patient–physician relationship plays. Snyderman and Weil further clarify that integrated medicine encompasses a holistic approach to preventative healthcare, covering all areas of lifestyle, including food, exercise, stress management, and emotional wellbeing. This patient-centered attitude encourages doctors to take a holistic approach, acknowledging patients as complex individuals with physical, emotional, and spiritual aspects, while also empowering patients to take an active role in their healthcare. In the end, integrated medicine reinterprets the role of the doctor, moving beyond only providing therapeutic help to become mentors, role models and mentors who inspire and support patients on their journey towards optimal wellness.

3.4 Global Regulatory Frameworks for Medicinal Plants

A global phenomenon is the extensive use of medicinal plants for primary healthcare. An astounding 80% of people in underdeveloped nations depend on medicinal plants for their medical requirements, according to the World Health Organization (WHO) (WHO 2013). The usage of medicinal plants is becoming more and more common in more affluent countries; thus, this trend is not just happening in underdeveloped ones. According to some estimates, up to 30% of people in nations like the US and the UK use medicinal plants (Cordell 2015). The dependence on medicinal plants for primary healthcare is astounding on a worldwide basis. The primary source of healthcare for almost 5.1 billion people globally is medicinal plants or medicinal plant products. This emphasizes the importance of medicinal plants in addressing the healthcare needs of a substantial proportion of the global population.

3.4.1 Challenges with Safety and Quality of Medicinal Plant Products

The escalating demand for medicinal plant products has sparked concerns regarding safety and quality. In response, the World Health Organization (WHO) has established rigorous standards to ensure the authenticity, purity, and efficacy of these products. These standards encompass quality assessment, good manufacturing practices (GMP), and good agriculture and collecting practices (GACP), providing a framework for the cultivation, harvesting, processing, and manufacturing of medicinal plant products (Sahoo and Manchikanti 2013).

3.4.2 Challenges in Quality Control

Medicinal plants are complicated goods, and the way they are made and processed can have an impact on safety, effectiveness, and quality control (WHO 2003, 2007a, b). In the absence of efficient quality management and regulation, patient outcomes may suffer from:

- Plant material that has been incorrectly identified or sourced; contaminated with pesticides, heavy metals, or microbial toxins.
- The use of synthetic medicines or other substances for adulteration.
- Inadequate manufacturing quality.
- Misleading or inaccurate information.

3.4.3 Safety Difficulties

Many reported concerns are caused by quality faults, however, there are additional dangers as well (Picking 2024):

- Medicinal plants direct adverse effects; reactions resulting from overdosing, improper dosage, or drug combinations.
- Idiosyncratic, allergic, or hypersensitive reactions.
- Mid-term and long-term harmful consequences, including organ damage or genotoxicity.
- Interactions between medicinal plants and drugs.

3.4.4 Regulatory Categories

To navigate the complex regulatory landscape for medicinal plants, it's essential to identify the primary categories that medicinal plant products fall into and the existing regulatory frameworks that govern them (Picking 2024). Medicinal plants are subject to varying classification systems and regulatory frameworks across different countries. The regulation of these plants is intricate, influenced by factors such as their intended use and geographical location. *Curcuma longa* is a prominent example that exemplifies this intricacy. Turmeric can be marketed and sold as a medicinal plant product, even though it is most known and used as a culinary spice in curry recipes around the world. In this context, turmeric may be linked to a variety of claims with differing degrees of clinical trial review, ranging from functional foods to over the counter (OTC) medications (Picking 2024).

3.4.4.1 Claim Categories

Many nations have set rules and regulations controlling health-related claims made about medicinal plant items in addition to regulatory categories. These statements may be found on product labels, websites, and pamphlets that go with them. A medical plant product's regulatory category is frequently determined by the kind and scope of claims made about it. For example, nutritional claims about the abundance of vital elements in turmeric products may be made. High polyphenols in turmeric which are known to have anti-inflammatory qualities may also support claims that it is a functional food (Picking 2024). Additionally, certain medical claims may be made in support of turmeric products based on traditional use that has been documented, screenings using biological assays, and clinical studies showing how well a standardized formulation works to treat, cure, or prevent a specific disease, or restoring or modifying physiological function.

3.4.4.2 Pharmacopoeias

The term "pharmacopoeia" originates from the Greek word "pharmakopoiia," literally translating to "the art of preparing medicines." Essentially, pharmacopoeias function as authoritative, legally binding guides that furnish detailed descriptions of medicinal substances employed within national healthcare systems, as defined by the World Health Organization (WHO 2013). Pharmacopoeias that include medicinal plant listings serve a significant role in bridging the gap in the commercial availability of medicinal plants across nations with and without a documented history of traditional medicine. The Indian Ocean Islands and sub-Saharan Africa provide a stark illustration of this inequality. Even though there are over 60,000 plant species in these areas, only 83 of the 1100 most popular commercial medicinal plants worldwide come from these locations (Blumenthal 2011; Eloff et al. 2010). This is mostly because traditional medicine relies on oral traditions and lacks official pharmacopoeias. China and India, on the other hand, have become the two biggest exporters of raw medicinal plant material worldwide due to their long histories of traditional medicine and the usage of medicinal plants (Schippmann et al. 2002). This draws attention to the importance of pharmacopoeias in promoting the development and commercialization of medicinal plants.

3.4.4.3 Medicinal Plant Monographs

A monograph is a comprehensive, unique description of a medication that can be published alone or as part of pharmacopoeias. Medicinal substances or mixtures of compounds that make health or treatment claims are described in detail in monographs. These descriptions are drawn from a variety of sources, such as documented traditional use, scientific study, and other pertinent data. Guidelines for writing monographs on medicinal plants have been published by the World Health Organization (WHO) as part of its Traditional Medicine Program (Zhang 1998). In addition, the WHO has produced four volumes of its own monographs (WHO 1999), which are an invaluable resource for researchers, regulatory bodies, and medical professionals.

3.4.4.4 Essential Drug Lists

The WHO created a list of essential medications in 1977 with the goal of providing for the majority of a nation's population's fundamental medical needs. To maintain its applicability and efficacy, this list—known as the WHO Model List of Essential Medicines (EML)—is updated every 2 years. Senna is the only medical plant product now included in the EML (WHO 2021). Nonetheless, a growing number of nations have created their own lists of necessary medications, many of which include items from medicinal plants. China, India, Mali, and Thailand are among the 22 member states (16%) that have added medicinal plant items to their national essential medication lists, according to a WHO survey (2005). This pattern demonstrates the increasing awareness of the value of medicinal plants in addressing the world's healthcare concerns.

3.4.4.5 Traditional Knowledge and Traditional Use

According to the World Intellectual Property Organization (WIPO), "traditional knowledge" is a dynamic and ever-evolving collection of practices, knowledge, and skills that have been created, maintained, and passed down through the generations within a community. These practices are frequently firmly ingrained in the community's social, cultural, or spiritual identity. Traditional knowledge, which includes the collective knowledge and experience of indigenous and local communities regarding the use, conservation, and management of plant-based resources, is crucial to the international regulation of medicinal plant products because it emphasizes the need for fair recognition, protection, and benefit-sharing agreements. Regulators need to be able to tell the difference between items that have a long history of traditional usage and those that don't. This distinction is essential for maintaining the safe and efficient use of medicinal plant products while honoring the customs and cultural heritage of people all around the world (Picking 2024).

3.4.4.6 Pharmacovigilance

"The science and activities relating to the detection, assessment, understanding, and prevention of adverse drug reactions" is how the World Health Organization (WHO) defines pharmacovigilance (Picking 2024). Pharmacovigilance's main goal is to keep an eye on medications' safety, including that of medicinal plant products, after they are made accessible to patients and healthcare professionals. The goal of this continuous assessment is to find adverse events that might have gone undetected in clinical trials. Pharmacovigilance is essential for guaranteeing the safety of medicinal plant products, even though it was first created for pharmaceutical pharmaceuticals (Shaw et al. 2012). China set up a thorough system in 1989 to track adverse drug responses for both Western medications and goods used in traditional Chinese medicine (TCM). This system has produced important insights under the Chinese State Food and Drug Administration's supervision, showing that 10–15% of reported adverse drug responses are related to medications used in traditional Chinese medicine (Zhang et al. 2012).

There are currently several global pharmacovigilance reporting systems in use, such as the WHO's Program for International Drug Monitoring (WHO PIDM). The VigiBase database, which is carefully maintained by the Uppsala Monitoring Center in Sweden, is the foundation of the WHO PIDM. Since its launch in 1968, VigiBase, the world's largest database of its sort, has accumulated an amazing collection of more than 30 million reports of suspected adverse drug reactions from WHO PIDM member countries. VigiBase collects priceless adverse effects data from over 100 nations worldwide, including significant contributors like the United States and China, through its vast network of Poison Information Centers. New reports are frequently added to the database, guaranteeing prompt access to vital safety data. Nearly six million reports of adverse drug interactions were available in VigiBase as of December 2010, highlighting its significance as a vital resource for pharmacovigilance and drug safety monitoring. Interestingly, medicinal plant items were mentioned in 12,679 (0.2%) of these studies (Shaw et al. 2012; Zhang et al. 2012). This important resource enables the global monitoring of medicinal plant safety and facilitates the identification of potential risks associated with their use.

3.5 Market Trends and Economic Impact

Growing public interest in natural health products has led to a notable increase in demand for medicinal herbs during the past 20 years. In Europe, sales of medicinal plants increased by a staggering 25% between 2008 and 2013, whereas in China and India, sales doubled during the same time frame (Sofowora et al. 2013). According to FBI (2022) projections, the global market for medical plant products is expected to increase at an astonishing rate of 15–25% per year, reaching a stunning $400 billion by 2025. Data from the past shows how quickly this market has grown. The global market for medicinal plants, including raw materials and products made from them, was projected to be worth $65 billion in 2000 (WHO 2003). The global market for herbal medicines has reached a staggering $165.66 billion, with the World Health Organization (WHO) predicting a remarkable surge to $5 trillion by 2050 for herbal products and traditional medicine (Booker et al. 2012). This exponential growth in demand has spawned a lucrative industry, driving the relentless collection of plants for herbal medicine and plant-based products. However, the darker side of this booming market is the alarming overexploitation of numerous medicinal plant species, pushing them to the brink of extinction (Volenzo and Odiyo 2020). The unsustainable harvesting of these plants not only threatens their survival but also jeopardizes the long-term viability of the herbal medicine industry.

4 Therapeutic Applications of Medicinal Plants

People from a wide range of cultural backgrounds have used plants' therapeutic and medicinal qualities to promote health and wellness for thousands of years. With over 25% of all prescribed medications coming from plant-based sources, medicinal plants continue to be an essential part of contemporary medicine (Alonso Castro et al. 2012). This ongoing dependence on medicinal plants emphasizes how important they are as a wealth of pharmacologically active substances. In addition to being phytomedicines used to cure a variety of illnesses, medicinal plants also supply crucial chemical structures needed to create new antimicrobial drugs (Manikandan and Ramasubbu 2020). Traditional medicinal plants contain a variety of chemicals that are used to treat chronic and infectious ailments. In areas where the usage of traditional plants is still common, a significant amount of information has been gathered about the plants employed to treat various afflictions (Manikandan et al. 2017).

Medicinal plants have formed the backbone of traditional medicine since pre-Hispanic times, representing a vital component of humanity's collective knowledge (Heinrich et al. 1998). The therapeutic value of these plants resides in their secondary metabolites, which exhibit pharmacological properties, rendering them an indispensable source of bioactive compounds for diverse medicinal applications, including antitumor, antiviral, antiepileptic, antibiotic, anti-inflammatory, and antinociceptive treatments (Wang et al. 2019). Notable examples of medicinal plants include *Papaver somniferum*, which yields potent pain-relieving alkaloids such as codeine and morphine. Artemisia annua and *Cinchona ledgeriana* provide lifesaving antimalarial compounds. *Taxus brevifolia* and *Catharanthus roseus* have yielded anticancer treatments like paclitaxel and vinblastine/vincristine. *Dioscorea mexicana* serves as the source of diosgenin, a potent steroid utilized in oral contraceptives (Pirintsos et al. 2022; Howes et al. 2020; Shen et al. 2018) (Fig. 1.1; Table 1.1).

Thymus vulgaris holds a revered place in traditional medicine, with widespread cultivation in many countries, particularly in rural areas where herbal remedies are relied upon to treat a myriad of ailments. The plant is specifically valued for its efficacy in addressing inflammation-related conditions, including rheumatism, muscle swelling, insect bites, and pain management. In addition, modern pharmacological studies on thyme's essential oil have confirmed its therapeutic potential, revealing a range of bioactive compounds that exhibit anti-inflammatory, antioxidant, antibacterial, and antifungal properties, thereby validating its traditional uses and underscoring its potential for integration into contemporary medicine (Hosseinzadeh et al. 2015).

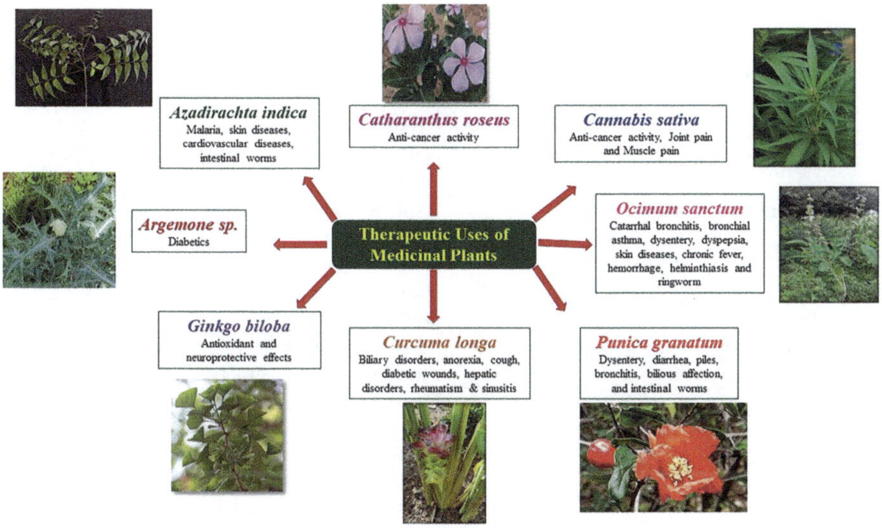

Fig. 1.1 Therapeutic uses of medicinal plants

Table 1.1 Therapeutic applications of medicinal plants

S. No.	Name of the plant	Medicinal uses	References
1.	*Aconitum heterophyllum*	Diarrhea, diabetes, cough, rheumatism, dyspepsia, stomachache, fever, digestive, and nervous system disorders	Kumar et al. (2019a, b)
2.	*Artocarpus heterophyllus*	Hypertension, anemia, asthma, dermatosis, and diarrhea	Moke et al. (2017)
3.	*Asparagus adscendes*	Diarrhea, gonorrhea, dysuria, weakness, lean and thinness, erectile dysfunction, piles, cough and dysentery	Das et al. (2023)
4.	*Azadirachta indica*	Malaria, skin diseases, cardiovascular diseases, intestinal worms	Satyanarayana et al. (2015)
5.	*Capsicum frutescens*	Diabetes, bronchitis, burning feet, arthritis, stomachache, diarrhea, and dysentery	Chattopadhyay and Bandyopadhyay (2005)
6.	*Centella asiatica*	Inflammation, diabetes, dysentery, hysteron-epilepsy, leprosy, rheumatism, dizziness, hemorrhoids, diarrhea, tuberculosis, skin lesions, and asthma	Abas et al. (2016)
7.	*Curcuma longa*	Biliary disorders, anorexia, cough, diabetic wounds, hepatic disorders, rheumatism, and sinusitis	Kumar and Sakhya (2013)
8.	*Eriobotrya japonica*	Headache, low back pain, phlegm, asthma, dysmenorrhea, cough, chronic bronchitis, diabetes, and skin diseases	Lin et al. (2018)
9.	*Gymnema sylvestre*	Anti-periodic, stomachic, laxative, diuretic, cough remedy, snakebite, biliousness, parageusia, and furunculosis	Mall et al. (2009)
10.	*Harungana madagascariensis*	Gastrointestinal disorders, cardiovascular disorders, malaria, leprosy, anemia, tuberculosis, fever, angina, nephrosis, dysentery, bleeding, piles syphilis, gonorrhea, and parasitic skin diseases	Tom et al. (2018)

(continued)

Table 1.1 (continued)

S. No.	Name of the plant	Medicinal uses	References
11.	*Musa paradisiaca*	Diarrhea, dysentery, intestinal lesions in ulcerative colitis, diabetes, sprue, uremia, nephritis, gout, hypertension, wound healing, inflammation, headache, and cardiac diseases	Ahmed et al. (2021)
12.	*Ocimum sanctum*	Catarrhal bronchitis, bronchial asthma, dysentery, dyspepsia, skin diseases, chronic fever, hemorrhage, helminthiasis, and ringworm	Hannan et al. (2014)
13.	*Punica granatum*	Dysentery, diarrhea, piles, bronchitis, bilious affection, and intestinal worms	Sadeghipour et al. (2014)
14.	*Terminalia arjuna*	Diabetes, cirrhosis, anemia, cardiovascular, and viral diseases	Thompson et al. (2014)
15.	*Withania somnifera*	Diabetes, cough and cold, insomnia, leprosy, bronchitis, asthma, tumors, tubercular glands, arthritis, nervous disorders	Sarangi et al. (2013)
16.	*Zingiber officinale*	Stomach ache, nausea, diarrhea, vomiting, joint and muscle pain, inflammatory diseases	Wattanathorn et al. (2011)
17.	*Allium sativum*	Antioxidant activity	Nwozo et al. (2023), Manikandan and Noel (2019), Ayyanar and Ignacimuthu (2009), Dhanaraj et al. (2025), and Paulraj et al. (2021)
18.	*Amaranthus hybridus*		
19.	*Anarcadium occidentale*		
20.	*Vernonia amygdalina*		
21.	*Leucas aspera*		
22.	*Ocimum basilicum*		
23.	*Sesamum prostratum*		
24.	*Momordica cymbalaria*		

(continued)

Table 1.1 (continued)

S. No.	Name of the plant	Medicinal uses	References
25.	*Acalypha hispida*	Anti-inflammatory activity	Cabral et al. (2016), Barbosa et al. (2016), Rosas et al. (2015), Mostofa et al. (2017), Zahidin et al. (2017), Siraj et al. (2016), Rios et al. (2017), Ortiz et al. (2016), Kyei et al. (2016), Cruz et al. (2016), Funk et al. (2016), Zahidin et al. (2017), Meyanungsang et al. (2015), and Shubashini and Uma (2010)
26.	*Acalypha indica*		
27.	*Achyranthes aspera*		
28.	*Alternanthera brasiliana*		
29.	*Borreria verticillata*		
30.	*Himatanthus drasticus*		
31.	*Heliotropium indicum*		
32.	*Matricaria chamomilla*		
33.	*Mimosa tenuiflora*		
34.	*Momordica charantia*		
35.	*Myracrodruon urundeuva*		
36.	*Phyllanthus niruri*		
37.	*Ruellia asperula*		
38.	*Schinus terebinthifolius*		
39.	*Solanum paniculatum*		
40.	*Spondias mombin*		
41.	*Spondias tuberose*		
42.	*Spondias purpurea*		
43.	*Zingiber officinale*		
44.	*Acalypha indica*		
45.	*Erythrina variegate*		
46.	*Phyllanthus maderaspatensis*		
47.	*Cardiospermum halicacabum*		
48.	*Glycyrrhiza glabra*	Antimicrobial activity	Sultan et al. (2014), Álvarez-Martínez et al. (2021), Manikandan and Ramasubbu (2020), Packiyalakshmi et al. (2017), Manikandan et al. (2021), Manikandan et al. (2022), Manikandan and Ramasubbu (2023), Zahidin et al. (2017), Manikandan et al. (2020a, b), Ramasubbu et al. (2020), Mina and Mina (2017), and Surendran et al. (2020)
49.	*Astragalus membranaceus*		
50.	*Echinacea* spp.		
51.	*Nigella sativa*		
52.	*Allium sativum*		
53.	*Zingiber officinale*		
54.	*Camellia sinensis*		
55.	*Myrtus communis*		
56.	*Mentha piperita*		
57.	*Thymus daenensis*		
58.	*Pimenta dioica*		
59.	*Cuminum cyminum*		
60.	*Memecylon heyneanum*		
61.	*Millettia peguensis*		
62.	*Acalypha indica*		
63.	*Cymbopogon citrates*		
64.	*Eugenia rottleriana*		
65.	*Garcinia imberti*		
66.	*Melaleuca citrina*		
67.	*Garcinia travancorica*		
68.	*Orthosiphon aristatus*		
69.	*Syzygium densiflorum*		

(continued)

Table 1.1 (continued)

S. No.	Name of the plant	Medicinal uses	References
70.	*Withania somnifera*	Anticancer activity	Umadevi et al. (2013), Heijden et al. (2004), Pushkarev et al. (2008), Liu et al. (2014), Padhye et al. (2009), Nabandith et al. (2004), Manikandan and Ramasubbu (2017), and Zahidin et al. (2017)
71.	*Catharanthus roseus*		
72.	*Cannabis sativa*		
73.	*Garcinia indica*		
74.	*Syzygium aromaticum*		
75.	*Garcinia* sp.		
76.	*Acalypha indica*		
77.	*Argemone Mexicana*	Antidiabetic activity	Yin et al. (2008), Yin et al. (2012), Murugesan et al. (2024), Manikandan et al. (2019), Raju et al. (2015), Shaheen et al. (2014), Kichu et al. (2015), Morvin Yabesh et al. (2014), and Divya et al. (2024)
78.	*Berberis aquifolium*		
79.	*B. aristata*		
80.	*B. vulgaris*		
81.	*Coptis chinensis*		
82.	*Eschscholzia californica*		
83.	*Hydrastis Canadensis*		
84.	*Tinospora cordifolia*		
85.	*Xanthorhiza simplicissima*		
86.	*Phellodendron amurense*		
87.	*Mimosa pudica*		
88.	*Tinospora cordifolia*		
89.	*Chamaecostus cuspidatus*		
90.	*Pterocarpus marsupium*		
91.	*Momordica cymbalaria*		
92.	*Pisonia grandis*		
93.	*Parthenium integrifolium*		
94.	*Cadaba fruticosa*		
95.	*Acalypha indica*		
96.	*Albizia lebbeck*		
97.	*Vitex negundo*		
98.	*Senna auriculata*		
99.	*Gymnema sylvestre*		
100.	*Pedalium murex*		
101.	*Stevia rebaudiana*		

4.1 Antioxidant Properties

Antioxidants, which are abundant in plants, are essential for protecting against illnesses brought on by free radicals (Ofoedu et al. 2022). Plants mostly produce these antioxidant chemicals as secondary metabolites, which are vital for human health but not necessary for the survival of the plant. The non-nutritive parts of plants known as phytochemicals, or "plant chemicals," have been shown to provide a variety of health advantages and the ability to fend against disease (Coulibaly et al. 2014). It has been demonstrated that these bioactive substances provide protection against a number of chronic illnesses, underscoring the significance of plant-based diets and the potential of phytochemicals in illness prevention and management.

The antioxidant qualities of several plants such as *Vernonia amygdalina, Amaranthus hybridus, Anarcadium occidentale*, and *Allium sativum*, were investigated by Nwozo et al. (2023). The findings showed that these plant extracts had strong DPPH radical scavenging activity, high FRAP values, and high total phenol and flavonoid content, all of which were indicative of considerable antioxidant activities. These results imply that these plant extracts' antioxidant properties, which offer defenses against oxidative stress in live cells, may be a factor in their possible health advantages.

4.2 Anti-inflammatory Properties

Anti-inflammatory medications play a vital role in mitigating tissue damage and enhancing patient comfort by modulating the pathophysiological process of inflammation. Given the vast array of plant species available for investigation, a collaborative, multidisciplinary approach is essential for the discovery of novel, naturally occurring anti-inflammatory compounds. By integrating knowledge from diverse disciplines, researchers can identify and develop new, potent molecules, ultimately leading to the creation of innovative anti-inflammatory therapies that address unmet medical needs and improve patient outcomes (Nunes et al. 2020).

Numerous secondary metabolites have been found in species belonging to the Anacardiaceae family, such as *Schinus terebinthifolius, Myracrodruon urundeuva, Spondias mombin, Spondias tuberosa*, and *Spondias purpurea*, according to phytochemical studies. The main bioactive substances that give these species their anti-inflammatory qualities are phenols, triterpenes, flavonoids, and cinnamic acid, which are abundant in these species, according to research (Cabral et al. 2016; Barbosa et al. 2016). The anti-inflammatory characteristics of the Euphorbiaceae family, which includes species like *Acalypha indica, Acalypha hispida*, and *Phyllanthus niruri*, are attributed to their abundance of phenolic compounds, saponins, tannins, and triterpenes (Mostofa et al. 2017; Zahidin et al. 2017; Siraj et al. 2016). Furthermore, studies have found anti-inflammatory compounds in a wide range of species from different families, such as *Achyranthes aspera*, Alternanthera *brasiliana, Borreria verticillata, Himatanthus drasticus, Heliotropium indicum, Matricaria chamomilla, Momordica charantia, Mimosa tenuiflora, Ruellia asperula, Solanum paniculatum*, and *Zingiber officinale* (Ortiz et al. 2016; Rios et al. 2017).

4.3 Antimicrobial Properties

Angiosperm plants are important sources of phytochemical compounds that play a major role in human healthcare maintenance (Gurusamy et al. 2022). Known for their antiviral, antibacterial, and antifungal qualities, medicinal plants are used in both conventional and contemporary treatments all over the world. In drug discovery, phytochemical substances originating from these plants are essential for creating analogs of pharmaceutical products (Gurusamy et al. 2024; Surendran et al. 2020). The powerful effects of several herbs used in traditional medicine against a variety of pathogenic bacteria have been well recognized (Cook 1996). Our knowledge of the functional characteristics of medicinal plants is being quickly expanded by ongoing research projects, which are offering a foundation for the creation of around 25–50% of already marketed medications (Conly and Johnston 2005). These plants' antibacterial qualities vary, but the majority have synergistic actions (Ventola 2015), which lessen the negative side effects of synthetic medications (Schcolnik-Cabrera 2023). Furthermore, it has been discovered that several plants function as quorum quenchers (Sun and Shahrajabian 2023), underscoring their potential for medicinal use.

Medicinal plants have a long history of effectively treating microbial diseases. These include *Allium sativum*, *Astragalus membranaceus*, *Camellia sinensis*, *Echinacea* spp., *Glycyrrhiza glabra*, *Nigella sativa*, and *Zingiber officinale*. If properly studied and used, these plants have strong immune-boosting qualities and show promise in the fight against bacterial infections (Sultan et al. 2014). Essential oils extracted from plants high in polyphenols and terpenes have a major antibacterial mode of action that involves interfering with the construction and function of bacterial cell membranes. For example, it has been demonstrated that essential oils derived from plants including *Myrtus communis, Mentha piperita, Thymus daenensis, Pimenta dioica*, and *Cuminum cyminum* can interfere with the bacterial cells' plasma membrane, which in turn inhibits the bacteria (Álvarez-Martínez et al. 2021). This demonstrates the potential of these plants as organic antimicrobials, which calls for more research into their possible uses in medicine.

4.4 Anticancer Properties

For decades, herbal treatments have been a crucial component of medical care in developing countries, and their use is still common today. The natural antibacterial qualities of plants are one of the main factors influencing their usage in medicine. As studies have advanced, the focus has switched to investigating the possible properties and uses of terrestrial plant extracts in the creation of drugs based on nanomaterials, with an emphasis on the treatment of cancer. This is noteworthy since many plant species have already been used to prevent and treat cancer, and studies have found that certain species have anticancer capabilities. Medicinal herbs have long been used to cure cancer; *Curcuma longa, Zingiber officinale, Withania somnifera, Catharanthus roseus, Glycyrrhiza glabra*, and *Cannabis sativa* are a few examples of plants that have shown promise as therapeutic effects and have been employed in traditional herbal medicine in underdeveloped nations (Ochwang'i et al. 2014; Costa-Lotufo et al. 2005).

4.5 Antidiabetics Properties

Natural products, particularly those derived from plants, play a vital role in the development of novel medications, serving as a rich source of promising lead candidates (Sharifi-Rad et al. 2018; Salehi et al. 2018). The accessibility, affordability, and favorable safety profile of plant-based preparations make them an indispensable component of many therapeutic regimens, especially in rural areas where access to conventional medications may be limited (Arya et al. 2011). Furthermore, plants offer a vast array of bioactive compounds, characterized by potent pharmacological properties and a lack of adverse effects, rendering them an attractive option for the development of innovative treatments (Mishra et al. 2018; Abdolshahi et al. 2018). *Lagerstroemia speciosa, Aloe vera, Cinnamomum* sp., *Momordica charantia, Theobroma cacao, Capparis* sp., *Trigonella foenum-graecum, Coffea arabica, Allium sativum, Gymnema* sp., *Psidium guajava, Salvia officinalis, Urtica* sp., *Glycine max, Camellia sinensis, Juglans regia, Curcuma longa* and *Ilex paraguariensis* are among the many medicinal plants that have been used traditionally to treat diabetes and its related comorbidities, according to Rios et al. (2017). With an emphasis on compounds of high therapeutic interest, such as amorfrutins, berberine, fukugetin, gurmarin, gymnemic acids, honokiol, palmatine, phlorizin, and trigonelline which have demonstrated promise in reducing complications related to diabetes, the authors also explored the mechanisms of action of natural products as antidiabetic agents.

5 Phytochemical and Pharmacological Aspects of Medicinal Plants

Plants have phytochemicals in their roots, stems, leaves, flowers, fruits, and seeds, among other components (Costa et al. 1999). Notably, the outer layers of plant tissues are where a large number of phytochemicals, particularly color molecules, tend to congregate. However, the amounts of phytochemicals can fluctuate greatly throughout plant types and are also affected by growth conditions, cooking methods, and processing (King and Young 1999). Although there are supplemental forms of phytochemicals, the available data indicates that they could not offer the same health advantages as phytochemicals derived from food. The phytochemical composition of medicinal plants represents a broad repository of bioactive or functional compounds central to their therapeutic potential. These compounds include steroids, terpenoids, flavonoids, sulfides, polyphenols, lignans, alkaloids, saponins, polysaccharides, proteins, and enzymes, each of which contributes uniquely to medicinal properties (Banwo et al. 2021; Yu et al. 2021; Samtiya et al. 2021).

Table 1.2 Bioactive compounds found in medicinal plants and its biological properties

S. No.	Name of the plant	Bioactive compounds	Biological properties	References
1.	*Ajuga turkestanica*	Ecdysterone, cyasterone, ajugalactone, ajugasterone, α-ecdysone, turkesterone	Anabolic activity and stimulate gallbladder contraction to promote bile flow	Abdukadirov et al. (2004, 2005)
2.	*Allium cepa*	Flavonoids	Antibacterial	Cushnie and Lamb (2005)
3.	*Allium sativum*	Allicin, sulfur compounds, and flavonoids	Anticardiovascular, antioxidant and antiplatelet properties	Chang et al. (2016)
4.	*Andrographis paniculata*	Andrographolides and flavonoids	Immunomodulatory activity	Chang et al. (2016)
5.	*Aspergillus* spp.	Polyphenols	Antifungal	Hammer et al. (2003)
6.	*Bacopa monnieri*	Bacosides	Neuroprotective	Wu et al. (2008)
7.	*Berberis* spp	Alkaloids	Antibacterial	Vuddanda et al. (2010)
8.	*Camellia sinensis*	Catechins	Antimicrobial, anti-inflammatory, and antioxidant	Cushnie and Lamb (2005)
9.	*Camptotheca acuminate*	Alkaloids	Anticancer	Li and Weng (2017)
10.	*Catharanthus roseus*	Alkaloids	Anticancer	Ferlay et al. (2013)
11.	*Crataegus* spp.	Flavonoids, procyanidins, and triterpenoids	Anti-inflammatory, antioxidant, and anticardiovascular	Li et al. (2017)
12.	*Curcuma longa*	Curcumin	Antioxidant, neuroprotective, anti-inflammatory, and anti-apoptotic activity	Ahmed et al. (2017)
13.	*Ferula pallida*	Sesquiterpene phenylpropanoid derivatives, pallidones, and sesquiterpene chromone derivatives	Antioxidant, skin wounds treatment	Kim et al. (2003)
14.	*Ginkgo biloba*	Flavonoids	Antioxidant and neuroprotective effects	Cui et al. (2018)
15.	*Glycyrrhiza glabra*	Guaiacylpropan-1-ol, cresol, Guaiacol, Guaiacylethane,	Antiviral, anti-inflammatory, antioxidant	Mamadalieva et al. (2013)
16.	*Hypericum perforatum essential oil,*	α-pinene, Caryophyllene oxide, β-caryophyllene, Spathulenol	Anti-inflammatory, antimicrobial, diarrhea	Kholmatov and Kosimov (1994)

(continued)

Table 1.2 (continued)

S. No.	Name of the plant	Bioactive compounds	Biological properties	References
17.	*Hyssopus seravschanicus*	Carvacrol, Myrcene, Sabinene, p-cymene, β-pinene, 1,8-cineole, Pinocamphone,	Anti-inflammatory, rheumatism, gastrointestinal	Dzhumaev et al. (1990)
18.	*Melaleuca alternifolia*	Polyphenols	Antifungal	Hammer et al. (2003)
19.	*Origanum tyttanthum*	β-D-glucoside, Carvacrol, Coumarins, Oleanolic, Thymol, Rosmarinic acids, Ursolic, Thymol, Naringenin, eryodictiol	Anti-tuberculosis and antimicrobial activity	Baser et al. (2002) and Bigović et al. (2011)
20.	*Origanum vulgare*	Polyphenols	Antifungal	Hammer et al. (2003)
21.	*Panax ginseng*	Ginsenosides	Immunomodulatory activity	Kumar et al. (2019a, b)
22.	*Punica granatum*	Flavonoids	Antiviral	Wang et al. (2009)
23.	*Silene wallichiana*	Viticosterone E, viticosterone E22-O-benzoate, 20-hydroxyecdysone-22-benzoate, 20-hydroxyecdysone 2-deoxy-20-hydroxyecdysone 2-deoxyecdysone,	Anabolic adaptogenic, tonic, antioxidant, antimicrobial	Madziga et al. (2010)
24.	*Taxus brevifolia*	Terpenoids	Anticancer	Wani et al. (1971)
25.	*Ungernia victoris*	Galanthamine, Haemanthidine, Hippeastrine, Lycorine, Narwedine, Nortazettine, Tazettine, Ungerine and Ungeridine,	Bronchitis, ulcers, poliomyelitis, neurological diseases	Berkov et al. (2009)

Among these, phenolic compounds, flavonoids, and alkaloids are particularly abundant in traditional medicinal plants and are often associated with their antioxidant, anti-inflammatory, and antimicrobial activities (Roy et al. 2022; Nwozo et al. 2023) (Table 1.2; Fig. 1.2).

5.1 Alkaloids

Higher plants, especially dicotyledonous species, are the primary sources of alkaloids, with only occasional occurrences in lower plant forms. Within these plants, alkaloids can be distributed throughout the entire organism or concentrated in

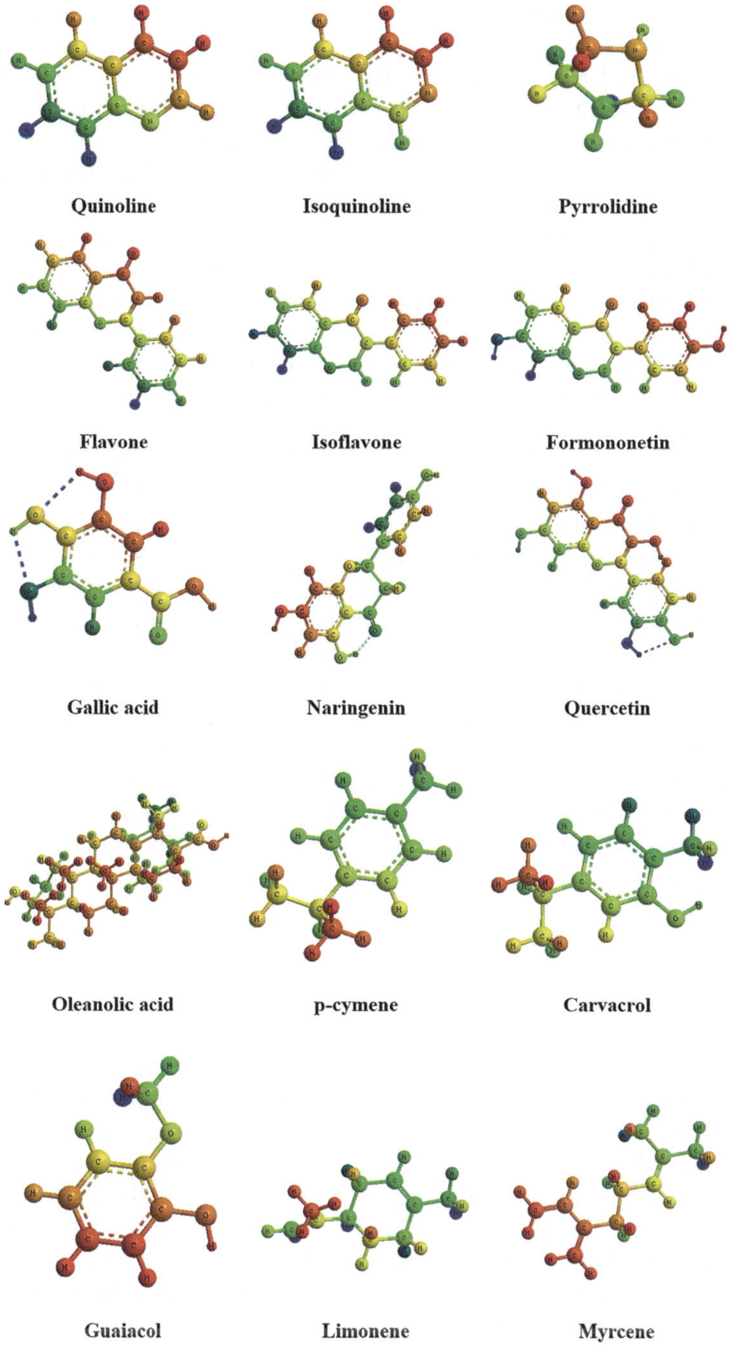

Fig. 1.2 Chemical structure of some bioactive compounds from medicinal plants

specific organs, such as roots, leaves, or seeds. Despite the unclear functions of some alkaloids, research has revealed their vital importance in plant defense and survival strategies. Specifically, alkaloids act as potent deterrents against herbivores, provide protection against pathogens, and inhibit the growth of competing plant species, ultimately contributing to the plant's ecological success (Cushnie et al. 2014). The possible health advantages of plant alkaloids have been the main subject of much research and use in traditional medicine throughout different locations. Pharmacognosy and pharmaceutical applications have placed a lot of emphasis on this field of study. For example, a thorough analysis of 320 plant species from 33 families in Uzbekistan showed a notable alkaloid content (Aripov 1995). In addition, through this study, more than 540 new compounds were identified from 292 species, including isoquinoline, quinoline, indole, amaryllis, steroid, quinazolone, sulfur-containing, pyrrolizidine, diterpenes and pyrrolidine (Aripov 1995).

The plant species *Ungernia victoris*, which is native to the Fergana Valley in western Tien Shan, is unique to the Gissar and Alay mountains. Numerous alkaloids have been found in this plant's leaves and bulbs, according to research (Berkov et al. 2009). Additionally, galacturonic anhydride was successfully extracted and identified from the plant's leaves in a study conducted by Malikova and Rakhmanberdyeva (2013). Furthermore, it has been discovered that a variety of aporphine alkaloids, such as lysicamine, O-methylmoschatoline, lanuginosine, liriodenine, and proaporphine (Nmethylcrotsparine, remerine), are present in the tulip tree (Ziyaev et al. 1973). Several studies have reported the isolation and identification of various alkaloids from plants. For instance, Tulyaganov and Kozimova (2005) identified dehydroschoberine, nitraraine, nitraramine, schoberine, sibiridine, peganol and vasicinone, in *Nitraria schoberi*. *Convolvulus subhirsutus* was found to contain convolamine, convolvine, convolidine, phyllalbine and other alkaloids (Gapparov et al. 2007). Similarly, *Convolvulus pseudocanthabrica* was reported to have convolvine, convolvidine, convolamine, convolicine, and phyllalbine (Gapparov and Aripova 2011).

5.2 Flavonoids

The plant world is home to flavonoids, a broad class of secondary metabolites with varying phenolic structures (Xia et al. 2013). There is a lot of interest in the possible therapeutic uses of flavonoids because many of them have been shown to have anticancer effects (Kim et al. 2003). The Fabaceae family is notably abundant in flavonoids, with multiple species exhibiting an impressive range of these substances. For example, a variety of flavonoids, such as flavones, isoflavones, formononetin, chrysoeriol, apigenin, luteolin, thermopsoside, and cynaroside, have been identified in *Thermopsis alterniflora* (Satimov et al. 2003). Similarly, flavonoids such as apigenin, luteolin, quercetin, and cinaroside have been found in *Vicia subvillosa* (Yuldashev et al. 2007). *Ammothamnus lehmannii* has been discovered to contain a diverse array of flavonoids, including ammothamnidin, cynaroside, lehmannin, isoquercetin, luteolin, and quercetin (Bakirov et al. 1987). Flavonoids are abundant in the genus *Scutellaria* (Malikov and Yuldashev 2002). It has been shown that different species in this genus contain a wide variety of flavonoids. Baicalein, chrysin, and 5,7-dihydroxy-2′-methoxyflavone, for instance, are found in *S. cordifrons* and *S. phyllostachya* (Siddikov et al. 2006). Scutellarin, chrysin, apigenin, and cynaroside are found in other species, including *S. ramosissima* and *S. immaculata* (Mamadalieva et al. 2011). Furthermore, it has been discovered that *Scutellaria ocellata* and *S. nepetoides* contain flavonoids such as oroxylin A, wogonin, and apigenin (Yuldashev and Karimov 2001), whereas *S. phyllostachya* has a variety of glycosides, such as wogoniside and chrysin 7-O-β-D-methyl glucuronide (Siddikov et al. 2007).

5.3 Phenolic Compounds

Numerous pharmacological effects, such as anti-cancer, anti-inflammatory, cardiovascular, osteoporosis, diabetic mellitus, and neuroprotective qualities, have been demonstrated for phenolic compounds, a class of significant secondary metabolites present in medicinal plants (Li et al. 2013; Daglia 2012). Notably, fruits and a variety of drinks, including tea, coffee, and red wine, which are consumed in large numbers around the world, are the main sources of these advantageous plant chemicals in the human diet. The relevance of a balanced, phytochemical-rich diet is highlighted by the potential benefits of including these phenolic-rich foods and beverages in one's diet for the prevention and treatment of a few chronic illnesses. The phenolic components of *Euphorbia ferganensis*, a medicinally significant plant, were examined by Abdulladzhanova et al. (2001). They found that the plant's air-dried raw tissue contains approximately 9.1% polyphenols. Quercitrin, gallic acid, kaempferol-3-O-glucoside, quercetin, 1-O-galloyl-4,6-trihydroxy-3,4,31-trimethoxydiphenoyl-β-D-glucose, 3-O-geraniin, 1,2,3-tri-O-galloyl-β-D-glucose, and 2,3-di-O-galloyl-β-D-glucose were the nine chemicals that made up the polyphenols. A rich source of phenolic compounds, *Bergenia hissarica* also contains aloe-emodin, aloe-emodin8-O-β-glucoside, physeion, emodin 1-O-β-D-glucopyranoside and chrysophanein (Yuldashev et al. 1993). A variety of bioactive substances, including ptyrosol, salidroside, gallic acid, and epigallocatechin-3-O-gallate, have been discovered to be present in other native plants with therapeutic value, such as *Rhodiola litvinovii* (Melikuziev et al. 2013). Similarly, gallic acid, ellagic acid, and quercetin-3-O-glucoside have been discovered to be present in *Geranium saxatile*, indicating the potential medicinal use of these plants (Siddikov et al. 2011).

5.4 Glycosides

A sugar moiety, usually D-glucose, L-rhamnose, or L-fructose, and an aglycone, which can be any of a variety of natural products, such as flavonoids, terpenes, or other bioactive compounds, make up the diverse class of natural compounds known as glycosides (Zahmanov et al. 2015). Numerous glycosides from different plant species have been found and isolated by researchers, demonstrating the great chemical variety of these substances. For example, Takeda et al. (1999) reported that a variety of bitter secoiridoid glycosides, such as sweroside, gentiopicroside, and olivierosides, were identified in Gentiana olivieri. These results underline the significance of further phytochemical studies on glycosides derived from plants, which could result in the identification of new bioactive substances with possible medicinal uses. A novel phenolic glucoside was also extracted from *Origanum tyttanthum* by them (Takeda et al. 2008). According to other research, *Astragalus unifoliolatus* has triterpene glycosides, including cyclounifolioside A and oleanolic acid (Kucherbaev et al. 2002), whereas *Ferula varia* contains sesquiterpene lactone glucosides (Kurimoto et al. 2012). Glycosides are versatile compounds with a wide range of applications, including their use as purgatives, cardiotonics, antidepressants, coloring agents, diuretics, flavoring agents, and antifungal, antidiabetic, antipyretic, stomachic, anthelmintic, and antirheumatic agents. The potent sources of glycosides are often found in various parts of plants that have been utilized for their medicinal properties (Bhattacharya and Chakraborty 2024).

5.5 Terpenoids

Monoterpenoids are the main constituents of essential oils, which are known for their aromatic qualities. Terpenoids are a broad class of naturally occurring molecules that are present in many different types of plants. Angelica species, *Ferula gummosa*, cypress, hyssop, frankincense, juniper, myrtle, peppermint, pine, rose, and spruce are among the many plant species that contain these compounds in high concentrations. Essential oils, especially those derived from plant leaves, can include complex combinations of monoterpene hydrocarbons, including alpha and beta pinene, Δ3-carene, limonene, and myrcene. On the other hand, more specific monoterpenoids are frequently found in essential oils from flowers and seeds. It's interesting to note that different plant species can display enantiomer diversity in optically active compounds. For instance, *Pinus palustris* contains (+)-α-pinene, although (−)-β-pinene is also present in *Pinus caribaea* and *P. pinaster*. Similarly, S-(+)-linalool is found in coriander, whereas R-(−)-linalool is present in *Cinnamomum camphora* (Singh 2007).

Terpenoids have exhibited a remarkably broad spectrum of antimicrobial activity, effectively inhibiting the growth of a wide range of pathogenic bacteria, encompassing both Gram-positive and Gram-negative strains. The lipophilic nature of terpenoids enables them to readily penetrate the bacterial cell wall and membrane, leading to a disruption of membrane integrity and potential. This disruption triggers a cascade of cellular events, including the leakage of essential cellular contents, denaturation of cytoplasmic proteins, and inactivation of critical cellular enzymes. Ultimately, these events culminate in the death of the bacterial cell, underscoring the potent antimicrobial properties of terpenoids (Raut and Karuppayil 2014). Recent research on terpenoids, especially those present in medicinal plants, has greatly advanced our knowledge of their wide range of biological activity and important medical applications. Numerous medicinal benefits, such as anticancer, antimalarial, anti-inflammatory, antimicrobial, and antidiabetic actions, have been discovered for terpenoids. They have also been demonstrated to reduce blood sugar, prevent and treat cardiovascular illnesses, and encourage transdermal absorption. Additionally, several terpenoids have neuroprotective, anti-aging, immunomodulatory, antioxidant, and insecticidal properties. Notably, several terpenoids have shown great promise as therapeutic agents by being extensively utilized in clinical practice, including paclitaxel and artemisinin (Yang et al. 2020).

Terpenoids have been shown to have strong anticancer effects, mostly by preventing the growth of tumor cells and causing necrosis and death. Reductions in oxidative stress and inflammation are frequently linked to their anticancer effects. Notably, some terpenoids, such as eugenol, thymol, and β-caryophyllene, can cause oxidative stress in cancer cells only while sparing healthy cells. Terpenoids are important because they disturb the membrane potential of cancer cells, especially in mitochondrial membranes, inhibit angiogenesis and the PI3K/Akt/NF-κB pathway, and trigger apoptosis by caspase activation. Additionally, it has been demonstrated that terpenoids, when combined with chemotherapeutic drugs, increase their effectiveness. For example, β-caryophyllene increases the anticancer action of paclitaxel by facilitating its uptake by cancer cells (Lesgards et al. 2014). Particularly, the anticancer effect of eugenol has been thoroughly investigated. At 0.5 μM, it has been shown to cause cervical cancer cells to undergo apoptosis without harming healthy cells. Furthermore, eugenol stops melanoma cell lines from growing by 50% after a day. When 125 mg/kg of eugenol was administered twice a week to a melanoma xenograft model, the tumor size decreased by 40% in comparison to control mice (Lesgards et al. 2014).

6 Success Stories and Case Studies

The use of medicinal plants continues to grow globally as part of traditional and modern healthcare systems. Success stories illustrate the integration of traditional medicinal plants into clinical practice and their global impact. One example is the development of artemisinin-based therapies for malaria, derived from the *Artemisia annua* plant. Artemisinin was discovered through traditional Chinese medicine (TCM) practices. It has revolutionized malaria treatment protocols worldwide. It is a trusted solution in the treatment of malaria. This has proven to save millions of lives. Efforts to ensure sustainable harvesting and cultivation of *Artemisia annua* have become a priority to meet global demands and support long-term availability (Aremu et al. 2024).

In sub-Saharan Africa, community-based cultivation initiatives have been established to conserve high-demand species like *Harpagophytum procumbens* (devil's claw). These programs not only preserve biodiversity but also strengthen local economies. The sustainable harvesting practices for devil's claw have set a precedent for combining conservation with economic development (El Allaoui et al.

2024). Traditional systems like Ayurveda and TCM are increasingly gaining global recognition. A prominent example is the widespread use of turmeric (*Curcuma longa*) in Ayurveda. Known for its anti-inflammatory and antioxidant properties, turmeric has undergone extensive scientific validation. This has lead to its incorporation into dietary supplements and pharmaceuticals. Curcumin, the active compound in turmeric, has been explored for its potential therapeutic effects on chronic diseases, including arthritis and cancer (Aremu et al. 2024; El Allaoui et al. 2024). Ethnobotanical studies indicate the extensive knowledge held by indigenous communities. For example, the Amazon rainforest in South America has been a good source of plant-based treatments, such as quinine extracted from Cinchona bark for malaria. This traditional remedy has informed the development of modern antimalarial drugs and indicates the importance of preserving indigenous knowledge. Advances in DNA barcoding have refined the identification and classification of medicinal plants, which enhances their pharmaceutical potential. For example, studies on *Zanthoxylum nitidum* revealed it to be a complex of four distinct species used for treating rheumatism rather than a single species. Similarly, genetic analysis of Corydalis species in East Asia, traditionally used for pain relief, uncovered that taxonomically distinct species were biologically synonymous. These findings emphasize the necessity of molecular investigations to align taxonomy with biochemical properties, improving drug development and quality control (Khoja et al. 2024; Davis and Choisy 2024).

Phylogenomic studies have further illuminated the origins and evolutionary trajectories of medicinal plants. Research on *Cannabis sativa* and *Erythroxylum coca* has provided insights into their domestication, spread, and the selection of traits for medicinal use. Ancient DNA analyses combined with radiocarbon dating have been instrumental in tracing historical usage patterns. This approach has identified plants with significant therapeutic potential, including compounds for antimalarial and cardiovascular treatments (Khoja et al. 2024). Traditional knowledge, when integrated with modern chemotaxonomy and phylogenomics, has advanced the discovery and conservation of medicinal plants. For instance, geographical mapping and ecological role analysis of *Panax* species (used for cancer and diabetes therapies) and *Dioscorea* species (a source of contraceptives) have offered critical insights. Predictive models assessing these plants' adaptability to environmental changes provide valuable data for sustainable use and biodiversity preservation (Khoja et al. 2024; Davis and Choisy 2024).

7 Challenges and Future Directions

Medicinal plants have served as a foundation for healthcare systems, yet their effective integration into modern medicine faces challenges. Overharvesting and biodiversity loss pose significant threats to the sustainability of medicinal plant populations. Species such as *Prunus africana*, valued for treating prostate disorders, and *Taxus brevifolia*, the source of the cancer drug paclitaxel, have been overexploited to critical levels. These practices endanger individual species and destabilize ecosystems. This limits the discovery of novel pharmaceutical agents. Sustainable harvesting techniques, community-based cultivation initiatives, and the establishment of gene banks are essential measures to address these challenges and preserve biodiversity (Howes et al. 2020).

Another issue is the lack of consistent global regulation and standardization in the production of herbal supplements. The absence of harmonized Good Manufacturing Practices (GMP) results in variable product quality and safety. This discrepancy undermines consumer trust and poses potential health risks. Comprehensive regulatory frameworks that include quality control parameters and standardized testing protocols are urgently needed to ensure the safe and effective use of herbal products. Funding for research into chemotaxonomy and the identification of chemical markers is equally important to improve the profiling of complex plant extracts and their therapeutic potential (Wishart et al. 2021; Sahu et al. 2021).

The integration of traditional knowledge with modern scientific approaches is another area that requires attention. Traditional healers possess insights into the medicinal properties of plants, often encompassing a holistic understanding of health that extends beyond physical ailments to spiritual and environmental dimensions. However, ethical concerns surrounding the access and sharing of this knowledge persist, particularly regarding financial benefits and intellectual property rights (Ola 2022; Che et al. 2024). Establishing clear benefit-sharing frameworks that are supported by international agreements such as the Nagoya Protocol can help address these challenges while promoting equitable partnerships between traditional knowledge holders and researchers (Ambler et al. 2021).

Research on underexplored flora offers potential for discovering new nutraceuticals and pharmaceuticals. Despite their promise, many plant species remain understudied due to limited resources and institutional support. Governments and funding organizations should prioritize research into the lesser-known species while encouraging collaborations between traditional medicine practitioners, botanists, and pharmacologists. Such efforts could also include promoting ecotourism, showcasing the medicinal plant heritage of specific regions, and providing economic incentives for local communities to conserve their natural resources.

The safety and efficacy of medicinal plants remain areas of concern, as some bioactive compounds may have toxic effects. Comprehensive studies on the potential adverse effects of natural products are necessary for their safe incorporation into modern healthcare systems. Furthermore, integrating these findings into permissive health insurance policies can enhance public access to validated traditional medicines. Policymakers must also address the complex challenge of respecting traditional knowledge while aligning it with scientific validation and modern regulatory models.

Future directions should focus on leveraging advancements in technology and science to address these challenges. For example, omics technologies, including genomics, proteomics, and metabolomics, offer tools to uncover the mechanisms of action of bioactive compounds and improve drug development. Phylogenomic studies can identify "hot spots" of biodiversity with high therapeutic potential that guide conservation efforts and drug discovery initiatives. Additionally, the World Health

Organization's strategic frameworks can serve as a guide for integrating traditional medicine into global healthcare systems, ensuring a balance between respecting traditional practices and meeting modern scientific standards. By addressing these challenges through interdisciplinary collaboration, effective policy development, and sustainable practices, medicinal plants can continue to play a vital role in healthcare innovation while preserving their cultural and ecological significance for future generations.

8 Conclusions

For thousands of years, medicinal plants have played a crucial role in human healthcare by providing a wealth of bioactive chemicals with potential for therapeutic use. Since plants are the source of about 25% of all prescribed drugs worldwide, their significance in primary healthcare cannot be disputed. Their antibacterial, anti-inflammatory, and antioxidant qualities are attributed to phytochemicals like terpenoids, alkaloids, and flavonoids. Medicinal plants have enormous potential for healthcare innovation, despite obstacles including overharvesting and worries about their efficacy and toxicity. Future paths should concentrate on interdisciplinary cooperation, sustainable practices, and efficient policy development in order to address these issues. The cultural and ecological importance of medicinal plants can be preserved while new therapies are discovered through the fusion of traditional knowledge and contemporary scientific methods.

References

Abas F, Khatib A, Perumal V, Suppaiah V, Ismail A, Hamid M, Lajis NH (2016) Metabolic alteration in obese diabetes rats upon treatment with *Centella asiatica* extract. J Ethnopharmacol 180:60–69

Abdolshahi A, Naybandi-Atashi S, Heydari-Majd M, Salehi B, Kobarfard F, Ayatollahi SA, Sharifi-Rad J (2018) Antibacterial activity of some Lamiaceae species against *Staphylococcus aureus* in yoghurt-based drink (Doogh). Cell Mol Biol 64(8):71–77

Abdukadirov IT, Khodzhaeva MA, Turakhozhaev MT, Mamatkhanov AU (2004) Carbohydrates from *Ajuga turkestanica*. Chem Nat Compd 40(1):85–86

Abdukadirov IT, Yakubova MR, Nuriddinov KR, Mamatkhanov AU, Turakhozhaev MT (2005) Ecdysterone and turkesterone in *Ajuga turkestanica* determined by HPLC. Chem Nat Compd 41(4):475–476

Abdulladzhanova NG, Mavlyanov SM, Dalimov DN (2001) Phenolic compounds of Euphorbia ferganensis B. Fedtsch. Chem Natural Comp 37(2):193–194

Ahmed T, Gilani AH, Abdollahi M (2017) Curcumin and other dietary polyphenols: potential mechanisms of neuroprotection and for the treatment of neurodegenerative diseases. Mol Neurobiol 54(6):4096–4112

Ahmed OM, Abd El-Twab SM, Al-Muzafar HM, Amin KA, Abdel Aziz SM, Abdel-Gabbar M (2021) *Musa paradisiaca* L. leaf and fruit peel hydroethanolic extracts improved the lipid profile, glycemic index and oxidative stress in nicotinamide/streptozotocin-induced diabetic rats. Vet Med Sci 7(2):500–511

Aiyegoro OA, Akinpelu DA, Afolayan AJ, Okoh AI (2008) Antibacterial activities of crude stem bark extracts of *Distemonanthus benthamianus* Baill. Afr J Biotechnol 8:356–361
Alonso Castro A, Maldonado Miranda JJ, Zarate Martinez A, del Jacobo Salcedo MR, Fernández Galicia C, Figueroa Zuñiga AL et al (2012) Medicinal plants used in the Huasteca Potosina, México. J Ethnopharmacol 143:292–298
Álvarez-Martínez FJ, Barrajón-Catalán E, Herranz-López M, Micol V (2021) Antibacterial plant compounds, extracts and essential oils: an updated review on their effects and putative mechanisms of action. Phytomedicine 90:153626
Ambler J, Dearden PK, Wilcox P, Hudson M, Tiffin N (2021) Including digital sequence data in the Nagoya protocol can promote data sharing. Trends Biotechnol 39(2):116–125
Aremu AO, Luo B, Mussarat S (2024) Medical ethnobotany. BMC Complement Med Ther 24(1):1–2
Aripov KN (1995) The chemistry of plant substances in Uzbekistan. Chem Nat Compd 31(1):2–15
Arya V, Gupta V, Ranjeet K (2011) A review on fruits having anti-diabetic potential. J Chem Pharm Res 3:204–212
Ayyanar M, Ignacimuthu S (2009) Herbal medicines for wound healing among tribal people in Southern India: ethnobotanical and scientific evidences. Int J Appl Res Nat Prod 2(3):29–42
Azaizeh H, Saad B, Khalil K, Said O (2006) The state of the art of traditional Arab herbal medicine in the Eastern region of the Mediterranean: a review. Evid Based Complement Alternat Med 3(2):229–235
Bakirov EK, Yusupova SS, Sattikulov A, Vdovin AD, Malikov VM, Yagudaev MR (1987) Flavonoids of *Ammothamnus lehmannii*. Structure of lehmannin and of ammothamnidin. Chem Nat Compd 23:429–435
Banwo K, Olojede AO, Adesulu-Dahunsi AT, Verma DK, Thakur M, Tripathy S, Utama GL (2021) Functional importance of bioactive compounds of foods with potential health benefits: a review on recent trends. Food Biosci 43:101320
Barbosa HM, Nascimento JN, Araujo TA, Duarte FS, Albuquerque UP, Vieira JR, Lira EC (2016) Acute toxicity and cytotoxicity effect of ethanolic extract of *Spondias tuberosa* Arruda bark: hematological, biochemical and histopathological evaluation. An Acad Bras Cienc 88:1993–2004
Baser KHC, Ozek T, Nuriddinov HR, Demirci AB (2002) Essential oils of two *Hypericum* species from Uzbekistan. Chem Nat Compd 38(1):54–57
Berkov S, Georgieva L, Kondakova V, Atanassov A, Viladomat F, Bastida J, Codina C (2009) Plant sources of galanthamine: phytochemical and biotechnological aspects. Biotechnol Biotechnol Equip 23(2):1170–1176
Berndtson K (1998) Integrative medicine: business risks and opportunities. Phys Exec 24(6):22–26
Bhattacharya S, Chakraborty S (2024) A review on medicinal plants and its importance from glycosides. Int J Res Appl Sci Eng Technol 12(X):715–729
Bigović D, Savikin K, Janković T, Menković N, Zdunić G, Stanojković T, Djurić Z (2011) Antiradical and cytotoxic activity of different *Helichrysum plicatum* flower extracts. Nat Prod Commun 6(6):819–822
Blumenthal M (2011) African herbal pharmacopoeia – book review. HerbalGram 90:71–72
Bojadzievski P (1992) The health services in Bitola through the centuries. Bitola: Society of science and art, 15–27
Booker A, Johnston D, Heinrich M (2012) Value chains of herbal medicines—research needs and key challenges in the context of ethnopharmacology. J Ethnopharmacol 140(3):624–633
Buerki RA, Higby GJ (2006) Dosage forms and basic preparations: history. In: Encyclopedia of pharmaceutical technology. Informa Healthcare, London, pp 948–974
Cabral B, Siqueira EM, Bitencourt MA, Lima MC, Lima AK, Ortmann CF, Zucolotto SM (2016) Phytochemical study and anti-inflammatory and antioxidant potential of *Spondias mombin* leaves. Rev Bras 26(3):304–311
Chang Q, Zuo Z, Harrison F, Chow MS (2016) Hawthorn. J Clin Pharmacol 56(6):660–670
Chatard JA (1908) Avicenna and Arabian medicine. Bull Johns Hopkins Hosp 19:S157–S160

Chattopadhyay RR, Bandyopadhyay M (2005) Effect of *Azadirachta indica* leaf extract on serum lipid profile changes in normal and streptozotocin-induced diabetic rats. Afr J Biomed Res 8(2):101–104

Che CT, George V, Ijinu TP, Pushpangadan P, Andrae-Marobela K (2024) Traditional medicine. In: *Pharmacognosy*. Academic, pp 11–28. https://doi.org/10.1016/B978-0-443-18657-8.00037-2

Chinsembu KC, Hedimbi M (2010) Ethnomedicinal plants and other natural products with anti-HIV active compounds and their putative modes of action. Int J Biotechnol Mol Biol Res 1(6):74–91

Conly JM, Johnston BL (2005) Where are all the new antibiotics? The new antibiotic paradox. Can J Infect Dis Med Microbiol 16(3):159

Cook HJ (1996) Physicians and natural history. In: *na*, pp 91–105

Cordell GA (2015) Ecopharmacognosy and the responsibilities of natural product research to sustainability. Phytochem Lett 11:332–346

Costa MA, Xia ZQ, Davin LB, Lewis NG (1999) Toward engineering the metabolic pathways of cancer-preventing lignans in cereal grains and other crops. In: Phytochemicals in human health protection, nutrition, and plant defense. Kluwer Academic/Plenum Publishers, New York, pp 67–87

Costa-Lotufo LV, Khan MTH, Ather A, Wilke DV, Jimenez PC, Pessoa C, de Moraes MO (2005) Studies of the anticancer potential of plants used in Bangladeshi folk medicine. J Ethnopharmacol 99(1):21–30

Coulibaly AY, Hashim R, Sulaiman SF, Sulaiman O, Ang LZP, Ooi KL (2014) Bioprospecting medicinal plants for antioxidant components. Asian Pac J Trop Med 7(Suppl 1):S553–S559. https://doi.org/10.1016/S1995-7645(14)60289-3

Cruz MP, Andrade CM, Silva KO, de Souza EP, Yatsuda R, Marques LM, Clemente-Napimoga JT (2016) Antinoceptive and anti-inflammatory activities of the ethanolic extract, fractions and flavones isolated from *Mimosa tenuiflora* (Willd.) Poir (Leguminosae). PLoS One 11(3):e0150839

Cui Q, He R, Tang L (2018) Flavonoids, neuroprotection, and cognition. In: Flavonoids – from biosynthesis to human health. IntechOpen, pp 123–143

Cushnie TT, Lamb AJ (2005) Antimicrobial activity of flavonoids. Int J Antimicrob Agents 26(5):343–356

Cushnie TT, Cushnie B, Lamb AJ (2014) Alkaloids: an overview of their antibacterial, antibiotic-enhancing and antivirulence activities. Int J Antimicrob Agents 44(5):377–386

Daglia M (2012) Polyphenols as antimicrobial agents. Curr Opin Biotechnol 23(2):174–181

Das KSP, Gupta RK, Raut G, Patel BP (2023) Traditional uses, phytochemistry and pharmacology of *Asparagus adscendens* Roxb.: a review. J Ayurveda Campus 4(1):39–48

Davis CC, Choisy P (2024) Medicinal plants meet modern biodiversity science. Curr Biol 34(4):R158–R173. https://doi.org/10.1016/j.cub.2023.12.038

Dervendzi V (1992) Contemporary treatment with medicinal plants. Tabernakul, Skopje, pp 5–43

Dhanaraj FI, Kalimuthu JK, Balamurugan PS, Subramani P, Katerere DR, Gurusamy M (2025) Investigating the phytochemical profile and antioxidant activity of different solvent extracts of *Sesamum prostratum* Retz. Plants 14(4):519

Dimitrova Z (1999) The history of pharmacy. St Clement of Ohrid, Sofija, pp 13–26

Divya C, Paul M, Lata C, Manikandan G, Ramasubbu R (2024) Ethnomedicinal survey of the Paliyar tribe: a case study of Kadamalaikundu. Theni District Int J Anthropol Ethnol 8(1):15

Dzhumaev KK, Zenkevich IG, Tkachenko KG, Tsiburskaya IA (1990) Essential oil of the leaves of *Hyssopus seravschanicus* from South Uzbekistan. Chem Nat Compd 1:123–125

El Allaoui H, El Ahmadi K, El Abdouni A, Dira I, El Bastrioui M, Bouhrim M, Haboubi K (2024) Trends and insights in medicinal plant extract research: a ten-year bibliometric and visualization study. Horticulturae 10(11):1163. https://doi.org/10.3390/horticulturae10111163

Eloff JN, Gurib-Fakim A, Phillips LD (2010) African herbal pharmacopoeia. Brendler T (ed) vol 53. Association for African Medicinal Plants Standards, Port Louis

Ferlay J, Steliarova-Foucher E, Lortet-Tieulent J, Rosso S, Coebergh JWW, Comber H, Bray F (2013) Cancer incidence and mortality patterns in Europe: estimates for 40 countries in 2012. Eur J Cancer 49(6):1374–1403
Fortune Business Insights (2022) Herbal medicine market size & COVID-19 impact analysis retrieved 30 October 2022. Market Res Rep Food Suppl Herbal Med 2022:2022–2029. https://www.fortunebusinessinsights.com/herbal-medicine-market-106320
Funk JL, Frye JB, Oyarzo JN, Chen J, Zhang H, Timmermann BN (2016) Anti-inflammatory effects of the essential oils of ginger (*Zingiber officinale* roscoe) in experimental rheumatoid arthritis. Pharma Nutr 4(3):123–131
Gapparov AM, Aripova SF (2011) Alkaloids from the aerial part and roots of *Convolvulus pseudocanthabrica* indigenous to Uzbekistan. Chem Nat Compd 47:673–674
Gapparov AM, Razzakov NA, Aripova SF (2007) Alkaloids of *Convolvulus subhirsutus* from Uzbekistan. Chem Nat Compd 43:291–292
García de Alba García JE, Ramírez Hernández BC, Robles Arellano G, Zañudo Hernández J, Salcedo Rocha AL, García de Alba Verduzco JE (2012) Conocimiento y uso de las plantas medicinales en la zona metropolitana de Guadalajara. Desacatos 39:29–44. México: CIESAS. https://www.redalyc.org/pdf/139/13923111003.pdf
Gorunovic M, Lukic P (2001) Pharmacognosy. Gorunovic M, Beograd, pp 1–5
Gurib-Fakim A (2006) Medicinal plants: traditions of yesterday and drugs of tomorrow. Mol Asp Med 27(1):1–93
Gurusamy M, Subramanian V, Raju R (2022) Diversity and conservation status of flora in Pilavakkal Dam foothills of Western Ghats, Tamil Nadu, India. Indones J For Res 9(2):215–237
Gurusamy M, Charlie K, Palanisamy K, Chelladurai D, Katerere DR (2024) An attempt to conserve a vulnerable tree species of *Santalum album* L. through micropropagation. Indones. J For Res 11(1):81–89
Hammer KA, Carson CF, Riley TV (2003) Antifungal activity of the components of *Melaleuca alternifolia* (tea tree) oil. J Appl Microbiol 95(4):853–860
Hannan J, Ojo O, Rokeya L, Khaleque J, Akhter M, Flatt P, Abdel-Wahab Y (2014) Actions underlying antidiabetic effects of *Ocimum sanctum* leaf extracts in animal models of type 1 and type 2 diabetes. Vet Med Sci 7:500–511
Heijden R, Jacobs D, Snoeijer W, Hallard D, Verpoorte R (2004) The *Catharanthus* alkaloids: pharmacognosy and biotechnology. Curr Med Chem 11(5):607–628
Heinrich M, Ankli A, Frei B, Wiemann C, Sticher O (1998) Medicinal plants in Mexico: healer's consensus and cultural importance. Soc Sci Med 47:1859–1871
Heinrich M, Williamson EM, Gibbons S, Barnes J (2012) Fundamentals of pharmacognosy and phytotherapy. Elsevier Health Sci (E-book)
Hosseinzadeh S, Jafarikukhdan A, Hosseini A, Armand R (2015) The application of medicinal plants in traditional and modern medicine: a review of *Thymus vulgaris*. Int J Clin Med 6(9):635–642
Howes MJR, Quave CL, Collemare J, Tatsis EC, Twilley D, Lulekal E, Nic Lughadha E (2020) Molecules from nature: reconciling biodiversity conservation and global healthcare imperatives for sustainable use of medicinal plants and fungi. Plants People Planet 2(5):463–481. https://doi.org/10.1002/ppp3.10138
Hsu E, Barrett R (2008) Traditional Asian medical systems
Kassaye KD, Amberbir A, Getachew B, Mussema Y (2006) A historical overview of traditional medicine practices and policy in Ethiopia. Ethiop J Health Dev 20(2):127–134
Kew KM, Quinn M, Quon BS, Ducharme FM (2016) Increased versus stable doses of inhaled corticosteroids for exacerbations of chronic asthma in adults and children. Cochrane Database Syst Rev 6:CD007524. https://doi.org/10.1002/14651858.CD007524.pub4
Khoja AA, Waheed M, Haq SM, Bussmann RW (2024) The role of plants in traditional medicine and current therapy: a case study from North part of Kashmir Himalaya. Ethnobot Res Appl 27:1–23. https://doi.org/10.32859/era.27.5.1-23
Kholmatov KK, Kosimov AI (1994) Medicinal plants. Izd-vo Ibn-Sina, Tashkent, 298 p

Kichu M, Malewska T, Akter K, Imchen I, Harrington D, Kohen J, Vemulpad SR, Jamie JF (2015) An ethnobotanical study of medicinal plants of Chungtia village, Nagaland, India. J Ethnopharmacol 166:5–17. https://doi.org/10.1016/j.jep.2015.02.053

Kiguba R, Ononge S, Karamagi C, Bird SM (2016) Herbal medicine use and linked suspected adverse drug reactions in a prospective cohort of Ugandan inpatients. BMC Complement Altern Med 16:1–8

Kim DO, Chun OK, Kim YJ, Moon HY, Lee CY (2003) Quantification of polyphenolics and their antioxidant capacity in fresh plums. J Agric Food Chem 51(22):6509–6515

King AMY, Young G (1999) Characteristics and occurrence of phenolic phytochemicals. J Am Diet Assoc 99(2):213–218

Kopp J, Wang GY, Horch RE, Pallua N, Ge SD (2003) Ancient traditional Chinese medicine in burn treatment: a historical review. Burns 29(5):473–478

Kovacevic N (2000) Fundamentals of pharmacognosy. Personal edition, Beograd, pp 170–171

Kucherbaev KJ, Uteniyazov KK, Kachala VV, Saatov Z, Shashkov AS (2002) Triterpene glycosides from plants of the *Astragalus* genus. Structure of cyclounifolioside A from *Astragalus unifoliolatus*. Chem Nat Compd 38:175–178

Kumar N, Sakhya SK (2013) Ethnopharmacological properties of *Curcuma longa*: a review. Int J Pharm Sci Res 4(1):103–112

Kumar JS, Jain R, Jain N (2019a) Antiobesity potential of *Aconitum heterophyllum* roots. Int J Pharm Bio Sci 10(3):65–74

Kumar S, Patra A, Mandal S, Biswas RR (2019b) *Andrographis paniculata* (Burm. f.) Wall. ex Nees: a comprehensive review on ethnopharmacological, phytochemical, and pharmacological studies. Chin. J Nat Med 17(9):681–711

Kurimoto SI, Suzuki K, Okasaka M, Kashiwada Y, Kodzhimatov OK, Takaishi Y (2012) New sesquiterpene lactone glucosides from the roots of *Ferula varia*. Phytochem Lett 5(4):729–733

Kyei S, Koffuor GA, Ramkissoon P, Ameyaw EO, Asiamah EA (2016) Anti-inflammatory effect of *Heliotropium indicum* Linn on lipopolysaccharide-induced uveitis in New Zealand white rabbits. Int J Ophthalmol 9(4):528

Lesgards JF, Baldovini N, Vidal N, Pietri S (2014) Anticancer activities of essential oil constituents and synergy with conventional therapies: a review. Phytother Res 28(10):1423–1446

Li H, Weng JR (2017) Bioactivities of essential oil from *Cinnamomum camphora* and its effect on DNA damage and cancer cell proliferation. J Food Drug Anal 25(4):797–805

Li S, Li SK, Gan RY, Song FL, Kuang L, Li HB (2013) Antioxidant capacities and total phenolic contents of infusions from 223 medicinal plants. Ind Crop Prod 51:289–298

Li X, Wu Q, Xie Y et al (2017) The effects of ginsenosides on the immune response. J Ginseng Res 41(4):450–458

Lin CH, Shih ZZ, Kuo YH, Huang GJ, Tu PC, Shih CC (2018) Antidiabetic and antihyperlipidemic effects of the flower extract of *Eriobotrya japonica* in streptozotocin-induced diabetic mice and the potential bioactive constituents in vitro. J Funct Foods 49:122–136

Liu H, Schmitz JC, Wei J, Cao S, Beumer JH, Strychor S et al (2014) Clove extract inhibits tumor growth and promotes cell cycle arrest and apoptosis. Oncol Res 21(5):247

Madziga HA, Sanni S, Sandabe UK (2010) Phytochemical and elemental analysis of *Acalypha wilkesiana* leaf. J Am Sci 6(11):510–514

Maizes V, Schneider C, Bell I, Weil A (2002) Integrative medical education: development and implementation of a comprehensive curriculum at the University of Arizona. Acad Med 77(9):851–860

Malikov VM, Yuldashev MP (2002) Phenolic compounds of plants of the *Scutellaria* L. genus. Distribution, structure, and properties. Chem Nat Compd 38:358–406

Malikova MK, Rakhmanberdyeva RK (2013) Pectinic substances from leaves of *Ungernia victoris*. Chem Nat Compd 49:596–598

Mall GK, Mishra PK, Prakash V (2009) Antidiabetic and hypolipidemic activity of *Gymnema sylvestre* in alloxan-induced diabetic rats. Glob J Biotechnol Biochem 4(1):37–42

Mamadalieva NZ, Herrmann F, El-Readi MZ, Tahrani A, Hamoud R, Egamberdieva DR et al (2011) Flavonoids in *Scutellaria immaculata* and *S. ramosissima* (Lamiaceae) and their biological activity. J Pharm Pharmacol 63(10):1346–1357

Mamadalieva N, Egamberdieva D, Tiezzi A (2013) In vitro biological activities of the components from *Silene wallichiana*. Med Aromat Plant Sci Biotechnol 7(1):1–6

Manikandan G, Noel SD (2019) Antioxidant activity and phytochemical analysis of methanolic leaf extract of *Leucas aspera* (Wild) Link. Int J Pharm Biol Sci 9(2):320–325

Manikandan G, Ramasubbu R (2017) A note on the population status and threats on two endemic and endangered species of *Garcinia* of Agasthyamalai Biosphere Reserve, Tamil Nadu, India. J Threat Taxa 9(10):10839–10845

Manikandan G, Ramasubbu R (2020) Antimicrobial activity of leaf extracts of *Memecylon heyneanum* Benth. ex Wight & Arn.: an endemic tree species of Southern Western Ghats. Adv Zool Bot 8(3):258–268

Manikandan G, Ramasubbu R (2023) GC/MS analysis of phytoconstituents and antimicrobial evaluation leaf extract of *Garcinia imberti* Bourd. Arab J Med Aromat Plants 9(3):134–159

Manikandan G, Vimala Rani A, Divya C, Ramasubbu R (2017) GC-MS analysis of phytochemical constituents in the petroleum ether leaf extracts of *Millettia peguensis*. Int Res J Pharm 8(9):144–150. https://doi.org/10.7897/2230-8407.089170

Manikandan G, Pandiselvi P, Sobana N, Murugan M (2019) GC-MS analysis of chemical constituents in the methanolic tuber extract of *Momordica cymbalaria* Hook. F Int Res J Pharm 10(1):135–140

Manikandan G, Saranya M, Gayathri S (2020a) GC-MS analysis of phyto-constituents of the essential oil from the leaves of *Melaleuca citrina* (Curtis) Dum. Cours Adv Zool Bot 8(3):87–98

Manikandan G, Divya C, Ramasubbu R (2020b) Antimicrobial properties of various leaf extracts of *Syzygium densiflorum* Wall. Ex Wight & Arn.: an endemic and endangered tree species of Western Ghats. In: Prakash S, Manikandan G, Shanmugam SA (eds) A text book of social relationship between microbes, human and environment, pp 149–164

Manikandan G, Ragavi K, Gayathri S (2021) GC-MS analysis of bioactive compounds and antimicrobial potential of the essential oil of *Cymbopogon citratus* (DC.) Stapf. Res J Chem Environ 25(2):54–65

Manikandan G, Rajalakshmi A, Suganya P, Kumari RM, Katerere DR (2022) Chemical profiling, antimicrobial and anti-malarial activity of leaf essential oil of *Eugenia rottleriana* Wt. & Arn (Myrtaceae). Arab J Med Aromat Plants 8(3):57–76

Melikuziev FA, Nishanbaev SZ, Turgunov KK, Bobakulov KM (2013) Phenolic compounds from roots of *Rhodiola litvinovii*. Chem Nat Compd 49:349–350

Meyanungsang K, Teresa M, Kaisarun A, Imchawati I, David H, James K, Subramanyam RV, Joanne FJ (2015) An ethnobotanical study of medicinal plants of Chungtia village, Nagaland, India. J Ethnopharmacol 166:5–17. https://doi.org/10.1016/j.jep.2015.02.053

Milinčević V (ed) (1980) Chilandar medical codex N. 517. Narodna biblioteka Srbije

Mina EC, Mina JF (2017) Ethnobotanical survey of plants commonly used for diabetes in Tarlac of Central Luzon Philippines. IIUM Med J Malaysia 16(1). https://doi.org/10.31436/imjm.v16i1.354

Mishra AP, Saklani S, Salehi B, Parcha V, Sharifi-Rad M, Milella L et al (2018) *Satyrium nepalense*, a high altitude medicinal orchid of Indian Himalayan region: chemical profile and biological activities of tuber extracts. Cell Mol Biol 64(8):35–43

Moke LE, Koto-te-Nyiwa NGN, Bongo LM, Noté OP, Mbing JN, Mpiana PT (2017) *Artocarpus heterophyllus* Lam. (Moraceae): phytochemistry, pharmacology and future directions, a mini-review. J Adv Bot Zool 5(3):1–8

Morvin Yabesh E, Prabhu S, Vijayakumar S (2014) An ethnobotanical study of medicinal plants used by traditional healers in Silent Valley of Kerala, India. J Ethnopharmacol 154(3):774–789. https://doi.org/10.1016/j.jep.2014.05.004

Mostofa R, Ahmed S, Begum MM, Sohanur Rahman M, Begum T, Ahmed SU et al (2017) Evaluation of anti-inflammatory and gastric anti-ulcer activity of *Phyllanthus niruri* L. (Euphorbiaceae) leaves in experimental rats. BMC Complement Altern Med 17:1–10

Murugesan R, Stephen DN, Vairakannu T, Gurusamy M, Govindharajan S (2024) Ethno-medicinal survey of targeted tribes in Idukki District, Kerala. Int J Pharm Res Allied Sci 13(2–2024):58–79
Nabandith V, Suzui M, Morioka T, Kaneshiro T, Kinjo K, Matsumoto K, Akao Y, Iinuma M, Yoshimi N (2004) Inhibitory effects of crude alpha-mangostin, a xanthone derivative, on two different categories of colon preneoplastic lesions induced by 1,2-dimethylhydrazine in the rat. Asian Pac J Cancer Prev 5(4):433–438
Nikolovski B (1995) Essays on the history of health culture in Macedonia. Macedonian Pharmaceutical Association, Skopje, pp 17–27
Nunes CDR, Barreto Arantes M, de Faria M, Pereira S, Leandro da Cruz L, de Souza PM, Pereira de Moraes L et al (2020) Plants as sources of anti-inflammatory agents. Molecules 25(16):3726
Nwozo OS, Effiong EM, Aja PM, Awuchi CG (2023) Antioxidant, phytochemical, and therapeutic properties of medicinal plants: a review. Int J Food Prop 26(1):359–388. https://doi.org/10.1080/10942912.2022.2157425
Ochwang'i DO, Kimwele CN, Oduma JA, Gathumbi PK, Mbaria JM, Kiama SG (2014) Medicinal plants used in treatment and management of cancer in Kakamega County, Kenya. J Ethnopharmacol 151(3):1040–1055
Ofoedu CE, Ofoedu EO, Chacha JS, Owuamanam CI, Efekalam IS, Awuchi CG (2022) Comparative evaluation of physicochemical, antioxidant, and sensory properties of red wine as markers of its quality and authenticity. Int J Food Sci 1:8368992. https://doi.org/10.1155/2022/8368992
Ojha SN, Tiwari D, Anand A, Sundriyal RC (2020) Ethnomedicinal knowledge of a marginal hill community of Central Himalaya: diversity, usage pattern, and conservation concerns. J Ethnobiol Ethnomed 16:1–21
Ola K (2022) Role of traditional knowledge in the COVID-19 battle. J World Intellect Prop 25(2):279–291. https://doi.org/10.1111/jwip.12220
Ortiz MI, Fernández-Martínez E, Soria-Jasso LE, Lucas-Gómez I, Villagómez-Ibarra R, González-García MP et al (2016) Isolation, identification and molecular docking as cyclo-oxygenase (COX) inhibitors of the main constituents of *Matricaria chamomilla* L. extract and its synergistic interaction with diclofenac on nociception and gastric damage in rats. Biomed Pharmacother 78:248–256
Packiyalakshmi D, Athilakshmi P, Gayathri S, Karthiga P, Thiri Bhuvaneswari R, Manikandan G (2017) Antimicrobial potential of different solvents leaf extract of *Millettia peguensis* against selected pathogens. Pharm Innov J 6(10):119–124
Padhye S, Ahmad A, Oswal N, Sarkar FH (2009) Emerging role of Garcinol, the antioxidant chalcone from *Garcinia indica* Choisy and its synthetic analogs. J Hematol Oncol 2:1–13
Patwardhan B, Warude D, Pushpangadan P, Bhatt N (2005) Ayurveda and traditional Chinese medicine: a comparative overview. Evid-Based Complement Alternat Med 2(4):465–473
Paulraj S, Pandi S, Rajalakshmi A, Gurusamy M (2021) *Momordica cymbalaria* plant tubers using silver nanoparticles synthesis and applied for the biological activity. J Sci Res 65(5):151–154
Pelagic V (1970) Pelagic folk teacher. Freedom, Beograd, pp 500–502
Pharmacopoeia B (2007) British Pharmacopoeia Commission. A-290, 4
Picking D (2024) The global regulatory framework for medicinal plants. In: Pharmacognosy. Academic, pp 769–782
Pirintsos S, Panagiotopoulos A, Bariotakis M, Daskalakis V, Lionis C, Sourvinos G et al (2022) From traditional ethnopharmacology to modern natural drug discovery: a methodology discussion and specific examples. Molecules 27(13):4060
Pushkarev VM, Kovzun OI, Tronko MD (2008) Antineoplastic and apoptotic effects of cannabinoids. N-acylethanolamines: protectors or killers. Exp Oncol 30(1):6–21
Qureshi RA, Ghufran MA (2007) Indigenous knowledge of selected medicinal wild plants of District Attock, Punjab, Pakistan. Pak J Bot 39(7):2291–2299
Raju R, Nambi SK, Gurusamy M (2015) In vitro propagation of *Pisonia grandis* R. Br.: an indigenous vegetable and promising medicinal plant. Phytomorphology 65:133–138
Ramasubbu R, Manikandan G, Kala NS (2020) Phytochemical composition and antimicrobial properties of leaf essential oil of *Garcinia imberti* Bourd. and *G. travancorica* Bedd. Int J Adv Sci Technol 29(2):1082–1092

Raut JS, Karuppayil SM (2014) A status review on the medicinal properties of essential oils. Ind Crop Prod 62:250–264

Rios R, Silva HBF, Carneiro NVQ, Costa RS, Carneiro TCB, Marques CR et al (2017) Anti-inflammatory activity of Jurubeba (*Solanum paniculatum* L.) through reducing the T-bet and GATA3 gene expression, in vitro. J Allergy Clin Immunol 139(2):AB268

Rosas EC, Correa LB, de Almeida PT, Costa TEMM, Mazzei JL, Heringer AP et al (2015) Anti-inflammatory effect of *Schinus terebinthifolius* Raddi hydroalcoholic extract on neutrophil migration in zymosan-induced arthritis. J Ethnopharmacol 175:490–498

Routh H, Bhowmik K (1999) Traditional Indian medicine in dermatology. Clin Dermatol 17(1):41–47

Roy A, Khan A, Ahmad I, Alghamdi S, Rajab BS, Babalghith AO et al (2022) (2022) flavonoids a bioactive compound from medicinal plants and its therapeutic applications. Biomed Res Int 1:5445291. https://doi.org/10.1155/2022/5445291

Saad B, Azaizeh H, Abu-Hijleh G, Said O (2006) Safety of traditional Arab herbal medicine. Evid-Based Complement Alternat Med 3(4):433–439

Sadeghipour A, Eidi M, Ilchizadeh Kavgani A, Ghahramani R, Shahabzadeh S, Anissian A (2014) Lipid lowering effect of *Punica granatum* L. peel in high lipid diet fed male rats. Evid-Based Complement Alternat Med 2014(1):432650

Sahoo N, Manchikanti P (2013) Herbal drug regulation and commercialization: an Indian industry perspective. J Altern Complement Med 19(12):957–963

Sahu AK, Padhy RK, Das D, Gautam A (2021) Improving financial and environmental performance through MFCA: a SME case study. J Clean Prod 279:123751. https://doi.org/10.1016/j.jclepro.2020.123751

Said HM (1980) History and philosophy of science: proceedings of the international congress of the history and philosophy of science, Islamabad, 8–13 December, 1979

Salehi B, Kumar NVA, Şener B, Sharifi-Rad M, Kılıç M, Mahady GB et al (2018) Medicinal plants used in the treatment of human immunodeficiency virus. Int J Mol Sci 19(5):1459

Samtiya M, Aluko RE, Dhewa T, Moreno-Rojas JM (2021) Potential health benefits of plant food-derived bioactive components: an overview. Foods 10(4):839. https://doi.org/10.3390/foods10040839

Sarangi A, Jena S, Sarangi AK, Swain B (2013) Anti-diabetic effects of *Withania somnifera* root and leaf extracts on streptozotocin induced diabetic rats. J Cell Tissue Res 13(1):3597

Satimov GB, Mamatkhanov AU, Kotenko LD, Turakhodzhaev MT, Madrakhimov SN, Faizieva SK et al (2003) A total flavonoid extract from *Thermopsis altherniflora*: development of the production technology, creation of a medicinal preparation, and evaluation of hypolipidemic activity. Pharm Chem J 37(4):42–44

Satyanarayana K, Sravanthi K, Shaker IA, Ponnulakshmi R (2015) Molecular approach to identify antidiabetic potential of *Azadirachta indica*. J Ayurveda Integr Med 6(3):165–174

Scarborough J (1978) Theophrastus on herbals and herbal remedies. J Hist Biol 11:353–385

Schcolnik-Cabrera A (2023) Current approaches to overcome antimicrobial resistance. Curr Med Chem 30(1):3–4

Schippmann U, Leaman DJ, Cunningham AB (2002) Impact of cultivation and gathering of medicinal plants on biodiversity: global trends and issues. Biodivers Ecosyst Approach Agric Forestry Fish:769–782

Shaheen S, Ramzan S, Haroon N, Hussain K (2014) Ethnopharmacological and systematic studies of selected medicinal plants of Pakistan. Pak J Sci 66(2):175–180

Sharifi-Rad M, Salehi B, Sharifi-Rad J, Setzer WN, Iriti M (2018) *Pulicaria vulgaris* Gaertn. essential oil: an alternative or complementary treatment for Leishmaniasis. Cell Mol Biol 64(8):18–21

Shaw D, Graeme L, Pierre D, Elizabeth W, Kelvin C (2012) Pharmacovigilance of herbal medicine. J Ethnopharmacol 140(3):513–518

Shen L, Xu J, Luo L, Hu H, Meng X, Li X, Chen S (2018) Predicting the potential global distribution of diosgenin-contained *Dioscorea* species. Chin Med 13:1–10

Shubashini SK, Uma S (2010) Ethnobotanical documentation of a few medicinal plants in the Agasthiayamalai region of Tirunelveli District, India. Ethnobotanical Leaflets 2:6
Siddikov GU, Yuldashev MP, Batirov EK, Abdullaev SV (2006) Flavonoids from *Scutellaria cordifrons* and *S. phyllostachya* roots. Chem Nat Compd 42(3):356
Siddikov GU, Yuldashev MP, Abdullaev SV (2007) Flavonoids from *Scutellaria phyllostachya* roots. Chem Nat Compd 43(3):324
Siddikov DR, Nishanbaev SZ, Narbutaeva DA, Vdovin AD, Abdullaev ND (2011) Polyphenols from *Geranium saxatile*. Chem Nat Compd 47:630–631
Singh G (2007) Chemistry of terpenoids and carotenoids. Discovery Publishing House, pp 1–286
Siraj MA, Shilpi JA, Hossain MG, Uddin SJ, Islam MK, Jahan IA, Hossain H (2016) Anti-inflammatory and antioxidant activity of *Acalypha hispida* leaf and analysis of its major bioactive polyphenols by HPLC. Adv Pharm Bull 6(2):275
Snyderman R, Weil AT (2002) Integrative medicine: bringing medicine back to its roots. Arch Intern Med 162(4):395–397
Sofowora A, Ogunbodede E, Onayade A (2013) The role and place of medicinal plants in the strategies for disease prevention. Afr J Tradit Complement Altern Med 10(5):210–229
Sovilj M, Spasojević M (2001) Production and application of essential oils from the domestic medicinal plant. PTEP 5(1–2):34–38
Sultan MT, Buttxs MS, Qayyum MMN, Suleria HAR (2014) Immunity: plants as effective mediators. Crit Rev Food Sci Nutr 54(10):1298–1308
Sun W, Shahrajabian MH (2023, 1845) Therapeutic potential of phenolic compounds in medicinal plants—natural health products for human health. Molecules 28(4)
Surendran A, Ramasubbu R, Gurusamy M (2020) Chemical signatures of essential oil: an alternative tool for the identification of the species of *Syzygium* Gaertn. of Tamil Nadu. Indian Forester 146:806
Svarstad H, Dhillion SS (2000) Responding to bioprospecting. From biodiversity in the South to medicines in the North
Sykiotis GP, Kalliolias GD, Papavassiliou AG (2006) Hippocrates and genomic medicine. Arch Med Res 37(1):181–183
Tabuti JR, Kukunda CB, Kaweesi D, Kasilo OM (2012) Herbal medicine use in the districts of Nakapiripirit, Pallisa, Kanungu, and Mukono in Uganda. J Ethnobiol Ethnomed 8:1–15
Takeda Y, Masuda T, Honda G, Takaishi Y, Ito M, Ashurmetov OA et al (1999) Secoiridoid glycosides from *Gentiana olivieri*. Chem Pharm Bull 47(9):1338–1340
Takeda Y, Tomonari M, Arimoto S, Masuda T, Otsuka H, Matsunami K et al (2008) A new phenolic glucoside from an Uzbek medicinal plant, *Origanum tyttanthum*. J Nat Med 62:71–74
Thompson HAJ, Ojo O, Flatt PR, Abdel-Wahab Y (2014) Aqueous bark extracts of *Terminalia arjuna* stimulates insulin release, enhances insulin action and inhibits starch digestion and protein glycation in vitro. Austin J Endocrinol Diabetes 1:1–6
Thorwald J (1991) Power and knowledge of ancient physicians. August Cesarec, Zagreb, pp 10–255
Tom ENL, Nankia FD, Nyunaï N, Thernier CG, Demougeot C, Dimo T (2018) Myocardial potency of aqueous extract of *Harungana madagascariensis* stem bark against isoproterenol-induced myocardial damage in rats. Universal J Pharm Res 3:17–24
Tucakov J (1948) Pharmacognosy. Academic Books, Beograd, pp 8–21
Tulyaganov TS, Kozimova NM (2005) Alkaloids from *Nitraria schoberi*. O-acetylnitraraine. Chem Nat Compd 41(5):578
Umadevi M, Kumar K, Bhowmik D, Duraivel S (2013) Traditionally used anticancer herbs in India. J Med Plants Stud 1(3):56–74
Ventola CL (2015) The antibiotic resistance crisis: part 1: causes and threats. Pharm Ther 40(4):277
Volenzo T, Odiyo J (2020) Integrating endemic medicinal plants into the global value chains: the ecological degradation challenges and opportunities. Heliyon 6(9):e04970
Vuddanda PR, Chakraborty S, Singh S (2010) Berberine: a potential phytochemical with multispectrum therapeutic activities. Expert Opin Investig Drugs 19(10):1297–1307

Wang GF, Shi LP, Ren YD, Liu QF, Liu HF, Zhang RJ et al (2009) Anti-hepatitis B virus activity of chlorogenic acid, quinic acid and caffeic acid in vivo and in vitro. Antivir Res 83(2):186–190

Wang J, Xu J, Gong X, Yang M, Zhang C, Li M (2019) Biosynthesis, chemistry, and pharmacology of polyphenols from Chinese *Salvia* species: a review. Molecules 24:155. https://doi.org/10.3390/molecules24010155

Wani MC, Taylor HL, Wall ME, Coggon P, McPhail AT (1971) Plant antitumor agents. VI. Isolation and structure of taxol, a novel antileukemic and antitumor agent from *Taxus brevifolia*. J Am Chem Soc 93(9):2325–2327

Wanjohi BK, Njenga EW, Sudoi V, Kipkore WK, Moore HL, Davies MI (2020) Ecological knowledge of indigenous plants among the Marakwet community (Embobut Basin), Elgeyo Marakwet County (Kenya). Ethnobot Res Appl 20(1):1–16

Wattanathorn J, Jittiwat J, Tongun T, Muchimapura S, Ingkaninan K (2011) *Zingiber officinale* mitigates brain damage and improves memory impairment in focal cerebral ischemic rat. Evid Based Complement Altern Med 1:429505

Wiart C (2007) Ethnopharmacology of medicinal plants: Asia and the Pacific. Springer Science & Business Media

Wink M (2010) Introduction: biochemistry, physiology and ecological functions of secondary metabolites. In: Annual plant reviews volume 40: biochemistry of plant secondary metabolism. Wiley-Blackwell, Chichester, pp 1–19

Wishart DS, Bartok B, Oler E, Liang KY, Budinski Z, Berjanskii M et al (2021) MarkerDB: an online database of molecular biomarkers. Nucleic Acids Res 49(D1):D1259–D1267. https://doi.org/10.1093/nar/gkaa1067

World Health Organization (1999) WHO monographs on selected medicinal plants, vol 2. World Health Organization

World Health Organization (2003) WHO guidelines on good agricultural and collection practices [GACP] for medicinal plants. World Health Organization

World Health Organization (2005) National policy on traditional medicine and regulation of herbal medicines: report of a WHO global survey. World Health Organization

World Health Organization (2007a) WHO guidelines on good manufacturing practices (GMP) for herbal medicines. World Health Organization

World Health Organization (2007b) WHO guidelines for assessing quality of herbal medicines with reference to contaminants and residues. World Health Organization

World Health Organization (2013) World Health Organization traditional medicine strategy 2014–2023. WHO, Geneva, pp 1–78

World Health Organization (2021) WHO model list of essential medicines. Retrieved 15/06/2022, from https://www.who.int/groups/expert-committee-on-selection-and-use-of-essential-medicines/essential-medicines-lists

World Health Organization (WHO) (2013) Review of world pharmacopoeias. In: International meeting of world pharmacopoeias, p 20

Wu ZM, Yin XX, Ji L, Gao YY, Pan YM, Lu Q, Wang JY (2008) Ginkgo biloba extract prevents against apoptosis induced by high glucose in human lens epithelial cells. Acta Pharmacol Sin 29(9):1042–1050

Xia J, Gao J, Inagaki Y, Kokudo N, Nakata M, Tang W (2013) Flavonoids as potential anti-hepatocellular carcinoma agents: recent approaches using HepG2 cell line. Drug Discov Ther 7(1):1–8

Xu J, Yang Y (2009) Traditional Chinese medicine in the Chinese health care system. Health Policy 90(2–3):133–139

Yang W, Chen X, Li Y, Guo S, Wang Z, Yu X (2020) Advances in pharmacological activities of terpenoids. Nat Prod Commun 15(3):1934578X20903555

Yin J, Xing H, Ye J (2008) Efficacy of berberine in patients with type 2 diabetes mellitus. Metabolism 57(5):712–717

Yin J, Ye J, Jia W (2012) Effects and mechanisms of berberine in diabetes treatment. Acta Pharm Sin B 2(4):327–334

Yu M, Gouvinhas I, Rocha J, Barros AI (2021) Phytochemical and antioxidant analysis of medicinal and food plants towards bioactive food and pharmaceutical resources. Sci Rep 11(1):10041. https://doi.org/10.1038/s41598-021-89437-4

Yuldashev MP, Karimov A (2001) Flavonoids of *Scutellaria ocellata* and *S. nepetoides*. Chem Nat Compd 37:431–433

Yuldashev MP, Batirov EK, Malikov VM (1993) Anthraquinones of *Bergenia hissarica*. Chem Nat Compd 29(4):543–544

Yuldashev MP, Muminova BA, Drenin AA, Botirov EK (2007) Flavonoids from the aerial part of *Vicia subvillosa*. Chem Nat Compd 43:34–36

Zahidin NS, Saidin S, Zulkifli RM, Muhamad II, Ya'akob H, Nur H (2017) A review of *Acalypha indica* L. (Euphorbiaceae) as traditional medicinal plant and its therapeutic potential. J Ethnopharmacol 207:146–173

Zahmanov G, Alipieva K, Denev P, Todorov D, Hinkov A, Shishkov S et al (2015) Flavonoid glycosides profiling in dwarf elder fruits (*Sambucus ebulus* L.) and evaluation of their antioxidant and anti-herpes simplex activities. Ind Crop Prod 63:58–64

Zhang X (1998) Regulatory situation of herbal medicines: a worldwide review, vol 26. World Health Organization, pp 223–228

Zhang L, Yan J, Liu X, Ye Z, Yang X, Meyboom R et al (2012) Pharmacovigilance practice and risk control of traditional Chinese medicine drugs in China: current status and future perspective. J Ethnopharmacol 140(3):519–525

Ziyaev R, Abdusamatov A, Yunusov SY (1973) Alkaloids of *Liriodendron tulipifera*. Chem Nat Compd 9:475–476

Safety, Toxicity, and Regulation in Use of Medicinal Plants

2

Somvir Singh, Arti Thakur, Vishav Kiran, Kanika Dulta, Arti Jamwal Sharma, and Mainak Deb Choudhury

Abstract

Medicinal plants contain various bioactive compounds that are significant in both traditional and contemporary therapeutic approaches, addressing a wide array of diseases. In addition to the benefits and historical application of herbal medicines, significant issues regarding safety, toxicity, and regulatory frameworks related to medicinal herbs exist. This chapter assesses the critical factors affecting the safe utilization of medicinal plants, encompassing potential adverse effects, challenges related to standardization and quality control, and regulatory obstacles across different global regions. Alkaloids, glycosides, tannins, and heavy metals are substances that raise toxicological concerns due to their potential to damage vital organs, disrupt cellular functions, or induce cancer when ingested in excessive amounts or over prolonged durations. Additionally, environmental factors, such as soil quality, climate, and toxin exposure significantly influence the toxicity levels of plants. This chapter emphasizes the importance of

S. Singh (✉) · M. D. Choudhury
Department of Biosciences, University Institute of Biotechnology, Chandigarh University Gharuan, Mohali, Punjab, India
e-mail: samrajput0002@gmail.com

A. Thakur
Department of Botany, Shoolini Institute of Life Sciences and Business Management, Solan, India

V. Kiran
LR Institute of Pharmacy Jabil Kyar, Solan, India

K. Dulta
Center of Advanced Innovation Technologies, VSB-Technical University of Ostrava, Ostrava, Poruba, Czech Republic

A. J. Sharma
Department of Biosciences, Career Point University, Hamirpur, Himachal Pradesh, India

K. Kuca et al. (eds.), *Medicinal Plants and Their Bioactives in Human Diseases*,
https://doi.org/10.1007/978-3-032-01356-9_2

recent scientific advancements, such as metabolomics, genomics, and analytical chemistry, in developing standardized methods for the identification and quantification of bioactive constituents in medicinal plants. Moreover, studies grounded in evidence regarding the pharmacokinetics, pharmacodynamics, and toxicity of medicinal plants are crucial for the revision and enhancement of regulatory standards. The use of medicinal plants in healthcare systems offers numerous advantages; however, stringent safety and regulatory measures are essential to safeguard patient health. This chapter emphasizes the necessity of a comprehensive strategy that integrates stringent regulatory measures with scientific research to facilitate safe therapeutic applications while safeguarding public health.

Keywords

Medicinal plants · Bioactive compounds · Safety · Toxicity · Regulation · Standardization

1 Introduction

Prehistoric societies, including the Egyptians, Greeks, Chinese, and Indians, have used medicinal plants (Pan et al. 2014). Early mentions of the healing qualities of plants may be found in writings, such as Huangdi Neijing (ancient China) and Ebers Papyrus (ancient Egypt). Both the Ayurveda system of medicine in India and the Traditional Chinese Medicine (TCM) system rely profoundly on the therapeutic qualities of plants (Roy Upton 2015). Herbalists and apothecaries in the Western world formalized the use of medicinal plants; many of these plant-based treatments were eventually embraced and developed into contemporary pharmacology (David et al. 2015). Medicinal plants have always exhibited encouraging good qualities all across the world in traditional medicine systems (Leonti and Casu 2013). Their different natural chemical components show different therapeutic qualities and can be used to create contemporary drugs (David et al. 2015). Compiling the knowledge of all medicinal systems gives us an elaborated profile of all medicinal plants, including its pharmacological as well as healing qualities. All have studied and used the conventional therapeutic use of these plants: Egyptians, Greeks, Chinese, and Indians. Ayurveda in India greatly depends on the therapeutic advantages of plants. Practitioners in the modern world employ these plants (Ramawat and Goyal 2008).

According to the World Health Organization (WHO), traditional medicine is "the sum total of the knowledge, skills, and practices based on the theories, beliefs, and experiences inherent to diverse cultures, whether explainable or not, used in the conservation of health as well as in the prevention, diagnosis, improvement, or treatment of physical and psychological illness" (WHO 2019). Since ancient times, medicinal plants have been a vibrant source of both restorative and protective medical therapy preparations for humans (Gurib-Fakim 2006). Modern ethnobotanical studies are quite important for the preservation of ancient medicinal cultures as well as for knowledge of historical and cultural shifts. The protection of traditional medicinal plant resources depends on this as well (Leonti 2011). Many individuals nowadays look for herbal treatments either in addition to or in place of traditional medicine. In both Western and non-Western health practices, phytotherapy has grown increasingly popular (Sendker and Sheridan 2017). Health food stores and pharmacies reflect this trend with their stocks of tinctures, teas, and supplements. Natural medicines, including herbs, vitamins, minerals, and various other supplements, became a momentous and broadly used stream universal. Their three primary categories are physical treatments, such as exercise, yoga, bright light, and acupuncture; nutraceutical (vitamins and minerals), such as vitamin B12, niacin, magnesium, omega-3 fatty acids, and herbal preparations, such as St. John's wort, roseroot, and saffron (Klemow et al. 2011).

Though modern medicine has developed, herbal products still satisfy therapeutic needs in clinical and traditional environments. Aromatherapy, herbalism, gemmotherapy, herbal medicine in pharmacy, Chinese herbal medicine, and so on are among the kinds of phytotherapy seen in CAM (Parfait and Lawrence 2023). For millennia, aromatherapy has been used. Ancient civilizations in China, India, Egypt, and other places included fragrant plant elements in resins, balms, and oils. Aromatherapy is the practice of using aromatic substances made up of essential oils and other aroma complexes (Kenner and Requena 2001). Aromatherapy either has no medical evidence that it can treat, prevent, or cure any condition. Herbalism is the preparation and sale of medicinal plants or derivative products (Hoffmann 2003). Designing placebo-controlled studies is difficult since the goal of aromatherapy is the scent of the products (Graham et al. 2003). Herbal preparations may not be suitable for some people who take other drugs, including hormonal contraceptives, or have serious health issues like liver or kidney disease, or those who are going to have surgery, pregnant or nursing women, the elderly, and children (Williamson 2003).

2 Importance of Medicinal Plants in Healthcare

A range of plants employed in herbal medicine, "herbology," or "herbal medicine" constitutes the term "medicinal plant." It is the study of such practices as well as the use of plants for medical purposes. Since ancient times, medicinal plants have been quite important in human health; some of these plants are seen as essential nutritional sources. Among these plants are ginger, green tea, walnuts, aloe, pepper, and turmeric. Among other things, some plants and their derivatives are major sources of active chemicals used in aspirin and toothpaste; medicinal plants are believed to be a rich source of components that can be used to create pharmacopoeial, non-pharmacopoeial, or synthetic drugs; the use of medicinal plants is considered to be quite safe as there are no or very few adverse effects; and medicinal plants are thought to provide a special and renewable resource for the discovery of possible novel medications and biological entities due to the structural and biological diversity of its elements (Adhami et al. 2018). Food additives from medicinal plants during the span of a person's life, nutrition, and diet are major factors in promoting and maintaining health. A plant-based diet may help to prevent and treat chronic diseases, such as diabetes, heart disease, hypertension, obesity, chronic inflammation, and cancer (WHO 2019).

3 Need for Safety and Toxicity Evaluation

Bhowmik et al. (2009) claims that several herbal medicines are either ineffective or have some side effects, and therefore rigorous safety and toxicity assessments of medicinal plants are essential. For instance, a clinical trial on concentrated ginger extract, which is used to treat various infections, showed negative gastrointestinal side effects, such as mouth irritation, heartburn, diarrhea, dyspepsia or nausea, and eructation. Several in vitro studies have also shown that ginger can suppress thromboxane production, therefore, delaying platelet aggregation and extending bleeding time. This poses particular risks for individuals with injuries and pregnant women, as prolonged bleeding time can lead to serious complications. Moreover, some studies have suggested that ginger consumption during pregnancy may increase the risk of fetal death, reabsorption, and interfere with the normal implantation process. Another example is Cinchona, also known as quina or red cinchona, which is a valuable source for quinine- a well-known antimalarial compound. Despite its efficacy in malaria, Cinchona bark has an intolerable toxicity risk from quinine overdoses despite its efficacy, quinine carries significant toxicity risks when consumed in excessive doses. Quinine showed several side effects, such as allergic responses, cardiotoxicity, anemia, and hypoglycemia, and a condition known as cinchonism, which encompasses symptoms, such as headaches, visual problems, tinnitus, dizziness, hearing loss, diarrhea, and nausea. Most of these negative effects are associated with the optimized quinine dose (Sarecka-Hujar and Szulc-Musioł 2022). Table 2.1 listed some toxic effects and symptoms of medicinal plants.

4 Regulatory Challenges in Herbal Medicine

Although herbal medicines are quite important, in most situations human health control remains difficult. A study revealed that Malawi lacked sufficient significant legislative tools to control herbal medicines. Unlike Nigeria, they lacked a national program, a national institute, and a national pharmacopeia, which complicated matters for the Malawi herbal industry (Mponda et al. 2025). National drug authorities' ignorance of herbal medicines, inadequate assessment techniques, and national drug authorities' ignorance of herbal medicines all contribute to the delay in the development of national policies, laws, and regulations for traditional medicines, modern/alternative medicines, and herbal medicines. Standardization difficulties are another obstacle; for safe and effective use of herbal drugs, consistency in composition and biological activity are fundamental criteria. Herbal medications, on the other hand, often fall short of this criterion because of several issues, including plant identification challenges, genetic variation, differences in growing conditions,

Table 2.1 Common toxic effects and symptoms of medicinal plants

Sr. No.	Plant	Toxic effect	Symptoms	References
1.	*Atropa belladonna*	Neurotoxicity	Hallucinations, dry mouth, dilated pupils	Mills and Bone (2000)
2.	*Digitalis* spp.	Cardiotoxicity	Arrhythmias, nausea, blurred vision	Schmidt et al. (2005)
3.	*Aristolochia* spp.	Nephrotoxicity, carcinogenicity	Kidney failure, urothelial carcinoma	Jarup (2003)
4.	*Ricinus communis*	Cytotoxicity (Ricin toxin)	Severe vomiting, diarrhea, multi-organ failure	Klich 2007
5.	*Aconitum* spp.	Cardiotoxicity, neurotoxicity	Hypotension, paresthesia, arrhythmias	Izzo (2005)
6.	*Hypericum perforatum*	Photosensitivity	Skin rash, redness, burning sensation	Schmidt et al. (2005)
7.	*Pyrrolizidine alkaloids*	Hepatotoxicity	Liver damage, jaundice, abdominal pain	Ncube et al. (2008)
8.	*Ephedra* spp.	Cardiovascular toxicity	Hypertension, tachycardia, stroke	Chan (2003), Schmidt et al. (2005)
9.	*Cinnamomum cassia*	Hepatotoxicity (Coumarin)	Liver damage, fatigue, abdominal pain	Ncube et al. (2008), Jarup (2003)
10.	*Camellia sinensis*	Iron absorption inhibition	Anemia, fatigue	Calixto (2005)
11.	*Crotalaria* spp.	Hepatotoxicity (Alkaloids)	Liver cirrhosis, ascites	Jarup (2003), Schmidt et al. (2005)
12.	*Taxus* spp.	Cardiotoxicity (Taxine alkaloids)	Bradycardia, cardiac arrest	Calixto (2005)
13.	*Nerium oleander*	Cardiotoxicity (Cardiac glycosides)	Nausea, vomiting, arrhythmias	Izzo (2005)
14.	*Glycyrrhiza glabra*	Mineralocorticoid toxicity	Hypertension, hypokalemia	Tilburt and Kaptchuk (2008)
15.	*Chelidonium majus*	Hepatotoxicity	Jaundice, elevated liver enzymes	Calixto (2005)
16.	*Ruta graveolens*	Photosensitivity, neurotoxicity	Skin blisters, dizziness, seizures	Klich (2007)
17.	*Abrus precatorius*	Cytotoxicity (Abrin)	Nausea, organ failure, death	Schmidt et al. (2005)
18.	*Senna* spp.	Gastrointestinal toxicity	Diarrhea, abdominal cramping	Calixto (2005), Järup (2003)
19.	*Colchicum autumnale*	Cytotoxicity	Vomiting, diarrhea, multi-organ failure	Izzo (2005)

(continued)

Table 2.1 (continued)

Sr. No.	Plant	Toxic effect	Symptoms	References
20.	*Solanum nigrum*	Neurotoxicity	Delirium, convulsions, coma	Ncube et al. (2008)
21.	*Zingiber officinale (excess)*	Gastrointestinal discomfort	Nausea, heartburn	Schmidt et al. (2005)
22.	*Allium sativum (raw excess)*	Hemolysis in predisposed individuals	Fatigue, anemia, jaundice	Calixto (2005)
23.	*Curcuma longa (excess)*	Gastrointestinal discomfort	Bloating, nausea	
24.	*Papaver somniferum*	Respiratory depression (Opioids)	Drowsiness, coma, death	Mills and Bone (2013)
25.	*Erythroxylum coca*	Cardiovascular and CNS toxicity	Tachycardia, hyperthermia, seizures	Schmidt et al. (2005)
26.	*Cassia occidentalis*	Hepatotoxicity, myopathy	Muscle pain, jaundice, liver failure	Ncube et al. (2008)
27.	*Datura stramonium*	Neurotoxicity	Hallucinations, dry mouth, convulsions	Izzo (2005)
28.	*Cyclamen* spp.	Gastrointestinal toxicity	Vomiting, diarrhea, convulsions	Schmidt et al. (2005)
29.	*Viscum album*	Cardiovascular toxicity	Hypotension, bradycardia	
30.	*Claviceps purpurea*	Neurotoxicity (Ergot alkaloids)	Seizures, hallucinations, gangrene	

diversity in harvesting techniques and extract processing, and insufficient knowledge of active pharmacologic principles (Kumar et al. 2016).

Another difficulty in herbal medicines is safety, and it is a key component of quality control. Most of the issues connected to the use of traditional and herbal medicines seem to come from the classification of many of these products as foods or dietary supplements in certain countries. Evidence of the quality, efficacy, and safety of these herbal medicines is therefore not necessary prior to marketing. Similarly, quality tests and manufacturing standards may be less strict or regulated; in certain instances, traditional health practitioners may not be certified or licensed (Mills and Bone 2013). As a result, the safety of herbal and traditional medicines has drawn great attention from national health authorities as well as the general people. If an herbal remedy is effective, quality assurance is required to guarantee that the product has the desired effects. Quality assurance is crucial even without efficacy data since plant contamination is a major concern. Common adulterants include botanicals, hazardous metals, bacteria, microbial toxins, pesticides, and fumigation chemicals. Thus, a key factor of safety is quality (Parveen et al. 2015).

The complicated structure of herbal medicines is a significant obstacle for clinical trials. The researchers, the manufacturers, and the regulatory agencies have to use strict scientific methods and clinical trials to guarantee the quality and lot-to-lot consistency of the traditional herbal products, therefore overcoming this issue, earning public confidence, and bringing herbal products into the mainstream of today's health care system. The quality and lot-to-lot consistency of the products depend mostly on the quality control of source materials and their manufacturing into the final products since the identities of the final products are not well defined and there are practically no purification steps involved in the production of herbal products. Modern technology allows for monitoring of the quality and consistency of the heterogeneous herbal products. The preferred way to demonstrate the safety and efficacy of a therapeutic product is by means of a well-designed clinical trial (WHO 2014). Pharmacovigilance issues: The most prevalent misconception about herbal medicines is that they are very safe and hence can be used safely by the patient on his or her own, without a doctor's prescription. Often with disappointing end results, side effects, or undesired after-effects, this belief has driven large-scale self-medication by people all over the world. Developing pharmacovigilance practices for herbal medicines calls for large-scale awareness of individuals. The variety of herbal medicine increases the difficulties of herbal pharmacovigilance, including fundamental issues, such as determining the most suitable herb naming system (botanical, common, pharmaceutical name, or herbal drug name) and verification of the botanical identity of the herbal components (Zhang et al. 2014).

5 Toxicity in Medicinal Plants

Despite their significance to human health, many medicinal plants contain toxic compounds that warrant careful consideration. Lack of knowledge regarding usage and components in herbal medicines has made people to overlook its toxicity. Most people consider herbal medicines as being safe; however, clinical evidence indicates that they can be just as harmful as pharmaceutical drugs, particularly when consumed in large doses or by vulnerable individuals. This section explores the various types of toxic compounds found in medicinal plants, the factors influencing their toxicity, and the potential for herb-drug interactions and associated adverse effects (Di Nunzio et al. 2017).

5.1 Alkaloids

A group of nitrogen-containing chemicals found in plants, fungi, bacteria, and animals, alkaloids are the alkaloid structure's nitrogen atom presence that gives the compound drug qualities. While hydrogen in primary and secondary amines can serve as a proton donor, the nitrogen's lone pair electrons can accept the protons. Mostly in the higher plants, particularly in plant roots, fruits, and stem, alkaloids are found (Shi and Vakoc 2014). Natural alkaloids: future viewpoints, biological functions, and fundamental features. Both plants and animals depend on alkaloids significantly. The alkaloid derived from the beta-carboline group, for example, has been used as an anti-HIV, anti-parasitic, and antibacterial. Anticancer medications for leukemias, lymphomas, testicular cancer, breast cancer, lung cancers, and Kaposi's sarcoma have been made from the *Vinca* alkaloid (vinblastine and vincristine) isolated from *Catharanthus roseus*, berberine isolated from *Coptis* spp., and taxol isolated from *Taxus bravifolia*. Alkaloids, however, include some chemicals harmful to people; plants create alkaloids based on their needs. Certain manufactured alkaloids have been shown to harm people and animals. Hepatotoxic pyrrolizidine alkaloids (PAs), for instance, are ester alkaloids made of necine and necic acid. These alkaloids are present in various plant families such as Asteraceae, Boraginaceae and Fabaceae. The presence of a double bond at the necine moiety classifies them as toxic alkaloids. Usually, the harmful effects of PAs are linked to the conversion rate of PAs by oxidases, organism vulnerability, dosage amount, and exposure time. In biological systems, the oxidases can convert PAs to reactive pyrroles, which subsequently, the product of catalytic reaction can interact with nucleic acids and proteins. Some examples of PAs are riddelliine, retrosine, monocrotaline, symphytine, lasiocarpine, heliotrine, senkirkine, petasitenine, and indolizidine (Adibah and Azzreena 2019).

5.2 Glycosides

A glycoside is a chemical compound consisting of a sugar moiety bound to a non-sugar component (aglycone) through a glycosidic bond. Chemically, they are hydroxyls of a sugar that can create ethers with other alcohols or esters with acids. Glycosides are the molecules in which a sugar component is attached to some other non-sugar component. In living things, glycosides serve many vital functions. If these chemicals are required, plants keep significant chemicals as inactive glycosides; when the glycosides are exposed to water and an enzyme, the sugar component is broken off, therefore releasing the chemical for use. Though, if they are consumed either raw or processed poorly, their consumption might be harmful because of cyanogenic glycosides, linamarin in cassava, and taxiphillin in bamboo shoots, which break down to generate hydrogen cyanide. Sufficient processing to decompose the cyanogenic glycoside and eliminate the resulting hydrogen cyanide can greatly lower the possible toxicity of bamboo shoots and cassava. Traditional preparation methods show the need for elimination of the natural poisons before consumption. Grown for its enlarged starch-packed roots, cassava has roughly 30% starch and very little (Zaman and Azzeme 2019).

5.3 Tannins

Tannins are polyphenolic compounds found in various plants, including those used in herbal medicine. While they have antioxidant and anti-inflammatory properties, high concentrations can be toxic. Tannins can cause gastrointestinal irritation, nausea, and vomiting and may interact with medications, reducing their absorption

(Sharma et al. 2021). Toxic effects of tannins include: Gastrointestinal problems: tannins can cause stomach upset, nausea, and vomiting due to their astringent properties; liver damage prolonged exposure to high levels of tannins has been linked to liver damage and cancer; and finally Interference with medication absorption: tannins may reduce the absorption of certain medications, such as iron supplements (Pizzi 2021).

5.4 Heavy Metal, Pesticide Residue, and Microbial Contamination

Herbal plants can become toxic due to heavy metals like lead (Pb), cadmium (Cd), mercury (Hg), and arsenic (As) that can enter through soil, water, and air pollution (Rahman and Singh 2019). The release of waste from factories, mining, and smelting is the major cause of plant toxicity. Heavy metal toxicity leads to various health risks, including gastrointestinal cancer, fragile immune systems, mental growth retardation, malnutrition, and neurological and cardiovascular diseases. Poor agricultural practices like improper and excess use of pesticides, fertilizers, and wastewater irrigation introduce metals into the plant chain. Excessive spraying of pesticides on plants increases its accumulation (Yadav et al. 2023). They enter the soil and disrupt the plant uptake. Some pesticides remain settled on the surface of plants, and a lack of washing or post-harvest treatment retains residues. Some examples of toxic pesticides include aldrin, dieldrin, DDT, parathion, and endosulfan (Yadav et al. 2023). Microbial contamination in herbal plants leads to spoilage, low efficacy, impurity, and potential health risks. Some examples of toxic pesticides include aldrin, dieldrin, DDT, parathion, and endosulfan (Ovuru et al. 2024). Microbial contamination results from various reasons, such as irrigation from contaminated water and cultivation in contaminated soil. Improper drying, packaging, and storage promote microbial growth, and airborne microbes, dust, and humidity subsidize contamination.

5.5 Factors Influencing Toxicity

Naturally, most herbal medicines are not toxic; however, different factors may influence the toxicity of herbal medicines. Some of these factors and their influence on toxicity of herbal medicines, such as dose, preparation, storage, and individual variability, are discussed below:

5.5.1 Dose

A dose is the amount of medication recommended for consumption at a given time. Bose et al. (2021) addresses various elements, including self-medication, whereby dose might affect toxicity. Herbal medications can be bought off the market without a prescription. Many people often take traditional plant-based treatments like neem and tulsi as first therapy before seeing the doctors. Herbal medications are also sometimes promoted as miracle cures for several conditions people are reluctant to see a doctor about. Many people turn to herbal medicine when they have given up on the allopathic system of treatment or cannot tolerate the side effects. Many patients use herbal treatments alongside their usual allopathic drugs without their doctors' knowledge. To remain healthy, many also use herbal treatments prophylactically. All these factors could help herbal drug and remedy self-medication. Unqualified, untrained doctors can also cause incorrect dosages. Across many nations, untrained or unqualified doctors are recommending herbal treatments. Although many nations have laws and rules, enforcing them on herbal practitioners is quite challenging. Usually, medical professionals with extensive knowledge of human anatomy, drug pharmacology, and years of experience are permitted to practice herbal therapy in developed countries. Many respected universities around the world are providing degrees in alternative systems of medicine. Nearly half of the real herbal practitioners still lack formal education or training. They gain this knowledge and information from their ancestors and seniors. These practitioners are unaware themselves of the negative side effects herbal drugs could cause (Cragg and Newman 2001).

Improper intake could affect herbal medicine toxicity; doses of conventional allopathic drugs are set after years of testing and trials. The dose fixation of the allopathic drugs is mostly governed by age or body weight. Herbal medications, on the other hand, are usually exempt from such rigorous testing or regulatory processes. Sometimes herbal treatments are taken as dietary supplements; no clear dose plan is being described. Furthermore, incorrect intake might result from inadequate or improper packaging as well as lack of a cup or spoon for precise herbal drug dosing. Often, the pharmaceutical companies involved in the production and promotion of herbal medications mislead patients by asserting their product to be free of side/adverse effects. At times, people are driven to overdose themselves in order to obtain rapid relief from their suffering. Unaware of the cumulative toxicity the active herbal medications may cause on the human body, some individuals keep using such drugs over months and years (Bose et al. 2021).

5.5.2 Preparation

The preparation of herbal medicines can also be one more element affecting their toxicity. For example, many herbal medications are sold without guaranteeing consistent quality under subpar product. They fall short of the required criteria. The cause of this is insufficient testing prior to the marketing of the completed goods. Sometimes herbal recipes lack the active components stated on the labels. Adulteration could follow from wrong identification and authentication of the plant sample (Cragg and Newman 2001). Whether intentional or not, the above-mentioned mistakes could even cause complete absence of the bioactive phytol components. Intentional contamination could also result in poor quality of herbal products. Improper storage of the plant sample of the finished product could also cause efficacy loss. Unclaimed on the labels, for instance, earthy materials, minerals, and heavy metals, among other things, may also be present. Toxins or pesticides used in plant farming might possibly clarify the presence of dangerous elements in the dietary supplement. Individual variation, such as pregnant women, may also influence the toxicity of herbal remedies; non-pregnant people may experience various side effects from the same medicinal plant (Maffe et al. 2013).

5.5.3 Storage

The toxicity of herbal drugs can be greatly impacted by storage conditions. Temperature, humidity, and exposure to light are some examples of variables that might result in chemical reactions that change the herbal product's composition and possibly increase its toxicity. For example, poor storage might result in contamination from pesticides, heavy metals, or microbes like fungi that cause toxicity. Furthermore, harmful substances may be created because of the breakdown of plant parts or metabolites. Standardization and quality control procedures are necessary to guarantee the safety and authenticity of herbal therapeutic products, as well as to reduce these hazards (Jitareanu et al. 2022).

5.5.4 Herb–Drug Interactions and Adverse Effects

Given the frequency of herbal supplement use among people alongside conventional pharmacological treatments, interactions between herbal drugs and conventional medications raise particular worry. Interactions might happen through several channels, including pharmacokinetic and pharmacodynamic processes. Some herbs, like St. John's Wort (*Hypericum perforatum*), can activate the cytochrome P450 enzyme system, therefore, lowering plasma levels of co-administered drugs, including warfarin and oral contractive (Jia et al. 2022). Other herbs, such as *Ginkgo biloba*, could block the same enzyme system, therefore increasing the bleeding risk when used with anticoagulant drugs (Izzo and Ernst 2009). Abebe (2002) claims that the negative effects connected to herbal products range from moderate to severe. Some herbs, such as kava (*Piper methysticum*), have been linked to liver damage and other major health issues. Other herbs, such as ephedra (*Ephedra sinica*), have been connected to cardiovascular problems, including hypertension and arrhythmias. To reduce related concerns and offer knowledge on possible interactions and side effects, healthcare professionals should know their patients' use of herbal medicines. Doing so will help them to enable the safe and efficient use of herbal supplements.

6 Methods for Assessing Safety and Toxicity

6.1 Cytotoxicity Assays

Evaluating the negative consequences of medicinal plant extracts on both normal and malignant cells requires cytotoxicity research. These tests measure a drug's capacity to cause or stop cell death. Commonly used techniques are the MTT assay, which evaluates mitochondrial activity as a sign of cell viability, and the Trypan Blue Exclusion Test, which identifies dead cells depending on membrane integrity. Lysosomal activity and membrane damage are also assessed using the neutral red uptake and lactate dehydrogenase (LDH) leakage tests, respectively. Rapid, affordable, consistent results provided by these tests are necessary for the initial screening of plant safety profiles (Mosmann 1983; Fotakis and Timbrell 2006). Desired sensitivity, cell type, and the plant extract's character will determine the choice of test.

6.2 Mutagenicity Assays

Mutagenicity testing identifies substances that can induce genetic mutations, which may lead to carcinogenesis. The most widely used assay for this purpose is the Ames test, which uses histidine-dependent strains of *Salmonella typhimurium* to detect mutations that restore the bacteria's ability to synthesize histidine. This test can be performed with or without metabolic activation (using the S9 liver fraction) to better simulate in vivo conditions. It is widely used to screen plant extracts for potential mutagenic compounds prior to conducting further in vivo studies (Maron and Ames 1983). The Ames test is sensitive, easy to perform, and relatively inexpensive, making it a reliable first-line assay in toxicological studies of medicinal plants.

6.3 Genotoxicity Assays

Genotoxicity tests evaluate if a chemical damages cellular genetic material, therefore possibly causing mutations, chromosomal fragmentation, or other genomic instabilities. An exact technique for spotting DNA strand breakage in single cells is the comet test, sometimes known as single-cell gel electrophoresis. Often used, the micronucleus assay finds chromosomal fragments or whole chromosomes lagging during mitosis, therefore generating small nuclei in daughter cells. Particularly with regard to long-term medical use, these tests are vital for assessing the genotoxic capacity of plant extracts (Tice et al. 2000; Fenech 2000). By improving mutagenicity tests and providing a more in-depth knowledge of the processes behind DNA, damage caused by phytochemicals, in vitro genotoxicity testing helps to clarify mutagenicity tests.

6.4 Acute Toxicity Studies

Acute toxicity studies are conducted to evaluate the harmful effects of a single dose or multiple doses administered within a 24-hour period. These studies help determine the median lethal dose (LD_{50}), which is the amount required to cause death in 50% of the test population, typically rodents. Observation periods typically last 14 days, during which signs of toxicity, behavioral changes, mortality, and gross pathological changes are recorded. These studies provide preliminary safety data essential for setting dose ranges in further toxicological assessments and clinical trials. The OECD guideline 423 (Acute Oral Toxicity—Acute Toxic Class Method) is widely adhered to for standardization purposes. Medicinal plant extracts, as complex mixtures of bioactive compounds, require assessments to confirm that they do not produce significant toxic effects at therapeutic doses (OECD 2001; Parasuraman 2011).

6.5 Sub-Chronic Toxicity Profile

Sub-chronic toxicity studies are designed to evaluate the effects of repeated exposure to a substance over an intermediate duration, typically 28 to 90 days. These studies are crucial for understanding the cumulative toxic effects of medicinal plant extracts when used over a moderate period, reflecting more realistic therapeutic scenarios. Parameters assessed include changes in body weight, food and water intake, hematological and biochemical profiles, as well as histological examinations of vital organs such as the liver, kidneys, and heart. Respectively, the OECD recommendations 407 and 408 offer protocols for 90-day oral toxicity studies and 28-day ones. For plants with possible long-term medicinal application, these studies are particularly crucial since they guarantee no sublethal toxicities build up over time (OECD 2008; Mukinda and Syce 2007).

6.6 Chronic Toxicity Studies

Chronic toxicity studies evaluate the health impacts associated with extended exposure to a medicinal plant extract, typically conducted over a duration of 6 months or longer. These studies are essential for assessing the potential carcinogenicity, reproductive toxicity, or systemic organ damage linked to long-term consumption. Comprehensive monitoring includes clinical signs, mortality rates, hematological and biochemical indices, organ weights, and thorough histopathological evaluations. Chronic studies are crucial for determining the No-Observed-Adverse-Effect Level (NOAEL), which is fundamental in assessing safe exposure levels for humans. Due to the growing interest in herbal preparations for chronic conditions, long-term studies are essential for establishing regulatory safety profiles (WHO 2004; Wagstaff 2010).

7 Clinical Trials Using Medicinal Plants

Establishing the safety and efficacy of therapeutic herbs for human use depends on clinical research. These studies are phased: Phase I looks at safety and dose in a small group of healthy volunteers; Phase II assesses efficacy and side effects in a larger sample of patients; Phase III verifies therapeutic benefits and monitors adverse reactions in a more general population. Because of their complicated phytochemical makeup and differences in plant sources, preparation techniques, and dosages, medicinal plant-derived products sometimes present particular difficulties in clinical research. Reproducibility and regulatory approval depend on strict standardization and good clinical practice (GCP) policies. Integrating contemporary trial design frameworks, botanical clinical studies have to also consider ethnopharmacological knowledge and traditional use patterns. To close the gap between conventional use and evidence-based therapy, the World Health Organization (WHO) stresses the importance of scientifically sound clinical assessment of herbal medicines (WHO 2000; Ekor 2014).

8 Future Perspectives and Innovations

Advanced analytical techniques are essential for establishing the safety of medicinal plants since they guarantee efficacy, purity, and regulatory compliance. These processes help to discover contaminants, standardize herbal formulations, and identify active components. Modern analytical techniques help to ensure the safety, purity, and efficacy of medicinal plants before they are used in pharmaceuticals, nutraceuticals, and herbal formulations (Garg 2020). The popularity of herbal products also raises questions about their safety and control. Certain plants could reveal their negative side effects with prescribed medications. Currently, regulatory bodies like the European Medicines Agency and the U.S. Food and Drug Administration are working to improve control and verify the safety of herbal plants and supplements (Ekor 2014). Traditionally used in human health practices, medicinal plants are a major component of contemporary well-being and medicine. Though the focus on sustainable and scientifically validated therapies is increasing, plants are still a vital component of both conventional and modern healthcare systems. Though their use in conventional medicine is anticipated to rise, several studies on the safety and effectiveness of plant-based medications offer fresh prospects for the treatment of several ailments (Davis and Choisy 2024). Despite the long history of traditional medical use of many medicinal plants, modern scientific research is necessary to confirm their efficacy and safety. Regulation incorporates evidence-based practices while confirming that traditional knowledge is valued for the benefit of human health (Wang et al. 2023).

9 Conclusions

Medicinal plants hold immense potential in modern healthcare, but their widespread use necessitates a robust framework for safety, toxicity assessment, and regulatory compliance. Scientific advancements, coupled with stringent regulations, will help bridge the gap between traditional knowledge and evidence-based medicine. Future research should prioritize standardized methodologies, clinical validation, and sustainable practices to ensure the safe and effective integration of medicinal plants into mainstream healthcare. A collaborative approach between researchers, policymakers, and healthcare professionals will be key to harnessing the full potential of medicinal plants while minimizing risks.

References

Abebe H (2002) Potential Entomopathogenic Fungi for the Control of Macrotermes subhyalinus (Isoptera: Termitidae) Ph D Thesis, Hannover

Adhami M, Haghdoost AA, Sadeghi B, Malekpour Afshar R (2018) Candidate miRNAs in human breast cancer biomarkers: a systematic review. Breast Cancer 1:198–205

Adibah KZ, Azzreena MA (2019) Plant toxins: alkaloids and their toxicities. GSC Biol Pharm Sci 6:21–29

Bhowmik D, Chiranjib, Dubey P, Chandira M, Kumar S (2009) Herbal drug toxicity and safety evaluation of traditional medicines. Arch Appl Sci Res 1(2):32–56

Bose S, Datta R, Kirlin W (2021) Toxicity studies related to medicinal plants. In: Evidence based validation of traditional medicines: a comprehensive approach. Springer Singapore, Singapore, pp 621–647

Calixto JB (2005) Twenty-five years of research on medicinal plants in Latin America: a personal view. J Ethnopharmacol 100:131–134

Chan K (2003) Some aspects of toxic contaminants in herbal medicines. Chemosphere 52:1361–1371

Cragg GM, Newman DJ (2001) Natural product drug discovery in the next millennium. Pharm Biol 39:8–17

David B, Wolfender JL, Dias DA (2015) The pharmaceutical industry and natural products: historical status and new trends. Phytochem Rev 14:299–315

Davis CC, Choisy P (2024) Medicinal plants meet modern biodiversity science. Curr Biol 34:R158–R173

Di Nunzio M, Valli V, Tomás-Cobos L, Tomás-Chisbert T, Murgui-Bosch L, Danesi F, Bordoni A (2017) Is cytotoxicity a determinant of the different in vitro and in vivo effects of bioactives? BMC Complement Altern Med 17:1–4

Ekor M (2014) The growing use of herbal medicines: issues relating to adverse reactions and challenges in monitoring safety. Front Pharmacol 14:177

Fenech M (2000) The in vitro micronucleus technique. Mutat Res/Fundam Mol Mech Mutagen 455:81–95

Fotakis G, Timbrell JA (2006) In vitro cytotoxicity assays: comparison of LDH, neutral red, MTT and protein assay in hepatoma cell lines following exposure to cadmium chloride. Toxicol Lett 160:171–177

Garg R (2020) A textbook of pharmacognosy & phytochemistry-I Edition 2020. Sia Publishers & Distributers

Graham PH, Browne L, Cox H, Graham J (2003) Inhalation aromatherapy during radiotherapy: results of a placebo-controlled double-blind randomized trial. J Clin Oncol 21:2372–2376

Gurib-Fakim A (2006) Medicinal plants: traditions of yesterday and drugs of tomorrow. Mol Asp Med 27:1–93

Hoffmann D (2003) Medical herbalism: the science and practice of herbal medicine. Simon and Schuster

Izzo AA (2005) Herb–drug interactions: an overview of the clinical evidence. Fundam Clin Pharmacol 19:1–6

Izzo AA, Ernst E (2009) Interactions between herbal medicines and prescribed drugs: an updated systematic review. Drugs 69:1777–1798

Jarup L (2003) Hazards of heavy metal contamination. British medical bulletin 68(1):167–82

Jia E, Lu C, Lin G, Ge G (2022) Phamaconekits of herbal medicines and herb- drug interactions. Front Pharmacol 13:1107777

Jitareanu A, Trifan A, Vieriu M, Caraba L (2022) Current trends in toxicity assessment of herbal medicines: a narrative review. Processes 11(1):83

Kenner D, Requena Y (2001) Botanical medicine: a European professional perspective. Paradigm publications

Klemow KM, Bartlow A, Crawford J, Kocher N, Shah J, Ritsick M (2011) Herbal medicine: biomolecular and clinical aspects, vol 2. CRC Press, pp 211–218
Klich MA (2007) *Aspergillus flavus*: the major producer of aflatoxin. Mol Plant Pathol 8:713–722
Kumar A, Venkatesh MP, Kumar PT (2016) Regulations and challenges of herbal medicines in Russia. Int J Ayurvedic Herb Med 6:2149–2161
Leonti M (2011) The future is written: impact of scripts on the cognition, selection, knowledge and transmission of medicinal plant use and its implications for ethnobotany and ethnopharmacology. J Ethnopharmacol 134:542–555
Leonti M, Casu L (2013) Traditional medicines and globalization: current and future perspectives in ethnopharmacology. Front Pharmacol 4:92
Maffè S, Paffoni P, Colombo L, Davanzo F, Dellavesa P, Cucchi L, Zenone F, Paino AM, Bergamasco L, Signorotti F, Parravicini U (2013) Herbs and cardiotoxic effects. G Ital Cardiol 14:445–455
Maron DM, Ames BN (1983) Revised methods for the Salmonella mutagenicity test. Mutat Res/Environ Mutagen Relat Subj 113:173–215
Mills S, Bone K (2000) Principles and practice of phytotherapy. Modern herbal medicine. Churchill Livingstone, Edinburgh.
Mills S, Bone K (2013) Principles and practice of phytotherapy: modern herbal medicine, 2nd edn. Churchill Livingstone
Mosmann T (1983) Rapid colorimetric assay for cellular growth and survival: application to proliferation and cytotoxicity assays. J Immunol Methods 65:55–63
Mponda JS, Muula AS, Choko A, Ajuwon AJ, Moody JO (2025) Traditional medicine regulation status and challenges in Malawi and Nigeria. BMC Complement Altern Med 25:53
Mukinda JT, Syce JA (2007) Acute and chronic toxicity of the aqueous extract of *Artemisia afra* in rodents. J Ethnopharmacol 112:138–144
Ncube NS, Afolayan AJ, Okoh AI (2008) Assessment techniques of antimicrobial properties of natural compounds of plant origin: current methods and future trends. Afr J Biotechnol 7:1797
OECD (2001) OECD guidelines for the testing of chemicals: acute oral toxicity – acute toxic class method (test no. 423)
OECD (2008) Repeated dose 90-day oral toxicity study in rodents (test no. 408)
Ovuru KF, Izah SC, Yasmin H, Enerijiofi KE, Das M, Ogwu MC (2024) Microbial contaminants of herbal remedies: health risks and sustainable quality control strategies. In: Herbal medicine phytochemistry: applications and trends. Springer International Publishing, Cham, pp 1571–1600
Pan SY, Litscher G, Gao SH, Zhou SF, Yu ZL, Chen HQ, Zhang SF, Tang MK, Sun JN, Ko KM (2014) Historical perspective of traditional indigenous medical practices: the current renaissance and conservation of herbal resources. J Evid Based Complement Altern Med 2014:525340
Parasuraman S (2011) Toxicological screening. J Pharmacol Pharmacother 2:74–79
Parfait BT, Lawrence MT (2023) Pharmacognosy, phytotherapy and modern medicine. Int Jr Infect Dis Epidemol 23:1
Parveen A, Parveen B, Parveen R, Ahmad S (2015) Challenges and guidelines for clinical trial of herbal drugs. J Pharm Bioallied Sci 7:329–333
Pizzi A (2021) Tannins medical/pharmacological and related applications: a critical review. Sustain Chem Pharm 22:100481
Rahman Z, Singh VP (2019) The relative impact of toxic heavy metals (THMs) (arsenic (As), cadmium (Cd), chromium (Cr) (VI), mercury (Hg), and lead (Pb)) on the total environment: an overview. Environ Monit Assess 191:1–21
Ramawat KG, Goyal S (2008) The Indian herbal drugs scenario in global perspectives. In: Bioactive molecules and medicinal plant, vol 16. Springer, Berlin, Heidelberg, pp 325–347
Roy Upton RH (2015) Traditional herbal medicine, pharmacognosy, and pharmacopoeial standards: a discussion at the crossroads. In: Evidence-based validation of herbal medicine. Elsevier, pp 45–85
Sarecka-Hujar B, Szulc-Musioł B (2022) Herbal medicines-are they effective and safe during pregnancy? Pharmaceutics 14:171

Schmidt MS, Prisinzano TE, Tidgewell K, Harding W, Butelman ER, Kreek MJ, Murry DJ (2005) Determination of Salvinorin A in-body fluids by high performance liquid chromatography–atmospheric pressure chemical ionization. J Chromatogr B 818:221–225

Sendker J, Sheridan H (2017) History and current status of herbal medicines. In: Toxicology of herbal products. Springer, Cham, pp 11–27

Sharma K, Thakur IS, Kaushik G (2021) Occurrence and distribution of pharmaceutical compounds and their environmental impacts: a review. Bioresour Technol Rep 16:100841

Shi J, Vakoc CR (2014) The mechanisms behind the therapeutic activity of BET bromodomain inhibition. Mol Cell 54:728–736

Tice RR, Agurell E, Anderson D, Burlinson B, Hartmann A, Kobayashi H, Sasaki YF (2000) Single cell gel/comet assay: guidelines for in vitro and in vivo genetic toxicology testing. Environ Mol Mutagen 35:206–221

Tilburt JC, Kaptchuk TJ (2008) Herbal medicine research and global health: an ethical analysis. Bull World Health Organ 86:594–599

Wagstaff DJ (2010) Chronic toxicity studies. In: Hayes AW (ed) Principles and methods of toxicology, 5th edn. CRC Press, pp 1143–1165

Wang J, Chitsaz F, Derbyshire MK, Gonzales NR, Gwadz M, Lu S, Marchler GH, Song JS, Thanki N, Yamashita RA, Yang M (2023) The conserved domain database in 2023. Nucleic Acids Res 51:D384–D388

WHO (2000) General guidelines for methodologies on research and evaluation of traditional medicine. World Health Organization, Geneva

WHO (2004) Guidelines on safety monitoring of herbal medicines in pharmacovigilance systems. World Health Organization, Geneva

Williamson EM (2003) Drug interactions between herbal and prescription medicines. Drug Saf 26:1075–1092

World Health Organization (2014) Guidelines on clinical evaluation of traditional medicines in the WHO African region. World Health Organization Regional Office for Africa

World Health Organization (2019) WHO global report on traditional and complementary medicine 2019. World Health Organization

Yadav P, Singh RP, Gupta RK, Pradhan T, Raj A, Singh SK, Pandey KD, Kumar A (2023) Contamination of soil and food chain through wastewater application. In: Advances in chemical pollution, environmental management and protection, vol 9. Elsevier, pp 109–132

Zhang J, Wider B, Shang H, Li X, Ernst E (2014) Quality of herbal medicines: challenges and solutions. Complement Ther Med 20:100–106

Mechanisms of Hepatoprotection: Key Phytochemicals and Pathways

3

Ambreen Malik Uttra, Urooj Iqbal, and Sumera Qasim

Abstract

The liver is a crucial organ that regulates the body's metabolism, detoxification, and immunity, among various other functions. Chronic liver diseases (CLDs) present considerable health concern, especially in the developing countries, leading to around two million deaths each year. Conventional treatment therapies often entail significant side effects, highlighting an urge to introduce safer alternatives. Plant-based bioactive phytochemicals are grabbing interest due to their hepatoprotective properties, providing a natural way to mitigate liver damage caused by various hepatotoxins, drug-induced injuries, and chronic liver disease. Phytochemicals like flavonoids, saponins, alkaloids, polyphenols, and terpenoids directly attenuate reactive oxygen species (ROS) through the nuclear factor erythroid 2–related factor 2 (Nrf2) signaling pathway. Moreover, some phytochemicals downregulate nuclear factor-kappa B (NF-κB), hepatic stellate cells (HSCs), and transforming growth factor-beta (TGF-β) as a hepatoprotection mechanism. Furthermore, modulation of mitochondrial and death receptor-mediated apoptotic pathways, autophagy, and AMP-activated protein kinase (AMPK) pathways

A. M. Uttra
Department of Pharmacology, College of Pharmacy, University of Sargodha, Sargodha, Pakistan
e-mail: ambreen.malik@uos.edu.pk

U. Iqbal
Department of Pharmacology, College of Pharmacy, University of Sargodha, Sargodha, Pakistan

Primary and Secondary Health Care Department, Lahore, Punjab, Pakistan

S. Qasim (✉)
Department of Pharmacology, College of Pharmacy, Jouf University, Sakaka, Saudi Arabia

K. Kuca et al. (eds.), *Medicinal Plants and Their Bioactives in Human Diseases*,
https://doi.org/10.1007/978-3-032-01356-9_3

by phytochemicals can contribute to CLD management. This chapter provides an in-depth review of the natural hepatoprotective mechanisms that may protect and rebuild liver health, paving the way for plant-based treatment options for CLD.

Keywords

Hepatoprotective · Phytochemicals · Oxidative stress · Liver injury pathways · Inflammation · Fibrosis

1 Introduction

The liver performs several important tasks that help us to keep healthy, from macronutrient metabolism and protein synthesis to bile production and vitamin storage. Chronic liver disease (CLD) leads to a progressive decline in these functions, most often due to alcoholic misuse, inflammation, and metabolic disorders leading to fibrosis and eventually cirrhosis (Yu et al. 2024). CLD involves inflammation, cell death, and dysbiosis of gut microbiota related to chronic inflammation, resulting in exacerbated gut barrier dysfunction and impaired liver function. The growing global incidence of CLD contributes to approximately two million deaths worldwide annually. Therefore, liver diseases are now becoming more popular and a major public health concern due to poor lifestyles, environmental pollutants, and hereditary factors (Juanola et al. 2021).

Conventional therapies for chronic liver diseases face several limitations that hinder their effectiveness and accessibility. A significant challenge is the lack of effective pharmacotherapies, particularly for conditions like primary biliary cirrhosis (PBC) and alcoholic liver disease, where existing treatments such as ursodeoxycholic acid and corticosteroids have shown suboptimal efficacy and limited innovation over decades (Petrasek et al. 2015). Whereas antiviral treatments for hepatitis C virus (HCV) show a success rate of approximately 50% in clinical trials, with even lower rates in diverse patient populations, compounded by barriers such as patient preferences and comorbidities (Morrill et al. 2005). Additionally, treatments like ursodeoxycholic acid (UDCA) for cholestatic liver diseases leave a significant portion of patients at risk for disease progression, highlighting the need for more personalized and multi-targeted therapeutic strategies (Wagner and Fickert 2020). Furthermore, the systemic side effects and non-specific drug delivery associated with conventional therapies limit their efficacy, as seen in the challenges faced by patients with metabolic dysfunction-associated steatotic liver disease (MASLD) regarding antiretroviral therapy (Heya et al. 2023; Kalopitas et al. 2024). Overall, drawbacks surrounding the conventional treatment modalities advocate the need for alternative strategies such as phytoconstituent-based therapies, which may improve the treatment efficacy in chronic liver ailments.

Hepatoprotective agents of plant origin have emerged as extremely significant in light of the increasing incidence of liver injuries due to synthetic drug treatment and due to toxins. Various medicinal plants, such as *Silybum marianum, Amaranthus spinosus, Glycyrrhiza glabra, Moringa oleifera, Garcinia mangostana, Andrographis paniculata, , Mangifera indica,* various Phyllanthus, Curcuma, and Citrus species, have been recognized for their hepatoprotective properties, primarily attributed to their rich phytochemical profiles, including flavonoids, terpenoids, and phenolic compounds (Gonfa et al. 2024; Widyawati et al. 2024). These compounds exhibit several pharmacological attributes, including antioxidant and anti-inflammatory activity and ability to stabilize liver cell membranes, which minimizes hepatotoxicity (Służały et al. 2024; Motwani et al. 2021). Phytoconstituents are attracting more attention in the management of liver dysfunctions due to their cost-effectiveness and lower toxicity than synthesized drugs (Minnady et al. 2022). Although the market for herbal medications is $83 billion globally, which indicates the rising popularity and extensive use of herbal medicines (Shikh et al. 2024). Overall, hepatoprotective strategies from plant-based sources translate into a holistic approach to liver rescue, as the myriad of bioactive components available from nature are used to combat against liver toxicity (Siddiqui et al. 2024).

2 Phytochemicals with Hepatoprotective Properties

Phytochemicals, or bioactive compounds from plant sources, are the compounds that plants create for their own protection. Many come from foodstuffs such as whole grains, vegetables, fruits, seeds, herbs, and nuts. Phytochemicals are categorized into two groups, which include primary metabolites that comprises of chlorophylls, nucleic acids, sugars, nucleic acids, and proteins; while secondary metabolites include flavonoids, alkaloids, lignans, curcumins, terpenes, steroids, saponins, glucosides, and phenolics. Other activities associated with these secondary metabolites include antidiarrheal, antiallergic, antimicrobial, anthelmintic, antioxidant, antiviral and antispasmodic activities (Kumar et al. 2023; Mendoza and Silva 2018). Due to their diversity of mechanisms of action, medicinal plant-derived phytochemicals are considered a potential reservoir of hepatoprotective activities (Gonfa et al. 2024). Hepatoprotective phytochemicals and potential mechanisms have been summarized in Table 3.1.

3 Mechanisms of Hepatoprotection

3.1 Oxidative Stress and Antioxidant Defense

In humans, oxidative stress through imbalance between oxidative species production and hepatic antioxidant defense system, is the main cause of hepatic injury. Asymmetry towards ROS leads to overproduction of ROS, crucial mediators of hepatic injury in a host of liver disorders including sepsis-induced liver injury, MAFLD (metabolic associated fatty liver disease) as well as chronic liver disease including fibrosis and cirrhosis (Su et al. 2024; Seidita et al. 2024; Machado et al. 2023). ROS play a pivotal role in liver injury via activating pro-inflammatory

Table 3.1 Phytochemicals targeting chronic hepatic disorders

Classification of Phytocompounds	Phytocompounds	Mechanism/target	Hepatoprotective effects	References
Flavonoids	Luteolin	Inhibition of apoptosis, and suppression of ferroptosis through the stimulation of GPX4 and SLC7A11	Protecting the liver by alleviating cell morphology, reducing serum AST, ALT, and ALP levels, enhancing cell viability	Han et al. (2024)
	Quercetin	Inhibition of GSDMD cleavage in macrophages, blocking ASC assembly, and preventing NOD-like receptor protein 3 (NLRP3) and AIM2 inflammasome formation	Reduces macrophage pyroptosis, thereby alleviating tissue necrosis and inflammation during hepatic ischemia-reperfusion injury	Lin et al. (2024)
	Apigenin and chrysin	Enhance intracellular glutathione levels and inhibit oxidative stress through ERK2 Nrf2 ARE signaling pathways	Reduce oxidative damage and promotes antioxidant gene transcription	Özcan et al. (2020)
	Cyanidin	Enhances Nrf2 signaling and inhibits NF-κB pathways	Alleviates liver injury induced by CCl_4 via anti-inflammatory and anti-oxidative attributes	Wang et al. (2022)
	Butin	Modulates/decreased expression of IL6, TNF-α and IL-1β and lipid peroxidation	Reduced hepatic markers and improved histopathological outcomes against D-galactosamine mediated rats liver damage	Althurwi et al. (2023)
	Morin	Up regulates the SOD, GPx, CAT levels, as well as Nrf2 and HO-1 pathways modulation	Modulates oxidative stress and hepatic inflammation	Kızıl et al. (2023)

Saponins	Quillaja saponaria bark saponin	Modulates Bcl-2 and Bax and ameliorates inflammation via Nrf-2 and HO-1 pathways.	Suppression of oxidative stress and reduction of apoptosis.	Abdel-Reheim et al. (2022)
	Glycyrrhizin	Nrf2/HO-1 signaling pathway activation, inhibition of autophagy, oxidative stress, inflammation, and apoptosis	Reduction of inflammatory markers and improved antioxidant enzyme activities in HIRI (hepatic ischemia-reperfusion injury) rats	Kou et al. (2020)
	Gypenosides	Activation of the farnesoid X receptor (FXR) signaling pathway and inhibition of HSC activation	Improved liver lipid and bile acid metabolism reducing extracellular matrix deposition and ameliorates hepatic fibrosis	Li et al. (2022a, 2023b)
	Tea saponins	Inhibition of NF-KB pathway	Alleviating liver inflammation via reduction of TNF-α and interleukins expression	Wu et al. (2022)
	Diosgenin	Nrf2 pathway activation	Enhances antioxidant defenses by alleviating oxidative stress along with ferroptosis in hepatocytes	Zhang et al. (2024a)
	Ginsenosides RD	Inhibits STAT3-mediated pyroptosis	Reduces inflammation and liver damage	Li et al. (2024a)
Alkaloids	Berberine	Enhances endogenous antioxidants levels and suppresses NF-κB, TNF-α signaling	Alleviates oxidative stress mediated inflammation	Phogat et al. (2023)
	Betaine	Enhances AMPK and ACC phosphorylation, reducing ER stress, apoptosis	Inhibits steatosis and improves lipid metabolism in NAFLD	Seo et al. (2024)
	Palmatine	Autophagy modulation via the AMPK/mTOR signaling pathway	Maintains cellular homeostasis and prevents hepatocyte apoptosis	Lin et al. (2022)
	Glaucine	Inhibits the TGFβ-induced phosphorylation of Smad2	Reduces collagen, fibronectin, and alpha-SMA expression in hepatocytes.	Azamov et al. (2023)
	Protopine	Inhibits MAPK and NF-κB signaling	Reduces pro-inflammatory cytokines and certain enzymes levels including COX-2 and MMP-9, thereby attenuating inflammation in hepatic cells	Kim et al. (2022)
	Sophocarpine	Inhibits NF-κB signaling pathway Inhibits PI3K/AKT pathway	Reduced pro-inflammatory cytokines expression (IL-1β, IL-6, and TNF-α). Reduces apoptosis with improved cell survival	Jiang et al. (2018) and Chen et al. (2023)

(continued)

Table 3.1 (continued)

Classification of Phytocompounds	Phytocompounds	Mechanism/target	Hepatoprotective effects	References
Polyphenols	Resveratrol	Decreases ER (endoplasmic reticulum) stress and NF-κB pathway inhibition	Reduces collagen deposition and inflammation in hepatic fibrosis	Ma et al. (2022)
	Curcumin	Targets TLR4/NF-κB and TGF-β/Smad3 signaling pathways	Suppresses inflammation in NAFLD to hepatocellular carcinoma	Li et al. (2024c)
	Epigallocatechin-3-Gallate	Inhibition of pro-inflammatory and fibrotic signaling pathways, such as TGFβ/p-ERK/p-Smad1/2	Enhances antioxidant defenses, reduces oxidative stress and inflammation	Mostafa-Hedeab et al. (2022)
	Ellagic Acid	Nrf2 pathway activation	In liver cells enhances NQO1, CAT, and GPX1 expression, thereby reducing oxidative stress	Yu et al. (2024)
	Caffeic acid	Nrf2 pathway activation	Enhances heme oxygenase-1 (HO-1) and glutamate-cysteine ligase (GCL) levels	Yang et al. (2019)
	Chlorogenic Acid	Nrf2/HO-1 activation and JNK signaling pathway inhibition	Mitigates oxidative stress, enhances the cellular antioxidant response and reduction of ER stress and intracellular free radicals	Wang et al. (2023) and Li et al. (2023c)
Terpenoids	Glycyrrhizic acid	Mitigates thioacetamide-induced hepatic fibrosis, reduces oxidative stress and fibrotic markers	Inhibition of HMGB1-Beclin 1 autophagy pathway	Nassar et al. (2024)
	Geraniol	Nrf2/HO-1 signaling pathway activation	Enhances antioxidant defenses by increasing glutathione levels and reducing oxidative stress markers like malondialdehyde	El-Emam et al. (2020)
	Curcumol	Hits P53-TFR1-FTH1 signaling axis	Inhibits angiogenesis and potentially mitigating liver fibrosis	Wang et al. (2024)
	Carnosic Acid	Activates AMP-dependent kinase (AMPK)	Suppresses gluconeogenic and lipogenic gene expressions against FLD (fatty liver disease)	Hasei et al. (2021)
	Limonene	Down-regulates lipogenic transcription factors (FASN and SREBP1) as well as pro-inflammatory cytokines levels (IL-6, IL-1β, and TNF-α), while up-regulates NRF2/HO-1 pathway	Enhances liver protection by mitigating oxidative stress, reducing lipid accumulation and stamping down inflammation	Li et al. (2022b) and Amini et al. (2020)
	α-Humulene	Targets Akt signaling pathway.	Inhibits cell proliferation and induces apoptosis hepatocellular carcinoma.	Chen et al. (2019)

mechanisms including NF-κB and NLRP3 inflammasome which are involved in the synthesis and release of cytokines and fibrogenic factors. This activates HSC, which gives rise to fibrosis and cirrhosis (Ramos-Tovar and Muriel 2023). Oxidative stress causes worsening of mitochondrial dysfunction and stimulates lipid peroxidation, leading to liver injury (Banerjee et al. 2023).

Enzymatic and non-enzymatic mechanisms oppose detrimental effects of ROS-induced hepatocyte injury. Antioxidant enzymes, including CAT (catalase), GST (glutathione S-transferase), SODs (superoxide dismutase), GPs (glutathione peroxidases), and quinone reductase (QR), are indispensable in the antioxidant defense system. GSH and NADPH are non-enzymatic antioxidants and integral components of antioxidant defense mechanism. A range of non-enzyme antioxidants (i.e., phytochemicals), such as α-tocopherol and ascorbic acid, also play a major role in ROS detoxification (Venmathi Maran et al. 2022). To minimize oxidative stress, systems of antioxidant defense, including endogenous and exogenous antioxidants/supplements (e.g., polyphenols, vitamins and dietary changes), are required. The Mediterranean Diet, involving high intake of extra virgin olive oil, improves antioxidant capacity by decreasing liver steatosis (Seidita et al. 2024). In addition, the up-regulation of antioxidant response elements via the activation of Nrf2 provides another protective mechanism against oxidative damage (Ramos-Tovar and Muriel 2023).

Oxidative stress disrupts redox balance and activates inflammatory pathways and acts as a driving force behind the progression of liver diseases such as non-alcoholic fatty liver disease (NAFLD), alcoholic liver disease (ALD), and hepatocellular carcinoma (HCC) (Fernández-Iglesias et al. 2017; Ezhilarasan et al. 2020; Allameh et al. 2023). Many phytochemicals, including polyphenols such as quercetin, resveratrol, and curcumin, proved a potential restorative principle to reverse oxidative damage by modulating ROS and improving antioxidant defense (Machado et al. 2023). In particular, curcumin has been known to decrease the proinflammatory cytokines and lower the lipid peroxidation by inducing the antioxidant proteins, such as SOD, Nrf2, and GSH (Farzaei et al. 2018). Furthermore, the ethanolic extract of Polygonum multiflorum Thunb. (PME) has hepatoprotective effects via Nrf2 activation and regulation of bile acid transporters, reducing oxidative stress-induced liver injury (Lin et al. 2018).

In NAFLD, many antioxidants like silymarin and resveratrol modulate oxidative stress and inflammation; however, their clinical efficacy remains unclear (Ezhilarasan and Lakshmi 2022; Ezhilarasan and Langeswaran 2024). Owing to its metabolic and detoxification function, however, the liver is highly susceptible to oxidative stress, which necessitates a well-functioning antioxidant defense system that is key to maintaining hepatic health and preventing pathological conditions (Allameh et al. 2023). Consequently, there is ongoing hope that promoting exogenous antioxidant defenses through dietary and pharmacological approaches will be a useful intervention in liver disease with an oxidative stress component (Aghemo et al. 2022).

3.2 Nrf2 Pathway Activation

In mammals, a complex antioxidant system has evolved to ensure redox homeostasis in the liver. Disruption of this homeostasis leads to the release of ROS, such as superoxide anion, hydrogen peroxide, alkoxyl radical, hydroxyl radical, peroxyl radical, and singlet oxygen, which can intrinsically activate HSCs, which are a known critical component in the pathogenesis of liver diseases as well as other chronic and degenerative disorders (Arroyave-Ospina et al. 2021). With chronic liver disease, an increase in ROS occurs to oxidize membranes. Lipid peroxides are then transformed to aldehydes, including malondialdehyde (MDA) and 4-hydroxynonenal (4-HNE), which cause additional mitochondrial and DNA damage. This cycle leads to a buildup of dysfunctional mitochondria and DNA mutation. NRF2 (nuclear factor erythroid 2–related factor 2) is an essential transcription factor that mediates cellular defense against oxidative stress induced by a high amount of ROS. As shown in Fig. 3.1, under physiological conditions, the activity

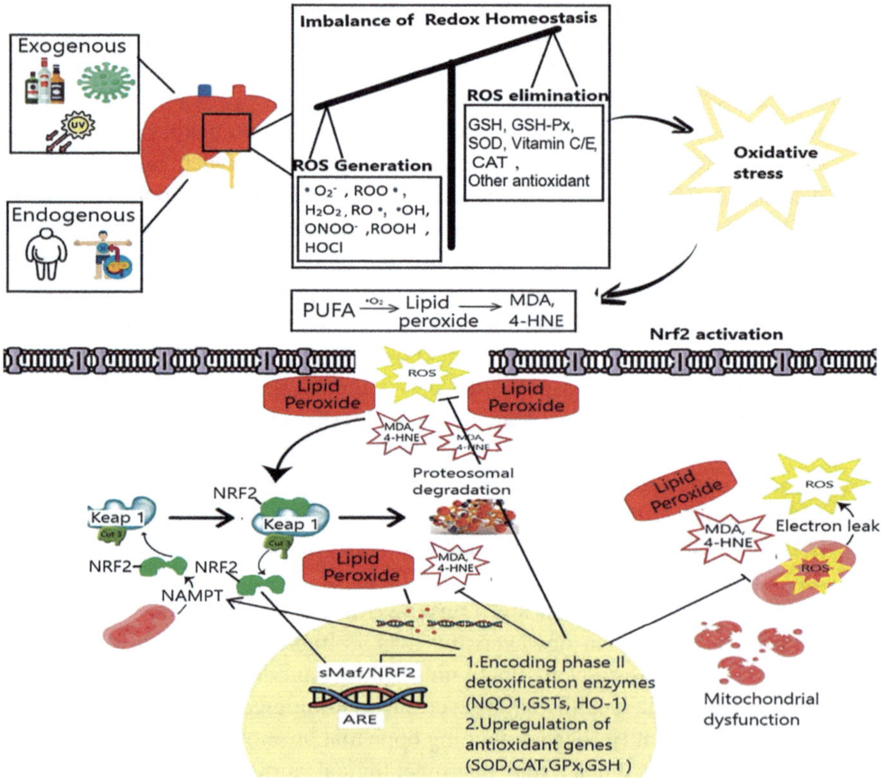

Fig. 3.1 Oxidative stress and Nrf2 activation in chronic liver disease. Key elements include Polyunsaturated fatty acids (PUFAs), small musculoaponeurotic fibrosarcoma oncogene homolog (sMAF), Nicotinamide Phosphoribosyl transferase (NAMPT), antioxidant response elements (AREs), reactive oxygen species (ROS), Kelch-like ECH-associated protein 1 (KEAP1), glutathione peroxidase (GPx), reduced glutathione (GSH), NAD(P)H quinone oxidoreductase 1 (NQO1), superoxide dismutase (SOD), catalase (CAT), glutathione S-transferases (GSTs), and heme oxygenase-1 (HO-1)

of NRF2 is tightly regulated through multiple posttranslational modifications. KEAP1 (Kelch-like ECH-associated protein 1) is a redox sensor that forms an E3 ubiquitin ligase complex with CULLIN3 that induces the ubiquitinoylation and degradation of NRF2. This tight control of NRF2 activity prevents uncontrolled activation of cytoprotective genes (Park et al. 2023).

Under physiological conditions, NRF2 exists in a complex with Keap1-Cul3 that leads to its proteasomal degradation. NRF2-activator drugs and oxidative stress modify KEAP1. NRF2 is accumulated and translocated into the nucleus, makes a heterodimer with small musculoaponeurotic fibrosarcoma oncogene homolog (sMAF), and binds to AREs located in promoters of target genes to activate transcription (Shaw and Chattopadhyay 2020). NRF2 activation drives a transcriptional program that includes many cytoprotective genes, such as phase II detoxification genes, including heme oxygenase-1 (HO-1), glutathione S-transferases (GSTs), and NAD(P)H quinone oxidoreductase 1 (NQO1), which are necessary for neutralizing toxic compounds and reactive metabolites. NRF2 then enters the nucleus, where detoxifying genes regulated by the antioxidant response element (ARE) are up-regulated, including for reduced glutathione (GSH), catalase (CAT), glutathione peroxidase (GPx), and superoxide dismutase (SOD), which all serve to neutralize lipid peroxides and aldehydes, lower ROS levels, and inhibit lipid peroxidation (Ramos-Tovar and Muriel 2020). Nicotinamide Phosphoribosyl transferase (NAMPT) is a protein that acts within mitochondria, which in a feedback loop increases the activity of NRF2 and mitochondrial function. It is helpful in maintaining homeostasis and cellular adaptation toward different stressors by decreasing HSCs activation, involving both antioxidant/lipid metabolism, protein degradation, and regulation of inflammation and other pathways. Therefore, it is important for liver health and the prevention of oxidative damage (Zhou et al. 2022a).

Flavonoids, polyphenols, and specific compounds such as sulforaphane, curcumin, and resveratrol are grouped among the most effective phytochemicals that target the Nrf2 pathway. Fruits and vegetables are rich sources of flavonoids and have been extensively investigated for their Nrf2 pathway-modulating ability to enhance antioxidant and anti-inflammatory responses, which play a key role in liver disease progression, particularly cirrhosis and fibrosis (Zhou et al. 2022b; Sharma et al. 2024). The most common and well-known Nrf2 activating compound is sulforaphane, derived from broccoli, which disrupts the interaction of Keap1 and Nrf2 and increases expression of cytoprotective genes (Dinkova-Kostova et al. 2017). Curcumin and resveratrol are both naturally occurring polyphenols that have been demonstrated to remodel the DNA methylation patterns of Nrf2, thereby reversing the epigenetic silencing and augmenting its antioxidative and chemo preventive functions (Divyajanani et al. 2023; Belka et al. 2024). Fermented *Syzygium samarangense* leaves (FWAL) have demonstrated protective effects against acetaminophen (APAP)-induced liver damage through the upregulation of Nrf2 and its downstream targets (e.g., HO-1 and NQO1) in addition to modulation of epigenetic factors, such as DNMTs and HDACs, to further promote Nrf2 expression (Huang et al. 2024). Also, the mixture of flavolignans derived from milk thistle known as silymarin has been recognized to be an Nrf2 bio activator having hepatoprotective role through modulating Nrf2 signaling pathway (Vargas-Mendoza et al. 2020). Such natural compounds may have a leading role in the prevention and treatment of chronic liver diseases through a targeted modulation of the Nrf2 pathway.

3.3 Anti-inflammatory Mechanisms

Inflammation is an evolutionary conserved survival mechanism that is a response to the presence of pathogens, but also to chemical or mechanical stress. that damages tissues. Overexpression of pro-inflammatory cytokines, ROS, and chemokines mainly produced by macrophages and various immune cells can induce critical cellular damage and play an essential role in the development of many inflammatory diseases (Zlatanova et al. 2024). Chronic inflammation is an important factor in the pathogenesis and progression of many chronic liver diseases and is a link between metabolic dysregulation and disease development (Tanwar et al. 2020). Kupffer cells and neutrophils are the main cellular participants in the liver injury inflammatory response in the pathological process (Zhang et al. 2019). NF-κB is a central mediator of inflammatory responses and a regulator of innate and adaptive immunity. NF-κB protein family consists of NF-κB1 (p105/p50), NF-κB2 (p100/p52), RelA (p65), RelB, and c-Rel. These subunits go on to form either homodimers or heterodimers, which is how NF-κB acts. Once generated, NF-κB moves to the nucleus and interacts with the consensus sequence on DNA, called the κB site, to drive target gene transcription. When NF-κB is not activated, NF-κB is maintained in the cytoplasm in an inactive form bound to IκB proteins in the cytosol, including IκBα, IκBβ, and IκBε. The activation of NF-κB can follow either the classical or non-classical pathways, as illustrated in Fig. 3.2 (Blaj et al. 2023).

The classical pathway, mostly driven by the RelA-p50 heterodimer can be prompted by inflammatory stimuli, including lipopolysaccharide, inflammatory cytokines (TNF-α, IL-1, IL-6), and physical/chemical stimuli. Stimulation of the classical pathway activates transforming growth factor-beta-activated kinase 1 (TAK1), which then phosphorylates the IκB kinase (IKK) complex, which in turn phosphorylates IκBα, causing its ubiquitination and degradation. As a result, the RelA-p50 dimer is released, translocates into the nucleus, and regulates various pro-inflammatory genes expressions (Sun 2017; Li et al. 2023a). Other than that, the non-classical pathway, which is activated by a number of different types of extracellular stimuli (e.g., B cell-activated receptor (BAFF-R), lymphotoxin β receptor (LTβR), and CD40-activated NIK kinase), has lesser activation stimulation. The N-terminal of p100 displays self-inhibition in the resting state, preventing the transcriptional activity of p100/Rel B (Kendellen et al. 2014). After stimulation of cells, this pathway activates NF-κB-inducing kinase (NIK), which phosphorylates IKKα. Upstream, IKKα (IκB kinase α) activates and subsequently pathways lead to phosphorylation of the C-terminal region of p100, culminating in the generation of p52. After a sequential series of phosphorylations, the p52-RelB dimer translocates into the nucleus, initiating transcription of immune cell developmental target genes of NF-κB (Yang et al. 2024). In chronic liver diseases, the activation of kupffer cells

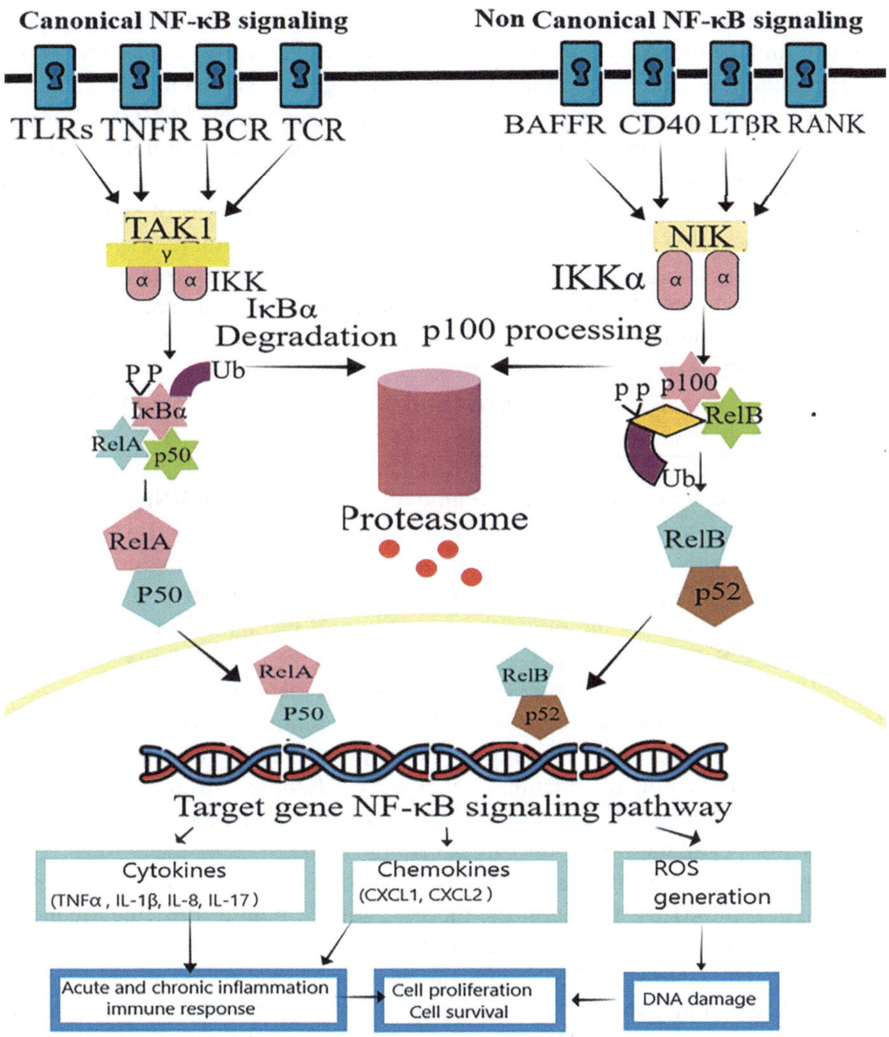

Fig. 3.2 Canonical and non-canonical NF-κB signaling pathways. Receptors such as toll-like receptor (TLR), tumor necrosis factor receptor (TNFR), B cell receptor (BCR), and T cell receptor (TCR) activate the canonical pathway. Subsequent to activation of this cascade, phosphorylation and degradation of inhibitory protein IκB take place, and NF-κB gets activated by release from the complex with IκB, displacement, and translocation into the nucleus. The non-canonical pathway depends on triggering the NF-κB2 (p100)/RelB complex via the receptor activator of nuclear factor kappa (RANK), cluster of differentiation 40 (CD40), lymphotoxin-β receptor (LTβR), and B cell activating factor receptor (BAFFR). This cascade culminates in phosphorylation of NIK, then IKKα. The p52-RelB heterodimer is now activated and translocates to the nucleus. NF-κB activation governs multiple cellular responses through the regulation of gene expression associated with cytokines, chemokines, and others

triggers an NF-κB signaling pathway (downstream of the previous signaling pathways) to induce the expression of pro-inflammatory cytokines, such as TNF-α and IL-1β, chemokines, adhesion molecules, and other genes, which ultimately leads to cellular process changes and promotes liver injury (Peng et al. 2020). Such proinflammatory factors ultimately activate macrophage inflammatory protein 2 (MIP-2), which brings neutrophils to clump and adhere at the site of the injury. Degranulation of neutrophils recruited to the injury site releases great amounts of proteases and ROS and extends the inflammatory response, recruiting other immune system members (liver cell damage). NF-κB pathway inhibition plays a key role in HIRI, and can effectively inhibit the expression level of intercellular adhesion molecule-1 protein, cytokine-induced neutrophil chemoattractant, macrophage inflammatory protein 2, and TNF-α (Deng and Zhou 2024).

Several phytochemicals have been identified to modulate the NF-κB inflammatory pathway in liver disorders, offering potential therapeutic benefits. Resveratrol, a small polyphenol, has been shown to protect against liver injury induced by Toxoplasma gondii infection by intervening in the HMGB1/TLR4/NF-κB signaling pathway, thereby reducing inflammatory markers and liver damage (Lu et al. 2021). Likewise, isoflavones of *Radix Astragali*, including methylnissolin and methylnissolin-3-O-β-D-glucoside exerted the effect of hepatoprotective and anti-inflammatory actions by inhibiting the expression of pro-inflammatory cytokines, including iNOS, COX-2, IL-1β, IL-6, and TNF-α, via the NF-κB signaling pathway (Aobulikasimu et al. 2023). Other therapeutic agents that have been investigated in NAFLD effectively include silymarin, silibinin, curcumin, resveratrol, berberine, and quercetin, which modulate various pathways but mainly the NF-κB pathway, leading to a reduction in inflammatory markers (Ezhilarasan and Langeswaran 2024). Likewise, the garcinol of the *Garcinia dulcis* inhibits the secretion of pro-inflammatory cytokines (TNF-α, Il-6, and IL-1β) by modulating the NF-κB signaling pathway, preventing the translocation of pNF-κB from the cytosol to the nucleus (Chantree et al. 2023). Furthermore, the traditional Chinese medicine Huangqin decoction has been reported to alleviate hepatic inflammation in NAFLD via inhibition of the TLR4/NF-κB/NLRP3 pathway, with baicalein, baicalin, and liquiritin showing high affinity for this pathway (Yan et al. 2023). Garlic has several phytochemicals that are involved in the interaction process of the NF-κB inflammation pathway by affecting its signaling molecules. Inhibition of NF-κB nuclear translocation, which is important for the transcription of pro-inflammatory cytokines, is one of the main mechanisms. For example, water-soluble garlic polysaccharides were shown to inhibit the phosphorylation of IκB-α that results in the inactivation and nucleus translocation of NF-κB in lipopolysaccharide (LPS)-induced macrophages and to decrease the production of pro-inflammatory cytokines, such as IL-6, TNF-α, and IL-1β (Shao et al. 2023). Combined, these results suggest the ability of several phytochemicals to modulate the NF-κB pathway, as well as their potential for alleviating inflammation in liver disorders, highlighting their promise for the treatment of such chronic hepatic conditions.

3.4 Anti-fibrotic Pathways

In the healthy liver, HSCs are in a quiescent, non-replicative state, in which the cells have a star-shaped morphology and intracellular lipid droplet stores arise from retinyl palmitate vitamin A. HSC activation is a central step in the development of liver fibrosis and cirrhosis. Once activated, HSCs lose intracellular lipid droplets, adopt a myofibroblast-like morphology, and produce high levels of α-smooth muscle actin (α-SMA) and extracellular matrix (ECM), which result in liver fibrosis (Patidar et al. 2024). TGF-β is activated from the deposits in the ECM and expressed and released from various cell types during both acute and chronic liver injury (Tsuchida and Friedman 2017). Damage associated with the liver signals the release of TGF-β molecules from a large latent complex (LLC) following integrins binding to latent accessory protein (LAP). As depicted in Fig. 3.3, binding of free TGF-β to the TGF-β type II receptor (TβRII) leading to TGF-β type I receptor (TβRI) recruitment. Subsequently, TβRII phosphorylates TβRI at a Gly-Ser–rich (GS) domain to induce a conformational switch in TβRI and potentiate their binding/phosphorylation of its substrates, i.e., the SMAD2 and SMAD3 proteins (also known as receptor-activated SMADs or R-SMADs). Phosphorylation of the SMAD proteins at their C-terminal domain leads to the formation of SMAD2/3/4 heterocomplexes, which then translocate to the nucleus where they bind DNA and regulate transcription of various target genes such as αSMA and CTGF.

There are two noteworthy features of SMAD2: On the one hand, it does not bind to DNA, and on the other, SMAD3 has a low affinity for DNA binding. As a result, SMAD2/3/4 complexes typically attract additional transcriptional coactivators to stabilize transactivation complexes. Second, R-SMADs are capable of inducing various TGF-β target genes without SMAD4. Other TGF-β-activated signaling pathways or non-canonical TGF-β, SMAD-independent pathways, as seen here with MAPK, mTOR, PI3K/AKT, and Rho/GTPase pathways (Dewidar et al. 2019; Fabregat and Caballero-Díaz 2018). The TGF-β/Smads signaling pathway is particularly significant, with TGF-β1 being a potent pro-fibrotic cytokine that promotes HSC activation through pathways involving Smad3, Stat3, and Akt, as well as the epigenetic reader BrD4, which is crucial for profibrotic gene expression (Zhang et al. 2024b, Xu et al. 2023).

Phytochemicals have shown potential in inhibiting TGF-β signaling, a critical pathway in liver fibrogenesis, by targeting various mechanisms involved in HSC activation and ECM production (Fig. 3.3). Quercetin promotes significant reduction of hepatic fibrogenesis by blockade of TGF-β/Smad3 signaling pathway and decreases expression of fibrotic markers (α-SMA, collagen1α1) in HSCs (Shakerian et al. 2022). Resveratrol acts the most through the activation of an SIRT1, a NAD+-dependent deacetylase, which inhibits the TGF-β/Smad signaling pathway. Such inhibition is performed by increasing the upregulation of Smad7, an axis antagonist of TGF-β signaling, and decreasing the levels of TGF-β1, p-Smad3 and α-SMA biomarkers for fibrotic changes (Xia et al. 2024). In addition, the ellagitannins pedunculagin, casuarinin, and gallic acid from *Osbeckia octandra* have also been reported to attenuate liver fibrogenesis via the TGF- β/SMAD signaling pathway

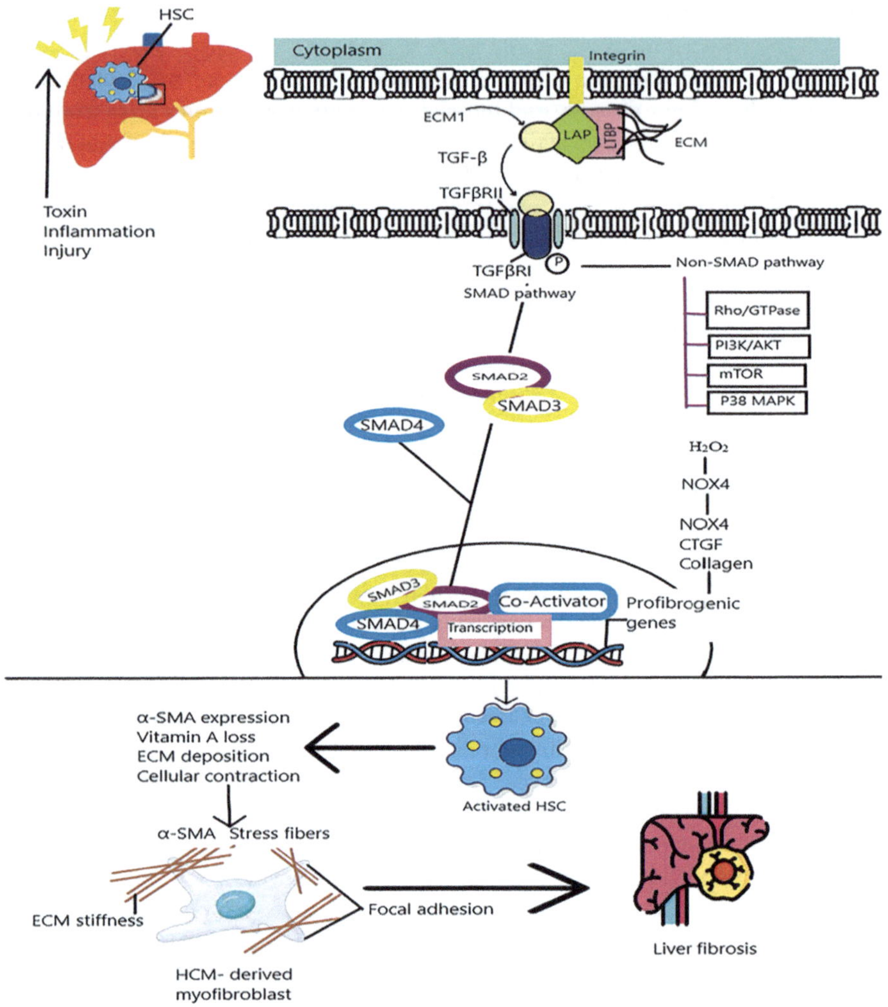

Fig. 3.3 Mechanism of liver fibrosis. Liver fibrosis is characterized by the activation of HSCs via TGF-β signaling, with Smad-dependent and Smad-independent pathways involved. Activated HSCs exhibit a phenotypic switch marked by loss of cytoplasmic lipid droplets, the acquisition of fibroblast-like morphology, and upregulation of α-SMA and ECM proteins

(Bogahawaththa et al. 2023). Emulating this, neo agaro oligosaccharides have also been demonstrated to block TGF-β/Smad signaling and reduce fibrogenic gene expression and collagen deposition in both animal and cell culture models (Yang et al. 2021). Moreover, the NPLC0393 extracted from *Gynostemma pentaphyllum* could regulate the TGF-β1/NDRG2/MAPK axis to inhibit the activation of HSC and exert an antifibrotic effect (Huang et al. 2020). Furthermore, TGF-β1/miR-195/Smads signaling pathway activates oxymatrine obtained from the *Sophorae Flavescentis Radix* that inhibited HSC activation effectively in a dose- and time-dependent manner (Gu et al. 2019). Overall, such plant-derived bioactive compounds provide a perspective way to develop novel treatments for liver fibrosis via targeting the TGF-β signaling pathway.

3.5 Role of Autophagy in Hepatoprotection

Autophagy is the process through which damaged organelles, misfolded proteins, and other cellular debris are degraded and recycled. This process is critical to cell-cell homeostasis and governs cell survival under stress conditions. There are three main types of autophagy: macro autophagy, micro autophagy, and chaperone-mediated autophagy. Of these, macro autophagy is the most well-characterized and involves the formation of double-membraned vesicles known as autophagosomes that enclose cytoplasmic contents destined to be degraded in lysosomes. Mechanistic target of rapamycin (mTOR), AMP-activated protein kinase (AMPK), and Beclin-1; are a few signaling pathways that regulate autophagy. These pathways are vital for sensing energy and nutrient status within the cell and adapting autophagic activity to preserve cell integrity (Gómez-Virgilio et al. 2022).

In the liver, autophagy is a protective process against different types of cellular stress such as oxidative stress, endoplasmic reticulum stress and lipid overload. The liver's role in processing and detoxifying harmful substances makes it especially prone to damage, and autophagy is key to alleviating some of these effects. Autophagy can remove damaged mitochondria via mitophagy, which decreases reactive oxygen species (ROS) production, thus providing protection at the mechanistic level. Through a specific type of autophagy called lipophagy, it can eliminate excess stored lipids, thus avoiding lipotoxicity and maintaining lipid homeostasis. Moreover, the mechanism of autophagy also plays a protective impact on hepatocytes by eliminating aggregated proteins and damaged organelles to prevent cell death including apoptosis and necrosis (Czaja et al. 2013). These processes work together to support the liver's adaptation to metabolic stress and recovery from injury (Fig. 3.4).

Disruption of autophagy contributes to the pathophysiology of various liver diseases, such as non-alcoholic fatty liver disease (NAFLD), alcoholic liver disease (ALD), and liver fibrosis. In the case of NAFLD, failing autophagy initiates

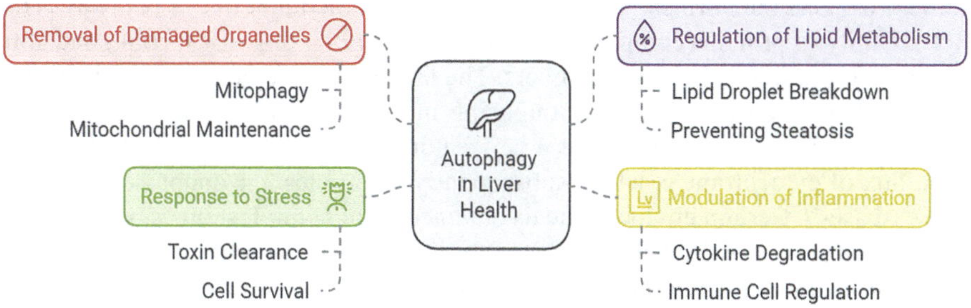

Fig. 3.4 Hepatoprotective role of autophagy during stress environment

lipid accumulation and inflammation, promoting the progression of the disease (Shen et al. 2024). Likewise, in ALD, inadequate autophagy leads to the buildup of toxic lipid metabolites and ROS, further perpetuating hepatocyte injury. Autophagy has a role in hepatic stellate cell (HSC) activation, which leads to oversecretion of extracellular matrix that is associated with liver fibrosis (Yu et al. 2022). Therapeutic potential autophagic reactivation in such settings has been associated with dampened disease severity. For instance, enhancing autophagy can improve lipid clearance, reduce oxidative stress, and suppress inflammatory responses, offering a multi-faceted approach to liver protection.

Several agents have demonstrated hepatoprotective effects through the modulation of autophagy. Everolimus, an mTOR inhibitor, has been shown to activate autophagy, conferring protection against ischemia-reperfusion injury in the liver (Lee et al. 2016; Zhu et al. 2022). Similarly, betaine, a natural compound, promotes autophagy to alleviate non-alcoholic fatty liver disease in mice, primarily through its role in reducing oxidative stress and lipid accumulation (Chen et al. 2022). Rutin (*Flos Sophorae Immaturus*) and curcumin induce NHSC (non-chemically induced HSC) autophagy by stimulating fatty acids by regulating the PI3K/AKT/mTOR pathway, resulting in the inhibition of NHSC activation (Chang et al. 2023). These results suggest the contribution of targeting autophagy to a therapeutic strategy for liver diseases, providing new insights into the pharmacological and natural approaches for hepatoprotection.

3.6 Apoptosis and Cell Survival Pathways in Hepatoprotection

Apoptosis, the mechanism of programmed cell death, is important for controlling hepatic homeostasis and clearing damaged or dysfunctional hepatocytes. Nevertheless, dysregulated apoptosis is involved in a range of liver pathologies such as hepatitis, NAFLD, and drug-induced liver injury. There are two main mechanisms for the regulation of apoptotic pathways in the liver: the intrinsic (mitochondrial) pathway and the extrinsic (death receptor-mediated) pathway. Both pathways result in the activation of caspases, a family of proteases that carry out apoptosis through the cleavage of cellular proteins, ultimately resulting in cell death (Cao et al. 2016).

Mitochondrial dysfunction that occurs by oxidative stress, toxic insult, or DNA damage triggers the intrinsic pathway. An important mediator of this pathway is the Bcl-2 family of proteins comprised of both pro-apoptotic (e.g., Bax, Bak) and anti-apoptotic (e.g., Bcl-2, Bcl-xL) members. The fate of hepatocytes is decided because of the balance between these two opposing members. Mitochondrial outer membrane permeabilization (MOMP) is a key feature of intrinsic apoptosis and enables the efflux of cytochrome c, which subsequently assembles the apoptosome to activate caspase-9. Certain phytochemicals obtained from natural sources, such as silymarin from milk thistle, have exhibited the property of stabilizing mitochondrial membranes and upregulating anti-apoptotic proteins, which provide protection against hepatocyte death (Karimi et al. 2011).

Conversely, the extrinsic pathway is activated upon the engagement of death ligands (i.e., FasL, TNF-α) to the associated death receptors on the external surface of the cell. This triggers formation of the death-inducing signaling complex (DISC) and activation of caspase-8 that can directly activate executioner caspases, such as caspase-3. However, depending on the circumstances, excessive activation of this pathway may aggravate tissue damage in liver diseases, such as hepatitis and cholestasis (Guicciardi and Gores 2010). Some phytochemicals, such as quercetin and curcumin, show the ability to downregulate death receptor expression and inhibit the activity of caspase-8, which provides hepatoprotective effects (Li et al. 2024b; Sohn et al. 2021). A second major pathway involved in the maintenance of hepatocyte survival is the phosphoinositide-3-kinase/AKT signaling (PI3K/Akt) cascade that stimulates cell survival and proliferation and actively inhibits apoptotic pathways. For example, activated Akt phosphorylates and inactivates pro-apoptotic proteins such as Bad, shifting the balance towards cell survival. Furthermore, Akt upregulates expression of anti-apoptotic factors such as Bcl-2 and encourages cellular repair systems. The active components (ginsenosides) of ginseng are known to activate the PI3K/Akt pathway, thereby preventing oxidative and inflammatory stressor-induced apoptosis of hepatocytes (Ghafouri-Fard et al. 2022). Phytochemicals that display hepatoprotective properties and modulate such apoptosis and survival pathways often assist in the resumption of liver functionality alongside prevention from further damage. For instance, grape-derived resveratrol reverses mitochondrial apoptosis-driven damage by activating SIRT1 and AMPK, which in turn inhibit pro-apoptotic pathways and stimulate mitochondrial biogenesis and cellular repair (Wu et al. 2024). Such dual action enhances the liver's resilience to stress and improves overall recovery. The regulation of apoptosis by phytochemicals thus represents a promising avenue for therapeutic intervention in liver diseases.

4 Conclusions

Chronic liver diseases (CLDs) are one of the significant health challenges worldwide with a significant burden of diseases due to oxidative stress, inflammation, and fibrosis. Conventional treatments are restricted by limited efficacy and systemic adverse effects, which motivates the exploration of phytochemicals for potential clinical use as an alternative. These bioactive compounds derived from plants, including flavonoids, polyphenols, and terpenoids, have been shown to possess hepatoprotective effects through alterations in oxidative stress, modulation of relevant signaling pathways (such as Nrf2 and NF-κB), inhibition of fibrosis via TGF-β/Smads pathways, and enhancement of autophagy. Their multi-functional properties provide a holistic approach to tackling the multi-mechanistic pathogenesis of liver injury. This chapter explores the target molecular pathways of phytochemicals and their natural interventions for promoting liver health. Phytochemicals show great potential in treating CLDs based on their antioxidant, anti-inflammatory, and anti-fibrotic activities. This is, of course, a preliminary study, and further research is necessary to transform these results into clinically applicable therapies.

References

Abdel-Reheim AM, Ashour AA, Khattab AM, Gaafar AA (2022) *Quillaja saponaria* bark saponin attenuates methotrexate-induced hepatic oxidative stress, inflammation and associated liver injury in rats. J Appl Pharm Sci 12:129–141

Aghemo A, Alekseeva OP, Angelico F, Bakulin IG, Bakulina NV, Bordin D, Bueverov AO, Drapkina OM, Gillessen A, Kagarmanova EM (2022) Role of silymarin as antioxidant in clinical management of chronic liver diseases: a narrative review. Ann Med 54:1548–1560

Allameh A, Niayesh-Mehr R, Aliarab A, Sebastiani G, Pantopoulos K (2023) Oxidative stress in liver pathophysiology and disease. Antioxidants 12:1653

Althurwi HN, Altharawi A, Alharthy KM, Albaqami FF, Alzarea SI, Al-Abbasi FA, Nadeem MS, Kazmi I (2023) Butin prevents liver damage triggered by D-galactosamine via regulating lipid peroxidation and proinflammatory cytokines in rodents. J King Saud Univ Sci 35:102934

Amini R, Asle-Rousta M, Aghazadeh S (2020) Hepatoprotective effect of limonene against chronic immobilization-induced liver damage in rats. Naunyn Schmiedeberg's Arch Pharmacol 393:2053–2059

Aobulikasimu N, Zheng D, Guan P, Xu L, Liu B, Li M, Huang X, Han L (2023) The anti-inflammatory effects of isoflavonoids from *Radix Astragali* in hepatoprotective potential against LPS/D-gal-induced acute liver injury. Planta Med 89:385–396

Arroyave-Ospina CJ, Wu Z, Geng Y, Moshage H (2021) Role of oxidative stress in the pathogenesis of non-alcoholic fatty liver disease: implications for prevention and therapy. Antioxidants 10:174

Azamov B, Lee K-M, Hur J, Muradillaeva S, Shim W-S, Lee C, Song P (2023) Oxoglaucine suppresses hepatic fibrosis by inhibiting TGFβ-induced Smad2 phosphorylation and ROS generation. Molecules 28:4971

Banerjee P, Gaddam N, Chandler V, Chakraborty S (2023) Oxidative stress-induced liver damage and remodeling of the liver vasculature. Am J Pathol 193:1400

Belka M, Gostyńska-Stawna A, Stawny M, Krajka-Kuźniak V (2024) Activation of Nrf2 and FXR via natural compounds in liver inflammatory disease. Int J Mol Sci 25:11213

Blaj AL, Cucu IA, Tamba IB, Turliuc DM (2023) The role of the NF-κB pathway in intracranial aneurysms. Brain Sci 13:1660

Bogahawaththa S, Kawamura T, Bandaranayake U, Hirakawa T, Yamada G, Ishino H, Hirohashi T, Kawaguchi S-I, Wijesundera KK, Wijayagunawardane MP (2023) Identification and mechanistic investigation of ellagitannins from *Osbeckia octandra* that attenuate liver fibrosis via the TGF-β/SMAD signaling pathway. Biosci Biotechnol Biochem 87:1295–1309

Cao L, Quan X-B, Zeng W-J, Yang X-O, Wang M-J (2016) Mechanism of hepatocyte apoptosis. J Cell Death 9:JCD.S39824

Chang J, Huang C, Li S, Jiang X, Chang H, Li M (2023) Research progress regarding the effect and mechanism of dietary polyphenols in liver fibrosis. Molecules 29:127

Chantree P, Martviset P, Thongsepee N, Sangpairoj K, Sornchuer P (2023) Anti-inflammatory effect of garcinol extracted from *Garcinia dulcis* via modulating NF-κB signaling pathway. Nutrients 15:575

Chen H, Yuan J, Hao J, Wen Y, Lv Y, Chen L, Yang X (2019) α-Humulene inhibits hepatocellular carcinoma cell proliferation and induces apoptosis through the inhibition of Akt signaling. Food Chem Toxicol 134:110830

Chen W, Xu M, Xu M, Wang Y, Zou Q, Xie S, Wang L (2022) Effects of betaine on non-alcoholic liver disease. Nutr Res Rev 35:28–38

Chen X-J, He M-H, Zhang X, Liao S-G, Zeng Z, Li Z, Tu B, Long Q-D, Lin Y (2024) Kurarinol A from *Sophora flavescens* as potential anti-liver fibrosis agent that inhibits the activation of LX-2 cells by regulating TGF-β/Smads signaling pathway

Chen Y, Lei Y, Wang H, Wang L, Xu J, Wang S, Yu M, Peng Z, Xiao F, Tian D (2023) Sophoricoside attenuates autoimmune-mediated liver injury through the regulation of oxidative stress and the NF-κB signaling pathway. Int J Mol Med 52:1–13

Czaja MJ, Ding W-X, Donohue TM, Friedman SL, Kim J-S, Komatsu M, Lemasters JJ, Lemoine A, Lin JD, Ou J-HJ (2013) Functions of autophagy in normal and diseased liver. Autophagy 9:1131–1158

Deng R-M, Zhou J (2024) Targeting NF-κB in hepatic ischemia–reperfusion alleviation: from signaling networks to therapeutic targeting. Mol Neurobiol 61:3409–3426

Dewidar B, Meyer C, Dooley S, Meindl-Beinker N (2019) TGF-β in hepatic stellate cell activation and liver fibrogenesis—updated 2019. Cells 8:1419

Dinkova-Kostova TA, Fahey WJ, Kostov VR, Kensler WT (2017) KEAP1 and done? Targeting the NRF2 pathway with sulforaphane. Trends Food Sci Technol 69:257–269

Divyajanani S, Harithpriya K, Ganesan K, Ramkumar K, Mohanram K (2023) Dietary polyphenols remodel DNA methylation patterns of NRF2 in chronic disease. Nutrients 15:3347

El-Emam ZS, Soubh AA, Al-Mokaddem KA, Abo El-Ella MD (2020) Geraniol activates Nrf-2/HO-1 signaling pathway mediating protection against oxidative stress-induced apoptosis in hepatic ischemia-reperfusion injury. Naunyn Schmiedeberg's Arch Pharmacol 393:1849–1858

Ezhilarasan D, Lakshmi T (2022) A molecular insight into the role of antioxidants in nonalcoholic fatty liver diseases. Oxidative Med Cell Longev 2022:9233650

Ezhilarasan D, Langeswaran K (2024) Hepatocellular interactions of potential nutraceuticals in the management of inflammatory NAFLD. Cell Biochem Funct 42:e4112

Ezhilarasan D, Rajeshkumar S, Lakshmi T (2020) The role of oxidative stress in chronic liver diseases. In: Role of oxidative stress in pathophysiology of diseases. Springer, Singapore, pp 13–25

Fabregat I, Caballero-Díaz D (2018) Transforming growth factor-β-induced cell plasticity in liver fibrosis and hepatocarcinogenesis. Front Oncol 8:357

Farzaei MH, Zobeiri M, Parvizi F, El-Senduny FF, Marmouzi I, Coy-Barrera E, Naseri R, Nabavi SM, Rahimi R, Abdollahi M (2018) Curcumin in liver diseases: a systematic review of the cellular mechanisms of oxidative stress and clinical perspective. Nutrients 10:855

Fernández-Iglesias A, Hide D, Gracia-Sancho J (2017) Oxidative stress in liver diseases. In: Gastrointestinal tissue. Academic, pp 125–140

Ghafouri-Fard S, Balaei N, Shoorei H, Hasan SMF, Hussen BM, Talebi SF, Taheri M, Ayatollahi SA (2022) The effects of ginsenosides on PI3K/AKT signaling pathway. Mol Biol Rep 49:6701–6716

Gómez-Virgilio L, Silva-Lucero MDC, Flores-Morelos DS, Gallardo-Nieto J, Lopez-Toledo G, Abarca-Fernandez AM, Zacapala-Gómez AE, Luna-Muñoz J, Montiel-Sosa F, Soto-Rojas LO (2022) Autophagy: a key regulator of homeostasis and disease: an overview of molecular mechanisms and modulators. Cells 11:2262

Gonfa YH, Bachheti A, Semwal P, Rai N, Singab AN, Bachheti RK (2024) Hepatoprotective activity of medicinal plants, their phytochemistry, and safety concerns: a systematic review. Z Naturforsch C 80

Gu Y, Lu J, Sun W, Jin R, Ohira T, Zhang Z, Tian X (2019) Oxymatrine and its metabolite matrine contribute to the hepatotoxicity induced by radix Sophorae tonkinensis in mice. Exp Ther Med 17:2519–2528

Guicciardi ME, Gores GJ (2010) Apoptosis as a mechanism for liver disease progression. Semin Liver Dis 30:402–410

Han Z, Batudeligen CH, Narisu A, Xu Y, Xue L (2024) Luteolin attenuates CCl4-induced hepatic injury by inhibiting ferroptosis via SLC7A11. BMC Complement Med Ther 24:193

Hasei S, Yamamotoya T, Nakatsu Y, Ohata Y, Itoga S, Nonaka Y, Matsunaga Y, Sakoda H, Fujishiro M, Kushiyama A (2021) Carnosic acid and carnosol activate AMPK, suppress expressions of gluconeogenic and lipogenic genes, and inhibit proliferation of HepG2 cells. Int J Mol Sci 22:4040

Heya MS, García-Ponce R, Soto BAM, Verde-Star MJ, Soto-Domínguez A, García-Hernandez DG, Saucedo-Cárdenas O, Hernández-Salazar M, Guillén-Meléndez GA (2023) Green alternatives in treatment of liver diseases: the challenges of traditional medicine and green nanomedicine. Chem Biodivers 20:e202300463

Huang H, Wang K, Liu Q, Ji F, Zhou H, Fang S, Zhu J (2020) The active constituent from Gynostemma pentaphyllum prevents liver fibrosis through regulation of the TGF-β1/NDRG2/MAPK axis. Front Genet 11:594824

Huang SH, Lin ZH, Wu TY, Wei GJ, Tung YC, Su ZY (2024) Fermented *Syzygium samarangense* leaves mitigate oxidative stress-triggered hepatotoxicity through Nrf2 activation and epigenetic regulation. Food Biosci 57:103518

Jiang Z, Meng Y, Bo L, Wang C, Bian J, Deng X (2018) Sophocarpine attenuates LPS-induced liver injury and improves survival of mice through suppressing oxidative stress, inflammation, and apoptosis. Mediat Inflamm 2018:5871431

Juanola O, Martínez-López S, Francés R, Gómez-Hurtado I (2021) Non-alcoholic fatty liver disease: metabolic, genetic, epigenetic and environmental risk factors. Int J Environ Res Public Health 18:5227

Kalopitas G, Arvanitakis K, Tsachouridou O, Malandris K, Koufakis T, Metallidis S, Germanidis G (2024) Metabolic dysfunction-associated steatotic liver disease in people living with HIV—limitations on antiretroviral therapy selection. Life 14:742

Karimi G, Vahabzadeh M, Lari P, Rashedinia M, Moshiri M, Garbhapu A, Marzban M (2011) "Silymarin", a promising pharmacological agent for treatment of diseases. Iran J Basic Med Sci 14:308–317

Kendellen FM, Bradford WJ, Lawrence LC, Clark SK, Baldwin SA (2014) Canonical and non-canonical NF-κB signaling promotes breast cancer tumor-initiating cells. Oncogene 33:1297–1305

Kim M, Kim H, Kim H (2022) Anti-inflammatory effect of protopine through MAPK and NF-κB signaling regulation in HepG2 cell. Molecules 27:4601

Kızıl HE, Caglayan C, Darendelioğlu E, Ayna A, Gür C, Kandemir FM, Küçükler S (2023) Morin ameliorates methotrexate-induced hepatotoxicity via targeting Nrf2/HO-1 and Bax/Bcl2/Caspase-3 signaling pathways. Molecular Biology Reports, 50(4):3479-3488

Kou X, Zhu J, Xie X, Hao M, Zhao Y (2020) The protective effect of glycyrrhizin on hepatic ischemia-reperfusion injury in rats and possible related signal pathway. Iran J Basic Med Sci 23:1232

Kumar A, Nirmal P, Kumar M, Jose A, Tomer V, Oz E, Proestos C, Zeng M, Elobeid T, Sneha K (2023) Major phytochemicals: recent advances in health benefits and extraction method. Molecules 28:887

Lee S, Kim K, Kim O, Lee S, Kim S (2016) Activation of autophagy by everolimus confers hepatoprotection against ischemia–reperfusion injury. Am J Transplant 16(7):2042–2054

Li H, Niu X, Zhang D, Qu M-H, Yang K (2023a) The role of the canonical NF-κB signaling pathway in the development of acute liver failure. Biotechnol Genet Eng Rev 39(1):775–795

Li H, Xi Y, Liu H, Xin X (2022a) Gypenosides ameliorate high-fat diet-induced non-alcoholic steatohepatitis via farnesoid X receptor activation. Front Nutr 9:914079

Li W, Kong Y, Zhang C (2024a) Ginsenoside RD prevents acute liver injury in mice by inhibiting STAT3-mediated NLRP3/GSDMD activation. J Biochem Mol Toxicol 38(1):e23825

Li H, Wang H, Yang A, Xue M, Wang J, Lv Q, Liu J, Hu L, Zhang Y, Wang X (2023b) Gypenosides synergistically reduce the extracellular matrix of hepatic stellate cells and ameliorate hepatic fibrosis in mice. Molecules 28(13):5448

Li J, Ge H, Xu Y, Xie J, Karim N, Yan F, Mo J, Chen W (2023c) Chlorogenic acid alleviates oxidative damage in hepatocytes by regulating miR-199a-5p/GRP78 axis. Food Biosci 53:102595

Li L, Lei X, Chen L, Ma Y, Luo J, Liu X, Xu X, Zhou G, Feng X (2024b) Protective mechanism of quercetin compounds against acrylamide-induced hepatotoxicity. Food Sci Human Wellness 13(1):225–240

Li Y, Deng X, Tan X, Li Q, Yu Z, Wu W, Ma X, Zeng J, Wang X (2024c) Protective role of curcumin in disease progression from non-alcoholic fatty liver disease to hepatocellular carcinoma: a meta-analysis. Front Pharmacol 15:1343193

Li Y, Yang M, Lin H, Yan W, Deng G, Ye H, Shi H, Wu C, Ma G, Xu S (2022b) Limonin alleviates non-alcoholic fatty liver disease by reducing lipid accumulation, suppressing inflammation and oxidative stress. Front Pharmacol 12:801730

Lin E-Y, Chagnaadorj A, Huang S-J, Wang C-C, Chiang Y-H, Cheng C-W (2018) Hepatoprotective activity of the ethanolic extract of *Polygonum multiflorum* Thunb. against oxidative stress-induced liver injury. Evid Based Complement Alternat Med 2018:4130307

Lin GS, Zhao MM, Fu QC, Zhao SY, Ba TT, Yu HX (2022) Palmatine attenuates hepatocyte injury by promoting autophagy via the AMPK/mTOR pathway after alcoholic liver disease. Drug Dev Res 83(7):1613–1622

Lin J, Li F, Jiao J, Qian Y, Xu M, Wang F, Sun X, Zhou T, Wu H, Kong X (2024) Quercetin, a natural flavonoid, protects against hepatic ischemia–reperfusion injury via inhibiting Caspase-8/ASC dependent macrophage pyroptosis. J Adv Res 70:555

Lu JM, Jin GN, Lu YN, Zhao XD, Lan HW, Mu SR, Piao LX (2021) Resveratrol modulates Toxoplasma gondii infection induced liver injury by intervening in the HMGB1/TLR4/NF-κB signaling pathway. European J of Pharmacology, 910:174497

Ma Z, Sheng L, Li J, Qian J, Wu G, Wang Z, Zhang Y (2022) Resveratrol alleviates hepatic fibrosis associated with decreased endoplasmic reticulum stress-mediated apoptosis and inflammation. Inflammation 45(2):812–823

Machado FI, Miranda GR, Dorta JD, Rolo PA, Palmeira MC (2023) Targeting oxidative stress with polyphenols to fight liver diseases. Antioxidants 12(6):1212

Mendoza N, Silva EME (2018) Introduction to phytochemicals: secondary metabolites from plants with active principles for pharmacological importance. In: Phytochemicals: source of antioxidants and role in disease prevention, vol 25. IntechOpen, London, pp 1–5

Minnady M, Jayapal G, Poochi S, Nethaji P, Mathalaimuthu B (2022) Hepatoprotective effect of indigenous medicinal plants: a review. Indian J Pharm Sci 84(4):1116

Morrill JA, Shrestha M, Grant RW (2005) Barriers to the treatment of hepatitis C: patient, provider, and system factors. J Gen Intern Med 20(8):754–758

Mostafa-Hedeab G, Hassan ME, Halawa TF (2022) Epigallocatechin gallate ameliorates tetrahydrochloride-induced liver toxicity in rats via inhibition of TGFβ/p-ERK/p-Smad1/2 signaling, antioxidant, anti-inflammatory activity. Saudi Pharm J 30(10):1293–1300

Motwani H, Gadhavi H, Mangukia N, Patel SK, Rawal RM, Solanki HA (2021) Hepatoprotective plants role in human health: a cross-kingdom review. J Med Plants Stud 9(6):41–51

Nassar SZ, Soliman N, Abdulmalek SA, Aly RG, Elatrebi S, Allam EA (2024) Glycyrrhizic acid protects against thioacetamide-induced hepatic fibrosis in rats by inhibition of HMGB1 and its crosstalk with autophagy-related protein Beclin-1. Bull Egypt Soc Physiol Sci 44(1):256–271

Özcan Ö, Aldemir O, Karabulut B (2020) Flavones (apigenin, luteolin, chrysin) and their importance for health. Mellifera 20(1):16–27

Park J-S, Rustamov N, Roh Y-S (2023) The roles of NFR2-regulated oxidative stress and mitochondrial quality control in chronic liver diseases. Antioxidants 12(10):1928

Patidar P, Hirani N, Bharti S, Baig SM (2024) Key regulators of hepatic stellate cell activation in alcohol liver disease: a comprehensive review. Int Immunopharmacol 141:112938

Peng C, Ouyang Yaobin L, Nonghua LN (2020) The NF-κB signaling pathway, the microbiota, and gastrointestinal tumorigenesis: recent advances. Front Immunol 11:1387

Petrasek J, Iracheta-Vellve A, Saha B, Satishchandran A, Kodys K, Fitzgerald KA, Kurt-Jones EA, Szabo G (2015) Metabolic danger signals, uric acid and ATP, mediate inflammatory cross-talk between hepatocytes and immune cells in alcoholic liver disease. J Leukoc Biol 98(2):249–256

Phogat A, Singh J, Kumar V, Malik V (2023) Berberine mitigates acetamiprid-induced hepatotoxicity and inflammation via regulating endogenous antioxidants and NF-κB/TNF-α signaling in rats. Environ Sci Pollut Res 30:87412–87423

Ramos-Tovar E, Muriel P (2020) Molecular mechanisms that link oxidative stress, inflammation, and fibrosis in the liver. Antioxidants 9:1279

Ramos-Tovar E, Muriel P (2023) Free radicals, antioxidants, nuclear factor-E2-related factor-2 and liver damage. Vitamins and Hormones. Elsevier

Seidita A, Cusimano A, Giuliano A, Meli M, Carroccio A, Soresi M, Giannitrapani L (2024) Oxidative stress as a target for non-pharmacological intervention in MAFLD: could there be a role for EVOO? Antioxidants 13:731

Seo J, Kwon D, Kim SH, Byun MR, Lee YH, Jung YS (2024) Role of autophagy in betaine-promoted hepatoprotection against non-alcoholic fatty liver disease in mice. Curr Res Food Sci 8:100663

Shakerian E, Akbari R, Mohammadtaghvaei N, Gahrooie M, Afarin R (2022) Quercetin reduces hepatic fibrogenesis by inhibiting TGF-β/Smad3 signaling pathway in LX-2 cell line. Jundishapur J Natl Pharm Prod 17

Shao X, Li J, Zhang H, Zhang X, Sun C, Ouyang X, Wang Y, Wu X, Chen C (2023) Anti-inflammatory effects and molecular mechanisms of bioactive small molecule garlic polysaccharide. Front Nutr 9:1092873

Sharma R, Majee C, Mazumder R, Padhi S, Khan F, Pal RS (2024) Insight into the regulation of Nrf2/Keap1 pathway by flavonoids as an approach for treatment of liver diseases: a review. Indian J Pharm Educ Res 58(special issue):s40–s57

Shaw P, Chattopadhyay A (2020) Nrf2–ARE signaling in cellular protection: mechanism of action and the regulatory mechanisms. J Cell Physiol 235:3119–3130

Shen Q, Yang M, Wang S, Chen X, Chen S, Zhang R, Xiong Z, Leng Y (2024) The pivotal role of dysregulated autophagy in the progression of non-alcoholic fatty liver disease. Front Endocrinol 15:1374644

Shikh VE, Khaytovich DE, Tsvetkov ND (2024) Herbal extracts in hepatoprotection: antioxidant and immunomodulatory effects. A review. Terapevticheskii Arkhiv 96:836–845

Siddiqui A, Bansal P, Pentela B, Mazumder A, Tyagi PK, Kapoor B (2024) Protective effects of herbal agents against hepatorenal toxicity: a review. J Natl Remedies 24:1621–1632

Służały P, Paśko P, Galanty A (2024) Natural products as hepatoprotective agents—a comprehensive review of clinical trials. Plants 13:1985

Sohn SI, Priya A, Balasubramaniam B, Muthuramaligam P, Sivasankar C, Selvaraj A, Valliammai A, Jothi R, Pandian S (2021) Biomedical applications and bioavailability of curcumin—an updated overview. Pharmaceutics 13:2102

Su H, Xu T, Sun R, Xiang Y, Gao Y, Zai K, Fang S (2024) Research progress of oxidative stress in sepsis-associated liver injury

Sun SC (2017) The non-canonical NF-κB pathway in immunity and inflammation. Nat Rev Immunol 17:545–558

Tanwar S, Rhodes F, Srivastava A, Trembling MP, Rosenberg MW (2020) Inflammation and fibrosis in chronic liver diseases including non-alcoholic fatty liver disease and hepatitis C. World J Gastroenterol 26:109

Tsuchida T, Friedman SL (2017) Mechanisms of hepatic stellate cell activation. Nat Rev Gastroenterol Hepatol 14:397–411

Vargas-Mendoza N, Morales-González Á, Morales-Martínez M, Soriano-Ursúa MA, Delgado-Olivares L, Sandoval-Gallegos EM, Madrigal-Bujaidar E, Álvarez-González I, Madrigal-Santillán E, Morales-Gonzalez JA (2020) Flavolignans from silymarin as Nrf2 bioactivators and their therapeutic applications. Biomedicine 8:122

Venmathi Maran BA, Iqbal M, Gangadaran P, Ahn BC, Rao PV, Shah MD (2022) Hepatoprotective potential of Malaysian medicinal plants: a review on phytochemicals, oxidative stress, and antioxidant mechanisms. Molecules 27:1533

Wagner M, Fickert P (2020) Drug therapies for chronic cholestatic liver diseases. Annu Rev Pharmacol Toxicol 60:503–527

Wang B, Cui S, Mao B, Zhang Q, Tian F, Zhao J, Tang X, Chen W (2022) Cyanidin alleviated CCl4-induced acute liver injury by regulating the Nrf2 and NF-κB signaling pathways. Antioxidants 11:2383

Wang J, Huang N, Zhao T, Wang L, Zheng Y, Xiao H (2024) Effects of curcumol on ferroptosis and tube forming ability of hepatic sinus endothelial cells

Wang Q, Liu T, Koci M, Wang Y, Fu Y, Ma M, Ma Q, Zhao L (2023) Chlorogenic acid alleviated AFB1-induced hepatotoxicity by regulating mitochondrial function, activating Nrf2/HO-1, and inhibiting noncanonical NF-κB signaling pathway. Antioxidants 12:2027

Widyawati R, Yuniarti M, Lukiswanto W, Sektiari B (2024) The role of oligomeric-grape seed proanthocyanidine extract (O-GSPE) on cholestasis liver damage in animal model. Res J Pharm Technol 17:3793–3800

Wu S-K, Wang L, Wang F, Zhang J (2024) Resveratrol improved mitochondrial biogenesis by activating SIRT1/PGC-1α signal pathway in SAP. Sci Rep 14:26216

Wu Z, Sun L, Chen R, Wen S, Li Q, Lai X, Zhang Z, Cao F, Sun S (2022) Chinese tea alleviates CCl4-induced liver injury through the NF-κB or Nrf2 signaling pathway in C57BL-6J mice. Nutrients 14:972

Xia Y, Luo Q, Gao Q, Huang C, Chen P, Zou Y, Chen X, Liu W, Chen Z (2024) SIRT1 activation ameliorates rhesus monkey liver fibrosis by inhibiting the TGF-β/Smad signaling pathway. Chem Biol Interact 394:110979

Xu F, Lu S, Pan N, Zhao F, Jia X, Wang S, Zhang Y, Zhou Y (2023) Bromodomain protein 4 is a key molecular driver of TGFβ1-induced hepatic stellate cell activation. Biochim Biophys Acta, Mol Cell Res 1870:119569

Yan B-F, Wang Y, Wang W-B, Ding X-J, Wei B, Liu S-J, Fu T-M, Chen L, Zhang J-Z, Liu J (2023) Huangqin decoction mitigates hepatic inflammation in high-fat diet-challenged rats by inhibiting TLR4/NF-κB/NLRP3 pathway. J Ethnopharmacol 303:115999

Yang S-Y, Pyo C-M, Nam M-H, Lee K-W (2019) ERK/Nrf2 pathway activation by caffeic acid in HepG2 cells alleviates its hepatocellular damage caused by t-butylhydroperoxide-induced oxidative stress. BMC Complement Altern Med 19:13

Yang D, Su J, Chen Y, Chen G (2024) The NF-κB pathway: key players in neurocognitive functions and related disorders. Eur J Pharmacol:177038

Yang JH, Ku SK, Cho IJ, Lee JH, Na C-S, Ki SH (2021) Neoagarooligosaccharide protects against hepatic fibrosis via inhibition of TGF-β/Smad signaling pathway. Int J Mol Sci 22:2041

Yu C, Xu Y, Zhao M, Song P, Yu J (2024) New insights into mechanism of ellagic acid alleviating arsenic-induced oxidative stress through MAPK/Keap1-Nrf2 signaling pathway response, molecular docking and metabolomics analysis in HepG2 cells. Ecotoxicol Environ Saf 285:117029

Yu X, Elfimova N, Müller M, Bachurski D, Koitzsch U, Drebber U, Mahabir E, Hansen HP, Friedman SL, Klein S (2022) Autophagy-related activation of hepatic stellate cells reduces cellular miR-29a by promoting its vesicular secretion. Cell Mol Gastroenterol Hepatol 13(6):1701–1716

Zhang T, Guo J, Gu J, Chen K, Li H, Wang J (2019) Protective role of mTOR in liver ischemia/reperfusion injury: involvement of inflammation and autophagy. Oxidative Med Cell Longev 2019:7861290

Zhang X, Yin G, Chen S, Meng D, Yu W, Liu H, Wang L, Zhang F (2024a) Diosgenin ameliorating non-alcoholic fatty liver disease via Nrf2-mediated regulation of oxidative stress and ferroptosis. Diabetes Obes Metab 26. https://doi.org/10.1111/dom.15452

Zhang Y, Ren L, Tian Y, Guo X, Wei F, Zhang Y (2024b) Signaling pathways that activate hepatic stellate cells during liver fibrosis. Front Med 11:1454980

Zhou J, Zheng Q, Chen Z (2022a) The Nrf2 pathway in liver diseases. Front Cell Dev Biol 10:826204

Zhou Q, Zhang N, Hu T, Xu H, Duan X, Liu B, Chen F, Wang M (2022b) Dietary phenolic-type Nrf2-activators: implications in the control of toxin-induced hepatic disorders. Food Funct 13:5480–5497

Zhu L, Chen Y, Ding W, Duan Y, Sun D, Lu Y (2022) Caspase-3/Treg and PI3K/AKT/mTOR pathway is involved in liver ischemia reperfusion injury (IRI) protection by everolimus. Transpl Immunol 71:101541

Zlatanova M, Nešić A, Trbojević-Ivić J, Četić D, Gavrović-Jankulović M (2024) Targeting NF-κB signaling: selected small molecules downregulate pro-inflammatory cytokines in both food allergen and LPS-induced inflammation. Int J Mol Sci 25:5798

Plant-Based Remedies for Nonalcoholic Fatty Liver Disease

4

V. M. Ragitha, Parvathy Susheelan, P. S. Syananth, Biba Vikas, and Lekshmi K. Edison

Abstract

A prevalent liver condition known as nonalcoholic fatty liver disease (NAFLD) is a leading cause of cryptogenic cirrhosis and chronic liver disease, particularly in developed countries. NAFLD pathogenesis is multifactorial and involves critical processes such as resistance to insulin, lipid metabolism disruption, oxidative stress, inflammation, programmed cell death, and fibrosis. Collectively, these processes contribute to disease progression. The transition from simple steatosis (fatty liver) to nonalcoholic steatohepatitis (NASH) is often driven by inflammation triggered by lipid accumulation. Therefore, the effective amelioration of steatosis and inflammation is crucial for successful NAFLD therapy. In recent years, herbal medicines have garnered significant interest as potential treatments for NAFLD, primarily because of their wide availability, fewer side effects, and documented therapeutic mechanisms. Various herbal derivatives that have shown promise in targeting the underlying pathological mechanisms of NAFLD have been extensively studied. Garlic, milk thistle, red grapes, and ginger are widely used in the treatment of NAFLD. The combination of such botanicals has already been employed in the Ayurvedic system, and research has indicated that herbal

V. M. Ragitha (✉) · P. Susheelan
MagGenome Technologies Pvt Limited, Ernakulam, Kerala, India

P. S. Syananth
Department of Biological Sciences, P D Patel Institute of Applied Sciences, Charotar University of Science and Technology, Changa, Gujarat, India

B. Vikas
NextGen Precision Health, University of Missouri, Columbia, MO, USA

L. K. Edison (✉)
Department of Comparative, Diagnostic, and Population Medicine, College of Veterinary Medicine, University of Florida, Gainesville, FL, USA
e-mail: edison.le@ufl.edu

K. Kuca et al. (eds.), *Medicinal Plants and Their Bioactives in Human Diseases*,
https://doi.org/10.1007/978-3-032-01356-9_4

medicines can improve steatosis and inflammation through multiple mechanisms that influence lipid metabolism and inflammatory pathways. These herbal treatments have been evaluated in preclinical studies and clinical trials and have shown beneficial effects on liver health without significant adverse effects. This chapter highlights the protective effects of various herbs in their natural or medicinal forms and their mechanisms of action in ameliorating steatosis and inflammation. This emphasizes the importance of traditional herbal preparations, crude extracts from medicinal plants, and pure natural compounds as valuable resources for groundbreaking drug research and future advancement of NAFLD treatments.

Keywords

NAFLD · NASH · Plants · Preclinical studies · Pathophysiology · Treatment options

1 Introduction

Nonalcoholic fatty liver disease (NAFLD) is a condition in which fat accumulates in hepatocytes without any secondary cause of liver damage, such as alcohol consumption. Normally, fat deposition also occurs in healthy individuals, but when it exceeds a percentage of five, it is considered pathological (Sharma and John 2023). Fat deposition occurs in hepatocytes, which make up to 80% of the liver mass, and are the primary cells responsible for liver function. Thus, fat accumulation affects the liver's day-to-day activities, including the metabolism of nutrients, synthesis of serum components, and detoxification. Although the liver is the only organ with the ability to regenerate, fat deposition interferes with the stem cell niche and fails to regain its physiological activities. Lifestyle changes are considered a primary risk factor for fatty liver disease because, like many other organs, liver functions depend on our biological clock (Reinke and Asher 2015). The physiological activities of the liver change based on its systemic environment, and disturbances in feeding, fasting, and rest-activity cycles lead to the progression of liver damage. NAFLD is strongly associated with obesity. Thus, unhealthy food habits and lack of sleep ultimately create liver issues but can be reversed by correcting those issues if cirrhosis has not started. NAFLD can be classified into different grades based on its progression. Imaging systems and biopsies reveal the grade of fatty liver, which is divided into four categories, as described below (Han et al. 2023):

Grade 0: Normal, healthy liver (0–5% fat deposition).

Grade 1 (mild): Steatosis, infrequent zone 3 ballooning, intralobular inflammation, no/mild portal inflammation

Grade 2 (moderate): Fatty liver with noticeable cell swelling in zone 3, mild chronic inflammation with polymorphonuclear cell infiltration, and mild to moderate inflammation in the portal area.

Grade 3 (severe): Panacinar steatosis, cell swelling, and noticeable disruption of normal structure predominantly in zone 3 region, chronic inflammation with polymorphonuclear cells present, and mild to moderate inflammation in the portal region.

Grade 3 progresses into different stages of fibrosis, and ultimately liver cirrhosis and irreversible damage to the liver.

Changes up to Grade 2 can be reversed by the introduction of good lifestyle changes and drugs that helps to detoxify the liver. This chapter covers the effectiveness of plants for treating NAFLD and its associated symptoms, which have already been used in the Ayurvedic system of medicine. NAFLD, a chronic liver disorder, is suspected to be the major cause of liver transplantation by 2030 (Gurian et al. 2020). Steatosis should not be caused by alcohol, drugs, or viral-induced steatosis. Nonalcoholic fatty liver disease includes mainly two stages: nonalcoholic fatty liver (NAFL) and nonalcoholic steatohepatitis (NASH) (Friedman et al. 2018). NAFLD comprises a series of events in the liver, with a difference in the severity of injury, which results in fibrosis. NAFL is a hepatosteatosis, which refers only to fat accumulation without significant interference with the physiological activities of the liver (Gurian et al. 2020). Thus, NAFL patients are in the low-risk group compared to those with NASH (Singh et al. 2015). NASH is a condition in which significant inflammation is reported with prominent liver damage and is known as steatohepatitis. Steatohepatitis can progress to liver cirrhosis, liver failure, or hepatocarcinoma. Cardiovascular issues and malignancies are non-liver-related complications of NASH (Lindenmeyer and McCullough 2018; Rinella and Sanyal 2016).

1.1 Diagnosis of NAFL and Grades

As NAFL does not have any clinical symptoms, diagnosis only occurs when blood parameters are evaluated as part of other diseases (Reinshagen et al. 2023). Serum glutamic oxaloacetic transaminase (SGOT), serum glutamic pyruvic transaminase (SGPT), and liver enzymes were abnormal. Cholesterol levels are also high, which leads to a detailed diagnosis, including imaging. No treatment option is generally prescribed for NAFL patients, and adopting good lifestyle changes, including regular physical activity, optimum body weight, and a nutritious diet, can reverse the condition. NASH may show clinical symptoms such as elevated body temperature, difficulty in breathing, vomiting of blood, dark or black tarry stools, swollen abdomen, slurred speech, fatigue, and mental confusion (Khoonsari et al. 2017). As the liver has been linked to all metabolic activities, the symptoms are not normally associated only with the liver. Traditional medicine systems also mention the variation of symptoms associated with liver disease; sometimes, the symptoms may be related to other issues. Liver imaging and biopsy are the main diagnostic tools that confirm NAFLD and differentiate the different grades of progression.

1.2 Prevalence of NAFLD

Research has demonstrated that the burden of NAFLD and NASH has increased worldwide. A recent meta-analysis published by Yonossi et al. (2023) proved that the incidence of the disease has increased worldwide along with the increased

Table 4.1 Global prevalence of NAFLD and NASH in general population above 20+ age (2019) (Younossi et al. 2023; Wong et al. 2023)

Region	NAFLD prevalence (%)	NASH prevalence (%)
Global	32.16	5.27
North Africa & Middle East	42.62	5.85
North America and Australia	38.47	5.00
Latin America	34.45	7.11
Western Europe	32.47	4.02
East Asia	32.31	4.76
Asia Pacific	29.77	4.49
South Asia	29.29	5.42
South-East Asia	24.25	5.30

occurrence of type 2 diabetes and obesity. A comprehensive analysis of 92 studies with approximately nine million individuals globally confirmed that the prevalence of NAFLD was approximately 32.16% in 2019 (Younossi et al. 2023). The region- and age-wise differences in prevalence were also studied by the same group, and the findings are mentioned in the table below (Table 4.1). This study also showed that the pooled prevalence increased from 25.26% in 1990–2006 to 38.2% (33.72–42.89) in 2016–2019 (Younossi et al. 2023). Worldwide, the frequency of NASH in individuals with NAFLD is approximately 16.02%. Latin America had the highest prevalence of NASH (Table 4.1).

Research conducted by Shalimar et al. (2022) showed a trend of NAFLD in India. This meta-analysis revealed that the pooled prevalence of NAFLD in India was approximately 38.6% and 35.4% in adults and children, respectively (Shalimar et al. 2022). This study also found sex-based differences in the occurrence of NAFLD, with males showing a higher rate (39.4%) compared to females (35.4%). Some national and international studies have proven the relationship between NAFLD and conditions such as obesity, type 2 diabetes, and excessive nutrition. According to the World Health Organization, one out of eight people lives with obesity. The number of people with obesity has doubled and quadrupled among children (World Health Organization 2024). Even in India, approximately 20% of children are overweight, and studies have shown that weight gain during childhood is a higher risk for NAFLD than weight gain in adulthood (Shalimar et al. 2022). A strong significant relationship occurs between the high prevalence of NAFLD and NASH and the conditions of obesity and type 2 diabetes (Williamson et al. 2011; Leite et al. 2009; Targher et al. 2007; Lu et al. 2013).

1.3 Pathophysiology of NAFLD

Fat metabolism occurs in the liver. Disturbances in uptake and lack of physical activity led to the accumulation of fat in the liver. Many questions remain unanswered regarding fat deposition in the liver, but obesity still plays a significant role in the development of this disease. Changes in food habits disturb the normal flora of the intestine, which alters the biomolecules responsible for maintaining homeostasis of the alimentary canal. Many studies have been conducted to validate the dietary dependence of disease progression, mainly by using high amounts of fat and fructose (Erbas et al. 2018; Lau et al. 2017). An increased carbohydrate and fatty acid diet, de novo lipogenesis, and free fatty acids from adipocytes contribute to the deposition of lipids into hepatocytes (Donnelly et al. 2005). Fat accumulation does not progress to steatohepatitis if fat intake through the diet is reduced (Monetti et al. 2007; Harley et al. 2014).

Genetic factors are also significantly involved in the pathogenesis of NAFLD. Lipotoxicity can ultimately lead to inflammation and cirrhosis. Studies have shown that changes in the lipid metabolic pathway prevent glycogen synthesis in the liver, which causes lipid accumulation. Patatin-like phospholipase domain-containing protein 3 (PNPLA3) gene is located on the 22nd chromosome. The protein encoded by this gene has lipase activity, which breaks down triglycerides (Salari et al. 2021, BasuRay et al. 2017). The I148M variation in PNPLA3 changes its function and is more prone to associate with lipid molecules without lipase activity, thus decreasing fat metabolism and increasing fat deposition (BasuRay et al. 2017). The I148M variant reduces the serum anti-inflammatory molecule adiponectin, which also inhibits profibrotic hepatic stellate cells. Mutations in this gene are considered prominent risk factors for NAFLD (Sherman et al. 2023). Mutation in the second member of the transmembrane 6 superfamily, located on chromosome 19 (TM6SF2), is another gene that is correlated with the development of NAFLD (Kozlitina et al. 2014; Holmen et al. 2014). Studies have shown that this gene variant is actively involved in the progression of hepatic fibrosis and hepatocellular carcinoma (Luo et al. 2022). Lysosomal acid lipase (LAL) has been proven to promote NAFLD progression (Pericleous et al. 2017; Carter et al. 2019; Sookoian et al. 2020).

1.4 Current Treatment Options

The pathological effect of NAFLD ranges from NAFL (non-alcoholic fatty liver) with simple steatosis and without any inflammation to non-alcoholic steatohepatitis (NASH) with steatosis followed by hepatocyte ballooning, eventually resulting in fibrosis and cirrhosis, ultimately leading to hepatocellular carcinoma (Takahashi et al. 2015; Adams et al. 2006). Changes in lifestyle, including a balanced diet followed by regular physical activity, help to improve the disease condition. Treatment strategies for NAFLD mainly focus on identifying and treating associated metabolic conditions, such as diabetes, improving insulin resistance, and using hepatoprotective agents. Rosiglitazone and pioglitazone are thiazolidinediones and metformin, a biguanide, increases insulin sensitivity and has been shown to improve steatosis and aminotransferase levels in NAFLD patients (Ratziu et al. 2008; Torres et al. 2008). Homeopathy is one of the most popular holistic systems of medicine. This mode of treatment focuses on the natural healing mechanisms of the body. Some of the remedies in homeopathy for treating liver diseases include bryonia, chelidonium, podophyllum, lycopodium, sulfur, and phosphorous (Roohullah and Khan 2017). The use of plants for treating ailments has been reported, and modern science has mentioned the effectiveness of plants. Hence, this chapter investigates the plants and their bioactive compounds that are mainly used in traditional Ayurvedic systems at the same time as being validated by modern science for treating NAFLD and associated symptoms.

2 Medicinal Plants and Their Bioactive Compounds for NAFLD Treatment- Preclinical Studies

NAFLD is significantly appearing among young populations, and one added advantage is it can be reversed. Usage of hepatic protective plants fastens the recovery rate from NAFLD. The effectiveness of some of these plants is even evaluated by modern science also. The bioactive agents and mechanisms of action of some of those plants are discussed in Table 4.2.

Abelmoschus esculentus has been proven to have antisteatotic effects by various researchers. Prom-in et al. (2021) showed that polyphenols, mucilage polysaccharides, and glutathione present in *A. esculentus* reduced NASH progression in mouse models (Prom-in et al. 2021). Cell-based studies by Peng et al. showed that certain fractions of *A. esculentus* ameliorate hepatic lipogenesis and lipid uptake by suppressing dipeptidyl peptidase-4 (DPP-4), and its expression has been shown to cause metabolic syndrome through insulin resistance (Peng et al. 2022; Gabbia et al. 2024). The herb *A. millefolium* showed hepatoprotective effects against fatty liver disease by inhibiting diabetes mellitus. Studies have confirmed that *A. millefolium* extract reverses hyperglycemia, corrects abnormalities in lipid metabolism, and improves liver enzyme levels in diabetic models (Rezaei et al. 2020). Mouhid et al. (2019) proved that this plant extract reduces hypercholesterolemia and alleviates metabolic stress in mice with obesity induced by a high-fat diet (Mouhid et al. 2019). *Aegle marmelos* has been effective in ameliorating the deleterious effects of a diet induced by high fat in obese mice, in which the aqueous extract reverses serum total cholesterol and triglyceride levels (Sharshar et al. 2022).

Research by Emamat et al. (2020) demonstrated that people consuming Allium vegetables had a 64% lower risk than those who consumed less (Emamat et al. 2020). *Allium cepa* is another plant that reverses carbon tetrachloride-induced fatty liver disease in Wistar rats (Babatunde et al. 2023). It works by the reduction of liver enzyme activities and lipid profiles (cholesterol and triglycerides) and causes an increase in HDL-cholesterol and protein concentration, while others have linked the effect of fermented versus raw garlic against steatosis in the liver by reducing the fatty acids in the serum and normalizing the disproportion between hepatic lipid influx and efflux (Shojaei-Zarghani et al. 2022; Chen et al. 2014). Garlic products increase fecal lipid excretion and alleviate hepatic lipogenesis by downregulating the synthesis of fatty acids, sterol regulatory element-binding protein-1c, β-hydroxy β-methylglutaryl-CoA reductase, acetyl-CoA carboxylase, and stearoyl-coenzyme A desaturase 1 (Chen et al. 2014; Ha et al. 2015). Yu et al. (2021) and Xiao et al.

Table 4.2 List of plants scientifically proven for abolishing NAFLD in preclinical studies

Plant	Extract/bioactive agent	Mechanisms	References
Abelmoschus esculentus	Polyphenols, mucilage polysaccharides, glutathione	Ameliorates lipoenesis and lipid absorption	Prom-in et al. (2021)
A. millefolium	Plant extract	Reverses hyperglycemia, corrects abnormalities in lipid metabolism, and improves liver enzyme levels	Mouhid et al. (2019), Rezaei et al. (2020)
Aegle marmelos	Aqueous extract	reverses serum total cholesterol and triglyceride levels	Sharshar et al. (2022)
Allium cepa	S-allylmercaptocysteine	Antioxidant and antiinflammatory effect	Xiao et al. (2013), Yu et al. (2021)
Allium sativum	S-allyl-l-cysteine, E-Ajoene, Z-Ajoene,	Prevent fat biosynthesis and lower triglycerides	Jannah et al. (2021)
Aloe vera	*Aloe vera* extract	Increased activity of liver catalase and superoxide dismutase	Abubakar et al. (2022), Dibal et al. (2022)
Alstonia scholaris	Indole alkaloids	Promote lipolysis and prevent lipogenesis	Sun et al. (2022)
Amaranthus spinosus	Methanol and aqueous extracts	Anti-hepatosteatosis	Prince et al. (2021)
Andrographis paniculata	Andrographolide, isondrographolide, 14-deoxy-11,12-didehydroandrographolide, 14-deoxyandropholide, and 14-acetylandrographolide, Isoandrographolide	Prevent hyperlipidemia and fat accumulation in the liver	Toppo et al. (2017), Cabrera et al. (2017), Liu et al. (2020), Ran et al. (2023), Pipitone et al. (2023)
Apium graveolens	Flavonoids	Reverses fat-induced histological changes in the liver	He et al. (2019)
Argemone mexicana	Root extract	Improved obesity-related blood parameters	Morey et al. (2023)
Artemisia capillaris	Scoparone	Ameliorates NAFLD by targeting the PPARα signaling pathway	Huang et al. (2024)
Cajanus cajan	Stilbenoids (Cajanolactone)	Regulates genes associated with adipogenesis and fat deposition	Luo et al. (2020), (2021a, b).

(continued)

Table 4.2 (continued)

Plant	Extract/bioactive agent	Mechanisms	References
Capparis spinosa	Fruit extract	Reverses NAFLD by down regulating SREBP-1c and a PPARα-independent pathway	Akbari et al. (2020, 2022)
Cassia semen	Ethanol extract	Reverses NAFLD by regulating expression of enzymes associated with lipid metabolism	Meng et al. (2019)

(2013) studied the effect of S-allylmercaptocysteine, a water-soluble molecule derived from garlic and demonstrated its potential for the treatment of non-alcoholic fatty liver disease using in vivo animal model. *Allium sativum* contains components like S-allyl-l-cysteine, E-Ajoene, Z-Ajoene, which are believed to prevent fat biosynthesis and lower triglycerides in serum. A single bulb garlic extract was observed to reverse hepatic steatosis in a mouse model with a high-fat diet (Jannah et al. 2021). A meta-analysis study that dealt with the effect of garlic on NAFLD proved that *A. sativum* not only prevents NAFLD but also reverses the NAFLD associated changes in the liver, such as lower liver enzymes (alanine transaminase and aspatin transaminase) (Mardi et al. 2023). Hyperlipidemic patients received treatment with garlic (20 g) daily with 1 tbsp lemon juice, which significantly reduced the total cholesterol, LDL-cholesterol, and fibrinogen levels compared to the control groups (Mirunalini et al. 2011; Khan 2011; Aslani et al. 2016).

Aloe vera, a well-known plant with anti-inflammatory activity, has been reported to have a positive effect on NAFLD. Abubakar et al. (2022) and Dibal et al. (2022) examined the efficacy of *A. vera* extract in a high-fat fructose diet (HFFD) and high-fat diet-induced mice models. Dibal et al. showed that a high-fat fructose diet increased cholesterol, but in mice treated with *A. vera*, the cholesterol level was not significantly different from that of the control group (Abubakar et al. 2022). The hepatocytes in control as well as high-fat fructose + *A. vera* extract-treated mice were shown to have normal physiology (Abubakar et al. 2022). But the alanine aminotransferase (ALT) level in HFFD-treated mice was significantly higher when compared to the control group. *A. vera* extract consumption reversed the increased levels of albumin, and improved enzyme activities of superoxide dismutase and catalase in the mice group treated with HFFD. Dibal et al. observed mice treated with a high-fat diet showed elevated cholesterol, triglycerides, and low-density lipoprotein. *A. vera* extract could reverse these changes along with the increased activity of liver catalase and superoxide dismutase (Dibal et al. 2022). A normal liver was observed with *A. vera* treatment, whereas mice induced with a high-fat diet showed numerous hepatic vacuoles with lipids.

Alstonia scholaris contains indole alkaloids, and Sun et al. evaluated the effectiveness of these alkaloids in NAFLD using a high-fat mice animal model. The study showed that the alkaloids attenuate the expression of genes associated with the synthesis of lipids. But the treatment with these alkaloids up-regulates the enzymes associated with lipolysis (Sun et al. 2022) through the oxidation of fatty acids in the hepatocytes. Thus, *A. scholaris* has proven to have therapeutic properties for fatty liver treatment. *Amaranthus spinosus* is a common vegetable, and a study in a rat model showed that powdered supplementation with methanol and aqueous extracts reduced high-carbohydrate, high-fat-induced increases in fat deposition, body weight, organ weight, lipid levels, glycemic status, hepatic and renal biomarkers, hepatic antioxidant status, and hepatosteatosis (Prince et al. 2021). Creatine kinase, alkaline phosphatase, aspartate, and alanine aminotransferase levels were normalized by *A. spinosus* administration.

Andrographis paniculata is a widely used medicinal plant in Ayurveda medicine. Studies have shown that *A. paniculata* extract inhibited inflammatory markers like Tumor Necrosis Factor-α, and Interleukins—1 & 6, thereby reducing hepatic steatosis and stiffness in the liver (Hossain, 2024). Studies have shown that diterpenoids (andrographolide, isondrographolide, 14-deoxy-11,12-didehydroandrographolide, 14-deoxyandropholide, and 14-acetylandrographolide) are mostly responsible for the bioactivity of *A. paniculata* extract to regulate lipid metabolism, prevent inflammation, oxidative stress, and apoptosis (Liu et al. 2020; Toppo et al. 2017). 14-deoxy-11,12-didehydroandrographolide downregulating the expression of the inflammasome (nucleotide-binding domain, leucine-rich–containing family, pyrin domain–containing-3) (NLRP3) and caspase-1, which decrease cholesterol buildup. It limits the apoptosis encouraged by a high-fat, high-carbohydrate diet by reducing the ratio of caspase to pro-caspase 3 (Liu et al. 2020). Ran et al. evaluated and proved the efficacy of andrographolide in a mouse model using a high-fat diet, and their study showed that andrographolide exerts its antisteatotic effect by suppressing FATP2 (fatty acid transport protein 2)-mediated fatty acid uptake in liver cells (Ran et al. 2023). Another study showed that andrographolide reduced the expression of the NLRP3 inflammasome in a choline-deficient mouse model. Choline deficiency is widely used as an animal model for NAFLD, and this study proved that choline-deficient mice developed an increase in liver weight, liver fibrosis, and activation of inflammasomes (apoptosis-associated speck-like protein and NLRP3), and these changes were reversed by andrographolide. Additionally, andrographolide reduces cytokine and LPS-induced iNOS expression in primary hepatocytes (Cabrera et al. 2017). Pipitone et al., showed the combined effect of curcumin and andrographolide on high lipid content and reactive oxygen species (ROS) levels in an in vitro liver steatosis disease model in HepG2 cells. These compounds synergistically ameliorated hepatotoxic effects (Pipitone et al. 2023). Isoandrographolide, a derivative of andrographolide, prevents hyperlipidemia and fat accumulation in the liver, and the success rate and stability were much better than andrographolide (Toppo et al. 2017).

Apium graveolens, commonly known as celery, contains flavonoids, which are the major bioactive components. He et al. (2019) used a fat emulsion to induce hyperlipidemia in a rat model and studied the effectiveness of the flavonoids of *A. graveolens* on hyperlipidemia. The animals showed higher body weight and liver index, which were ameliorated by the flavonoid extract. The flavonoid extract restored the fat-induced histological changes in the liver (He et al. 2019). Zhao et al.

(2021) compared the effects of normal and fermented celery juice on obese mouse using a high-fat diet, and the study showed that fermented celery juice is superior to normal celery juice in inhibiting body weight gain, reducing dyslipidemia and hyperglycemia, and reversing visceral fat accumulation. Abd-Elrazek et al. (2022) studied the synergistic effect of curcumin and celery to ameliorate high-fat and fructose-induced changes in liver physiology, and the effect was further compared with silymarin, a reference drug. They established a disease model with liver enzyme elevation, hyperglycemia, hyperlipidemia, increased lysosomal enzyme activities, and hepatic malondialdehyde levels, along with a decline in hepatic glutathione levels and DNA fragmentation. The combined administration of celery and curcumin improved pathological changes better than silymarin (Abd-Elrazek et al. 2022).

Argemone mexicana is commonly called the Mexican poppy and was found to have anti-obesity activities in a high-fat diet-induced obese mice model. *A. mexicana* root extract reduced body weight and showed beneficial effects on obesity-related blood parameters, including total cholesterol, triglycerides, very low-density lipoprotein cholesterol, high-density lipoprotein cholesterol, low-density lipoprotein cholesterol, serum glutamic-oxaloacetic transaminase, serum glutamic pyruvic transaminase, total bilirubin, and blood glucose levels (Morey et al. 2023).

Artemisia capillaris is a plant renowned for its anti-inflammatory properties. Network pharmacology and molecular docking-based identification of molecules that are effective for NAFLD were evaluated by Lian et al., who found 51 ingredients targeting NAFLD-regulated genes in mice. This study also confirmed that the plant extract reduced lipid accumulation and inflammation in the liver induced by an HFD (Liang et al. 2024). NASH mice model was induced using a diet deficient in choline, L-amino acid defined, and high fat content, and ethanol extract of *A. capillaris* improves liver physiology by modulating the expression levels of antioxidant, lipolytic, and fibrosis-related genes, significantly improving plasma and lipid profiles, and significantly reducing CDAHFD-induced hepatic fibrosis (Erdenebileg et al. 2024). Huang et al. (2024) showed that scoparone, a natural compound derived from the plant, helped to improve NAFLD in high-fat diet-fed mice. Xu et al. (2025) established a methionine-choline-deficient diet that induced in vivo NAFLD in a mouse model, which exhibited lipid deposition and inflammatory responses. Other studies have also reported antilipidemic effects of plant extracts in rat and mouse models (Liang et al. 2022; Liu et al. 2016).

Cajanus cajan showed activity against fat deposition owing to its bioactive components. Stilbenoids (cajanolactone) from *C. cajan* reduced fat deposition in ovariectomized animals and diet-induced fat deposition in female mice by inhibiting lipogenic genes, ChREBP (carbohydrate response element-binding protein), and SREBP-1c (Sterol regulatory element-binding protein 1c), and stimulated apolipoprotein B mRNA expression (Luo et al. 2020, 2021a). Luo et al. (2021b) demonstrated that consumption of cajanolactone A results in a reduction in weight gain in female mice, and the mechanism was through downregulation of genes in the hypothalamus and the regulation of genes in adipose tissues involved in adipogenesis and fat deposition (Luo et al. 2021b).

Capparis spinosa improves steatohepatitis in rats induced by the administration of a high-fat diet. Studies have demonstrated different mechanisms for reducing NASH in rats through the regulation of genes associated with lipid physiology. Akbari et al. showed that rats consuming 20 mg/kg *C. spinosa* extract had reduced liver weight and index, lipids and glucose in serum, and liver enzymes (alanine aminotransferase, aspartate aminotransferase, etc.). The histological features of high-induced nonalcoholic steatohepatitis (NASH) were reversed by the extract (Akbari et al. 2020). Akbari et al. also demonstrated another mechanism of the extract in reducing NASH by downregulating SREBP-1c and a PPARα-independent pathway (Akbari et al. 2022). In 2024, Afarin et al. proved the synergistic effect of *C. spinosa* fruit extract and quercetin, and the combined effects reduced the body mass index, liver weight, abnormal liver enzymes (alanine aminotransferase and aspartate aminotransferase), triglycerides, and low-density lipids compared to those used alone. This synergistic effect also increases high-density lipids and reduces inflammatory markers (Afarin et al. 2024). *Cassia semen* has been reported to ameliorate NAFLD. Meng et al. evaluated the efficiency of dried ethanolic *C. semen* extract in increasing triglycerides, alanine transferase, aspartate transaminase, total cholesterol, superoxide dismutase, malondialdehyde, glutathione, tumor necrosis factor-α, interleukins (IL-6 & 8), and low-density lipoprotein receptor (LDL-R) mRNA levels in a high-fat containing diet-consuming rat NAFLD model (Meng et al. 2019).

3 Herbal Medicines and NAFLD: Clinical Supports

Studies focused on the effect of botanicals on NAFLD were not restricted to in vitro and in vivo animal studies. Clinical trials were also conducted to assess the ability of plant extracts to improve NAFLD-associated changes (Table 4.3). Sangouni et al. (2020a) performed a clinical trial using garlic powder, and their study showed that consumption of four tablets containing 400 mg of *Allium sativum* significantly reduced NAFLD-associated physiological changes in patients compared to patients receiving four tablets of 400 mg of starch daily for 12 weeks (Sangouni et al. 2020a). Another study from the same group confirmed that garlic consumption for 3 months (1600 mg/day) ameliorates metabolic syndrome-associated components. The outcomes included a significant reduction in diastolic and systolic blood pressure, waist circumference, triglyceride, γ-glutamyl transferase, fatty liver index, insulin, insulin resistance, and appetite and a significant increase in high-density lipoprotein (Sangouni et al. 2021). They also conducted another study, and garlic was shown to possess biological activities that ameliorate obesity-related physiological changes (Sangouni et al. 2020b).

Curcuma longa is a widely used plant not only in Ayurveda but also in other traditional medicine systems because of its medicinal properties. Curcumin is a major natural phenolic compound found in the rhizomes of turmeric. Clinical studies have confirmed that the consumption of curcuminoids and piperine has antidiabetic and hepatoprotective effects (Panahi et al. 2018, 2019; Mirhafez et al. 2021).

Table 4.3 List of plants scientifically proven for abolishing NAFLD in clinical studies

Plant	Extract/ bioactive agent	Results/key findings	References
Allium sativum	Powder	Ameliorates obesity-related physiological changes	Sangouni et al. (2020a, b), (2021)
Curcuma longa	Curcumin	Antidiabetic and hepatoprotective effects	Mirhafez et al. (2021), Panahi et al. (2018, 2019)
Piper nigrum	Piperine	Antidiabetic and hepatoprotective effects	Mirhafez et al. (2021)
Glycyrrhiza glabra	Root extract	Improves alanine aminotransferase and insulin resistance	Rostamizadeh et al. (2022)
Nigella sativa	Seed powder	Reduces C-reactive protein and nuclear factor kappa-B	Darand et al. (2019)
Portulaca oleracea	Seed powder, leaf extract	Improves metabolic profile and oxidative stress; reduces alanine and aspartate aminotransferases, γ-glutamyltransferase	Gheflati et al. (2019), Darvish Damavandi et al. (2021), Milkarizi et al. (2024)
Silybum marianum	Silybin	Reduces hypersensitive C-reactive protein and anti-inflammatory activity	Li et al. (2024), Shaker et al. (2025)
Pueraria lobata	Puerarin	Reduces hypersensitive C-reactive protein and anti-inflammatory activity	Li et al. (2024)
Salvia miltiorrhiza	Salvianolic acid	Reduces hypersensitive C-reactive protein and anti-inflammatory activity	Li et al. (2024)
Phyllanthus niruri	Powdered aerial part	Reduces total cholesterol	Li et al. (2024), Shaker et al. (2025)
Zingiber officinale	Rhizome extract	Liver transaminase levels	Ghoreishi et al. (2024), Attari et al. (2016)

Licorice (*Glycyrrhiza glabra*) has shown good results in clinical studies targeting NAFLD. Rostamizadeh et al. showed that daily consumption of 1000 mg of root extract for 12 weeks helps to reduce body weight and statistically improves alanine aminotransferase, insulin, insulin resistance, malondialdehyde serum levels, and ultrasonographic findings of liver steatosis, compared to the control group in a randomized double-blind clinical trial (Rostamizadeh et al. 2022). About 2 g of *Nigella sativa* seed consumption for 12 weeks reduced inflammatory biomarkers in patients with non-alcoholic fatty liver disease compared to the placebo group (starch 2 g/day). The levels of highly sensitive C-reactive protein and nuclear factor kappa-B decreased significantly only in Nigella seed powder (Darand et al. 2019).

Portulaca oleracea has shown successful results in clinical trials for ameliorating NAFLD-associated symptoms. One randomized clinical trial using the seeds of this plant, along with a low-calorie diet, improved the metabolic profile and oxidative stress (Gheflati et al. 2019). Darvish Damavandi et al. demonstrated that the consumption of leaf extract of the plant (300 mg/day) for 12 weeks led to a significant decrease in alanine and aspartate aminotransferases, γ-glutamyl transferase, fasting glucose, insulin resistance, triglycerides, and low-density lipoprotein (Darvish Damavandi et al. 2021). The same study showed no difference between the test and placebo-controlled groups in liver insulin, steatosis grade, total bilirubin, liver enzymes, lipid profile, and blood pressure between the test and placebo groups (Darvish Damavandi et al. 2021). Recently, a study from Milkarizi et al. (2024) reported the efficiency of Portulaca extract in ameliorating NAFLD-associated health issues, and this study also supported previous studies that showed improvements in liver enzymes (Milkarizi et al. 2024).

Silybum marianum is another plant with which clinical studies have evaluated the synergistic activity of three other plant species: *Pueraria lobata, Salvia miltiorrhiza, and Phyllanthus niruri*. Li et al. (2024) studied the effect of *S. marianum, S. miltiorrhiza, and P. lobata,* together on NAFLD patients (18–65 years). This study used tablets made of 23.0 mg silybin (*Silybum marianum*), 11.4 mg puerarin (*Pueraria lobata*), and 10.9 mg of salvianolic acid (*S. miltiorrhiza),* and tablets from each plant were consumed twice daily for 24 weeks along with placebo. After 24 weeks, the absolute lipid fat content and percentage of lipid fat content in the plant-extract-based tablet-treated group were compared to those in the placebo group, but the difference was not statistically significant. The extract intervention reduced hypersensitive C-reactive protein levels and increased the activity of aldehyde dehydrogenase (Li et al. 2024). Shaker et al. (2025) attempted another combination phase II, double blind, placebo-controlled, randomized, three-arm, interventional clinical trial with *S. marianum and P. niruri* (Shaker et al. 2025). This study used a capsule named Heptex made of powdered extract from the aerial part of *P. niruri* (200 mg) and powdered extract from the fruit of *S. marianum* (100 mg) with low (one capsule) and high doses (two capsules) for 36 weeks. Both low and high doses of capsules significantly reduced alanine transaminase levels, and only high doses reduced aspartate aminotransferase levels. Except for total cholesterol, there were no significant differences in the lipid profiles of other parameters between the three groups.

Zingiber officinale Roscoe is another plant that has shown beneficial effects on NAFLD in clinical studies. One study evaluated the effect of powdered *Zingiber officinale* (1000 mg/dose, twice daily for 3 months) in NAFLD patients with diabetes mellitus, and the treatment group showed a significant decrease in insulin resistance, serum insulin, body mass index, and waist and hip circumferences, and a significant increase in serum high-density lipids. There were no changes in inflammatory markers, but liver transaminase levels were reduced (Ghoreishi et al. 2024). Attari et al. (2016) also performed a study using ginger extract (2 g/day; 1 g/day) and starch as the placebo group. The consumption of ginger powder for 12 weeks improved the metabolic changes associated with obesity and decreased overall body weight.

4 Conclusions and Future Perspectives

The occurrence of NAFLD increased with increased incidence of obesity, and nowadays the incidence has alarmingly increased among young populations. Traditional medicine systems used the potential of plants for improving NAFLD and associated symptoms. Modern science also evaluated the effectiveness of plants using advanced technologies, and many of the plants proved to be promising candidates in cell lines and introduced animal models. Even though all plants are not good at improving all physiological issues associated with NAFLD, many of them function in different pathways to ameliorate the associated symptoms, such as lipidogenesis, fat accumulation prevention, anti-inflammatory activities, improved insulin sensitivity, etc. To an extent, many scientific studies further evaluated the effectiveness of plants in healthy and NAFLD patients. Even though these studies were only done in a limited number of patients, results showed a similar pattern for treating NAFLD, which means the effectiveness was not universal in different people nor was it equally effective in ameliorating all symptoms. These results make us focused on the activity of individual plants, which targets different symptoms associated with NAFLD and promising the future development of formulations targeting oxidative stress, inflammation, insulin resistance, and fibrosis associated with NAFLD. The effectiveness of these developing formulations should be evaluated using biomarkers associated with NAFLD rather than ameliorating clinical symptoms. These studies have to be done in much larger populations so that ethnobotanical data can be validated by genomics and personalized treatment for NAFLD.

References

Abd-Elrazek AM, Ibrahim SR, El-dash HA et al (2022) The ameliorative effect of *Apium graveolens* & curcumin against non-alcoholic fatty liver disease induced by high fructose-high fat diet in rats. Futur J Pharm Sci 8:26

Abubakar BM, Dibal NI, Attah MOO, Chiroma SM et al (2022) Exploring the antioxidant effects of *Aloe vera*: potential role in controlling liver function and lipid profile in high fat and fructose diet (HFFD) fed mice. Pharmacol Res - Mod Chin Med 4:100150

Adams LA, Angulo P et al (2006) Treatment of non-alcoholic fatty liver disease. Postgrad Med J 82:315–322

Afarin R, Hatami M, Monjezi S, Bineshfar F, Ahangarpour A et al (2024) Suppression of TGF-β/Smad3 signaling pathway by *Capparis spinosa* and quercetin in a rat model of nonalcoholic steatohepatitis. Iran J Basic Med Sci 27:1096–1104

Akbari R, Yaghooti H, Jalali MT, Khorsandi LS, Mohammadtaghvaei N et al (2020) *Capparis spinosa* improves the high fat diet-induced non-alcoholic steatohepatitis in rats: the possible role of FGF21. BMC Res Notes 13:356

Akbari R, Yaghooti H, Jalali MT, Khorsandi LS, Mohammadtaghvaei N et al (2022) *Capparis spinosa* improves non-alcoholic steatohepatitis through down-regulating SREBP-1c and a PPARα-independent pathway in high-fat diet-fed rats. BMC Res Notes 15:315

Aslani N, Entezari MH, Askari G, Maghsoudi Z, Maracy MR et al (2016) Effect of garlic and lemon juice mixture on lipid profile and some cardiovascular risk factors in people 30-60 years old with moderate hyperlipidaemia: a randomized clinical trial. Int J Prev Med 7:95

Babatunde PF, Adeniran OY, Metibemu DS, Bose OM et al (2023) Biochemical evaluation of the anti-hepatic steatosis activity of *Allium cepa* on carbon tetrachloride (CCl_4)-induced Wistar rats. IOSR J Biotechnol Biochem 10:1–7

BasuRay S, Smagris E, Cohen JC, Hobbs HH et al (2017) The PNPLA3 variant associated with fatty liver disease (I148M) accumulates on lipid droplets by evading ubiquitylation. Hepatology 66:1111–1124

Cabrera D, Wree A, Povero D et al (2017) Andrographolide ameliorates inflammation and fibrogenesis and attenuates inflammasome activation in experimental non-alcoholic steatohepatitis. Sci Rep 7:3491

Carter A, Brackley SM, Gao J, Mann JP et al (2019) The global prevalence and genetic spectrum of lysosomal acid lipase deficiency: a rare condition that mimics NAFLD. J Hepatol 70:142–150

Chen YC, Kao TH, Tseng CY, Chang WT, Hsu CL et al (2014) Methanolic extract of black garlic ameliorates diet-induced obesity via regulating adipogenesis, adipokine biosynthesis, and lipolysis. J Funct Foods 9:98–108

Darand M, Darabi Z, Yari Z, Saadati S, Hedayati M, Khoncheh A, Hosseini-Ahangar B, Alavian SM, Hekmatdoost A et al (2019) Nigella sativa and inflammatory biomarkers in patients with non-alcoholic fatty liver disease: results from a randomized, double-blind, placebo-controlled, clinical trial. Complement Ther Med 44:204–209

Darvish Damavandi R, Shidfar F, Najafi M, Janani L, Masoodi M, Akbari-Fakhrabadi M, Dehnad A et al (2021) Effect of Portulaca oleracea (purslane) extract on liver enzymes, lipid profile, and glycemic status in nonalcoholic fatty liver disease: a randomized, double-blind clinical trial. Phytother Res 35:3145–3156

Dibal NI, Buba F, Chiroma SM, Goni ZM, Kilobas HE, Sheriff H, Kuma JU, Kwaha TJ, Judge A, Muhammed A, Shuaibu HG, Falnyi ZG, Muhammad AA et al (2022) Aloe vera ameliorates hyperlipidemia, enhances endogenous antioxidant activity, and regulates liver function in high-fat diet fed BALB/c mice. Mediterr J Nutr Metab 15:505–515

Donnelly KL, Smith CI, Schwarzenberg SJ, Jessurun J, Boldt MD, Parks EJ et al (2005) Sources of fatty acids stored in liver and secreted via lipoproteins in patients with nonalcoholic fatty liver disease. J Clin Invest 115:1343–1351

Ebrahimzadeh Attari V, Ostadrahimi A, Asghari Jafarabadi M, Mehralizadeh S, Mahluji S et al (2016) Changes of serum adipocytokines and body weight following *Zingiber officinale* supplementation in obese women: a RCT. Eur J Nutr 55:2129–2136

Emamat H, Farhadnejad H, Tangestani H, Saneei Totmaj A, Poustchi H, Hekmatdoost A et al (2020) Association of allium vegetables intake and non-alcoholic fatty liver disease risk: a case-control study. Nutr Food Sci 50:1075–1083

Erbas O, Erdogan MA, Khalilnezhad A, Gürkan FT, Yiğittürk G, Meral A, Taskiran D et al (2018) Neurobehavioral effects of long-term maternal fructose intake in rat offspring. Int J Dev Neurosci 69:68–79

Erdenebileg S, Kim M, Nam Y, Cha KH, Le TT, Jung SH, Nho CW et al (2024) *Artemisia argyi* ethanol extract ameliorates nonalcoholic steatohepatitis-induced liver fibrosis by modulating gut microbiota and hepatic signaling. J Ethnopharmacol 333:118415

Friedman SL, Neuschwander-Tetri BA, Rinella ME, Sanyal AJ (2018) Mechanisms of NAFLD development and therapeutic strategies. Nat Med 24:908–922

Gabbia D, Zanotto I, Frion-Herrera Y, De Martin S et al (2024) Preliminary evidence of the anti-steatotic effect of *Abelmoschus esculentus* extracts in in vitro models of MASLD. Dig Liver Dis 56:S46

Gheflati A, Adelnia E, Nadjarzadeh A et al (2019) The clinical effects of purslane (*Portulaca oleracea*) seeds on metabolic profiles in patients with nonalcoholic fatty liver disease: a randomized controlled clinical trial. Phytother Res 33:1501–1509

Ghoreishi PS, Shams M, Nimrouzi M, Zarshenas MM, Lankarani KB, Fallahzadeh Abarghooei E, Talebzadeh M, Hashempur MH et al (2024) The effects of ginger (*Zingiber officinale* Roscoe) on non-alcoholic fatty liver disease in patients with type 2 diabetes mellitus: a randomized double-blinded placebo-controlled clinical trial. J Diet Suppl 21:294–312

Gurian E, Giraudi P, Rosso N, Tiribelli C, Bonazza D, Zanconati F, Giuricin M, Palmisano S, De Manzini N, Sergo V, Bonifacio A et al (2020) Differentiation between stages of non-alcoholic fatty liver diseases using surface-enhanced Raman spectroscopy. Anal Chim Acta 1110:190–198

Ha AW, Ying T, Kim WK et al (2015) The effects of black garlic (*Allium sativum*) extracts on lipid metabolism in rats fed a high fat diet. Nutr Res Pract 9:30–36

Han SK, Baik SK, Kim MY et al (2023) Non-alcoholic fatty liver disease: definition and subtypes. Clin Mol Hepatol 29(suppl):S5–S16

Harley IT, Stankiewicz TE, Giles DA, Softic S, Flick LM, Cappelletti M, Sheridan R, Xanthakos SA, Steinbrecher KA, Sartor RB, Kohli R, Karp CL, Divanovic S et al (2014) IL-17 signaling accelerates the progression of nonalcoholic fatty liver disease in mice. Hepatology 59:1830–1839

He U, Yang S, Zhang A, Zhang X, Sun J, Tian L et al (2019) Lipid-lowering and antioxidative effects of *Apium graveolens* L. root flavonoid extracts. RSC Adv 9:26757–26767

Holmen OL, Zhang H, Fan Y, Hovelson DH, Schmidt EM, Zhou W, Guo Y, Zhang J, Langhammer A, Løchen ML, Ganesh SK, Vatten L, Skorpen F, Dalen H, Zhang J, Pennathur S, Chen J, Platou C, Mathiesen EB, Wilsgaard T, Willer CJ et al (2014) Systematic evaluation of coding variation identifies a candidate causal variant in TM6SF2 influencing total cholesterol and myocardial infarction risk. Nat Genet 46:345–351

Hossain MS (2024) Evaluation of safety and efficacy of *Andrographis paniculata* extracts on hepatic steatosis and stiffness in patients with nonalcoholic fatty liver disease. J Clin Exp Hepatol 14:101963

Huang P, Yang L, Liu T, Jiang Y, Chen Z, Song H, Zheng P et al (2024) Scoparone alleviates nonalcoholic fatty liver disease by modulating the PPARα signaling pathway. Eur J Pharmacol 984:177033

Jannah FN, Anggraini PD, Gofur A, Lestari SR et al (2021) The prevention of single bulb garlic extract on non-alcoholic fatty liver disease (NAFLD) in high fat mice (Mus musculus) model. AIP Conf Proc 2353:030095

Khan R (2011) Effects of garlic on blood glucose levels and HbA1c in patients with type 2 diabetes mellitus. J Med Plant Res 5:2922–2928

Khoonsari M, Mohammad Hosseini Azar M, Ghavam R, Hatami K, Asobar M, Gholami A, Rajabi A, Safarnezhad Tameshkel F, Amirkalali B, Sohrabi M et al (2017) Clinical manifestations and diagnosis of nonalcoholic fatty liver disease. Iran J Pathol 12:99–105

Kozlitina J, Smagris E, Stender S, Nordestgaard BG, Zhou HH, Tybjærg-Hansen A, Vogt TF, Hobbs HH, Cohen JC et al (2014) Exome-wide association study identifies a TM6SF2 variant that confers susceptibility to nonalcoholic fatty liver disease. Nat Genet 46:352–356

Lau JK, Zhang X, Yu J (2017) Animal models of non-alcoholic fatty liver disease: current perspectives and recent advances. J Pathol 241:36–44

Leite NC, Salles GF, Araujo AL, Villela-Nogueira CA, Cardoso CR et al (2009) Prevalence and associated factors of non-alcoholic fatty liver disease in patients with type-2 diabetes mellitus. Liver Int 29:113–119

Li BY, Xi Y, Liu YP, Wang D, Wang C, Chen CG, Fang XH, Li ZX, Chen YM et al (2024) Effects of *Silybum marianum*, *Pueraria lobate*, combined with *Salvia miltiorrhiza* tablets on non-alcoholic fatty liver disease in adults: a triple-blind, randomized, placebo-controlled clinical trial. Clin Nutr ESPEN 63:2–12

Liang M, Huo M, Guo Y, Zhang Y, Xiao X, Xv J, Fang L, Li T, Wang H, Dong S, Jiang X, Yu W (2022) Aqueous extract of *Artemisia capillaris* improves non-alcoholic fatty liver and obesity in mice induced by high-fat diet. Front Pharmacol 13:1017329

Liang M, Dong S, Guo Y, Zhang Y, Xiao X, Ma J, Jiang X, Yu W (2024) Exploration of the potential mechanism of aqueous extract of *Artemisia capillaris* for the treatment of non-alcoholic fatty liver disease based on network pharmacology and experimental verification. J Pharm Pharmacol 76:1328–1339

Lindenmeyer CC, McCullough AJ (2018) The natural history of nonalcoholic fatty liver disease—an evolving view. Clin Liver Dis 22:11–21

Liu L, Zhao J, Li Y, Wan Y, Lin J, Shen A, Xu W, Li H, Zhang Y, Xu J, Peng J, Hong Z (2016) Artemisia capillaris formula inhibits hepatic steatosis via an miR-122-induced decrease in fatty acid synthase expression in vivo and in vitro. Sci Rep 13(6):4751–4758

Liu YT, Chen HW, Lii CK, Jhuang JH, Huang CS, Li ML et al (2020) A diterpenoid, 14-deoxy-11, 12-didehydroandrographolide, in Andrographis paniculata reduces steatohepatitis and liver injury in mice fed a high-fat and high-cholesterol diet. Nutrients 12(2):523

Luo ZH, Liu ZW, Mao Y, Shu R, Fu LC, Yang RY, Hu YJ, Shen XL et al (2020) Cajanolactone A, a stilbenoid from *Cajanus cajan*, prevents ovariectomy-induced obesity and liver steatosis in mice fed a regular diet. Phytomedicine 78:153290

Luo Z, Huang J, Li Z, Liu Z, Fu L, Hu Y, Shen X et al (2021a) Cajanolactone A, a stilbenoid from *Cajanus cajan* (L.) Millsp, prevents high-fat diet-induced obesity via suppressing energy intake. Front Pharmacol 12:695561

Luo ZH, Huang JW, Meng QQ, Wu HW, Yang RY, Fu LC, Hu YJ, Shen XL et al (2021b) Cajanolactone A, a stilbenoid from *Cajanus cajan*, inhibits energy intake and lipid synthesis/storage, and promotes energy expenditure in ovariectomized mice. Biomed Pharmacother 138:111491

Luo F, Oldoni F, Das A (2022) TM6SF2: a novel genetic player in nonalcoholic fatty liver and cardiovascular disease. Hepatol Commun 6(3):448–460

Lu WS, Sun RX, Gao YY, Wen JP, Pan RF, Li L, Wang J, Xian YX, Cao CX, Zheng M (2013) Nonalcoholic fatty liver disease and microvascular complications in type 2 diabetes. World J Gastroenterol 19(20):3134–3142

Mardi P, Kargar R, Fazeli R, Qorbani M (2023) Allium sativum: a potential natural compound for NAFLD prevention and treatment. Front Nutr 10:1059106

Meng Y, Liu Y, Fang N, Guo Y et al (2019) Hepatoprotective effects of *Cassia* semen ethanol extract on non-alcoholic fatty liver disease in experimental rat. Pharm Biol 57:98–104

Milkarizi N, Barghchi H, Belyani S, Bahari H, Rajabzade F, Ostad AN, Goshayeshi L, Nematy M, Askari VR (2024) Effects of Portulaca oleracea (purslane) on liver function tests, metabolic profile, oxidative stress and inflammatory biomarkers in patients with non-alcoholic fatty liver disease: a randomized, double-blind clinical trial. Front Nutr 11:1371137

Mirhafez SR, Dehabeh M, Hariri M, Farimani AR, Movahedi A, Naderan RD, Jamialahmadi T, Simental-Mendía LE, Sahebkar A (2021) Curcumin and piperine combination for the treatment of patients with non-alcoholic fatty liver disease: a double-blind randomized placebo-controlled trial. Adv Exp Med Biol 1328:11–19

Mirunalini S, Krishnaveni M, Ambily V (2011) Effects of raw garlic (Allium Sativum) on hyperglycemia in patients with type 2 diabetes mellitus. Pharmacol 2:968–974

Monetti M, Levin MC, Watt MJ, Sajan MP, Marmor S, Hubbard BK, Stevens RD, Bain JR, Newgard CB, Farese RV Sr, Hevener AL, Farese RV Jr (2007) Dissociation of hepatic steatosis and insulin resistance in mice overexpressing DGAT in the liver. Cell Metab 6(1):69–78

Morey P, Wankhade A, Paithankar V, Vyas JV (2023) Evaluation of antiobesity activity of Argemone mexicana Linn. in high-fat diet induced obesity in rats. Int J Res Publ Rev 4(5):6859–6866

Mouhid L, Gómez de Cedrón M, Quijada-Freire A, Fernández-Marcos PJ, Reglero G, Fornari T, Ramírez de Molina A (2019) Yarrow supercritical extract ameliorates the metabolic stress in a model of obesity induced by high-fat diet. Nutrients 12(1):72

Panahi Y, Khalili N, Sahebi E, Namazi S, Simental-Mendía LE, Majeed M, Sahebkar A (2018) Effects of curcuminoids plus piperine on glycemic, hepatic and inflammatory biomarkers in patients with type 2 diabetes mellitus: a randomized double-blind placebo-controlled trial. Drug Res 68(7):403–409

Panahi Y, Valizadegan G, Ahamdi N, Ganjali S, Majeed M, Sahebkar A (2019) Curcuminoids plus piperine improve nonalcoholic fatty liver disease: a clinical trial. J Cell Biochem 120(9):15989–15996

Peng CH, Ker YB, Li HH, Tsou SH, Lin CL, Huang CN (2022) Abelmoschus esculentus subfractions ameliorate hepatic lipogenesis and lipid uptake via regulating dipeptidyl peptidase-4-With improving insulin resistance. PLoS One 17(3):e0265444

Pericleous M, Kelly C, Wang T, Livingstone C, Ala A (2017) Wolman's disease and cholesteryl ester storage disorder: the phenotypic spectrum of lysosomal acid lipase deficiency. Lancet Gastroenterol Hepatol 2(9):670–679

Pipitone RM, Zito R, Lupo G, Javed A, La Mantia C, Di Maria G, Pratelli G, Di Salvo F, Fontana S, Pucci M, Carlisi D, Grimaudo S (2023) Curcumin and andrographolide co-administration safely prevent steatosis induction and ROS production in HepG2 cell line. Molecules 28(3):1261

Prince MRU, Zihad SMNK, Ghosh P, Sifat N, Rouf R, Al Shajib GM, Alam MA, Shilpi JA, Uddin SJ (2021) Amaranthus spinosus attenuated obesity-induced metabolic disorders in High-Carbohydrate-High-Fat Diet-Fed Obese Rats. Front Nutr 8:653918

Prom-in S, Wongwitwichot P, Wangpradit N, Chua KH, Kamisah Y, Kumar J, Kaewsrichan J (2021) Abelmoschus esculentus L. mitigates NAFLD pathologic phenotypes in C57BL/6J mouse model. IJPSR 12(11):5715–5722

Ran LS, Wu YZ, Gan YW et al (2023) Andrographolide ameliorates hepatic steatosis by suppressing FATP2-mediated fatty acid uptake in mice with nonalcoholic fatty liver disease. J Nat Med 77:73–86

Ratziu V, Giral P, Jacqueminet S, Charlotte F, Hartemann-Heurtier A, Serfaty L, Podevin P, Lacorte JM, Bernhardt C, Bruckert E, Grimaldi A, Poynard T, LIDO Study Group (2008) Rosiglitazone for nonalcoholic steatohepatitis: one-year results of the randomized placebo-controlled fatty liver improvement with Rosiglitazone Therapy (FLIRT) Trial. Gastroenterology 135(1):100–110

Reinke H, Asher G (2015) Circadian clock control of liver metabolic functions. Gastroenterology 150:574–580

Reinshagen M, Kabisch S, Pfeiffer AFH et al (2023) Liver fat scores for noninvasive diagnosis and monitoring of nonalcoholic fatty liver disease in epidemiological and clinical studies. J Clin Transl Hepatol 11:1212–1227

Rezaei S, Ashkar F, Koohpeyma F et al (2020) Hydroalcoholic extract of *Achillea millefolium* improved blood glucose, liver enzymes and lipid profile compared to metformin in streptozotocin-induced diabetic rats. Lipids Health Dis 19:81

Rinella ME, Sanyal AJ (2016) Management of NAFLD: a stage-based approach. Nat Rev Gastroenterol Hepatol 13:196–205

Roohullah S, Khan Z (2017) Review on non-alcoholic fatty liver disease and its homoeopathic treatment. Int J Res AYUSH Pharm Sci 1:184–192

Rostamizadeh P, Asl SMKH, Far ZG et al (2022) Effects of licorice root supplementation on liver enzymes, hepatic steatosis, metabolic and oxidative stress parameters in women with nonalcoholic fatty liver disease: a randomized double-blind clinical trial. Phytother Res 36:3949–3956

Salari N, Darvishi N, Mansouri K et al (2021) Association between PNPLA3 rs738409 polymorphism and nonalcoholic fatty liver disease: a systematic review and meta-analysis. BMC Endocr Disord 21:125

Sangouni AA, Mohammad Hosseini Azar MR, Alizadeh M (2020a) Effect of garlic powder supplementation on hepatic steatosis, liver enzymes and lipid profile in patients with non-alcoholic fatty liver disease: a double-blind randomised controlled clinical trial. Br J Nutr 124:450–456

Sangouni AA, Mohammad Hosseini Azar MR, Alizadeh M (2020b) Effects of garlic powder supplementation on insulin resistance, oxidative stress, and body composition in patients with non-alcoholic fatty liver disease: a randomized controlled clinical trial. Complement Ther Med 51:102428

Sangouni AA, Alizadeh M, Jamalzehi A et al (2021) Effects of garlic powder supplementation on metabolic syndrome components, insulin resistance, fatty liver index, and appetite in subjects with metabolic syndrome: a randomized clinical trial. Phytother Res 35:4433–4441

Shaker MK, Hassany M, Eysa B et al (2025) The activity of a herbal medicinal product of *Phyllanthus niruri* and *Silybum marianum* powdered extracts (Heptex®) in patients with apparent risk factors for nonalcoholic steatohepatitis: a phase II, multicentered, randomized, double-blind, placebo-controlled clinical trial. BMC Complement Med Ther 25:8

Shalimar EA, Bansal B et al (2022) Prevalence of non-alcoholic fatty liver disease in India: a systematic review and meta-analysis. J Clin Exp Hepatol 12:818–829

Sharma B, John S (2023) Nonalcoholic Steatohepatitis (NASH). StatPearls Publishing, Treasure Island. Available from: https://www.ncbi.nlm.nih.gov/books/NBK470243/

Sharshar AM, Ameen AA, El-Masry HG (2022) The bioactive role of aqueous extract of *Aegle marmelos* leaves on obese rats. Int J Health Sci 6(S9):620–629

Sherman DJ, Liu L, Mamrosh JL et al (2023) The fatty liver disease-causing protein PNPLA3-I148M alters lipid droplet-Golgi dynamics. bioRxiv. https://doi.org/10.1101/2023.10.13.562302

Shojaei-Zarghani S, Fattahi MR, Kazemi A et al (2022) Effects of garlic and its major bioactive components on non-alcoholic fatty liver disease: a systematic review and meta-analysis of animal studies. J Funct Foods 96:105206

Singh S, Allen AM, Wang Z et al (2015) Fibrosis progression in nonalcoholic fatty liver vs nonalcoholic steatohepatitis: a systematic review and meta-analysis of paired-biopsy studies. Clin Gastroenterol Hepatol 13:643–654

Sookoian S, Pirola CJ, Valenti L et al (2020) Genetic pathways in nonalcoholic fatty liver disease: insights from systems biology. Hepatology (Baltimore, Md.) 72:330–346

Sun SF, Zhong HJ, Zhao YL et al (2022) Indole alkaloids of Alstonia scholaris (L.) R. Br. alleviated nonalcoholic fatty liver disease in mice fed with high-fat diet. Nat Prod Bioprospect 12:14

Takahashi Y, Sugimoto K, Inui H et al (2015) Current pharmacological therapies for nonalcoholic fatty liver disease/nonalcoholic steatohepatitis. World J Gastroenterol 21:3777–3785

Targher G, Bertolini L, Rodella S et al (2007) Nonalcoholic fatty liver disease is independently associated with an increased incidence of cardiovascular events in type 2 diabetic patients. Diabetes Care 30:2119–2121

Toppo E, Darvin SS, Esakkimuthu S et al (2017) Effect of two andrographolide derivatives on cellular and rodent models of non-alcoholic fatty liver disease. Biomed Pharmacother 95:402–411

Torres DM, Harrison SA (2008) Diagnosis and therapy of nonalcoholic steatohepatitis. Gastroenterology 134:1682–1698

Williamson RM, Price JF, Glancy S et al (2011) Prevalence of and risk factors for hepatic steatosis and nonalcoholic fatty liver disease in people with type 2 diabetes: the Edinburgh Type 2 Diabetes Study. Diabetes Care 34:1139–1144

Wong VW, Ekstedt M, Wong GL et al (2023) Changing epidemiology, global trends and implications for outcomes of NAFLD. J Hepatol 79:842–852

World Health Organization (2024) One in eight people are now living with obesity. World Health Organization Available from: https://www.who.int/news/item/01-03-2024-one-in-eight-people-are-now-living-with-obesity

Xiao J, Ching YP, Liong EC et al (2013) Garlic-derived S-allylmercaptocysteine is a hepatoprotective agent in non-alcoholic fatty liver disease in vivo animal model. Eur J Nutr 52:179–191

Xu M, Ma L, Liang H et al (2025) Protective effects of small RNAs encapsulated in Artemisia capillaris-derived exosomes against non-alcoholic fatty liver disease. Front Pharmacol 15:1476820

Younossi ZM, Golabi P, Paik JM et al (2023) The global epidemiology of nonalcoholic fatty liver disease (NAFLD) and nonalcoholic steatohepatitis (NASH): a systematic review. Hepatology 77:1335–1347

Yu Q, Lee YY, Xia ZY et al (2021) S-allylmercaptocysteine improves nonalcoholic steatohepatitis by enhancing AHR/NRF2-mediated drug metabolizing enzymes and reducing NF-κB/IκBα and NLRP3/6-mediated inflammation. Eur J Nutr 60:961–973

Zhao D, Cao J, Jin H et al (2021) Beneficial impacts of fermented celery (Apium graveolens L.) juice on obesity prevention and gut microbiota modulation in high-fat diet fed mice. Food Funct 12:9151–9164

Phytochemicals in Prevention and Management of Cardiovascular Disease

5

Ananda Kumar Chettupalli, Aman Sisodia, S. K. Abdul Rahaman, Sarad Pawar Naik Bukke, and Narender Boggula

Abstract

Cardiovascular diseases, or CVDs, are a generic term for the numerous diseases that affect the coronary blood arteries and the heart. Phytochemicals of natural origin encompass a wide variety of vitamins, steroidal saponins, polyphenols, flavonoids, and organ sulfur compounds. A great number of bioactive chemicals obtained from terrestrial plants have been established to improve cardiovascular health and prevent cardiovascular disease. These include carotenoids, diosgenin, resveratrol, quercetin, isoflavones, catechin, sulforaphane, and tocotrienols. Cardiovascular disease is the global killer. Flavonoids are present in citrus fruits, onions, and cocoa and help in improving endothelial function, reducing blood pressure, and inhibiting platelet aggregation. Among all the studied phytochemical molecules, one well-studied molecule has been related to numerous health benefits, including cancer prevention and cardiovascular diseases. Other crucial substances that have been proven to be beneficial to many diseases are Brassica oleracea, curcumin, and berberine. Increasing data suggest that phytochemicals and plant-based whole meals are an effective and alternative therapy for lowering hyperlipidemia and CVD. Many foods and medicinal plants contain phytochemical antioxidants, which assist in the treatment and prevention of chronic oxidative stress disorders. Their intensive anti-inflammatory antioxidant and

A. K. Chettupalli (✉) · A. Sisodia · S. K. Abdul Rahaman
Department of Pharmacy, School of Medical & Allied Sciences, Greater Noida, Uttar Pradesh, India

S. P. N. Bukke
Department of Pharmaceutics and Pharmaceutical Technology, Kampala International University, Western Campus, Ishaka – Bushenyi, Uganda

N. Boggula
Department of Pharmaceutical Chemistry, Omega College of Pharmacy, Ghatkesar, Medchal, Telangana, India

K. Kuca et al. (eds.), *Medicinal Plants and Their Bioactives in Human Diseases*,
https://doi.org/10.1007/978-3-032-01356-9_5

free-radical scavenging properties result in protection against cardiovascular disorders. Herbal nanotechnology offers vast benefits over conventional herbal preparations and extracts of CVDs. In that case, improved solubility, bioavailability, protection against toxicity, enhanced pharmacological activity, lesser amounts of medication required, and enhanced stability are involved. This chapter puts forward phytochemicals as safe, natural, and cost-effective CVD preventive agents.

Keywords

Phytochemicals · Cardiovascular disease · Oxidative stress · Nanoparticles · Cardioprotection · Anti-inflammatory

1 Introduction

Cardiovascular diseases encompass a broad group of conditions such as high blood pressure, atherosclerosis, coronary artery disease (CAD), heart failure, and stroke. Some of the major factors responsible for CVDs are shown in Fig. 5.1. Despite great developments in medical science and therapeutic therapies, the global burden of CVDs remains huge and requires further study into new and complementary means

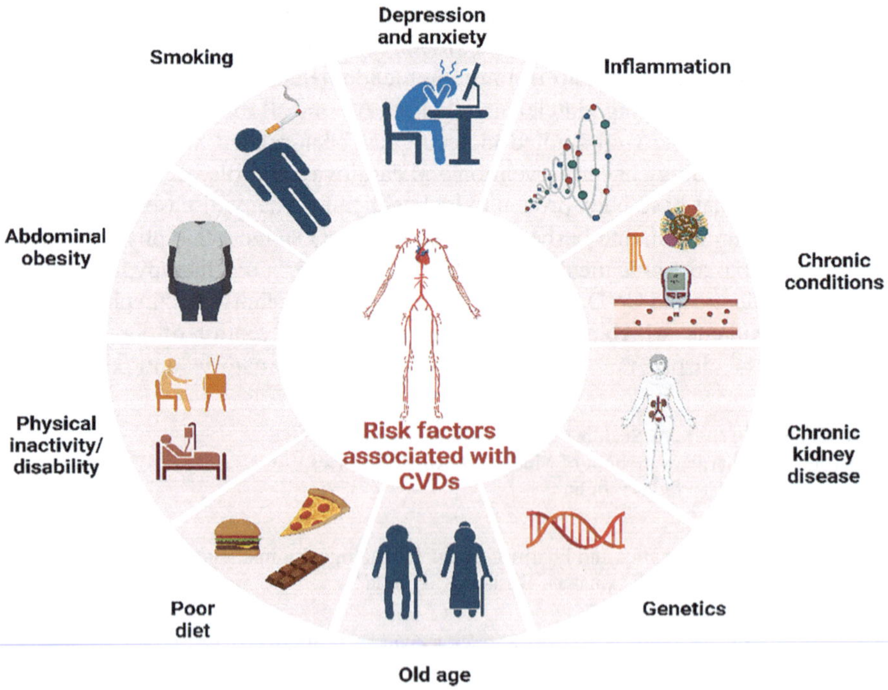

Fig. 5.1 Factors responsible for CVDs

of prevention and therapy (Chang et al. 2020). The area of phytoconstituents, which refers to plant-derived bioactive compounds, has materialized as the future study area, hence receiving considerable attention. Plant-based nutrients, such as flavonoids, have been found to have promising research for the alteration of heart-protective pathways and provide natural and sustainable responses to cardiovascular disease onset. This increased focus identifies the potential therapeutic repute of natural materials in cardiovascular health (Sharifi-Rad et al. 2020).

In the United States, one dies every 33 s from cardiovascular disease. According to 2016, India accounted for 63% of all NCD-related deaths with CVDs representing 27% of the deaths. The estimated number of deaths in India from CVD will rise from 2.26 million in 1990 to 4.77 million by 2020 (Huffman et al. 2011). Furthermore, CVDs account for 45% of fatalities in adults aged 40–69 (WHO 2016). Researchers and healthcare providers are interested in natural goods due to their diversity and often significant benefits that are valued in traditional medical systems (David et al. 2015). These products are attractive due to their potential to offer a holistic and tailored approach to cardiovascular health by tapping into the vast availability of bioactive substances and phytochemicals from nature (Mittal et al. 2024). This use of these natural medicines is strongly engraved in the communal understanding of traditional healing practices, and growing volumes of scientific research testify to their therapeutic usefulness (O'Connor 1995).

Other recommendations include the use of natural pigments produced from natural sources to treat cardiovascular diseases. The use of natural pigments in cardiovascular health is in line with our growing appreciation of the role of diet and lifestyle in heart outcomes. Colorful fruits and vegetables are not only pleasing to the eye but also contain a broad spectrum of bioactive chemicals such as carotenoids, flavonoids, and anthocyanins. These dyes impart bright colors to these foods, but they also impart several cardiovascular advantages. Flavonoids are the natural compounds of plant polyphenols of fruits, cereals, vegetables, roots, bark, flowers, stems, tea, and wine. Carotenoids, such as carotene, lycopene, lutein, cryptoxanthin, zeaxanthin, and astaxanthin are the most commonly occurring in diet; they contribute to the maintenance of cardiovascular disease (Shukla et al. 2010). To counter oxidative stress, nutritional components in food act as antioxidants, scavenging free radicals. Natural goods are important in the clinic, and more and more medicinal plants are being used as natural antioxidants. In addition to their antioxidant activity, natural commodities have other biological properties that reduce the risk of CVD (Chang et al. 2020). The purpose of this chapter is to examine the function of phytoconstituents and natural pigments in cardiovascular wellness, with particular emphasis on their potential in prevention and treatment processes. Knowledge about these bioactive molecules and how they work may give us great insights into formulating new, plant-derived therapeutic agents for the management of cardiovascular disorders.

2 Factors Associated with CVDs

The various factors associated with CVDs are discussed as below:

2.1 Ischemia Reperfusion Injury

Reoxygenation of tissue with a temporarily interrupted oxygen supply causes an increase in reactive oxygen species and oxidative damage, which also leads to further mitochondrial ultrastructure damage, function, and metabolism, known as ischemia-reperfusion injury (Wu et al. 2010). The flavonoid orientin is found in natural plant extracts. Orientin has been proven to possess anti-inflammatory, antitumor, and antioxidant properties. It also causes cardiomyopathy and protects against heart ischemia-reperfusion injury by increasing antioxidant defense (Wang et al. 2017). The dried leaves of a plant in the family Rosaceae have hawthorn leaf flavonoid that is known to have anti-inflammatory, antioxidant, and hypolipidemic effects. Hawthorn leaf flavonoids have been documented to enhance antioxidant enzymes and scavenger radicals, thereby reducing lipid peroxidation levels and protecting the endothelial lining of arteries against free oxygen radicals. Flavonoids isolated from hawthorn leaves can therefore prevent heart damage attributed to oxidative stress (Fu et al. 2013).

2.2 Hypertension

Hypertension is a clinical disease exhibited by elevated blood pressure levels. Besides the level of high blood pressure, there may be functional damage to the heart, brain, and kidneys. It has been suggested that reactive oxygen species are an important pathogenic mechanism for hypertension. Reactive oxygen species, in their role as signaling molecules, may regulate relaxation and expansion of smooth muscle cells within the vascular system and the endothelial system of blood vessels (Barbosa-Filho et al. 2006). Mulberries, peanuts, wine, and other plant-based foods contain naturally occurring polyphenols such as resveratrol. It can act as a natural antioxidant against cardiovascular disease conditions, which include treatment and prevention (Wang et al. 2017). Resveratrol, as shown in Table 5.2, which lowers blood pressure and reduces oxidative stress, also enhances the redox-protein capacity for modulating the redox environment of a cell (Theodotou et al. 2017). Tea polyphenols form the most prominent bioactive principles in tea, which help prevent or cure hypertension. Though tea polyphenols can immensely reduce abnormal blood pressures rises, however, they have virtually no effect upon the blood pressures level. Levels of oxidative damage are decreased, whereas levels of the antioxidant enzymes increased. Activity related to the inherent process (Faraji and Tarkhani 1999). Curcumin is a polyphenol compound derived from the rhizome of the turmeric plant. The therapeutic properties of turmeric are that it is anti-inflammatory, antioxidant, antifibrotic, and anticancer. Experiments show the potent antioxidant

Table 5.2 CVDs with mechanism targeted by phytochemicals and their sources

Cardiovascular disease	Targeted mechanism	Phytochemicals	Key benefits	Common sources	References
Hypertension	Nitric oxide production, RAAS modulation	Resveratrol, Anthocyanins	Vasodilation, reduced blood pressure	Grapes, berries	Najjar et al. (2021)
Atherosclerosis	LDL oxidation prevention, plaque stabilization	Curcumin, EGCG	Reduced plaque formation, improved vascular health	Turmeric, green tea	Zhang et al. (2021)
Congestive heart failure	Mitochondrial protection, reduced oxidative stress	Coenzyme Q10, quercetin, curcumin	Improved energy efficiency, reduced inflammation	Spinach, onions	Xu et al. (2024)
Hyperlipidemia	Cholesterol metabolism, lipid oxidation	Phytosterols, Berberine	Lower LDL levels, increased HDL	Nuts, barberry	Emamat et al. (2022)
Stroke	Platelet aggregation inhibition, endothelial protection	Salicylates, polyphenols	Reduced clot formation, enhanced recovery	Willow bark, dark chocolate	Duttaroy (2024)

activity of curcumin. Curcumin reduces cardiovascular resistance, increases vascular responsiveness, and averts the development and progression of hypertension through blood vessel lowering production of peroxides (Nakmareong et al. 2011).

2.3 Arrhythmia

Unusual sinoatrial node activation causes arrhythmias. Research showed that the imbalance of redox plays a vital role in the pathogenesis of arrhythmia. Several studies have shown tachyarrhythmia patients had much greater amounts of reactive oxygen species (MDA) and other markers of free radical damage in their blood than normal levels of SOD, TAC, GSH, and other antioxidant indicators (Brenyo and Aktas 2014). The rhizome of *Paeonia suffruticosa* contains an extract known as paeonol with anti-inflammatory, antioxidant, and antibacterial effects. Paeonol may protect against arrhythmias, shorten the duration of a conduction block or atrial fibrillation, and prevent myocardial ischemia during reperfusion injury (Lee 2015). Matrine is a natural alkaloid extracted from *Sophora flavescens*, a plant widely recognized for its medicinal properties. The matrine possesses anti-inflammatory, antiviral, and antiarrhythmic effects. The matrine physically can prevent the sodium ions from leaving the cardiac cell membrane to keep a normal heart rhythm going

(Zhou et al. 2014). By interfering with both the potassium and sodium ion transport channels of the cardiac membrane, it can also reduce the stress on the cardiomyocyte and promote irregular heartbeats (Zhou et al. 2022).

2.4 Heart Failure

Heart failure is the most common cause of hospital admission of cardiac patients, and it is considered a major public health problem. The symptom score could further classify heart failure as having either reduced or preserved ejection fraction (Li et al. 2015). Curcumin, (as shown in Table 5.2), also called diferuloylmethane, is a polyphenol that is extracted from the rhizomes of *Curcuma longa* L., or turmeric. Many studies conducted using HF experimental models have established that curcumin is effective. Curcumin enhances rat post-MI left ventricular systolic function when used with an ACE inhibitor. In their experiment, rats with moderate-sized MIs were subjected to sham surgery or left coronary artery ligation. After 1 week, rats were orally administered with enalapril, curcumin, or enalapril + curcumin for 6 weeks.(Morimoto et al. 2010) The evidence presented here, which consists of both in vitro and in vivo animal experiments, supports the notion that curcumin is active and helpful in treating heart failure (HF). It achieves this through several ways, such as increasing ventricular hypertrophy, controlling the expression of associated genes, and avoiding cardiomyocyte fibrosis (Choudhary 2011).

2.5 Hyperlipidemia

Hyperlipidemia, or high LDL cholesterol and low HDL cholesterol levels, is another significant risk factor for cardiovascular disease. Phytochemicals improve lipid homeostasis by reducing cholesterol absorption and enhancing lipid metabolism (Islam et al. 2021). Phytosterols (as shown in Table 5.2) found in nuts and seeds inhibit dietary cholesterol absorption and consequently lower serum LDL. Saponins found in most legumes enhance cholesterol excretion through bile acid binding; tocotrienols, on the other hand, which are abundant in palm oil and rice bran, decrease cholesterol synthesis and enhance lipid profiles (Bouchenak and Lamri-Senhadji 2013).

2.6 Atherosclerosis

Accumulation of plaques in the arterial walls is one of the main pathogenic factors of cardiovascular disease. Phytochemicals are rich in anti-inflammatory and anti-atherogenic properties, thereby reducing the plaque formation process. Turmeric has curcumin that reduces inflammation and inhibits smooth muscle cell development in arterial walls (Nimgulkar et al. 2015). Garlic-derived allicin inhibits LDL cholesterol oxidation, an essential step of plaque development, and works on improving endothelial functions (Fig. 5.1). In addition, quercetin, a flavonoid in onions and apples, also increases endothelial functions and inhibits foam cell growth, necessary for early atherosclerosis (Kitts 2005).

3 Cardiovascular Diseases and Medicinal Herbs

The major scientific basis for the development of new drugs and their natural sources is the compositional evaluation and chemical properties of plant-based substances. According to dose, most bioactive chemicals have good effects on human well-being and are extra nutritional components that are naturally present in foods in small levels (Chen et al. 2019). At least 5000 plant species have been used for over 3000 years and are known to possess therapeutic values. Some of these include Gaertn (milk thistle), *Silybum marianum* (L.), lily of the valley (*Convallaria majalis* L.), and borage (*Borago officinalis* L.) that are used to treat congestive heart failure (CHF) and affiliated cardiovascular diseases and arrhythmia. The bioactivities of these herbs are due to the phytochemicals they contain (Naveed et al. 2020). Some of the medicinal herbs used against cardiovascular diseases are discussed below:

3.1 Garlic (*Allium sativum* L.)

Garlic has been known to have positive and protective effects against infections, cardiovascular disease, and cancer. It is considered beneficial for lowering triglycerides and cholesterol, reducing blood pressure, and preventing platelet aggregation (Ahmed & Wang, 2021). Garlic may have anti-atherosclerotic benefits by reducing fat deposition in blood vessels and blocking oxidative and inflammatory processes (Sobenin et al. 2019).

3.2 Tea (*Camellia sinensis* L.)

It has been suggested that the consumption of tea helps in weight reduction, cancer, neurological problems, and improving memory and mental alertness (Chieng and Kistler 2022). Tea has been well known to be a good source of polyphenols, especially epigallocatechin gallate (EGCG), niacin, folic acid, riboflavin, pantothenic acid, potassium, magnesium, and manganese (Tijburg et al. 1997; Bhardwaj and Khanna 2013). Catechins exert protective effects against the vascular system for multiple reasons, such as antihypertensive, antioxidative, antiproliferative, antithrombogenic, anti-inflammatory, and lipid-lowering effects (Stangl et al. 2006).

3.3 Soybean (*Glycine max* L.)

It is believed that the use of medicinal soy helps with the symptoms of osteoporosis, cancer, cognitive impairments, and menopause infrequently. It seeks to minimize CVD through cholesterol and blood pressure levels (Chatterjee et al. 2018). According to several human studies, isoflavones may reduce CVD due to the following: decrease in thrombus formation, improved cerebrovascular reactivity, and raising of LDL resilience to oxidative stress (Martin et al. 2008).

3.4 Danshen (*Salvia miltiorrhiza* Bunge)

The use of Danshen as a form of herbal remedy was to treat the condition of CVD that include angina, myocardial infarction, coronary artery disease, high blood pressure, hyperlipidaemia, and cerebrovascular disease sickness (Cheng 2007). One of the most common components is Tanshinone IIA which is likely to have a favorable impact on hypertension, angina pectoris, MI, hyperlipidemia, and other diseases Gao et al. 2012). It may also help to minimize the occurrence of platelet-associated thromboembolic disease. The antioxidant properties of Salvia ketone IIA helps to protect the heart from perfusion injury (Ren-an et al. 2014). Additionally, it protects the heart's functionality and controls Ca^{2+} excess inside cells, which increases the danger of cell damage (Fang et al. 2018).

3.5 Hawthorn (Crataegus spp.)

Hawthorn extract is frequently used by herbalists in treating cerebral insufficiency, angina, CHF, and bradyarrhythmia. Due to its favorable inotropic and vasodilatory characteristics, hawthorn is believed to increase myocardial perfusion and reduce afterload (Melzer and Saller 2009). Since hawthorn potentiates the activity of digitalis, its concomitant use should be monitored closely for potential side effects. Additionally, hawthorn prevents the formation of thromboxane A2, which may pose a risk for bleeding in patients on anticoagulant or antiplatelet drugs (Orhan 2018).

3.6 Arjuna (*Terminalia arjuna*)

Terminalia arjuna is a widely known Ayurvedic medicinal plant used mostly for its cardioprotective activity. It has been traditionally employed to cure several cardiovascular diseases like hypertension, heart failure, and ischemic heart disease. (Ramesh and Palaniappan 2023) *T. arjuna* bark contains bioactive molecules like flavonoids, tannins, triterpenoids, and glycosides, all of which play a significant role in its cardioprotective actions.(Tahir et al. 2025).

4 The Role of Phytochemicals and Natural Products as Cardio-Preventive in Specific Cardiovascular Diseases

Plants naturally possess bioactive molecules, called phytochemicals, which confer several health benefits. They are crucial for human health and help in the prevention of disorders, especially CVDs (Altemimi et al. 2017). These compounds are interesting candidates for therapeutic use owing to their diverse biological properties involving cardioprotective, anti-inflammatory, and antioxidant activities (Adegbola et al. 2017). Phytochemicals inhibit plaque formation and enhance coronary circulation, thereby used to prevent CAD, which is characterized by the narrowing of the coronary arteries because of plaque deposition (Mir et al. 2024). Flavonoids prevent platelet aggregation, stabilize arterial plaques, and enhance endothelial function. Omega-3 fatty acids, although not strictly phytochemicals, help to decrease plaque formation and lower triglycerides, conferring substantial cardiovascular benefits (Kaur et al. 2024). In the case of a stroke caused by the disruption of blood flow to the brain, phytochemicals provide neuroprotective and cardiovascular benefits. *Ginkgo biloba* enhances cerebral circulation and decreases ischemic damage. It also protects neurons against oxidative damage (Prasad et al. 2022). Carotenoids present in colored fruits and vegetables help preserve vascular integrity and decrease oxidative stress, thereby lowering the risk of stroke (Orellana-Urzúa et al. 2023). Finally, some phytochemicals inhibit cardiac remodeling and protect mitochondrial activity and, thus, can help in the treatment of congestive heart failure (CHF), a disease wherein the heart is unable to pump blood sufficiently (Cicero and Colletti 2017). Ginseng includes ginsenosides, which reduce oxidative stress, improve mitochondrial efficiency, and boost cardiac output (Hyun et al. 2022). Although not a phytochemical, coenzyme Q10 enhances these effects by improving mitochondrial energy generation and minimizing oxidative damage in order to improve heart function (Franco et al. 2019).

Bioactive chemicals found in coffee include caffeine, diterpenes (cafestol and kahweol), and chlorogenic acid isomers.(Makiso et al. 2024) Polyphenols, flavonoids, and phytochemicals are all bioactive compounds which help in the prevention of cardiovascular disease (Din et al. 2023). They are more prone to manage the oxidation of low-density lipoproteins, which increases the amount of lipids in the circulation and controls the process of endothelial apoptosis (Jain et al. 2018). Many of the flavonoids commonly found in foods, including tea, coffee, spices, fruits, nuts, vegetables, and treatments of botanical origin, have been shown to decrease the incidence of CVD (Aron and Kennedy 2008). Plant sterols are bioactive compounds that act just like animal cholesterol. Plant sterols, an alkaloid that has a different side chain composition than cholesterol, have become quite popular since there is potential to reduce cholesterol in those products.(Nes 2011) Foods that are rich in plant sterols are an effective dietary supplement for lowering cardiovascular risk because they reduce blood levels of total cholesterol and low-density lipoprotein cholesterol by as much as 15% (Saleem et al. 2022). Some of the best-known plants are garlic, onion, leeks, and cruciferous vegetables including broccoli, cabbage, cauliflower, and kale. These plants have sulfur compounds that are often associated with a reduced risk of cardiovascular disease (Kris-Etherton et al. 2002). As it can scavenge free radicals and reduce inflammation, sulforaphane may find application in anti-CVD therapies. It can activate the Nrf2 basic leucine zipper transcription factor-related factor, as an anti-oxidative stress defense mechanism (Bachheti et al. 2022).

4.1 Flavonoids

These compounds are interesting candidates for therapeutic use owing to their diverse biological properties involving cardioprotective, anti-inflammatory, and antioxidant activities (Maleki et al. 2019). Some rich sources of quercetin in fruits include apples and berries (as shown in Table 5.1) that have been widely recognized to hinder inflammation and oxidative stress, prevent breakdown in arterial integrity, and prevent cardiovascular disease (Lakhanpal and Rai 2008). Catechins in cocoa and green tea elevate endothelial function, reduce oxidative damage, prevent atherosclerosis, and consequently contribute to good vascular health (Mangels and Mohler III 2017).

Table 5.1 Phytochemicals, their sources and properties related to CVD

Phytochemical	Source	Properties	References
Anthocyanins	Berries, cherries, and grapes	Antioxidant, anti-inflammatory, reduce blood pressure, protect against atherosclerosis	Garcia and Blesso (2021)
Carotenoids	Carrots, sweet potato, and tomatoes	Reduce oxidative stress and inflammation	Kaulmann and Bohn (2014)
Flavonoids	Citrus fruits, apples, onions, and tea	Antioxidant, anti-inflammatory activity, and help reduce blood pressure	Benavente-Garcia and Castillo (2008)
Curcumin	Turmeric	Antioxidant and anti-inflammatory activity, improve endothelial function	Karimian et al. (2017)
Betanin	Beets	Lower blood pressure, improve endothelial function and reduce the risk of heart disease	Mirmiran et al. (2020)
Cyanidin	Blackberries, blackcurrants, and cherries	Improve endothelial functions and reduce the risk of CVDs	Wang et al. (2021)
Ailanthone	Chinese tree of heaven	Reduce blood pressure and reduce the risk of CVDs	Andonova et al. (2024)
Resveratrol	Red grapes and red wine	Protect against atherosclerosis, reduce LDL cholesterol	Chassot et al. (2018)

4.2 Phenolic Acid

Phenolic acids are aromatic organic compounds that are characterized by containing a hydroxyl group in addition to having an aromatic ring. They have been most documented for their antioxidants, which prevent oxidative damage within cells by neutralizing free radicals (Kumar and Goel 2019). Whole grains, rice bran, and coffee are excellent sources of ferulic acid, a potent antioxidant capable of inhibiting inflammation and preventing lipid peroxidation (Alam 2019). Caffeic acid, present in coffee, a few fruits, and plants, helps to preserve good heart health since it reduces oxidative stress and controls cholesterol levels (Tajik et al. 2017).

4.3 Alkaloids

Alkaloids constitute a class of organic compounds that contain nitrogen and display an array of biological activities. Such properties include cardioprotective activity (Ushie et al. 2023). Berberine, a compound found in Berberi species of plants, has been known to regulate lipid metabolism, lower blood sugar, and improve cardiovascular health by reducing oxidative stress and inflammation (Ma et al. 2018).

4.4 Terpenoids

Terpenoids also have the names isoprenoids. They belong to hydrocarbons with multiplicity varieties of functional groups. It has several effects, which include lipid lowering, antioxidant, and anti-inflammatory (Kaur et al. 2019). Ginsenosides are present in ginseng have been shown to be effective in modulating the synthesis of nitric oxide (NO), lowering oxidative stress, and improving cardiac function, especially in illnesses such as heart failure and hypertension (Kim 2018).

4.5 Other Compounds

Other compounds, such as lignans, are rich in antioxidants and can be found in seeds such as flaxseed. They improve cardiovascular health by lowering cholesterol, enhancing lipid metabolism, and have anti-inflammatory properties (Imran et al. 2015). The chemical structures of the compounds that show therapeutic activity in cardiovascular disorders are shown in Fig. 5.2. Carotenoids are pigments having strong antioxidant activity, found in colorful fruits and vegetables, including bell peppers, tomatoes, and carrots (Fig. 5.2). They prevent atherosclerosis and other cardiovascular disorders through increased vascular health and protection against oxidative stress (Singh and Omer 2024). Some more compounds, their sources, and medicinal properties are mentioned in Table 5.1.

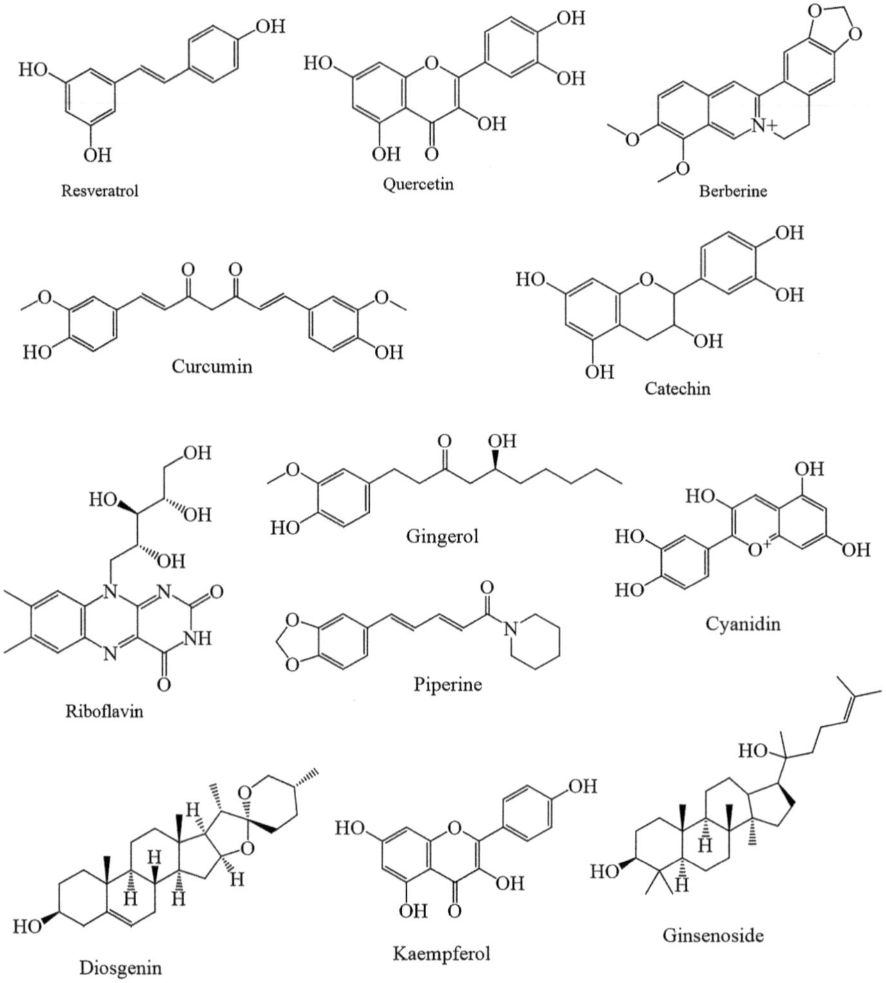

Fig. 5.2 Chemical structures of some phytocompounds associated with treatment of CVDs drawn using Chemdraw software

5 Molecular Mechanisms of Phytochemical Action

Phytochemicals provide cardioprotective effects via a variety of molecular pathways. This includes signaling pathway links (Upadhyay and Dixit 2015). Control of NO bioavailability, modulation of oxidative stress and mitochondrial function, and inhibition of proinflammatory cytokines. An understanding of these pathways has significant therapeutic promise for the prevention and treatment of cardiovascular disorders (Arandhara et al. 2024).

5.1 Interaction with Signaling Pathways

Phytochemicals influence critical cellular signaling pathways that regulate inflammation, oxidative stress, and apoptosis, all of which play important roles in CVD etiology (Guan et al. 2021).

5.1.1 NF-κB Pathway

NF-κB is a transcription factor that activates genes associated with inflammation and immunological responses (Lee and Burckart 1998). Curcumin and quercetin block NF-κB activation, leading to decreased production of pro-inflammatory cytokines and enzymes, such as TNF-α and COX-2 (Zhang et al. 2019).

5.1.2 MAPK Pathway

Mitogen-activated protein kinases regulate cellular responses to stress and inflammation (Cowan and Storey 2003). Catechins and resveratrol are chemicals that modulate MAPK signaling, resulting in reduced inflammatory responses and protection against endothelial dysfunction (Cowan and Storey 2003; Rahman et al. 2006).

5.2 Modulation of Nitric Oxide (NO) Bioavailability and Endothelial Function

Nitric oxide is a molecule of great importance in the vascular system. The molecule regulates vasodilation, blood pressure, and endothelial integrity. Phytochemicals improve NO production and bioavailability. This improves cardiovascular function (Bondonno et al. 2015). Green tea and cocoa contain flavonoids that stimulate endothelial nitric oxide synthase (eNOS) activity, which increases NO production. Phytochemicals protect NO from oxidative degradation by scavenging ROS, thus maintaining its vasodilatory effects (Duarte et al. 2014).

5.3 Influence on Inflammatory Cytokines

Such inflammatory cytokines are TNF-α and IL-6 and are involved extensively with CVDs. Phytochemicals thus suppress the synthesis and release of these cytokines, thereby minimizing inflammation (Zhang et al. 2019).

5.3.1 TNF-α

Curcumin and berberine suppress TNF-α production through inhibition of upstream signaling pathways, such as NF-κB. This suppresses arterial irritation and prevents plaque formation in atherosclerosis (Meng et al. 2022).

5.3.2 IL-6

Resveratrol and flavonoids decrease IL-6 expression, which reduces systemic inflammation and improves lipid metabolism (Peluso et al. 2013).

5.4 The Synergistic Effects of Phytochemicals and Conventional Therapies

When the established pharmacological drugs are combined with phytochemicals, the results of therapy may improve because of synergistic effects. Chemicals may enhance the pharmacodynamic effects of cardiovascular drugs, including statins, ACE inhibitors, and beta-blockers, thereby enhancing efficacy (Gao et al. 2022). Flavonoids, for example, may improve endothelial function and reduce oxidative stress, thus boosting the effectiveness of antihypertensive drugs (Maaliki et al. 2019). Antioxidant defense can assist to decrease drug-induced side effects, such as hepatotoxicity and muscle damage caused by statin treatment (Liu et al. 2019). An overall strategy for dealing with hard cardiovascular diseases, like atherosclerosis and congestive heart failure, involves targeting multiple pathways at one time (Ruparelia et al. 2017) (Table 5.2).

6 Challenges and Limitations

6.1 Bioavailability and Metabolism

One of the most significant obstacles in using phytochemicals for CVD prevention and management is bioavailability, which is defined as the extent and rate of entry of an active ingredient into systemic circulation, where it can exert its impact. Chemical structure, limited water solubility (for example, flavonoids and carotenoids), and first-pass metabolism in the liver can all reduce the efficiency of such chemicals. The variability in gut flora between people influences the conversion of phytochemicals into bioactive metabolites. Digestive enzymes, pH, and food matrix interactions (such as fat or protein content) can all have an impact on absorption and metabolism. Various research initiatives are currently focusing on enhancing bioavailability using innovative formulations such as nanoencapsulation, liposomes, and synthetic analogs (Nicolescu et al. 2023).

6.2 Variability in Natural Sources

Although phytochemicals are produced from plants, their content and concentration can vary greatly due to environmental, agronomic, and processing conditions. Plant phytochemical concentration is influenced by variations in soil quality, climate, and altitude, as well as cultivation strategies such as organic versus conventional farming. Post-harvest factors, including storage and transportation, may reduce their potency, and industrial processing processes (e.g., drying, frying, or extraction) can modify their bioactivity. This raises questions about the replicability and efficacy of phytochemicals, both in laboratory studies and as therapies (Kumar et al. 2023).

6.3 Safety and Dose

Although phytochemicals are naturally occurring and hence considered reasonably safe, concerns about safety and dose should be addressed with extreme caution. Excessive ingestion of certain phytochemicals, such as alkaloids or saponins, might cause gastrointestinal distress or liver damage. Interactions with medications complicate their use even further; for example, grapefruit flavonoids have been demonstrated to inhibit cytochrome P_{450} enzymes, impairing drug metabolism. Furthermore, individual heterogeneity, which includes hereditary characteristics and underlying health issues, influences how people react to phytochemicals. The lack of regulation in most over-the-counter supplements heightens the dangers of inconsistent doses and toxicity. To reduce these dangers, it is vital to define safe and effective dosage ranges through clinical trials (Bode and Dong 2015). The lack of consistency in phytochemical formulation and clinical application is a significant hurdle. Differences in extraction processes frequently result in variations in the purity and potency of active chemicals. Quality control difficulties in herbal supplement production led to variable amounts of bioactive components, making cross-study comparisons difficult. Furthermore, there is little regulatory control for phytochemical products. Many are classified as dietary supplements rather than pharmaceuticals, allowing them to undergo less stringent safety and efficacy testing. Another issue is that there are fewer large-scale, well-designed clinical trials on these medications, limiting the establishment of standard methods for their application. To maintain consistent quality and safety, researchers, regulatory agencies, and industry must work together. and effectiveness of phytochemical-based interventions to address these challenges (Yan et al. 2022). Table 5.3 shows the various challenges and clinical potentials of phytocompounds.

7 Future Perspectives

7.1 Exploration of Novel Phytochemicals

The discovery of new phytochemicals has enormous potential for enhancing the science of disease prevention and treatment, particularly in cardiovascular disorders. Though well-known plant chemicals such as flavonoids, polyphenols, and alkaloids have been linked to numerous cardiovascular benefits, there is a vast untapped reservoir of lesser-known plant species and their bioactive constituents. The new phytochemical search entails evaluating diverse plant species, such as indigenous traditional medicines utilized over time in different parts of the world (Shaito et al. 2020). There are some promising areas of research, such as genetically

Table 5.3 Challenges and clinical potentials of phytochemicals

Aspect	Challenges	Potential solutions	Examples of phytochemicals	References
Bioavailability	Low absorption, rapid metabolism	Use of nanoformulations, liposomal delivery	Curcumin. Resveratrol	Santos et al. (2011)
Clinical translation	Limited large scale clinical trials	Conducting more randomized controlled trails	Berberine, Anthocyanins	Ju et al. (2018)
Safety	Potential for herb–drug interaction	Careful dose optimization, toxicology studies	Garlic (Allicin), green tea, catechins	Rasheed et al. (2024)
Standardization	Variability in phytochemical content in sources	Developing standardized extracts and formulations	Flavonoids, Saponins	Kumar et al. (2025)
Cost and accessibility	High cost of extraction and purification	Use of sustainable and affordable sources	Plant sterols, Omega-3 fatty acids	Swetha and Mathanghi (2024)

modified plants, that potentially give higher concentrations of specific bioactive chemicals with enhanced pharmacokinetic qualities. Plant biotechnology has advanced to the point where plants can be genetically modified to produce more of specific phytochemicals like resveratrol or quercetin, which are widely recognized to have cardiovascular advantages. Targeting these molecules to increase their bioavailability can lead to more effective phytochemical-based medicines (Langston et al. 2023).

7.2 Advances in Drug Delivery Systems for Phytochemicals

The poor absorption, fast metabolism, and short half-lives of phytochemicals in the body frequently restrict their bioavailability and therapeutic efficiency. Drug delivery system advancements are thus crucial in overcoming these difficulties and enhancing phytochemical cardiovascular effects. Nanotechnology has emerged as a viable way to improve phytochemical delivery to target tissues. Such encapsulation, produced from liposomes, nanoencapsulation, and nanocarriers, allows phytochemicals to be protected against degradation, delivered in regulated release, and enhanced in their gastrointestinal absorption (Chavda et al. 2023). Liposomes, for example, are lipid-based vesicles that encapsulate hydrophobic phytochemicals, enhancing their solubility and absorption. Other interesting possibilities for improving the stability, bioavailability, and controlled release of phytochemicals are SLNs and NLC. These delivery methods can target specific tissues, such as the cardiovascular system, boosting the therapeutic benefit while reducing side effects (Pandey et al. 2022).

7.3 Role of Phytochemicals in Precision Medicine

As precision medicine becomes more prevalent in healthcare, there is a tremendous opportunity to adapt phytochemical medicines to individual patients based on their unique genetic, epigenetic, and phenotypic profiles. The efficacy of phytochemicals in cardiovascular disease can vary depending on genetic polymorphisms, dietary behaviors, and microbiome composition. Future study may focus on discovering biomarkers that predict the response to certain phytochemicals, enabling individualized therapy regimens for CVD patients (Giuliani et al. 2022). For example, some people's genetic variations metabolize phytochemicals in ways that reduce the compound's efficiency. As a result, pharmacogenomics may help identify which patients are most likely to respond to various plant-based medicines. Furthermore, the metabolism and bioavailability of phytochemicals by gut microbiota are considered crucial in precision medicine, given that variable compositions of the gut microbiota alter how individuals may respond to plant-based treatments (Kastenmüller et al. 2015).

7.4 Collaborative Approaches in Phytochemical Research

Given the complexities of the diseases and the various mechanisms by which phytochemicals exert their overall effects, several research efforts must be integrated to completely advance the effective creation of usable phytochemical-based therapeutics. Biologists, chemists, pharmacologists, and physicians must work together to elucidate the intricate relationships between phytochemicals and the human system (Majrashi et al. 2023). Biologists and chemists are needed to identify and describe novel phytochemicals, as well as comprehend their molecular and cellular modes of action. Pharmacologists can then examine the pharmacokinetics, pharmacodynamics, and potential adverse effects of these substances, while clinicians provide useful insights into the real-world use and clinical efficacy of phytochemical medicines (Nkwocha et al. n.d.).

8 Conclusions

The study of natural products in the context of cardiovascular health is a growing field with important implications for global health. Cardiovascular disease (CVD) prevalence remains a serious problem, necessitating innovative solutions beyond typical pharmaceutical interventions. This review looked at a wide range of natural chemicals, including herbal medicines and compounds originating from marine sources, and highlighted their potential as key components in the complex management and prevention of CVDs. Comprehensive interventions are required due to the intricate relationships between oxidative stress, inflammation, hypertension, and other cardiovascular risk factors. Fruits, vegetables, and medicinal herbs include natural pigments, antioxidants, and flavonoids that have been shown to be useful in decreasing oxidative stress, a key risk factor for cardiovascular disease. Their ability to regulate blood pressure and cholesterol profiles, together with their anti-inflammatory properties, makes them essential assets to cardiovascular health. Nano-phytomedicine research provides personalized treatment regimens and targeted medication delivery devices, adding a futuristic element to cardiovascular care. We are coming closer to a more focused and efficient treatment strategy with the synthesis of natural chemicals like quercetin and curcumin into nano-formulations, which show the potential for higher bioavailability and lower adverse effects. However, there are impediments to the potential field of natural products in cardiovascular treatment. Safety problems, limited scientific evaluation, and regulatory issues necessitate careful consideration and ongoing research. Informed patient-physician communication is critical for the successful integration of natural therapies into standard cardiology practice, which requires finding a balance between maximizing possible benefits and limiting dangers. Future advancements in cardiovascular treatment appear to be both exciting and transformative. The combination of genetics, technology, and natural goods enables personalized therapy based on each patient's unique genetic and metabolic profile. As knowledge and technology advance, the use of natural products in cardiovascular care is projected to grow in importance. This will assist to promote holistic approaches, reduce the global prevalence of heart-related illnesses, and provide people with more sophisticated and practical ways to maintain their cardiovascular health.

References

Adegbola P, Aderibigbe I, Hammed W, Omotayo T (2017) Antioxidant and anti-inflammatory medicinal plants have potential role in the treatment of cardiovascular disease: a review. Am J Cardiovasc Dis 7(2):19

Ahmed T, Wang C-K (2021) Black garlic and its bioactive compounds on human health diseases: a review. Molecules 26(16):5028

Alam MA (2019) Anti-hypertensive effect of cereal antioxidant ferulic acid and its mechanism of action. Front Nutr 6:121

Altemimi A, Lakhssassi N, Baharlouei A, Watson DG, Lightfoot DA (2017) Phytochemicals: extraction, isolation, and identification of bioactive compounds from plant extracts. Plants 6(4):42

Andonova T, Petkova Z, Teneva O, Antova G, Apostolova E, Naimov S, Mladenova T, Slavov I, Dimitrova-Dyulgerova I (2024) *Ailanthus altissima* seed oil—a valuable source of lipid-soluble components with DNA protective and antiproliferative activities. Foods 13(8):1268

Arandhara A, Saha D, Deka DJ, Deka M, Kumar Das B (2024) Redox imbalance and cardiovascular pathogenesis: exploring the therapeutic potential of phytochemicals. Curr Bioact Compd 20(9):55–78

Aron PM, Kennedy JA (2008) Flavan-3-ols: nature, occurrence and biological activity. Mol Nutr Food Res 52(1):79–104

Bachheti RK, Worku LA, Gonfa YH, Zebeaman M, Deepti, Pandey DP, Bachheti A (2022) [Retracted] prevention and treatment of cardiovascular diseases with plant phytochemicals: a review. Evid Based Complement Alternat Med 2022(1):5741198

Barbosa-Filho JM, Martins VKM, Rabelo LA, Moura MD, Silva MS, Cunha EVL, Souza MFV, Almeida RN, Medeiros IA (2006) Natural products inhibitors of the angiotensin converting enzyme (ACE): a review between 1980–2000. Rev Bras 16:421–446

Benavente-Garcia O, Castillo J (2008) Update on uses and properties of citrus flavonoids: new findings in anticancer, cardiovascular, and anti-inflammatory activity. J Agric Food Chem 56(15):6185–6205

Bhardwaj P, Khanna D (2013) Green tea catechins: defensive role in cardiovascular disorders. Chin J Nat Med 11:345

Bode AM, Dong Z (2015) Toxic phytochemicals and their potential risks for human cancer. Cancer Prev Res 8(1):1–8. https://doi.org/10.1158/1940-6207.CAPR-14-0160

Bondonno CP, Croft KD, Ward N, Considine MJ, Hodgson JM (2015) Dietary flavonoids and nitrate: effects on nitric oxide and vascular function. Nutr Rev 73(4):216–235

Bouchenak M, Lamri-Senhadji M (2013) Nutritional quality of legumes, and their role in cardiometabolic risk prevention: a review. J Med Food 16(3):185–198

Brenyo A, Aktas MK (2014) Review of complementary and alternative medical treatment of arrhythmias. Am J Cardiol 113(5):897–903

Chang X, Zhang T, Zhang W, Zhao Z, Sun J (2020) Natural drugs as a treatment strategy for cardiovascular disease through the regulation of oxidative stress. Oxidative Med Cell Longev 2020(1):5430407

Chassot LN, Scolaro B, Roschel GG, Cogliati B, Cavalcanti MF, Abdalla DSP, Castro IA (2018) Comparison between red wine and isolated trans-resveratrol on the prevention and regression of atherosclerosis in LDLr (−/−) mice. J Nutr Biochem 61:48–55

Chatterjee C, Gleddie S, Xiao C-W (2018) Soybean bioactive peptides and their functional properties. Nutrients 10(9):1211

Chavda VP, Nalla LV, Balar P, Bezbaruah R, Apostolopoulos V, Singla RK, Khadela A, Vora L, Uversky VN (2023) Advanced phytochemical-based nanocarrier systems for the treatment of breast cancer. Cancer 15(4):1023

Chen Z, Luo T, Zhang L, Zhou Z, Huang Y, Lu L, Yang Z, Wang L, Xian S (2019) A simplified herbal formula for the treatment of heart failure: efficacy, bioactive ingredients, and mechanisms. Pharmacol Res 147:104251

Cheng TO (2007) Cardiovascular effects of Danshen. Int J Cardiol 121(1):9–22

Chieng D, Kistler PM (2022) Coffee and tea on cardiovascular disease (CVD) prevention. Trends Cardiovasc Med 32(7):399–405

Choudhary MI (2011) Frontiers in cardiovascular drug discovery. Bentham Sci Publ 1:1–310

Cicero FG, Colletti A (2017) Nutraceuticals and dietary supplements to improve quality of life and outcomes in heart failure patients. Curr Pharm Des 23(8):1265–1272

Cowan KJ, Storey KB (2003) Mitogen-activated protein kinases: new signaling pathways functioning in cellular responses to environmental stress. J Exp Biol 206(7):1107–1115

David B, Wolfender J-L, Dias DA (2015) The pharmaceutical industry and natural products: historical status and new trends. Phytochem Rev 14:299–315

Din SRU, Saeed S, Khan SU, Kiani FA, Alsuhaibani AM, Zhong M (2023) Bioactive compounds (BACs): a novel approach to treat and prevent cardiovascular diseases. Curr Probl Cardiol 48(7):101664

Duarte J, Francisco V, Perez-Vizcaino F (2014) Modulation of nitric oxide by flavonoids. Food Funct 5(8):1653–1668

Duttaroy AK (2024) Functional foods in preventing human blood platelet hyperactivity–mediated diseases. Nutrients 6(21):3717

Emamat H, Zahedmehr A, Asadian S, Nasrollahzadeh J (2022) The effect of barberry (Berberis integerrima) on lipid profile and systemic inflammation in subjects with cardiovascular risk factors: a randomized controlled trial. BMC Complement Med Ther 22(1):59

Fang J, Little PJ, Xu S (2018) Atheroprotective effects and molecular targets of tanshinones derived from herbal medicine Danshen. Med Res Rev 38(1):201–228

Faraji MH, Tarkhani AHH (1999) The effect of sour tea (Hibiscus sabdariffa) on essential hypertension. J Ethnopharmacol 65(3):231–236

Franco R, Navarro G, Martínez-Pinilla E (2019) Hormetic and mitochondria-related mechanisms of antioxidant action of phytochemicals. Antioxidants 8(9):373

Fu J, Zheng Y, Li P, Li X, Shang X, Liu J (2013) Hawthorn leaves flavonoids decreases inflammation related to acute myocardial ischemia/reperfusion in anesthetized dogs. Chin J Integr Med 19:582–588
Gao S, Liu Z, Li H, Little PJ, Liu P, Xu S (2012) Cardiovascular actions and therapeutic potential of tanshinone IIA. Atherosclerosis 220(1):3–10
Gao Q, Feng J, Liu W, Wen C, Wu Y, Liao Q, Zou L, Sui X, Xie T, Zhang J (2022) Opportunities and challenges for co-delivery nanomedicines based on combination of phytochemicals with chemotherapeutic drugs in cancer treatment. Adv Drug Deliv Rev 188:114445
Garcia C, Blesso CN (2021) Antioxidant properties of anthocyanins and their mechanism of action in atherosclerosis. Free Radic Biol Med 172:152–166
Giuliani A, Montesanto A, Matacchione G, Graciotti L, Ramini D, Protic O, Galeazzi R, Antonicelli R, Tortato E, Bonfigli AR, Sabbatinelli J, Olivieri F (2022) The association between single nucleotide polymorphisms, including miR-499a genetic variants, and dyslipidemia in subjects treated with pharmacological or phytochemical lipid-lowering agents. Int J Mol Sci 23(10). https://doi.org/10.3390/ijms23105617
Guan R, Van Le Q, Yang H, Zhang D, Gu H, Yang Y, Sonne C, Lam SS, Zhong J, Jianguang Z (2021) A review of dietary phytochemicals and their relation to oxidative stress and human diseases. Chemosphere 271:129499
Huffman MD, Prabhakaran D, Osmond C, Fall CHD, Tandon N, Lakshmy R, Ramji S, Khalil A, Gera T, Prabhakaran P (2011) Incidence of cardiovascular risk factors in an Indian urban cohort: results from the New Delhi Birth Cohort. J Am Coll Cardiol 57(17):1765–1774
Hyun SH, Bhilare KD, In G, Park CK, Kim JH (2022) Effects of Panax ginseng and ginsenosides on oxidative stress and cardiovascular diseases: pharmacological and therapeutic roles. J Ginseng Res 46(1):33–38
Imran M, Ahmad N, Anjum FM, Khan MK, Mushtaq Z, Nadeem M, Hussain S (2015) Potential protective properties of flax lignan secoisolariciresinol diglucoside. Nutr J 14:1–7
Islam SU, Ahmed MB, Ahsan H, Lee YS (2021) Recent molecular mechanisms and beneficial effects of phytochemicals and plant-based whole foods in reducing LDL-C and preventing cardiovascular disease. Antioxidants 10(5):784
Jain S, Buttar HS, Chintameneni M, Kaur G (2018) Prevention of cardiovascular diseases with anti-inflammatory and anti-oxidant nutraceuticals and herbal products: an overview of preclinical and clinical studies. Recent Patents Inflamm Allergy Drug Discov 12(2):145–157
Ju J, Li J, Lin Q, Xu H (2018) Efficacy and safety of berberine for dyslipidaemias: a systematic review and meta-analysis of randomized clinical trials. Phytomedicine 50:25–34
Karimian S, Pirro M, Johnston TP, Majeed M, Sahebkar A (2017) Curcumin and endothelial function: evidence and mechanisms of protective effects. Curr Pharm Des 23(17):2462–2473
Kastenmüller G, Raffler J, Gieger C, Suhre K (2015) Genetics of human metabolism: an update. Hum Mol Genet 24(R1):R93–R101
Kaulmann A, Bohn T (2014) Carotenoids, inflammation, and oxidative stress—implications of cellular signaling pathways and relation to chronic disease prevention. Nutr Res 34(11):907–929
Kaur KK, Allahbadia G, Singh M (2019) Monoterpenes—a class of terpenoid group of natural products as a source of natural antidiabetic agents in the future—a review. CPQ Nutr 3(4):1–21
Kaur P, Dahiya R, Kaur G, Buttar HS, Wilson DW, Telessy IG (2024) Pathogenesis of atherosclerosis and coronary heart disease: epidemiology, diagnostic biomarkers and prevention by nutraceuticals, functional foods, and plant-derived therapies. In: Botanicals and natural bioactives: prevention and treatment of diseases, vol 2. Bentham Science Publisher, p 146
Kim JH (2018) Pharmacological and medical applications of Panax ginseng and ginsenosides: a review for use in cardiovascular diseases. J Ginseng Res 42(3):264–269
Kitts DD (2005) Antioxidant phytochemicals and potential synergistic activities at reducing risk of cardiovascular disease. In: Food-drug synergy and safety. CRC Press, pp 27–62
Kris-Etherton PM, Hecker KD, Bonanome A, Coval SM, Binkoski AE, Hilpert KF, Griel AE, Etherton TD (2002) Bioactive compounds in foods: their role in the prevention of cardiovascular disease and cancer. Am J Med 113(9):71–88

Kumar N, Goel N (2019) Phenolic acids: natural versatile molecules with promising therapeutic applications. Biotechnol Rep 24:e00370

Kumar A, Nirmal P, Kumar M, Jose A, Tomer V, Oz E, Proestos C, Zeng M, Elobeid T, Sneha V, Oz F (2023) Major phytochemicals: recent advances in health benefits and extraction method. Molecules 28(2):887

Kumar MM, Sharma A, Rinka Juneja DTP, Arif MM, Singh MAD, Kumar A (2025) Biological evaluation of herbal preparations and formulation standardization: a literature study. J Neonatal Surg 14(8s):273

Lakhanpal P, Rai DK (2008) Role of quercetin in cardiovascular diseases. Internet J Med Update 3(1):31–49

Langston FMA, Nash GR, Bows JR (2023) The retention and bioavailability of phytochemicals in the manufacturing of baked snacks. Crit Rev Food Sci Nutr 63(14):2141–2177

Lee S (2015) Computer simulation study of the potential anti-arrhythmic properties of Paeonol. J Physiol Pathol Korean Med 29(4):305–312

Lee J, Burckart GJ (1998) Nuclear factor kappa B: important transcription factor and therapeutic target. J Clin Pharmacol 38(11):981–993

Li L, Zhou X, Li N, Sun M, Lv J, Xu Z (2015) Herbal drugs against cardiovascular disease: traditional medicine and modern development. Drug Discov Today 20(9):1074–1086

Liu A, Wu Q, Guo J, Ares I, Rodríguez JL, Martínez-Larrañaga MR, Yuan Z, Anadón A, Wang X, Martínez MA (2019) Statins: adverse reactions, oxidative stress and metabolic interactions. Pharmacol Ther 195:54–84

Ma X, Chen Z, Wang L, Wang G, Wang Z, Dong X, Wen B, Zhang Z (2018) The pathogenesis of diabetes mellitus by oxidative stress and inflammation: its inhibition by berberine. Front Pharmacol 9:782

Maaliki D, Shaito AA, Pintus G, El-Yazbi A, Eid AH (2019) Flavonoids in hypertension: a brief review of the underlying mechanisms. Curr Opin Pharmacol 45:57–65

Majrashi TA, Alshehri SA, Alsayari A, Bin MA, Alrouji M, Alshahrani AM, Shamsi A, Atiya A (2023) Insight into the biological roles and mechanisms of phytochemicals in different types of cancer: targeting cancer therapeutics. Nutrients 15(7):1704

Makiso MU, Tola YB, Ogah O, Endale FL (2024) Bioactive compounds in coffee and their role in lowering the risk of major public health consequences: a review. Food Sci Nutr 12(2):734–764

Maleki SJ, Crespo JF, Cabanillas B (2019) Anti-inflammatory effects of flavonoids. Food Chem 299:125124

Mangels DR, Mohler ER III (2017) Catechins as potential mediators of cardiovascular health. Arterioscler Thromb Vasc Biol 37(5):757–763

Martin D, Song J, Mark C, Eyster K (2008) Understanding the cardiovascular actions of soy isoflavones: potential novel targets for antihypertensive drug development. Cardiovasc Haematol Disord Drug Targets 8(4):297–312

Melzer J, Saller R (2009) Proprietary herbal medicines in circulatory disorders: hawthorn, ginkgo, Padma 28. In: Herbal drugs: ethnomedicine to modern medicine. Springer, Berlin, pp 115–135

Meng T, Li X, Li C, Liu J, Chang H, Jiang N, Li J, Zhou Y, Liu Z (2022) Natural products of traditional Chinese medicine treat atherosclerosis by regulating inflammatory and oxidative stress pathways. Front Pharmacol 13:997598

Mir MA, Dar MA, Qadir A (2024) Exploring the landscape of coronary artery disease: a comprehensive review. Am J Biomed Pharm 1(1):9–22

Mirmiran P, Houshialsadat Z, Gaeini Z, Bahadoran Z, Azizi F (2020) Functional properties of beetroot (Beta vulgaris) in management of cardio-metabolic diseases. Nutr Metab 17:3

Mittal RK, Mishra R, Sharma V, Purohit P (2024) Bioactive exploration in functional foods: unlocking nature's treasures. Curr Pharm Biotechnol 25(11):1419–1435

Morimoto T, Sunagawa Y, Fujita M, Hasegawa K (2010) Novel heart failure therapy targeting transcriptional pathway in cardiomyocytes by a natural compound, curcumin. Circ J 74(6):1059–1066

Najjar RS, Schwartz AM, Wong BJ, Mehta PK, Feresin RG (2021) Berries and their polyphenols as a potential therapy for coronary microvascular dysfunction: a mini-review. Int J Mol Sci 22(7):3373

Nakmareong S, Kukongviriyapan U, Pakdeechote P, Donpunha W, Kukongviriyapan V, Kongyingyoes B, Sompamit K, Phisalaphong C (2011) Antioxidant and vascular protective effects of curcumin and tetrahydrocurcumin in rats with L-NAME-induced hypertension. Naunyn Schmiedeberg's Arch Pharmacol 383:519–529

Naveed M, Majeed F, Taleb A, Zubair HM, Shumzaid M, Farooq MA, Baig MMFA, Abbas M, Saeed M, Changxing L (2020) A review of medicinal plants in cardiovascular disorders: benefits and risks. Am J Chin Med 48(2):259–286

Nes WD (2011) Biosynthesis of cholesterol and other sterols. Chem Rev 111(10):6423–6451

Nicolescu A, Babotă M, Barros L, Rocchetti G, Lucini L, Tanase C, Mocan A, Bunea CI, Crişan G (2023) Bioaccessibility and bioactive potential of different phytochemical classes from nutraceuticals and functional foods. Front Nutr 10:1184535

Nimgulkar C, Ghosh S, Sankar AB, Uday KP, Surekha MV, Madhusudhanachary P, Annapurna BR, Raghu P, Bharatraj DK (2015) Combination of spices and herbal extract restores macrophage foam cell migration and abrogates the athero-inflammatory signalling cascade of atherogenesis. Vasc Pharmacol 72:53–63

Nkwocha CC, Ogugofor MO, Chukwuma IF (n.d.) Identification and characterization of phytochemicals and constituents in *Desmodium velutinum* stem using high performance liquid chromatography (HPLC). SSRN https://ssrn.com/abstract=4006028

O'Connor BB (1995) Healing traditions: alternative medicine and the health professions. University of Pennsylvania Press

Orellana-Urzúa S, Briones-Valdivieso C, Chichiarelli S, Saso L, Rodrigo R (2023) Potential role of natural antioxidants in countering reperfusion injury in acute myocardial infarction and ischemic stroke. Antioxidants 12(9):1760

Orhan IE (2018) Phytochemical and pharmacological activity profile of *Crataegus oxyacantha* L. (hawthorn)–a cardiotonic herb. Curr Med Chem 25(37):4854–4865

Pandey S, Shaikh F, Gupta A, Tripathi P, Yadav JS (2022) A recent update: solid lipid nanoparticles for effective drug delivery. Adv Pharm Bull 12(1):17–33

Peluso I, Raguzzini A, Serafini M (2013) Effect of flavonoids on circulating levels of TNF-α and IL-6 in humans: a systematic review and meta-analysis. Mol Nutr Food Res 57(5):784–801

Prasad DR, Maurya B, Mishra A (2022) Neuroprotective effect of *Ginkgo biloba* and its. Indopathy Neuroprotect Recent Adv:258. [Abbreviated source]

Rahman I, Biswas SK, Kirkham PA (2006) Regulation of inflammation and redox signaling by dietary polyphenols. Biochem Pharmacol 72(11):1439–1452

Ramesh P, Palaniappan A (2023) *Terminalia arjuna*, a cardioprotective herbal medicine–relevancy in the modern era of pharmaceuticals and green nanomedicine—a review. Pharmaceuticals 16(1):126

Rasheed H, Ahmed S, Sharma A (2024) Changing trends towards herbal supplements: an insight into safety and herb-drug interaction. Curr Pharm Biotechnol 25(3):285–300

Ren-an Q, Juan L, Chuyuan L, Wenjuan F, Chunyan H, Xuemei Y, Lin H, Hong N (2014) Study of the protective mechanisms of compound Danshen tablet against myocardial ischemia/reperfusion injury via the Akt-eNOS signaling pathway in rats. J Ethnopharmacol 156:190–198

Ruparelia N, Chai JT, Fisher EA, Choudhury RP (2017) Inflammatory processes in cardiovascular disease: a route to targeted therapies. Nat Rev Cardiol 14(3):133–144

Saleem A, Akhtar MF, Sharif A, Akhtar B, Siddique R, Ashraf GM, Alghamdi BS, Alharthy SA (2022) Anticancer, cardioprotective and anti-inflammatory potential of natural-sources-derived phenolic acids. Molecules 27(21):7286

Santos AC, Veiga F, Ribeiro AJ (2011) New delivery systems to improve the bioavailability of resveratrol. Expert Opin Drug Deliv 8(8):973–990

Shaito A, Thuan DTB, Phu HT, Nguyen THD, Hasan H, Halabi S, Abdelhady S, Nasrallah GK, Eid AH, Pintus G (2020) Herbal medicine for cardiovascular diseases: efficacy, mechanisms, and safety. Front Pharmacol 11:422

Sharifi-Rad J, Rodrigues CF, Sharopov F, Docea AO, Karaca AC, Sharifi-Rad M, Kahveci Karıncaoglu D, Gülseren G, Şenol E, Demircan E (2020) Diet, lifestyle and cardiovascular diseases: linking pathophysiology to cardioprotective effects of natural bioactive compounds. Int J Environ Res Public Health 17(7):2326

Shukla SK, Gupta S, Ojha SK, Sharma SB (2010) Cardiovascular friendly natural products: a promising approach in the management of CVD. Nat Prod Res 24(9):873–898

Singh A, Omer K (2024) Recent advancement in therapeutic activity of carotenoids. In: Dietary carotenoids—sources, properties, and role in human health. IntechOpen

Sobenin IA, Myasoedova VA, Iltchuk MI, Zhang DW, Orekhov AN (2019) Therapeutic effects of garlic in cardiovascular atherosclerotic disease. Chin J Nat Med 17(10):721–728

Stangl V, Lorenz M, Stangl K (2006) The role of tea and tea flavonoids in cardiovascular health. Mol Nutr Food Res 50(2):218–228

Swetha N, Mathanghi SK (2024) Towards sustainable omega-3 fatty acids production–a comprehensive review on extraction methods, oxidative stability and bio-availability enhancement. Food Chem Adv 4:100603

Tahir H, Akhtar MN, Bishoyi AK, Sanghvi G, Rasheed M, Afzaal M, Imran A, Saeed F, Khare N, Ahmed F (2025) Nutritional composition, phytochemical profile, extraction methods of bioactive components, and health benefits of *Terminalia arjuna* bark. eFood 6(2):e70038

Tajik N, Tajik M, Mack I, Enck P (2017) The potential effects of chlorogenic acid, the main phenolic components in coffee, on health: a comprehensive review of the literature. Eur J Nutr 56:2215–2244

Theodotou M, Fokianos K, Mouzouridou A, Konstantinou C, Aristotelous A, Prodromou D, Chrysikou A (2017) The effect of resveratrol on hypertension: a clinical trial. Exp Ther Med 13(1):295–301

Tijburg LB, Mattern T, Folts JD, Weisgerber UM, Katan MB (1997) Tea flavonoids and cardiovascular diseases: a review. Crit Rev Food Sci Nutr 37(8):771–785

Upadhyay S, Dixit M (2015) Role of polyphenols and other phytochemicals on molecular signaling. Oxidative Med Cell Longev 2015:504253

Ushie OA, Ochepo EO, Nkom PY, Ago MA, Gani J (2023) Medicinal uses of nitrogen group containing secondary metabolites. Trop J Sci Technol 4(1):71–82

Wang X, An F, Wang S, An Z, Wang S (2017) Orientin attenuates cerebral ischemia/reperfusion injury in rat model through the AQP-4 and TLR4/NF-κB/TNF-α signaling pathway. J Stroke Cerebrovasc Dis 26(10):2199–2214

Wang Y, Gallegos JL, Haskell-Ramsay C, Lodge JK (2021) Effects of chronic consumption of specific fruit (berries, citrus and cherries) on CVD risk factors: a systematic review and meta-analysis of randomised controlled trials. Eur J Nutr 60:615–639

WHO (2016) Cardiovascular diseases India. https://www.who.int/india/health-topics/cardiovascular-diseases

Wu P, Zhang Z, Wang F, Chen J (2010) Natural compounds from traditional medicinal herbs in the treatment of cerebral ischemia/reperfusion injury. Acta Pharm Sin B 31(12):1523–1531

Xu K, Qu L, Li H, Ren X, Yan N, Fu X (2024) Functional properties of dietary quercetin in cardiovascular health and disease. Food Front 5(5):1951–1967

Yan K, Cheng XJ, Bian GL, Gao YX, Li DQ (2022) The influence of different extraction techniques on the chemical profile and biological properties of *Oroxylum indicum*: multifunctional aspects for potential pharmaceutical applications. Evid Based Complement Altern Med 2022:8975320

Zhang L, Virgous C, Si H (2019) Synergistic anti-inflammatory effects and mechanisms of combined phytochemicals. J Nutr Biochem 69:19–30

Zhang S, Li L, Chen W, Xu S, Feng X, Zhang L (2021) Natural products: the role and mechanism in low-density lipoprotein oxidation and atherosclerosis. Phytother Res 35(6):2945–2967

Zhou Y, Wu Y, Deng L, Chen L, Zhao D, Lv L, Chen X, Man J, Wang Y, Shan H (2014) The alkaloid matrine of the root of *Sophora flavescens* prevents arrhythmogenic effect of ouabain. Phytomedicine 21(7):931–935

Zhou Y, Suo W, Zhang X, Lv J, Liu Z, Liu R (2022) Roles and mechanisms of quercetin on cardiac arrhythmia: a review. Biomed Pharmacother 153:113447

Clinical Trials and Future Directions in Herbal Treatments Against Hepatic and Cardiovascular Disease

6

Syed Naved Quadri, Sakshi Tiwari, Saman Fatima, Bilal Siddiqui, and M. Z. Abdin

Abstract

Hepatic and cardiovascular diseases (CVDs) remain major global health challenges; synthetic hepatoprotective agents used in clinical practices often lack therapeutic efficacy with undesirable side effects. Hepatic diseases are often associated with the progression of subclinical and clinical CVDs. It is evident from clinical data that coronary artery disease (CAD), arrhythmias, and stroke are more common in patients with hepatic impairments, which results in increased cardiovascular morbidity and mortality. The widespread availability, affordability, and evidence of efficacy in preclinical studies are the major reasons for the use of herbal medicines in previous times and their ever-increasing popularity in the current period. The available anecdotal information on the therapeutic benefits of herbal medicines cannot replace the rigor of clinical trials necessary to establish their safety and efficacy. This chapter aims to evaluate the efficacy of the herbal treatment as demonstrated in clinical trials. We have collected and tabulated the clinically evaluated hepatoprotective and CVD-protective herbals and their natural formulations with their current status in different trial stages. Literature was searched across three major databases PubMed, Web of Science, and Scopus to identify relevant clinical studies and ongoing current clinical trials evaluating the effectiveness and safety of herbal treatments. At present, berberine and silymarin have been approved by the US FDA to enter clinical phase IV stud-

S. N. Quadri · S. Tiwari · B. Siddiqui · M. Z. Abdin (✉)
Centre for Transgenic Plant Development, Department of Biotechnology, School of Chemical and Life Sciences, Jamia Hamdard, New Delhi, Delhi, India
e-mail: snquadri@jamiahamdard.ac.in; tiwarisakshi021@gmail.com; mohdbilalsiddiqui927@gmail.com; mzabdin@jamiahamdard.ac.in

S. Fatima
Amity Institute of Pharmacy (AIP), Amity University, Noida, Uttar Pradesh, India
e-mail: sfaima@amity.edu

K. Kuca et al. (eds.), *Medicinal Plants and Their Bioactives in Human Diseases*,
https://doi.org/10.1007/978-3-032-01356-9_6

ies. Compounds such as quercetin, resveratrol, allicin, and many others, offer promising cardiovascular protective effects. This chapter addresses the challenges of herbal treatments, particularly the poor bioavailability of active phytoconstituents, by utilizing nanoformulation-based strategies. To make herbal treatment more effective, the personalized, customized, and tailored herbal therapy approaches from a future development point of view have been discussed in detail.

Keywords

Cardiovascular Diseases (CVDs) · United States Food and Drug Administration (US-FDA) · Herbal Treatments · Phytoconstituents · Clinical Trials

1 Introduction

The liver and heart have long been known to interact with one another. In his well-known work "Canon" (The Law), Avicenna initially noted the interdependent effects of these two organs. This relationship was defined as "liver dryness" predominating over "heart wetness" and "heart warmth" predominating over "liver coolness and wetness." The liver's demeanor is "warm" and "wet" under normal circumstances, while the heart's is "warm" and "dry." Avicenna believed that an organ dysfunction or illness was described by a discrepancy between the body's temperaments (Khadem et al. 2013). Heart and liver disorders are major burdens on the healthcare system and major contributors to declining longevity and quality of life. According to projections from the Global Burden of Disease Study, 50 million individuals had hepatitis C and 254 million had hepatitis B in 2022. Every year, liver disease claims the lives of about two million people (liver cancer, cirrhosis, and viral hepatitis), which also constitutes 4% of all deaths globally (1 of every 25 deaths). According to this estimate, 600,000 to 900,000 fatalities are attributed to liver cancer (Devarbhavi et al. 2023). In recent decades, the prevalence of cardiovascular diseases (CVDs), which continue to be the leading cause of mortality worldwide, has dramatically increased. The Global Burden of Disease Study estimates that the number of deaths globally from CVDs rose from 12.4 million in 1990 to 19.8 million in 2022 due to growing populations and aging (Mensah et al. 2023). There is a clear correlation between liver disease and CVD. In addition to sharing several risk factors, including obesity, diabetes mellitus, and dyslipidemia, atherosclerotic cardiovascular disease (ASCVD) also advances in tandem with deteriorating metabolic circumstances among individuals suffering from non-alcoholic fatty liver disease (NAFLD) (Nakahara et al. 2014; Imajo et al. 2014). In contrary, persistent heart failure can cause liver damage through congestion, which in turn can result in cirrhosis and liver fibrosis. Additionally, it has been noted that several medications often used to treat CVD may potentially result in steatosis and liver damage (Di Pasqua et al. 2022). There are several different pathophysiological variables linked to fatty liver disease (FLD). These include oxidative stress, genetics, intestinal dysbiosis, inflammation, adipokines, and psychological stressors like despair and anxiety, among others, all of these are recognized indicators of CVD (Macavei et al. 2016). This suggests that additional study and actions are required to lessen the burden brought on by FLD. Low-grade systemic inflammation brought on by metabolic diseases like FLD and type 2 diabetes can result in several extrahepatic consequences, including an elevated risk for CVD (Gregor and Hotamisligil 2011).

A significant fraction of the global population receives their medical care from herbal medicines, which are also considered to be a part of many tribes' cultural traditions. Most pharmaceutically significant medications are plant-based. Medicines derived from plants are used extensively in both human and animal health systems worldwide. Over the past 10 years of the twentieth century, an increasing amount of biological, clinical, and phytochemical data indicates that plant-based herbal remedies are gaining popularity as a means of curing a range of ailments. There are several medicinal plants that are recognized to have cardioprotective properties, including *Tinospora cordifolia, Daucus carota Linn, Amaranthus viridis, Nerium oleander (NO) Linn, Hydrocotyle asiatica Linn, Ginkgo biloba, Terminalia arjuna, Curcuma longa, Anastatica hierochuntica*, and *Cichorium intybus*. Researchers have identified a significant number of phytochemicals from plant sources (Li et al. 2014a, b; Shah et al. 2019). Although phytoconstituents derived from plant products are thought to be the best and most proven source to produce new therapeutic agents, their widespread use for therapeutic purposes is still limited by their poor distribution, absorption, metabolism, and elimination, along with their limited toxicological characteristics. In conclusion, this chapter provides an in-depth analysis of herbal treatments with potential hepatoprotective and cardiovascular benefits, highlighting their progression through clinical trials. By systematically reviewing the evidence from major scientific databases, it underscores the growing acceptance of natural formulations, such as berberine and silymarin, which have advanced to late-stage clinical evaluations.

2 The Liver–Heart Axis: Pathophysiological Insights into Inflammation and Signaling Pathways

One important metabolic organ that may be responsible for chronic inflammation is the liver, which secretes cytokines, chemokines, and inflammatory markers into the bloodstream. Secondary cardiovascular adverse effects, including coagulation, increased plaque development, endothelial dysfunction, and altered vascular tone, may arise from this. Numerous inflammatory markers and cytokines, such as elevated hepatic expression of TNF-α, IL-6, CXCL10, and IL1RN, as well as serum levels of TNF-α, interleukin (IL)-6, CCL3, soluble intercellular adhesion molecule-1 (sICAM-1), and C-reactive protein (CRP), have been associated with FLD and an increased risk of CVD (Francque et al. 2016). Worsening inflammation and insulin resistance (IR) states are associated with worse cardiometabolic outcomes as NAFLD severity rises. Several findings showed that FLD was associated with a markedly elevated risk of coronary atherosclerosis. Because coronary artery calcification decreases vascular compliance, impairs myocardial perfusion, and increases the extent of atherosclerosis, it increases the risk of coronary artery disease (CAD), aberrant vasomotor responses, and long-term mortality when used as an independent indicator of cardiovascular complications (Sarwar et al. 2009). In a recent cross-sectional study, Chang et al. assessed the association between FLD of both alcoholic and non-alcoholic etiology and coronary artery calcium score (CAC) in 105,328 patients (Chang et al. 2019). The research revealed an association between an elevated CAC score and both ALD (alcohol liver disease) and NAFLD. The underlying pathophysiological mechanisms of congestive hepatopathy, which is typically subclinical, include elevated arterial oxygen saturation, declined hepatic blood flow, and elevated hepatic vein pressures (de Gonzalez and Lefkowitch 2017). The main cardiac conditions linked to congestive hepatopathy include cardiomyopathy, valvular diseases, left heart failure, and constrictive pericardial disease (Myers et al. 2003). The clinical and histological illness known as cardiogenic ischemic hepatitis, or acute cardiogenic liver injury, is typified by a sharp and brief increase in serum transaminase levels brought on by an abrupt decrease in cardiac output, which lowers hepatic blood flow. According to results from a retrospective investigation, acute liver injury can result from more than only the sudden drop in hepatic perfusion that occurs during cardiogenic shock. The research suggested that acute hepatic damage brought on by a hypotensive event is predisposed to by hepatic vein obstruction brought on by right-sided heart failure (Seeto et al. 2000). Although The mechanism behind the correlation between NAFLD and CVD and other cardiac problems remains unclear and may involve routes other than insulin resistance, such as oxidative stress, low-grade inflammation, and the impact of gut microbiome disturbances. NAFLD and the prognosis for CVD are strongly correlated. Although NAFLD is thought to promote CVD, it shares risk variables with CVD, including obesity, diabetes mellitus, and dyslipidemia. In addition to being linked to chronic inflammation, insulin resistance, coagulation problems, endothelial dysfunction, oxidative stress, dyslipidemia, and microbiome disruptions, NAFLD may also be linked to CVD (Ogresta et al. 2022).

Excess plasma lipid and high plasma insulin and glucose levels caused by insulin resistance and diabetes mellitus increase de novo lipogenesis and inhibit intracellular triglyceride hydrolysis. Triglyceride levels in the plasma rise as a result of this increase in hepatic triglyceride content. Furthermore, diminished membrane-bound low-density lipoprotein receptors and elevated intracellular cholesterol in the liver are linked to aberrant lipid and glucose metabolism in nonalcoholic fatty liver disease (NAFLD). These changes result in an increase in massive very-low-density lipoprotein and small dense low-density lipoprotein, which eventually accelerates atherosclerosis (Deprince et al. 2020). In addition to being a significant pathogenic factor linked to NAFLD, insulin resistance also results in uncontrolled lipolysis in fat cells and poor glucose transfer into skeletal muscle (Cobbina and Akhlaghi 2017). Super oxides and other reactive oxygen species cause oxidative stress that is derived from the liver. Oxidative stress triggers the production of proinflammatory cytokines (Rolo et al. 2012). Both the adipose and liver tissue enhance TNF-mRNA, which is linked to the phases of liver fibrosis. Toll-like receptor 2 triggers the inflammasome during the development of NAFLD (Rolo et al. 2012; Miura et al. 2013). Studies have shown that oxidative stress may contribute to the pathogenesis of CVD in people with FLD by causing aberrant cytokine production, such as TNF-α, IL-6, and CRP, which also aids in the progression of liver disease severity from simple steatosis to steatohepatitis (Bhatia et al. 2012). Furthermore, new research shows that immune cells have a major impact on the onset of HF. Immune cells that may influence the development or control of heart failure and cardiac remodeling in heart ischemia include clusters of differentiating 4 T cells or 8 T cells (Kumar et al. 2022). There have been reports of changes in circulating adipokines in metabolic syndromes and NAFLD, including elevated leptin levels or decreased adiponectin (Polyzos et al. 2016). Several adipokines, including apelin, omentin, and adiponectin, have been shown to exhibit cardioprotective and anti-inflammatory properties. Nevertheless, it has been found that the pro-inflammatory qualities of leptin, visfatin, resistin, and adipocyte fatty acid binding protein impair cardiovascular efficiency and raise the risk of CVD (Smekal and Vaclavik 2017). NAFLD is linked to ectopic fat reserves in the epicardium, pericardium, and myocardium, which are distinguished by unique metabolic hallmarks as shown by magnetic resonance spectroscopy (Lovric et al. 2018). According to a meta-analysis, there is a correlation between a higher frequency of atrial fibrillation and higher levels of proinflammatory indicators, including plasma IL-6 and CRP (Wu et al. 2013). The Canakinumab Anti-inflammatory Thrombosis Outcome Study (CANTOS) trial investigated if using a particular anti-inflammatory medication to reduce proinflammatory biomarkers reduces the risk of CVD events (Ridker et al. 2017). Commensals in the gastrointestinal tract transform some nutrients, including carnitine or choline, into trimethylamine (TMA), which flavin monooxygenases in the liver then metabolize to TMA N-oxide (TMAO). High circulating TMAO levels have been linked to negative CVD outcomes in numerous studies (Wang et al. 2011). Additionally, numerous meta-analyses have verified a substantial correlation between risk of CVD events and circulating TMAO levels (Qi et al. 2018). In a study evaluating 60 biopsy-proven NAFLD participants, higher TMAO levels were linked to a more severe case of NAFLD, while reduced betaine and betaine/choline ratios were seen (Chen et al. 2016). Despite the paucity of research on TMAO in NAFLD, it is well accepted that circulating TMAO levels are a crucial indicator of CVD, the primary cause of death for individuals with NAFLD. A recent study found that TMAO affects the heart's autonomic nerve system, which causes ischemia-induced ventricular arrhythmias (Meng et al. 2019). Furthermore, PERK (protein kinase RNA-like endoplasmic reticulum kinase), an endoplasmic reticulum stress kinase, might be involved. TMAO binds to the PERK receptor, activating PERK and inducing the transcription factor FoxO1, which is crucial in metabolic diseases (Chen et al. 2019). Interestingly, because TMAO increases NLRP3 and NF-κB, it may directly activate pro-inflammatory pathways, which in turn encourages vascular calcification (Zhang et al. 2020). NF-κB activation has been linked to a number of etiologies, including inflammation, thrombosis, ROS production, CM apoptosis, atherosclerosis, MI, cardiomyopathies, and cardiometabolic illnesses (such as diabetes) (Maier et al.

2012; Bonomini et al. 2015; Dai et al. 2017). Being generated via feedback loops, the intricate protein-protein interactions (PPIs) that take place between NF-κB components and the activators/repressors of its signaling pathway are noteworthy (Tieri et al. 2012).

By processing sterol regulatory element-binding protein-1 (SREBP-1), mTORC1 stimulates lipogenesis and protein synthesis by inhibiting the autophagic program (Laplante and Sabatini 2009) and inducing the transcription of acetyl-CoA carboxylase (ACC), fatty acid synthase (FASN), glycerol-3-phosphate acyltransferase (GPAT), and stearoyl-CoA desaturase-1 (SCD1) lipogenic genes (Deleyto-Seldas and Efeyan 2021). Additionally, AMPK, a crucial autophagy activator, is downregulated by mTORC1. Remarkably, Beclin1 has cardioprotective effects during the reperfusion stage, while mTORC1 controls the myocardial ischemia scenario (Ling et al. 2020). Recently, some researchers have used genetic engineering techniques to create a diet-induced obesity and NAFLD mice model and a liver-specific AMPKα1/α2 double knockout (Zhao et al. 2020). According to the study, AMPK knockout that is unique to the liver may exacerbate steatosis, fibrosis, inflammation, and lipid buildup. TUNEL staining revealed that AMPK knockout mice had a markedly higher number of apoptotic hepatocytes. The results lend more credence to AMPK's potential as a therapeutic and preventative strategy for NAFLD (Bai et al. 2016). Numerous factors may work alone or in concert, concurrently or sequentially, as part of the complicated pathophysiology of FLD. Sirt1 may regulate AMPK upstream kinases to activate AMPK, while activated AMPK may increase intracellular NAD+ levels through LKB1, starting the Sirt1 signaling pathway (Bai et al. 2016; Sharma et al. 2021). Recent studies of animal liver and cultured hepatocytes have shown that Sirt1 and lipin-1 are not only linked to the management of the inflammatory response but also to the lipid metabolism pathway that mediates FLD (You et al. 2017). Liver disease and CVD interact, so it's critical to evaluate both conditions. In a clinical environment, it might be challenging to differentiate their interactions and confounding effects because these diseases have comparable risk factors and fundamental pathology. Despite this restriction, the significance of liver and cardiovascular diseases has been growing in the contemporary era of aging populations. Given its significant influence on health prognosis, early detection and mitigation of CVD are particularly regarded as crucial.

3 Phytoconstituents with Hepatoprotective and Cardiovascular Protective Properties

Historically, medicines have frequently been derived from plants and herbs, as conventional extracts or as pure active substances. Naturally occurring substances are clearly a valuable source for developing novel medications that treat illnesses. Well-known medications derived from plants and herbs, among many others, include aspirin from the *Salix alba L.* tree, lovastatin from *Monascus purpureus L.*, reserpine from *Rauvolfia serpentina*, digoxin (cardiac glycoside) from *Digitalis purpurea*, ephedrine from *Ephedra sinica*, and taxol from *Taxus brevifolia* (Frishman

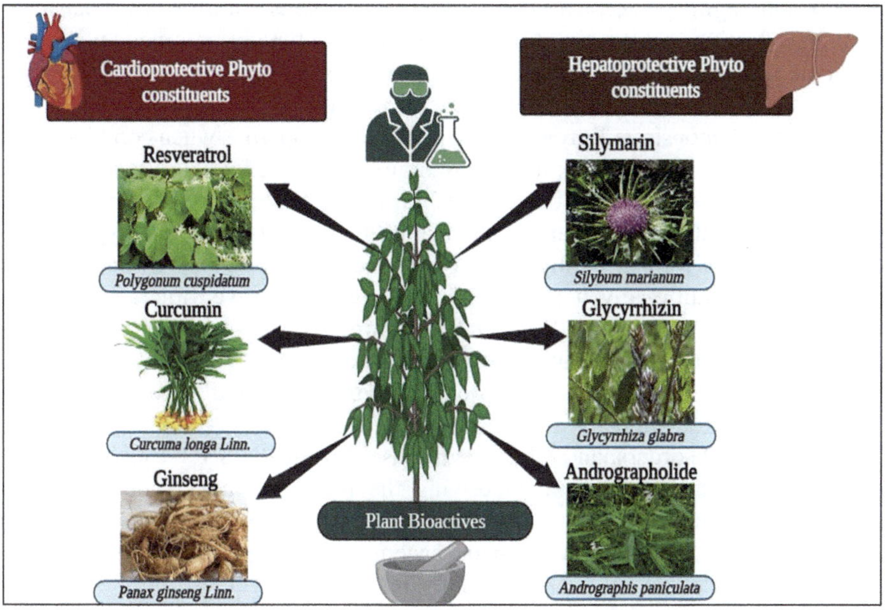

Fig. 6.1 Natural herbal compounds of therapeutic importance to combat the hepatic and cardiovascular diseases

et al. 2009; Cragg and Newman 2013) (Fig. 6.1). The fact that rural communities in emerging economies have turned to traditional herbal and plant-derived treatments to manage illnesses in the face of limited socioeconomic means further contributes to the growing interest in medicinal plant products. There are several reasons behind this, but primarily the fact that plant-based medications are less expensive and have less adverse effects (Tabassum and Ahmad 2011). Another factor contributing to the resurgence of interest in natural goods is that no synthetic drug screening database can match the biological functionality and structural variety of natural molecules (Davison and Brimble 2019).

4 Mechanistic Insights into Natural Compounds with Cardiovascular Protective Effects

Numerous bioactives included in herbal treatments provide them multidimensional cellular modes of action. Herbal medicines may possess diuretic, anti-inflammatory, anti-proliferative, vasorelaxant, and antioxidant properties. Moreover, medicinal herbs can suppress lipid peroxidation, endothelial dysfunction, platelet activation, generation of ROS, macrophage atherogenicity, and VSMC phenotypic switching. Many different types of CVD can be treated and managed using herbal remedies due to their broad spectrum of molecular and cellular targets. For instance, a variety of CVDs, such as myocardial infarction, atherosclerosis, coronary heart disease

(CHD), and angina pectoris, have been treated with *Salvia miltiorrhiza* (Danshen), sometimes known as red sage, an annual sage that has long been utilized in Chinese medicine (Gao et al. 2012). Danshen offers protection against the negative effects of homocysteine, which is a major risk factor for vascular disorders (Chan et al. 2004). Danshen has demonstrated strong anti-hypertensive properties when combined with *Pueraria montana var.* lobata, more commonly known as kudzu (Ng et al. 2011). Danshen capsules (1000 mg twice daily for 12 weeks) significantly reduced the SBP and pulse rate of individuals with uncontrolled mild to moderate hypertension undergoing conventional antihypertensive medication, according to one study in clinical trials. Additionally, it has been shown to be safe and well-tolerated in hypertension patients (Yang et al. 2012a, b). Resveratrol, or 3,5,4-trihydroxy-trans-stilbene, is a polyphenolic compound that occurs naturally in *Polygonum cuspidatum*, peanuts, berries, grapes, and their products, particularly red wine (Burns et al. 2002). Resveratrol interacts with several targets in age-related illnesses, cancer, heart and brain disorders, and other conditions. The AMPK/SIRT-1 pathway is the primary molecular mechanism underlying the biological effects of resveratrol. Indeed, in ECs and cardiac myoblasts, resveratrol decreases the generation of ROS and increases the intrinsic antioxidant systems, including superoxide dismutase enzymes (Spanier et al. 2009). Additionally, it lowers the synthesis of prostaglandin E2 and arachidonic acid. It counteracts the actions of the most significant molecules implicated in inflammation, including NF-κB, TNF-α, IL-6, monocyte chemotactic protein-1 (MCP-1), and inducible nitric oxide synthase (iNOS) activity. Additionally, it prevents the activity of phospholipase A2 and cyclooxygenase-2 (COX-2) (Moreno 2000; Yang et al. 2012a, b). A traditional example of an herb used to treat cardiovascular diseases, garlic (*Allium sativum*) is well-known for its many benefits against disorders linked to CVDs, including inflammation, oxidative stress, hypertension, and hyperlipidemia (Thomson et al. 2015). Garlic can help control atherosclerosis and hyperlipidemia by cholesterol and lowering LDL levels, lowering the amount of lipid in artery cells, and preventing the growth of VSMCs (Sun et al. 2018). Rosemary (*Rosmarinus officinalis*) demonstrates neuroprotection by preventing cerebral insufficiency linked to ischemic stroke, as evidenced by a reduction in the brain's restricted blood supply. Because of its anti-inflammatory effects, rosemary can lower the expression of mediators and enzymes that promote inflammation, including inducible iNOS and COX-2 (Seyedemadi et al. 2016). *Curcuma longa* has a naturally occurring phenolic component called curcumin (diferuloylmethane), which is extracted as a yellow pigment. Curcumin interacts with a variety of molecular targets, including Akt/Nrf2, AMPK/UCP2, MCP-1, IL-8, JNK, MAPK p38, ICAM-1, ERK, and Janus Kinase 2 (JAK2)/STAT3 (Parodi et al. 2006). Curcumin at a dose of 15 mg/kg appears to decrease xanthine oxidase, superoxide anion, and myeloperoxidase levels while increasing those of GST, SOD, catalase, GPx, and lipid peroxides. Furthermore, curcumin activates ARE in several cardiovascular cells exposed to 5–15 μM curcumin, making it a potent HO-1 stimulator in ECs (Motterlini et al. 2000). Numerous animal models have been used to study curcumin's impact in CVDs. For example, 1.66 mg of curcumin/kg prevented the development of aorta fatty streaks and had a hypolipidemic impact (Zingg et al.

2013). Curcumin's cardioprotective properties were demonstrated in rat models of myocardial ischemia (Pagliaro et al. 2015). The Chinese plant Huanglian contains an alkaloid called berberine (BBR), which was identified from *Hydrastis canadensis* and has been long used in traditional Chinese medicine (Liu et al. 2012). BBR may reduce inflammation by attenuating transcription activity, disrupting the MAPK pathways in an AMPK-dependent manner, inhibiting the Rho GTPase pathway, and inhibiting the NF-κB signaling pathway (Jeong et al. 2009). The activity of BBR in reversing the malfunctioning of endothelial progenitor cells (EPCs) was shown in in vitro investigations. Additionally, BBR has been demonstrated to lower cholesterol in both animals and hyperlipidemic individuals by mechanisms distinct from that of statins, including ERK pathway activation and increased LDLR expression on the surface of hepatocytes (Kong et al. 2004). Ginseng is known to have over 300 different bioactive compounds. The most bioactive substances that have been extracted from ginseng preparations are triterpene saponins called ginsenosides.

Ginseng has been utilized to treat hypertension in relation to CVDs. Ginseng's ability to improve vascular functioning results in hypotensive effects. Ginsenoside Rg3 activates eNOS, whereas NO-dependent vasorelaxation brought on by ginsenoside KRG enhances blood flow (Lee et al. 2016). Additionally, ginseng can increase NO production and eNOS expression. These impacts are mediated by the enhancement of eNOS dimer formation, a rise in NO production, and the suppression of arginase activity (Shin et al. 2013). The ginsenoside Rg3 can block Ca^{2+}-gated channels, activate $Ca2^{+}$-gated potassium channels, suppress ACE activity, and raise NO and cGMP levels (Park et al. 2014). Often called "lingzhi" or "reishi," Ganoderma lucidum is a variety of mushroom whose fruit body, spores, and mycelia are used to make a number of marketable G. lucidum products for medicinal purposes. Heart function may be negatively impacted by elevated blood pressure. Three peptides, known as ACE inhibitory peptides (ACEIPs), which can be utilized to treat hypertension, were isolated from G. lucidum and are QLVP, QDVL, and QLDL. QLVP has the ability to prevent ACE by interacting with its Gln242 and Lys472. Additionally, QLVP decreased the vasoconstrictor peptide endothelin-1's mRNA and protein expression in HUVECs and increased the phosphorylation of eNOS by angiotensin 1 in these cells (Wu et al. 2019).

5 Natural Compounds and Their Mechanistic Roles in Combating Hepatic Diseases

One of the body's primary organs, the liver is necessary for detoxification, metabolism, and the removal of numerous dangerous chemicals. Several medicinal herbs have been used over the years to treat a number of liver disorders. Several positive impacts on the liver are being found because of studies conducted utilizing a range of approaches on different plant species' entire extracts or purified active components. Silymarin, the active ingredient of *Silybum marianum (L.) Gaertn.*, commonly known as "milk thistle," is one of the ancient herbs and has been used extensively to treat liver conditions (Luo et al. 2018). Numerous liver disorders, including alcoholic liver disease, liver cirrhosis, and acute and chronic hepatitis brought on by viruses, drugs, or toxins, are treated with it. By limiting the increase of inducible NOS expression and preventing the metabolic activation of pyridine, silymarin guards against liver injury. Its suppression of lipid peroxidation and decrease in increased intracellular Ca^{2+} are principally responsible for its hepatoprotective effects (Zhang et al. 2013). Similarly, silymarin may prevent HCV-infected cells by inducing TNF-α to activate NF-κB and cause its nuclear translocation (Morishima et al. 2010). The traditional Chinese medication known as gancao is made from the dried roots of *Glycyrrhiza glabra* L., *Glycyrrhiza inflata* Bat., or Glycyrrhiza uralensis Fisch. Gancao is widely used to treat several liver conditions, such as viral hepatitis, liver fibrosis, hepatic steatosis, and liver damage (Li et al. 2014a, b). Because it inhibits TNF-α and caspase-3, glycyrrhizic acid (GA) has a hepatoprotective effect by reducing inflammation and apoptosis (PubChem 2024). The amount of cytochrome C released into the cytoplasm from the mitochondria is greatly decreased by GA. GA's anti-inflammatory properties stem from its suppression of myeloperoxide activity, TNF-α production, and NF-κB translocation into the nucleus. GA may have the ability to repair liver damage because it also raised the expression of proliferating cell nuclear antigen (PCNA) (El-Tahawy et al. 2011). The herbaceous plant *Andrographis paniculata Nees* contains two active chemical ingredients, andrographolide and neoandrographolide, which are widely known for their ability to manage liver disorders. The diterpene lactone class, which includes andrographolide, neoandrographolide, deoxy-andrographolide, panicoline, andrographine, and paniculide-A/B/C, is the primary active chemical ingredient that is extracted from the leaves (Prokunina-Olsson et al. 2013). In an animal model of chemical-induced liver injury, andrographolide reduces fibrosis, angiogenesis, and inflammation through anti-inflammatory and antioxidant pathways. Moreover, it promotes HO-1 through the p38 MAPK, and the MAPK/Nrf2 pathway exhibits anti-HCV properties (Lee et al. 2006). Moreover, andrographolide reduces the production of TNF-α and COX-2, reduces liver hypoxia, and mitigates hepatic apoptosis and fibrosis in rats by downregulating the expression of hypoxia-inducible genes, including vascular endothelial growth factor (VEGF) (Lee et al. 2015). *Phyllanthus niruri Linn.*, also referred to as "gale of the wind," contains phyllanthin, a strong hepatoprotective lignan that has traditionally been employed as a natural remedy for liver diseases, including jaundice (Ghosh et al. 2011). Additionally, these lignans regulate a "fatty liver" state and guard against alcohol-induced liver damage. Phyllanthus lignin's hepatoprotective benefits are brought about via its prevention of lipid peroxidation, superoxide, and hydroxyl radicals (Bhattacharjee and Sil 2007). The decoction of *Gegen qinlian* affects NAFLD via promoting PPAR-γ, which reduces inflammation and regulates lipid metabolism (Wang et al. 2015). *Alisma orientale (Sam.) Juz.* (Alismataceae), which inhibited de novo lipogenesis, promoted hepatic lipid outflow. Moreover, *A. orientalis (Sam.) Juz.* modified oxidative stress cytokines, fibrotic and inflammatory mediators, and ultimately affected

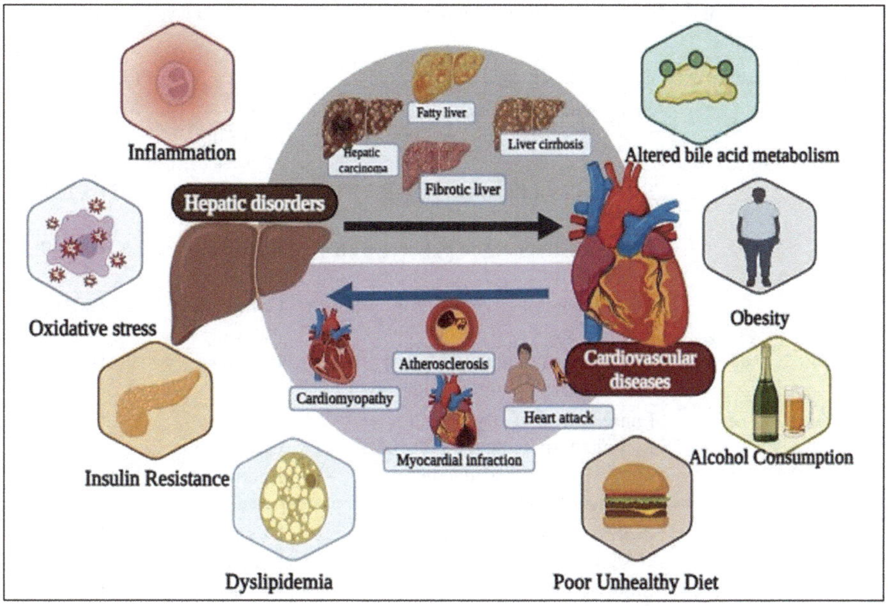

Fig. 6.2 Pathophysiology and the interlink in-between hepatic and cardiovascular diseases

panels for lipo-apoptosis and liver damage (Choi et al. 2019). The Asparagaceae plant, *Asparagus officinalis L.*, has a number of significant nutritional traits. With an IC_{50} of 45.52μg/mL, the fraction of green asparagus that was produced using ethyl acetate (EtOAc) significantly inhibits the development of murine t-HSC/Cl-6. Several compounds that were isolated from the EtOAc fraction were analyzed for their characteristics and potential to stop hepatic fibrosis (Zhong et al. 2015). Figure 6.2 shows the pathophysiology and the interlink in-between hepatic and cardiovascular diseases.

The development of liver disease is significantly influenced by oxidative damage. The role of oxidative stress in autoimmune syndrome-related liver disorders and viral hepatitis has been extensively researched and documented (Ezhilarasan 2018). The goal of the current chapter is to uncover the intricacies of medicinal plants by conducting extensive preclinical investigations and clinical studies to scientifically investigate the effectiveness of herbal-based medications and their safety (Giri et al. 2023). These methods eventually aid in standardizing the dosage schedule based on scientifically based findings and identifying an authentic pharmaceutical candidate and efficient therapeutic lead from natural sources, particularly plant origin. Most herbal remedies are now sold with the claims of preventing disease, promoting health, easing symptoms, and healing various illnesses. Table 6.1 listed some clinical trials for herbal medicines in treating hepatic diseases, while Table 6.2 listed some clinical trials for herbal medicines in treating cardiovascular diseases.

Table 6.1 Clinical trials for herbal medicines in treating hepatic diseases registered at ClinicalTrials.gov

Herbal compound	Title of the study	Status	Identifier
Silymarin	Silymarin in NAFLD	Not recruiting, phase 4	NCT02973295
	Silymarin for the treatment of NAFLD	Completed, phase 2	NCT02006498
	Effect of silymarin in patients with NAFLD	Recruiting, N/A	NCT03749070
	Phase I trial of silymarin for chronic liver diseases	Completed, phase1	NCT00389376
	Study to evaluate the effect of RGMA001on patients with NAFLD	Unknown, N/A	NCT01511523
	Phase II trial of silymarin for non-cirrhotic patients with NASH	Completed, phase 2	NCT00680407
	Milk Thistle Clinical Trial in Pediatric NAFLD	Recruiting	NCT06477146
Siliphos	A study of Siliphos in adults with NASH	Completed, phase2	NCT00443079
RGMA001 (Chinese Herbal Medicine)	Study to evaluate the effect of RGMA001on patients with NAFLD	Unknown, N/A	NCT01511523
Tablet of *silybum marianum, Pueraria lobate* and *salvia miltiorrhiza*	Effects of Tablets of Silybum Marianum, Pueraria Lobate and Salvia Miltiorrhiza on Fatty Liver	Unknown status	NCT05076058
Protandim (Blend of 5 herbal ingredients)	The effect of Protandim on NASH	Completed, N/A	NCT00977730
Resveratrol	Resveratrol for the treatment of NAFLD and insulin resistance in overweight adolescents	Completed, phase2/3	NCT02216552
	Long-term investigation of resveratrol on fat metabolism in obese Men with NAFLD	Completed, N/A	NCT01446276
	The effects of resveratrol supplement on biochemical factors and hepatic Fibrosis in patients with NASH	Completed, phase2/3	NCT02030977
	Resveratrol in patients with NAFLD	Completed, N/A	NCT01464801
	Potential beneficial effects of Resveratrol	Completed, N/A	NCT01150955
Berberine	Role of pioglitazone and berberine in the treatment of NAFLD	Completed, phase 2	NCT00633282
	Efficacy and safety of berberine in NASH	Recruiting, phase 4	NCT03198572
Curcumin	Curcumin supplement in NAFL patients with type2 diabetes	Completed, phase 2/3	NCT02908152
	The effect of curcumin on liver fat content in obese subjects	Not recruiting, N/A	NCT03864783

(continued)

Table 6.1 (continued)

Herbal compound	Title of the study	Status	Identifier
Ginger	Effect of Ginger Supplement on NAFL	Completed, phase 2/3	NCT02535195
	Effects of ginger on NAFLD in T2DM	Enrolling, phase1	NCT02289235
Shirazi's thyme	The effects of Zataria Multiflora Boiss. (Shirazi's thyme) on NAFLD	Completed, N/A	NCT02983669
Zhenzhu Tiaozhi capsules	Therapeutic effects of compound Zhenzhu Tiaozhi capsules in NAFLD	Recruiting, N/A	NCT03375580
Purified anthocyanin	Purified anthocyanin and NAFLD	Completed, phase1	NCT01940263
Bayberry juice	Clinical investigation on the effects of bayberry juice treatment in adult Subjects with features of fatty liver disease	Completed, N/A	NCT01707914
HepatoKeeper Herbal Essentials *(Schisandra chinensis, Paeonia lactiflora,* and *Punica granatum)*	Clinical Trials Using HepatoKeeper Herbal Essentials to Treat Non-alcoholic Fatty Liver Disease and Metabolic Factors	Completed, N/A	NCT06244550
Drug: Livitol-70	Herbal Supplements for Improvement of Liver Function in Participants with Alcoholic Liver Disease	Not yet recruiting	NCT03503708
Herbal therapy	Treatment of Chronic HCV Infected Egyptian Patients with Electromagnetic Waves and Herbal Therapy	Unknown status	NCT02215525
Chinese herbal medicine	Study on the Optimal Strategy for Acute-on-chronic Liver Failure with Integrative Treatment	Completed, N/A	NCT03577938
Herbal treatment	Herbal Treatment of Hepatitis C in Methadone Maintained Patients	Completed, N/A	NCT00010816
SHO-SAIKO-TO	SHO-SAIKO-TO for Patients with Chronic Hepatitis C: A Phase II Study	Completed, N/A	NCT00590564
Drug: diammonium glycyrrhizinate	Magnesium Isoglycyrrhizinate Followed by Diammonium Glycyrrhizinate and Combined with Entecavir in Chronic Hepatitis B (MAGIC-101)	Unknown status	NCT03349008
Artichoke extract	Effect of a Dietary Supplement on Lipid Pattern, Indexes of NAFLD and Systemic Inflammation in Healthy Subjects with Suboptimal LDL-C Levels	Not yet recruiting	NCT06403748

Table 6.2 Clinical trials for herbal medicines in treating cardiovascular diseases registered at ClinicalTrials.gov

Herbal compound	Title of the study	Status	Identifier
Herbal medicine Danshen-Gegen	Cardiovascular-Protective Effects of Herbal Medicine Danshen-Gegen	Completed, N/A	NCT01033630
Drug: XueBiJing injection (Xuebijing is a compound injection made from five Chinese herbs)	Additive Anti-inflammatory Action for Critically Ill Patients with Cardiovascular Surgery (Xuebijing) IV	Unknown status	NCT04993794
Solid extract of *Passiflora incarnata*	Short-term Effects of Passiflora and Meditation on Blood Pressure and Heart Rate	Unknown status	NCT03953469
Extract of *Ginkgo Biloba*	Extract of Ginkgo Biloba (EGB 761) and Vascular Function	Completed, N/A	NCT00029991
Drug: North American *Ginseng (Panax Quinquefolius)*	The Effect of North American Ginseng on Blood Pressure in Individuals with Hypertension	Completed, N/A	NCT00219960
Drug: Digoxin Drug: Hawthorn	Pharmacokinetic and Pharmacodynamic Interaction Study of Digoxin and Hawthorn	Completed, N/A	NCT00006330
Herbal therapy/ Botanical therapy (*Hibiscus sabdariffa*)	The Effect of *Hibiscus Sabdariffa L.* Tea on Blood Pressure	Completed, N/A	NCT00175110
Drug: MAK	Effects of Herbal Antioxidants on Cardiovascular Disease in Older Blacks	Completed, N/A	NCT00010725
Jing Si Herbal Tea	Effect of Jing Si Herbal Tea on Inflammation in Patients with Cardiovascular Disease	Unknown status	NCT05420987
Drug: Modified Wenxin Decoction	Clinical and Metabonomics Study on the Treatment of Heart Failure After Myocardial Infarction with Modified Wenxin Decoction	Unknown status	NCT04667598
Licorice extract	Licorice and Home Blood Pressure, Additional Study	Completed, N/A	NCT06019598

N/A not available
https://clinicaltrials.gov site was searched with keywords and the above data in Tables 6.1 and 6.2 were retrieved for the compounds/study in Clinical Trials

6 Future Directions in Herbal Treatment of Hepatic and Cardiovascular Diseases

The limited bioavailability of active phytoconstituents in herbal remedies for hepatic and cardiovascular illnesses has led to the development of novel techniques to boost their therapeutic efficiency. Nanotechnology-based delivery systems, including nano-emulsions, liposomes, and polymeric nanoparticles, are gaining popularity due to their potential to increase solubility, protect phytochemicals from degradation, and enable targeted administration to specific tissues. These systems improve the stability and bioavailability of hydrophobic phytoconstituents. In addition to nanotechnology, phytoconstituent structural modifications are being employed to improve their pharmacokinetic properties. Prodrug formation and the development of synthetic derivatives help to increase membrane permeability and metabolic stability, while bio-enhancers such as piperine reduce first-pass metabolism, allowing more of the active compound to reach the systemic circulation. Advanced formulation techniques play a crucial role in addressing bioavailability challenges. Self-emulsifying drug delivery systems (SEDDS) enhance the solubility of poorly water-soluble phytochemicals, while cyclodextrin complexes improve both their solubility and stability. Additionally, transdermal patches and buccal administration systems offer alternative delivery routes that bypass the gastrointestinal tract and first-pass metabolism, thereby increasing the bioavailability of active compounds (Buya et al. 2020). Furthermore, synergistic combinations of phytoconstituents, either with other natural compounds or with synthetic molecules, can enhance therapeutic outcomes through complementary mechanisms of action. Biotechnology approaches, such as metabolic engineering and microbial biotransformation, are being explored to optimize the production and conversion of bioactive compounds into more bioavailable forms. Finally, the success of these strategies depends on robust pharmacokinetic studies and well-designed clinical trials. These investigations are essential to understand absorption, distribution, metabolism, and excretion profiles, as well as to determine optimal dosing and efficacy. Together, these advanced approaches hold significant promise for overcoming the bioavailability barriers in herbal treatments and unlocking their full therapeutic potential for managing hepatic and cardiovascular diseases (Choudhury 2022).

The integration of personalized medicine and genomic profiling with innovative strategies to address low bioavailability is transforming the treatment landscape for hepatic and cardiovascular diseases. Genomic profiling allows for the identification of genetic variations that influence disease susceptibility, drug metabolism, and therapeutic response. This enables the development of tailored treatment strategies, including the use of bioactive herbal compounds that align with an individual's genetic makeup. By analyzing biomarkers and genetic markers, clinicians can design more effective, individualized therapies that optimize outcomes while minimizing adverse effects (Krzyszczyk et al. 2018). Personalizedmedicine (PM) is a novel approach that tailors therapeutic techniques to an individual's specific genetic, epigenomic, and proteomic characteristics. It emphasizes giving the appropriate treatment to the right individual at the right time. The fundamental issue in this profession is to determine the best effective therapy for a specific individual rather than depending on therapies established for the broader population. The primary idea of PM is to abandon the "one size fits all" model in favor of a personalized strategy that takes into account genetic diversity, environmental exposures, and lifestyle factors. PM's focus on prevention and therapy aims to improve disease prediction and prevention, allow for more accurate diagnoses, and offer focused, individualised remedies (Goetz and Schork 2018; Jatti Patil and Nagaraju 2022; Wang and Wang 2023).

When combined, genomic profiling and these innovative strategies enable a truly personalized approach. For instance, genomic data can identify individuals who may benefit most from specific bioactive compounds or delivery methods, while synergistic combinations of phytoconstituents and tailored biotechnology approaches, such as metabolic engineering and microbial biotransformation, further optimize therapeutic efficacy (Kussmann et al. 2023). Considering environmental and lifestyle factors alongside genetic data refines these personalized strategies, ensuring treatments are as effective as possible. Together, these advancements hold immense promise for revolutionizing the management of hepatic and cardiovascular diseases. They not only address the limitations of traditional treatments but also pave the way for developing customized herbal formulations and therapies designed to harmonize with everyone's unique genetic and environmental context (Gaudêncio et al. 2023).

Targeted herbal nanoformulations, prepared through green synthesis, represent a promising advancement in enhancing the efficacy of treatments for hepatic and cardiovascular diseases. Green synthesis leverages eco-friendly, sustainable methods that use plant extracts, biomolecules, or natural reducing agents to produce nanoparticles. These nanoparticles act as carriers for bioactive compounds, ensuring improved solubility, stability, and targeted delivery while minimizing toxicity and environmental impact. For hepatic diseases, such as liver fibrosis or non-alcoholic fatty liver disease (NAFLD), these formulations enable the precise delivery of herbal phytoconstituents like silymarin, curcumin, or resveratrol directly to liver cells (Bin-Jumah et al. 2020; Abuzeid et al. 2023). By enhancing bioavailability and bypassing first-pass metabolism, these nanoformulations ensure that the therapeutic compounds reach their target sites in optimal concentrations, thereby improving treatment outcomes. Similarly, in cardiovascular diseases, green-synthesized nanoparticles can encapsulate bioactive compounds like flavonoids, polyphenols, or terpenoids, which are known for their cardioprotective properties. These formulations can be engineered to target atherosclerotic plaques, reduce oxidative stress, or enhance endothelial function. For instance, nanoparticles loaded with herbal antioxidants can specifically target areas of inflammation in blood vessels, mitigating the progression of atherosclerosis or hypertension (Patra et al. 2018). The use of green synthesis ensures that the preparation process is free from harmful chemicals, making it safer for both patients and the environment. Additionally, these formulations can be tailored for controlled or sustained release, enabling long-lasting therapeutic effects with reduced dosing frequencies. The integration of such targeted herbal nanoformulations into clinical practice has the potential to revolutionize the treatment of hepatic and cardiovascular diseases, offering a sustainable, effective, and patient-centric approach (Khafaga et al. 2023; Al-darwesh et al. 2024).

7 Conclusions

Hepatic and cardiovascular diseases (CVDs) are closely interlinked health challenges, significantly contributing to global morbidity and mortality. The limitations of synthetic hepatoprotective agents, including insufficient efficacy and adverse effects, have driven a growing interest in herbal medicines. With their affordability, accessibility, and demonstrated potential in preclinical research, herbal treatments offer a promising alternative for managing hepatic diseases and associated CVDs.

This chapter highlights the findings of clinically evaluated herbal therapies, showcasing their role in addressing liver and cardiovascular conditions. Notable compounds such as berberine and silymarin have advanced to phase IV clinical trials, reflecting their therapeutic potential. Additionally, emerging evidence underscores the cardiovascular benefits of phytoconstituents like quercetin, resveratrol, and allicin. Despite these advancements, challenges such as the poor bioavailability of active herbal compounds remain significant barriers to their broader clinical application. Innovative strategies, including nanoformulation techniques, have been proposed to overcome these challenges and enhance the efficacy of herbal therapies. Furthermore, the future of herbal medicine lies in the development of personalized and tailored approaches, ensuring treatments are adapted to individual patient needs for maximum therapeutic benefit. In conclusion, while herbal medicines hold great promise, rigorous clinical trials, and advanced formulation strategies are essential to establish their safety, efficacy, and long-term viability in modern healthcare systems.

References

Abuzeid HM, Julien CM, Zhu L, Hashem AM (2023) Green synthesis of nanoparticles and their energy storage, environmental, and biomedical applications. Crystals 13(11):1576

Al-darwesh MY, Ibrahim SS, Mohammed MA (2024) A review on plant extract mediated green synthesis of zinc oxide nanoparticles and their biomedical applications. Results Chem 7:101368

Bai T, Yang Y, Yao YL, Sun P, Lian LH, Wu YL, Nan JX (2016) Betulin alleviated ethanol-induced alcoholic liver injury via SIRT1/AMPK signaling pathway. Pharmacol Res 105:1–12

Bhatia LS, Curzen NP, Byrne CD (2012) Nonalcoholic fatty liver disease and vascular risk. Curr Opin Cardiol 27(4):420–428

Bhattacharjee R, Sil PC (2007) Protein isolate from the herb, *Phyllanthus niruri* L. (Euphorbiaceae), plays hepatoprotective role against carbon tetrachloride induced liver damage via its antioxidant properties. Food Chem Toxicol 45(5):817–826

Bin-Jumah M, Al-Abdan M, Albasher G, Alarifi S (2020) Effects of green silver nanoparticles on apoptosis and oxidative stress in normal and cancerous human hepatic cells in vitro. Int J Nanomedicine 15:1537–1548

Bonomini F, Favero G, Rezzani R (2015) NF-κB—A key factor in atherogenesis and atheroprogression. In: Thrombosis, atherosclerosis and atherothrombosis—new insights and experimental protocols. IntechOpen

Burns J, Yokota T, Ashihara H, Lean ME, Crozier A (2002) Plant foods and herbal sources of resveratrol. J Agric Food Chem 50(11):3337–3340

Buya AB, Beloqui A, Memvanga PB, Préat V (2020) Self-nano-emulsifying drug-delivery systems: from the development to the current applications and challenges in oral drug delivery. Pharmaceutics 12(12):1194

Chan K, Chui SH, Wong DYL, Ha WY, Chan CL, Wong RNS (2004) Protective effects of Danshensu from the aqueous extract of *Salvia miltiorrhiza* (Danshen) against homocysteine-induced endothelial dysfunction. Life Sci 75(26):3157–3171

Chang Y, Ryu S, Sung KC, Cho YK, Sung E, Kim HN et al (2019) Alcoholic and non-alcoholic fatty liver disease and associations with coronary artery calcification: evidence from the Kangbuk Samsung Health Study. Gut 68(9):1667–1675

Chen S, Henderson A, Petriello MC, Romano KA, Gearing M, Miao J et al (2019) Trimethylamine N-oxide binds and activates PERK to promote metabolic dysfunction. Cell Metab 30(6):1141–1151

Chen YM, Liu Y, Zhou RF, Chen XL, Wang C, Tan XY et al (2016) Associations of gut-flora-dependent metabolite trimethylamine-N-oxide, betaine and choline with non-alcoholic fatty liver disease in adults. Sci Rep 6(1):19076

Choi E, Jang E, Lee JH (2019) Pharmacological activities of *Alisma orientale* against nonalcoholic fatty liver disease and metabolic syndrome: literature review. Evid Based Complement Alternat Med 2019(1):2943162

Choudhury A (2022) Potential role of bioactive phytochemicals in combination therapies against antimicrobial activity. J Pharmacopuncture 25(2):79–87

Cobbina E, Akhlaghi F (2017) Non-alcoholic fatty liver disease (NAFLD)—pathogenesis, classification, and effect on drug metabolizing enzymes and transporters. Drug Metab Rev 49(2):197–211

Cragg GM, Newman DJ (2013) Natural products: a continuing source of novel drug leads. Biochim Biophys Acta Gen Subj 1830(6):3670–3695

Dai X, Thiagarajan D, Fang J, Shen J, Annam NP, Yang Z et al (2017) SM22α suppresses cytokine-induced inflammation and the transcription of NF-κB inducing kinase (Nik) by modulating SRF transcriptional activity in vascular smooth muscle cells. PLoS One 12(12):e0190191

Davison EK, Brimble MA (2019) Natural product derived privileged scaffolds in drug discovery. Curr Opin Chem Biol 52:1–8

de Gonzalez AKK, Lefkowitch JH (2017) Heart disease and the liver: pathologic evaluation. Gastroenterol Clin N Am 46(2):421–435

Deleyto-Seldas N, Efeyan A (2021) The mTOR–autophagy axis and the control of metabolism. Front Cell Dev Biol 9:655731

Deprince A, Haas JT, Staels B (2020) Dysregulated lipid metabolism links NAFLD to cardiovascular disease. Mol Metab 42:101092

Devarbhavi H, Asrani SK, Arab JP, Nartey YA, Pose E, Kamath PS (2023) Global burden of liver disease: 2023 update. J Hepatol 79(2):516–537

Di Pasqua LG, Cagna M, Berardo C, Vairetti M, Ferrigno A (2022) Detailed molecular mechanisms involved in drug-induced non-alcoholic fatty liver disease and non-alcoholic steatohepatitis: an update. Biomedicine 10(1):194

El-Tahawy NF, Ali AH, Saied SR, Abdel-Wahab Z (2011) Effect of glycyrrhizin on lipopolysaccharide/D-galactosamine-induced acute hepatitis in albino rats: a histological and immunohistochemical study. Egypt J Histol 34(3):518–527

Ezhilarasan D (2018) Oxidative stress is bane in chronic liver diseases: clinical and experimental perspective. Arab J Gastroenterol 19(2):56–64

Francque SM, Van Der Graaff D, Kwanten WJ (2016) Non-alcoholic fatty liver disease and cardiovascular risk: pathophysiological mechanisms and implications. J Hepatol 65(2):425–443

Frishman WH, Beravol P, Carosella C (2009) Alternative and complementary medicine for preventing and treating cardiovascular disease. Dis Mon 55(3):121–192

Gao S, Liu Z, Li H, Little PJ, Liu P, Xu S (2012) Cardiovascular actions and therapeutic potential of tanshinone IIA. Atherosclerosis 220(1):3–10

Gaudêncio SP, Bayram E, Lukić Bilela L, Cueto M, Díaz-Marrero AR, Haznedaroglu BZ et al (2023) Advanced methods for natural products discovery: bioactivity screening, dereplication, metabolomics profiling, genomic sequencing, databases and informatic tools, and structure elucidation. Mar Drugs 21(5):308

Ghosh N, Ghosh R, Mandal V, Mandal SC (2011) Recent advances in herbal medicine for treatment of liver diseases. Pharm Biol 49(9):970–988

Giri BR, Baral R, Bhatt H, Khadka A, Tamrakar R, Timalsina G, Gyawali R (2023) Phytochemical screening, free-radical scavenging activity, in vitro alpha-amylase inhibitory activity, and in vivo hypoglycemic activity studies of several crude drug formulations based on selected medicinal plants of Nepal. Pharm Chem J 56(10):1369–1378

Glycyrrhizic acid, PubChem. Available at: https://pubchem.ncbi.nlm.nih.gov/compound/Glycyrrhizic-acid. Accessed December 26, 2024

Goetz LH, Schork NJ (2018) Personalized medicine: motivation, challenges, and progress. Fertil Steril 109(6):952–963

Gregor MF, Hotamisligil GS (2011) Inflammatory mechanisms in obesity. Annu Rev Immunol 29(1):415–445

Imajo K, Hyogo H, Yoneda M, Honda Y, Kessoku T, Tomeno W et al (2014) LDL-migration index (LDL-MI), an indicator of small dense low-density lipoprotein (sdLDL), is higher in non-alcoholic steatohepatitis than in non-alcoholic fatty liver: a multicenter cross-sectional study. PLoS One 9(12):e115403

Jatti Patil D, Nagaraju R (2022) Personalised precision medicine: a novel approach for oral cancer management

Jeong HW, Hsu KC, Lee JW, Ham M, Huh JY, Shin HJ et al (2009) Berberine suppresses proinflammatory responses through AMPK activation in macrophages. Am J Physiol Endocrinol Metab 296(4):E955–E964

Khadem E, Toosi MN, Ilkhani R (2013) Liver-heart inter-relationship in fatty liver disease based on the Avicenna's point of view. Iran J Public Health 42(6):648–649

Khafaga DSR, El-Khawaga AM, Elfattah Mohammed RA, Abdelhakim HK (2023) Green synthesis of nano-based drug delivery systems developed for hepatocellular carcinoma treatment: a review. Mol Biol Rep 50(12):10351–10364

Kong W, Wei J, Abidi P, Lin M, Inaba S, Li C et al (2004) Berberine is a novel cholesterol-lowering drug working through a unique mechanism distinct from statins. Nat Med 10(12):1344–1351

Krzyszczyk P, Acevedo A, Davidoff EJ, Timmins LM, Marrero-Berrios I, Patel M et al (2018) The growing role of precision and personalized medicine for cancer treatment. Technology (Singap World Sci) 6(3–4):79–100

Kumar V, Prabhu SD, Bansal SS (2022) CD4+ T-lymphocytes exhibit biphasic kinetics post-myocardial infarction. Front Cardiovasc Med 9:992653

Kussmann M, Abe Cunha DH, Berciano S (2023) Bioactive compounds for human and planetary health. Front Nutr 10:1193848

Laplante M, Sabatini DM (2009) An emerging role of mTOR in lipid biosynthesis. Curr Biol 19(22):R1046–R1052

Lee CY, Peng WH, Cheng HY, Chen FN, Lai MT, Chiu TH (2006) Hepatoprotective effect of *Phyllanthus* in Taiwan on acute liver damage induced by carbon tetrachloride. Am J Chin Med 34(3):471–482

Lee KH, Bae IY, Park SI, Park JD, Lee HG (2016) Antihypertensive effect of Korean red ginseng by enrichment of ginsenoside Rg3 and arginine–fructose. J Ginseng Res 40(3):237–244

Lee SJ, Kang JH, Iqbal W, Kwon OS (2015) Proteomic analysis of mice fed methionine and choline deficient diet reveals marker proteins associated with steatohepatitis. PLoS One 10(4):e0120577

Li GY, Gao HY, Huang J, Lu J, Gu JK, Wang JH (2014a) Hepatoprotective effect of *Cichorium intybus* L., a traditional Uighur medicine, against carbon tetrachloride-induced hepatic fibrosis in rats. World J Gastroenterol 20(16):4753

Li JY, Cao HY, Liu P, Cheng GH, Sun MY (2014b) Glycyrrhizic acid in the treatment of liver diseases: literature review. Biomed Res Int 2014(1):872139

Ling NX, Kaczmarek A, Hoque A, Davie E, Ngoei KR, Morrison KR et al (2020) mTORC1 directly inhibits AMPK to promote cell proliferation under nutrient stress. Nat Metab 2(1):41–49

Liu YX, Xiao CL, Wang YX, Li YH, Yang YH, Li YB et al (2012) Synthesis, structure–activity relationship and in vitro anti-mycobacterial evaluation of 13-n-octylberberine derivatives. Eur J Med Chem 52:151–158

Lovric A, Granér M, Bjornson E, Arif M, Benfeitas R, Nyman K et al (2018) Characterization of different fat depots in NAFLD using inflammation-associated proteome, lipidome and metabolome. Sci Rep 8(1):14200

Luo Q, Ding J, Zhu L, Chen F, Xu L (2018) Hepatoprotective effect of wedelolactone against concanavalin A-induced liver injury in mice. Am J Chin Med 46(4):819–833

Macavei B, Baban A, Dumitrascu DL (2016) Psychological factors associated with NAFLD/NASH: a systematic review. Eur Rev Med Pharmacol Sci 20(24) [no page range provided]

Maier HJ, Schips TG, Wietelmann A, Krüger M, Brunner C, Sauter M et al (2012) Cardiomyocyte-specific IκB kinase (IKK)/NF-κB activation induces reversible inflammatory cardiomyopathy and heart failure. Proc Natl Acad Sci 109(29):11794–11799

Meng G, Zhou X, Wang M, Zhou L, Wang Z, Wang M et al (2019) Gut microbe-derived metabolite trimethylamine N-oxide activates the cardiac autonomic nervous system and facilitates ischemia-induced ventricular arrhythmia via two different pathways. EBioMedicine 44:656–664

Mensah GA, Fuster V, Murray CJ, Roth GA, Global Burden of Cardiovascular Diseases and Risks Collaborators (2023) Global burden of cardiovascular diseases and risks, 1990–2022. J Am Coll Cardiol 82(25):2350–2473

Miura K, Yang L, van Rooijen N, Brenner DA, Ohnishi H, Seki E (2013) Toll-like receptor 2 and palmitic acid cooperatively contribute to the development of nonalcoholic steatohepatitis through inflammasome activation in mice. Hepatology 57(2):577–589

Moreno JJ (2000) Resveratrol modulates arachidonic acid release, prostaglandin synthesis, and 3T6 fibroblast growth. J Pharmacol Exp Ther 294(1):333–338

Morishima C, Shuhart MC, Wang CC, Paschal DM, Apodaca MC, Liu Y et al (2010) Silymarin inhibits in vitro T-cell proliferation and cytokine production in hepatitis C virus infection. Gastroenterology 138(2):671–681

Motterlini R, Foresti R, Bassi R, Green CJ (2000) Curcumin, an antioxidant and anti-inflammatory agent, induces heme oxygenase-1 and protects endothelial cells against oxidative stress. Free Radic Biol Med 28(8):1303–1312

Myers RP, Cerini R, Sayegh R, Moreau R, Degott C, Lebrec D, Lee SS (2003) Cardiac hepatopathy: clinical, hemodynamic, and histologic characteristics and correlations. Hepatology 37(2):393–400

Nakahara T, Hyogo H, Yoneda M, Sumida Y, Eguchi Y, Fujii H et al (2014) Type 2 diabetes mellitus is associated with the fibrosis severity in patients with nonalcoholic fatty liver disease in a large retrospective cohort of Japanese patients. J Gastroenterol 49:1477–1484

Ng CF, Koon CM, Cheung DWS, Lam MY, Leung PC, Lau CBS, Fung KP (2011) The anti-hypertensive effect of Danshen (Salvia miltiorrhiza) and Gegen (Pueraria lobata) formula in rats and its underlying mechanisms of vasorelaxation. J Ethnopharmacol 137(3):1366–1372

Ogresta D, Mrzljak A, Berkovic MC, Bilic-Curcic I, Stojsavljevic-Shapeski S, Virovic-Jukic L (2022) Coagulation and endothelial dysfunction associated with NAFLD: current status and therapeutic implications. J Clin Transl Hepatol 10(2):339

Pagliaro B, Santolamazza C, Simonelli F, Rubattu S (2015) Phytochemical compounds and protection from cardiovascular diseases: a state of the art. Biomed Res Int 2015(1):918069

Park JB, Kwon SK, Nagar H, Jung SB, Jeon BH, Kim CS et al (2014) Rg3-enriched Korean Red Ginseng improves vascular function in spontaneously hypertensive rats. J Ginseng Res 38(4):244–250

Parodi FE, Mao D, Ennis TL, Pagano MB, Thompson RW (2006) Oral administration of diferuloylmethane (curcumin) suppresses proinflammatory cytokines and destructive connective tissue remodeling in experimental abdominal aortic aneurysms. Ann Vasc Surg 20:360–368

Patra JK, Das G, Fraceto LF, Campos EVR, Rodriguez-Torres MDP, Acosta-Torres LS et al (2018) Nano based drug delivery systems: recent developments and future prospects. J Nanobiotechnol 16:1–33

Polyzos SA, Kountouras J, Mantzoros CS (2016) Adipokines in nonalcoholic fatty liver disease. Metabolism 65(8):1062–1071

Prokunina-Olsson L, Muchmore B, Tang W, Pfeiffer RM, Park H, Dickensheets H et al (2013) A variant upstream of IFNL3 (IL28B) creating a new interferon gene IFNL4 is associated with impaired clearance of hepatitis C virus. Nat Genet 45(2):164–171

Qi J, You T, Li J, Pan T, Xiang L, Han Y, Zhu L (2018) Circulating trimethylamine N-oxide and the risk of cardiovascular diseases: a systematic review and meta-analysis of 11 prospective cohort studies. J Cell Mol Med 22(1):185–194

Ridker PM, Everett BM, Thuren T, MacFadyen JG, Chang WH, Ballantyne C et al (2017) Antiinflammatory therapy with canakinumab for atherosclerotic disease. N Engl J Med 377(12):1119–1131

Rolo AP, Teodoro JS, Palmeira CM (2012) Role of oxidative stress in the pathogenesis of nonalcoholic steatohepatitis. Free Radic Biol Med 52(1):59–69

Sarwar A, Shaw LJ, Shapiro MD, Blankstein R, Hoffman U, Cury RC et al (2009) Diagnostic and prognostic value of absence of coronary artery calcification. JACC Cardiovasc Imaging 2(6):675–688

Seeto RK, Fenn B, Rockey DC (2000) Ischemic hepatitis: clinical presentation and pathogenesis. Am J Med 109(2):109–113

Seyedemadi P, Rahnema M, Bigdeli MR, Oryan S, Rafati H (2016) The neuroprotective effect of rosemary (*Rosmarinus officinalis* L.) hydro-alcoholic extract on cerebral ischemic tolerance in experimental stroke. Iran J Pharm Res: IJPR 15(4):875

Shah SMA, Akram M, Riaz M, Munir N, Rasool G (2019) Cardioprotective potential of plant-derived molecules: a scientific and medicinal approach. Dose-Response 17(2):1559325819852243

Sharma A, Anand SK, Singh N, Dwarkanath A, Dwivedi UN, Kakkar P (2021) Berbamine induced activation of the SIRT1/LKB1/AMPK signaling axis attenuates the development of hepatic steatosis in high-fat diet-induced NAFLD rats. Food Funct 12(2):892–909

Shin W, Yoon J, Oh GT, Ryoo S (2013) Korean red ginseng inhibits arginase and contributes to endothelium-dependent vasorelaxation through endothelial nitric oxide synthase coupling. J Ginseng Res 37(1):64

Smekal A, Vaclavik J (2017) Adipokines and cardiovascular disease: a comprehensive review. Biomed Pap Med Fac Univ Palacky Olomouc Czech Repub 161(1)

Spanier G, Xu H, Xia N, Tobias S, Deng S, Wojnowski L et al (2009) Resveratrol reduces endothelial oxidative stress by modulating the gene expression of superoxide dismutase 1 (SOD1), glutathione peroxidase 1 (GPx1) and NADPH oxidase subunit (Nox4). J Physiol Pharmacol 60(Suppl 4):111–116

Sun YE, Wang W, Qin J (2018) Anti-hyperlipidemia of garlic by reducing the level of total cholesterol and low-density lipoprotein: a meta-analysis. Medicine 97(18):e0255

Tabassum N, Ahmad F (2011) Role of natural herbs in the treatment of hypertension. Pharmacogn Rev 5(9):30

Thomson M, Al-Qattan KK, Js D, Ali M (2015) Anti-diabetic and anti-oxidant potential of aged garlic extract (AGE) in streptozotocin-induced diabetic rats. BMC Complement Altern Med 16:1–9

Tieri P, Termanini A, Bellavista E, Salvioli S, Capri M, Franceschi C (2012) Charting the NF-κB pathway interactome map. PLoS One 7(3):e32678

Wang RC, Wang Z (2023) Precision medicine: disease subtyping and tailored treatment. Cancer 15(15):3837

Wang YL, Liu LJ, Zhao WH, Li JX (2015) Intervening TNF-α via PPARγ with Gegenqinlian Decoction in experimental nonalcoholic fatty liver disease. Evid Based Complement Alternat Med 2015(1):715638

Wang Z, Klipfell E, Bennett BJ, Koeth R, Levison BS, DuGar B et al (2011) Gut flora metabolism of phosphatidylcholine promotes cardiovascular disease. Nature 472(7341):57–63

Wu NA, Xu B, Xiang Y, Wu L, Zhang Y, Ma X et al (2013) Association of inflammatory factors with occurrence and recurrence of atrial fibrillation: a meta-analysis. Int J Cardiol 169(1):62–72

Wu Q, Li Y, Peng K, Wang XL, Ding Z, Liu L et al (2019) Isolation and characterization of three antihypertension peptides from the mycelia of *Ganoderma lucidum* (Agaricomycetes). J Agric Food Chem 67(29):8149–8159

Yang L, Zhang J, Yan C, Zhou J, Lin R, Lin Q et al (2012a) Sirt1 regulates cd40 expression induced by TNF-α via NF-κB pathway in endothelial cells. Cell Physiol Biochem 30(5):1287–1298

Yang TY, Wei JCC, Lee MY, Chen CB, Ueng KC (2012b) A randomized, double-blind, placebo-controlled study to evaluate the efficacy and tolerability of Fufang Danshen (*Salvia miltiorrhiza*) as add-on antihypertensive therapy in Taiwanese patients with uncontrolled hypertension. Phytother Res 26(2):291–298

You M, Jogasuria A, Lee K, Wu J, Zhang Y, Kwang Lee Y, Sadana P (2017) Signal transduction mechanisms of alcoholic fatty liver disease: emerging role of lipin-1. Curr Mol Pharmacol 10(3):226–236

Zhang W, Hong R, Tian T (2013) Silymarin's protective effects and possible mechanisms on alcoholic fatty liver for rats. Biomol Ther 21(4):264

Zhang X, Li Y, Yang P, Liu X, Lu L, Chen Y et al (2020) Trimethylamine-N-oxide promotes vascular calcification through activation of NLRP3 (nucleotide-binding domain, leucine-rich-containing family, pyrin domain-containing-3) inflammasome and NF-κB (nuclear factor κB) signals. Arterioscler Thromb Vasc Biol 40(3):751–765

Zhao P, Sun X, Chaggan C, Liao Z, In Wong K, He F et al (2020) An AMPK–caspase-6 axis controls liver damage in nonalcoholic steatohepatitis. Science 367(6478):652–660

Zhong C, Jiang C, Xia X, Mu T, Wei L, Lou Y et al (2015) Antihepatic fibrosis effect of active components isolated from green asparagus (*Asparagus officinalis* L.) involves the inactivation of hepatic stellate cells. J Agric Food Chem 63(26):6027–6034

Zingg JM, Hasan ST, Meydani M (2013) Molecular mechanisms of hypolipidemic effects of curcumin. Biofactors 39(1):101–121

Antibacterial Phytochemicals: A New Era in Infection Management

7

Aqeela Ashraf, Hina Qaiser, Huma Shafique, Manzar Zahra, Laraib Zainab, and Hafsa Ahmad Qureshi

Abstract

Antimicrobial resistance (AMR) has already been declared as the next pandemic. This has led to a paradigm shift towards the use of natural plant-based products called phytochemicals. The pace of discovery of new antibiotics is slower as compared to the increasing trend of AMR. To combat this threat, many efforts are directed to study phytochemicals as a suitable candidate in place of antibiotics. Research has increasingly highlighted their potential not only as stand-alone treatments but also as complementary agents to conventional antibiotics, enhancing their efficacy and overcoming resistance mechanisms. Additionally, the use of phytochemicals offers a more sustainable and less toxic alternative to synthetic antibiotics, thereby paving the way for safer therapeutic options. This chapter summarizes various approaches adopted for the better selection of phytochemicals to be used as antibacterial agents. These include both *in-silico* studies involving dynamic simulations and molecular docking as well as in-vitro trials to study the efficacy of various phytocompounds against bacterial pathogens. Nonetheless, significant physicochemical barriers such as low bioavailability, pharmacokinetic properties of the agent, low solubility of phytocompounds, limited gastrointestinal absorption, low hydrophilicity, and slow intrinsic dissolution impede the effective use of these natural compounds as treatments for various infections. However, the use of nanoparticle-driven phytocompounds is

A. Ashraf (✉) · H. Shafique · L. Zainab · H. A. Qureshi
Department of Biology, Lahore Garrison University, Lahore, Pakistan
e-mail: aqeelamalik09@gmail.com

H. Qaiser
School of Biological Sciences, University of the Punjab, Lahore, Pakistan

M. Zahra
Department of Chemistry, Lahore Garrison University, Lahore, Pakistan

K. Kuca et al. (eds.), *Medicinal Plants and Their Bioactives in Human Diseases*,
https://doi.org/10.1007/978-3-032-01356-9_7

showing good results. As scientists continue to explore the rich diversity of plant secondary metabolites, the integration of antibacterial phytochemicals into clinical practice could herald a new era in infection management, promoting both health and environmental sustainability.

Keywords

Phytocompounds · Bacterial infections · Antimicrobial resistance · Nanoparticles · In silico study

1 Introduction

In the twenty-first century, the emergence of antimicrobial resistance (AMR) has evolved into a major worldwide health concern. It is expected that by 2050, AMR could result in ten million deaths and potentially drive 24 million individuals into extreme poverty by 2030 (Charani et al. 2021). The highest incidence of AMR infection is observed in infants under 1 year of age and in adults over 65 years of age (Fink et al. 2020). Despite advancements in immunization and chemotherapy, infectious diseases remain a major health concern. The inappropriate use of antibiotics has led to drug resistance, with bacteria developing various resistance mechanisms, rendering infectious diseases as one of the leading causes of morbidity (Subramaniam and Girish 2020). The rising incidence of antibiotic resistance by the bacteria undermines the efficacy of current therapeutic agents, resulting in prolonged illness, complex treatment regimens, and increased healthcare expenditure. Bacterial species exhibit rapid evolutionary adaptation to withstand adverse environmental conditions, antibiotic exposure, and host immune responses. The emergence of bacterial strains resistant to multiple drugs has been fueled by the overuse and misuse of antibiotics (Muteeb et al. 2023).

Phytochemicals present a potential alternative to complement conventional antibiotics, addressing the challenges posed by the rapid evolution of resistant microorganisms (AlSheikh et al. 2020). Several categories of plant-derived chemicals, such as alkaloids, phenols, coumarins, and terpenes, have shown notable antimicrobial properties against multi-drug resistant (MDR) microorganisms. These natural compounds interfere with key bacterial components, including membrane proteins, efflux pumps, biofilms, and quorum-sensing mechanisms, which play crucial roles in the emergence of antibiotic resistance. The complex composition of plant extracts, containing a diverse array of phytochemical components, makes it difficult for bacteria to quickly develop resistance mechanisms (Khameneh et al. 2021). This chapter discusses the growing concern of AMR and explores the potential of phytochemicals as alternative antibacterial agents.

2 The Evolution of Antibiotic Resistance: Pre, Present, and Beyond

In human history, various disease outbreaks have occurred. Differentiating between bacterial and viral origins for these epidemics is often challenging. Historical records from major civilizations, encompassing ancient Egypt, Mesopotamia, Babylon, Rome, China, India, and Islamic societies, contain accounts of widespread devastating illnesses (Huremović 2019). The early modern period witnessed numerous lethal epidemics and pandemics of predominantly bacterial origin. Figure 7.1 shows identified severe bacterial epidemics.

The introduction of antibiotics has markedly curtailed the proliferation of bacterial epidemics. During the antibiotic golden age (1930s–1960s) and the ensuing innovation plateau (1960s–2000s), pathogenic bacteria exerted a minimal discernible influence on human history. During this era, no significant bacterial pandemics were recorded, although several worldwide outbreaks of antibiotic-resistant bacteria occurred. Noteworthy examples include the 1964 typhoid outbreak in Aberdeen, Britain (Walker 1965), the 1979 anthrax incident in Sverdlovsk, Soviet Union (Marshall 1988), and the 2007 cholera epidemic in Baghdad (Khwaif et al. 2010). Since the 1960s, a trend has emerged: the rise of antibiotic resistance has been accompanied by a decline in the development of new antibiotic classes. This pattern poses a potential threat of resurgent epidemics and bacterial infections, which could have far-reaching consequences (Church and McKillip 2021). Presently, drug-resistant infections claim at least 700,000 lives worldwide annually, a figure projected to surge to ten million by 2050 if current trends persist (Church and McKillip 2021; Young and Craig 2020; Kawano 2021; Murray et al. 2022). To address this issue, the World Health Organization (WHO) has released a catalog of critically important antibiotic-resistant bacteria on a global scale, aiming to stimulate the exploration, invention, and creation of novel antibacterial medications. The majority of these organisms are classified as gram-negative bacteria (Tacconelli 2017).

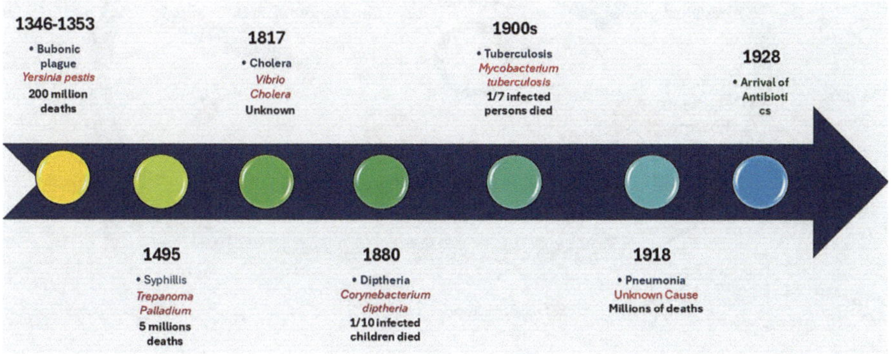

Fig. 7.1 Chronology of bacterial outbreaks from the beginning of the modern era until antibiotics were discovered

Numerous bacterial species discovered long ago that were successfully managed during the golden age of antibiotics have recently resurged as deadly pathogens worldwide.

Alexander Fleming, the Nobel laureate who discovered penicillin, cautioned in 1945 regarding the potential for antibiotic resistance due to misuse. His prediction was realized within a decade of widespread use (Hegab 2016). Nevertheless, advancements in medicinal chemistry have facilitated the rapid discovery of various classes of antibiotics with diverse bactericidal or bacteriostatic mechanisms. Unfortunately, the bacteria concurrently developed numerous resistance strategies (Fig. 7.2).

The majority of these novel antibiotics do not employ novel mechanisms of action but rather are variations or improvements in existing antibiotic classes (Boyd et al. 2021). Although these new antibiotics can treat many resistant bacteria, they are not effective against all strains. Considering the potential for some of these antibiotics to fail in regulatory approval and the inevitable development of resistance to those that succeed, it is evident that the current pipeline of antibiotics in development is insufficient to meet current and future patient needs (Kzhyshkowska et al. 2022). Over 70% of the bacteria responsible for severe infections are likely to show resistance to at least one commonly prescribed antibiotic (Overbye and Barrett 2005). Consequently, the combination of increasing antibiotic resistance and the paucity of new antibiotics presents a concern for the future of infectious disease treatment (Dougan et al. 2019).

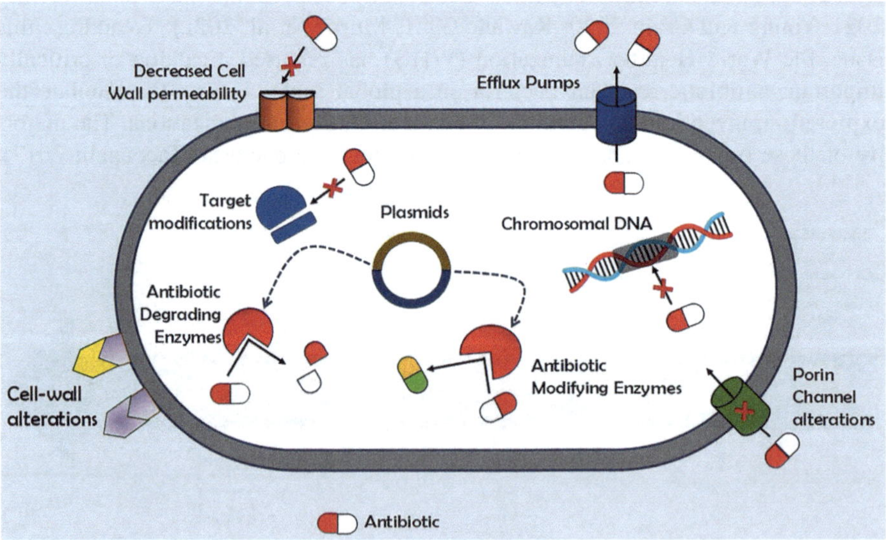

Fig. 7.2 Most significant bacterial resistance mechanisms against antibiotics

3 Rethinking Antibiotics: Challenges and Path Forward

Overuse of artificial antibiotics has been associated with considerable adverse effects and threatens microorganisms that are crucial for biogeochemical cycles. During the COVID-19 pandemic, hospitalized and immunocompromised patients have faced heightened risks, leading to an increase in secondary infections, often caused by antibiotic-resistant bacteria (Miethke et al. 2021). Two main factors contribute to antibiotic resistance: bacterial capabilities and human activity.

3.1 Bacterial Influence on Antibiotic Resistance

Bacteria, among the most successful microorganisms on Earth, have demonstrated remarkable resilience and adaptability across a wide range of environments. Their ability to adjust to various conditions has allowed bacteria to persist on the planet's surface for approximately 2 billion years (Schopf et al. 1965). Gaining a comprehensive insight into the molecular foundations of antibiotic resistance is crucial for the development of novel antibacterial compounds and innovative therapeutic approaches. Contemporary research has uncovered that bacterial organisms can evolve diverse resistance mechanisms, and the synergy of these mechanisms may give rise to superbugs or multidrug-resistant pathogens. Figure 7.2 demonstrates the various pathways through which bacteria can acquire resistance to antibacterial agents.

3.2 Human Influence on Antibiotic Resistance

The misuse and overuse of antibiotics are largely attributed to consumption patterns, improper and excessive usage, patient expectations, and human attitudes towards antibiotics (Byrne et al. 2019). The primary factor contributing to antibiotic resistance in human pathogens is self-medication. Research shows that while 20% of human antibiotic use occurs in hospital settings, 80% takes place in community environments. Significantly, it is estimated that up to 50% of antibiotic use in communities is considered dubious and unnecessary (Rather et al. 2017). Regrettably, humans extensively utilize antibiotics in livestock to promote growth and prevent disease. Numerous studies have demonstrated that antimicrobial resistance resulting from antibiotic use in animals adversely affects human health and promotes the development of resistant bacteria (Phillips et al. 2004). A comparable pattern has been noted in fish farming, where the extensive application of large amounts of antibiotics in fish food increases the likelihood of antibiotic remnants in fish meat and associated products (Cabello 2006). Additionally, antibiotic residues from drug manufacturers and the buildup of substantial quantities of discarded antibiotics in the ecosystem present a major environmental risk, considering that the worldwide antibiotic market utilizes 100,000–200,000 tons each year (Joakim Larsson 2014). Climate change may also contribute to the risk of the spread of pathogens. Shifting climatic factors can alter the distribution and abundance of organisms, leading to novel interactions and increased opportunities for the emergence of infectious diseases (Baker et al. 2022). As a result, various human activities have played a significant role in the development of antibiotic-resistant bacteria. These activities include the large-scale production of antibiotics, the accumulation of antibiotic residues in the environment, and the improper use of antibiotics in medical treatments, veterinary care, and agricultural practices.

4 Mechanistic of Antibacterial Phytoconstituents

Natural materials outperform synthetic treatments in scaffold variety, molecular complexity, stereochemical abundance, ring system diversity, and carbohydrate content (Vadhana et al. 2015).

The mechanism of antibacterial phytoconstituents has been illustrated below.

4.1 Protein Biosynthesis

Ribosomal subunits can be targeted by antibiotics to reduce protein synthesis and combat bacterial infections. DNA and RNA are crucial for protein production, and phytoconstituents targeting nucleic acids can have broad-spectrum antimicrobial effects (Othman et al. 2019). FtsZ (filamentous temperature-sensitive protein Z) is a protein involved in bacterial cell division, inhibitors target FtsZ to reduce bacterial proliferation (Barrows and Goley 2021). Inhibitors of FtsZ can reduce bacterial cell growth (Weart et al. 2007).

4.2 Bacterial Cell Membrane Permeability

Bacterial cell membranes protect the cell's environment and aid in cell identification (Mabhiza et al. 2016). Cell membrane-targeting antibiotics may work well against bacteria that are resistant. Bacteria classified as Gram-positive or Gram-negative have different membrane compositions. Cell membrane damage can reduce protection, hinder transport, and enhance conductivity (Kumari et al. 2020). Flavonoids, plant-derived compounds, have antibacterial properties against various bacterial diseases (Upadhyay et al. 2014). A study found that sanguinarine damages cell walls and membranes, increasing permeability and producing reactive oxygen species (Gu et al. 2023).

4.3 Inhibition of Efflux Pumps

Misuse of antibiotics has reduced antibiotic efficacy in MDR bacteria, leading to increased resistance (Hancock 2005). Inhibiting efflux pump genes can increase antibacterial properties in order to combat MDR bacteria (Shriram et al. 2018). Plant-based phytoconstituents like reserpine, carnosic acid, carnosol, lysergol, resveratrol, thymol, carvacrol, berberine, and palmatine have shown potential as efflux pump inhibitors, suppressing efflux pump activity in various bacterial species (Aghayan et al. 2017).

4.4 Biofilm Inhibition

Biofilms, bacterial colonies anchored in an extracellular matrix, contribute to antibiotic resistance. Quorum sensing (QS) is crucial for the development and stability of biofilms. Novel antibacterial medications can help stop MDR infections (Masák et al. 2014). Natural compounds from plants have shown antibacterial and antibiofilm properties (Fathi et al. 2021), inhibiting polymer matrix formation, cell adhesion, extracellular matrix synthesis, and virulence factor synthesis (Lu et al. 2019). Hordenine, parthenolide, cyclanidin, and ajoene suppress the expression of several QS-related genes and downregulate extracellular polymeric material, thus have shown antibacterial properties (Liu et al. 2021; Kalia et al. 2018; Gopu and Shetty 2016; Phipps et al. 2012).

5 Major Phytochemical Classes with Potent Antibacterial Activity

Nature's biodiversity provides diverse biological compounds, including secondary metabolites found in plants. These compounds, including alkaloids, polyphenols, organosulfur compounds, terpenes, tannins, and essential oils, have potential antimicrobial properties (Barbosa et al. 2020; Cappiello et al. 2020). Some of the major phytochemical classes have been described here.

5.1 Alkaloids

About 300 different plant families, bacteria, fungi, and animals contain alkaloids, which come in various secondary metabolites (Evans and Evans 2009). More than 18,000 distinct alkaloids have been found (Othman et al. 2019), they can exist as salts of organic acid or free bases (Robbers et al. 1996). Alkaloids are classified into true-, proto-, and pseudo-alkaloids based on their biosynthesis (Amirkia and Heinrich 2014). With respect to their chemical structure, they are divided into two groups: non-heterocyclic or proto-alkaloids and heterocyclic or real alkaloids (Evans and Evans 2009).

The biological activity of alkaloids, such as their antiviral, antibacterial, anticancer, and central nervous depressant qualities, has been thoroughly researched (Thawabteh et al. 2019). Recently, their antibacterial activity has been crucial in treating infectious diseases associated with MDR (Khameneh et al. 2015).

β-carboline alkaloids have higher selectivity for the human pathogen *S. aureus,* and their extracts from *Peganum harmala* seeds and roots showed good synergistic effects when co-administered with novobiocin, carbenicillin, and colistin (Darabpour et al. 2011). Two canthine-6-one type alkaloids from *Allium neapolitanum* demonstrated significant activity against MRSA and MDR *S. aureus* (O'Donnell and Gibbons 2007).

5.2 Flavonoids

Plants produce flavonoids, a class of phenolic chemicals, as secondary metabolites. They can be either glycosidic (connected to a sugar molecule, also called glycone) or non-glycosylated (aglycone) (Dias et al. 2021). These polyphenolic chemicals can exert antibacterial activity by inhibiting energy metabolism, nucleic acid assembly, and cytoplasmic membrane function (Donadio et al. 2021). Flavonoids may also overcome antibiotic resistance and increase the effectiveness of current drugs, making them a potential treatment for antibiotic-resistant illnesses (Shamsudin et al. 2022).

Flavonoids have been found to have varying levels of antibacterial activity. Researchers isolated two anti-methicillin-resistant *Staphylococcus aureus* (MRSA) agents from laurel leaves *(Laurus nobilis)*, kaempferol 3-O-α-l-(2′,4′-di-E-p-coumaroyl)-rhamnoside and kaempferol 3-O-α-l-(2′-Z-p-coumaroyl-4′-E-p-coumaroyl)-rhamnoside, which demonstrated superior antibacterial effectiveness. The structural characteristics of the compounds were found to be crucial for their antibacterial effect (Tsuchiya et al. 1996). Several flavonoids were extracted from *Paulownia tomentosa* fruit ethanol extract, showing equivalent antibacterial activity to commercially available medicines (Šmejkal et al. 2008). Dihydromorin and norartocarpetin were isolated from *Artocarpus heterophyllus* and were found to inhibit all tested strains more effectively than norartocarpetin (Septama et al. 2020).

5.3 Terpenoids

Large hydrocarbon groups known as terpenes primarily consist of 5-carbon isoprene (C5H8) units. Although larger chains like diterpenes and triterpenes also occur, the most prevalent terpenes are monoterpenes and sesquiterpenes, which vary in the amount of isoprene units they contain (Swamy et al. 2016). Many terpenes have weaker antibacterial properties (Nazzaro et al. 2013). Terpenoids are derived when terpenes undergo alteration, which includes addition or removal of functional groups (Barbieri et al. 2017). Thus, the antibacterial action of terpenoids is governed by their functional group (Griffin et al. 1999). Among the most researched terpenoids are citronella, geraniol, piperitone, linalyl acetate, thymol, menthol, carvacrol, linalool, and piperitone.

Terpenoids prevent microbial viability by interfering with oxygen intake and oxidative phosphorylation, which are required for energy generation (Shaw and Ingraham 1967). These mechanisms are essential for aerobic microbial growth and cell respiration. The research discovered terpenes that had considerable antibacterial action against specific bacterial strains. Thymol showed significant activity against *Bacillus cereus, Salmonella Typhimurium, Escherichia coli,* and *Staphylococcus aureus*. Eugenol had substantial efficacy against *S. aureus*. Significant effects were demonstrated by other oxygenated terpenes, such as l-carveol, citral, and trans-geraniol, especially against *S. aureus* and *E. coli* (Guimarães et al. 2019). In another study, salvipisone and aethiopinone (diterpenoids) were extracted from *Salvia sclarea* roots (Różalski et al. 2007). Against *S. aureus, Enterococcus faecalis,* and *S. epidermidis,* they exhibited antibacterial and antibiofilm action. The synergistic effects of salvipisone and aethiopinone against MRSA and methicillin-resistant *Staphylococcus epidermidis* (MRSE) were also evaluated when they were taken with antimicrobial medications such as oxacillin, vancomycin, and linezolid (Walencka et al. 2007).

5.4 Tannins

Tannins present in plants and fruits defend against pests and regulate plant development. Traditional medicine uses extracts, rich in tannins, which have potent antibacterial properties. Tannin-loaded nanoparticles and hydrogels have been shown to be effective in antibacterial applications (Bhalodia and Shukla 2011). Tannins are polyphenolic substances which are 500–30 kDa in molecular weights (Serrano et al. 2009). They are divided into two categories: hydrolysable tannins (HT) and condensed tannins (CT). These compounds possess significant antioxidant and antimicrobial properties, making them valuable for medicinal applications.

In a study, the antibacterial properties of tannins isolated from the stem bark and leaves of *Salix alba* were reported. *Staphylococcus aureus, Streptococcus pyogenes, Escherichia coli, Neisseria gonorrhoeae,* and *Shigella sonnei* were among the several bacterial strains that were more susceptible to the effects of bark tannins in methanolic extracts. Higher tannin concentrations and incubation periods resulted in larger inhibition zones, emphasizing their potential as natural antibacterial agents (Javed et al. 2020). Another investigation demonstrated that *Rubus eubatus* (blackberry) extract contains proanthocyanidins (PACs), which can inhibit the action of common oral bacteria, including *Porphyromonas gingivalis, Fusobacterium nucleatum,* and *Streptococcus mutans* (González et al. 2013). Moreover, lingonberry juice (*Vaccinium vitis-idaea L.*) was tested against *S. mutans* and *F. nucleatum*. After extracting its polyphenol-rich components, the anthocyanin and procyanidin fractions demonstrated high antibacterial activity, with the procyanidin fraction being particularly efficient against *S. mutans* (Riihinen et al. 2014).

5.5 Saponins

Saponins are a diverse collection of secondary metabolites characterized by a structure that includes a steroid or triterpenoid aglycone and one or more sugar chains (Moses et al. 2014). They are further classified into 11 major carbon-skeleton-based subgroups (Bailly and Vergoten 2020). These eleven subgroups include lupanes, steroids, lanostanes, cycloartanes, ursanes, tarax asteranes, oleananes, hopanes, dammaranes, and lanostanes (Kregiel et al. 2017). Many saponins are formed by combining glycones and carbohydrates at various attachment sites. Saponins are classified as steroidal or triterpenoid based on their aglycone chemical properties, known as sapogenin (Qasim et al. 2020). Saponin from *Camellia sinensis* (green tea) showed the highest antibacterial activity against *S. aureus* (Gram-positive) and moderate activity against *E. coli and S. enteritidis* (Gram-negative bacteria) (Khan et al. 2018). Many studies have discovered that *Paris polyphylla* is an effective medicinal plant with potent inhibitory effects on multidrug resistance treatment. *P. polyphylla* saponins also exhibit broad-spectrum antibacterial activity against acne-causing bacteria such as *Propionibacterium acnes, Staphylococcus aureus*, and *Staphylococcus epidermidis.* This suggests that *P. polyphylla* saponins have huge prospects in the pathophysiology of anti-acne treatments (Li et al. 2019). The biocidal nature of triterpenoid saponins involves membranolytic properties. The *Gnidia Glauca* (Fresen.) Gilg. leaf saponin fraction inhibited the growth of both Gram-negative pathogens, *X. oryzae* pv. *oryzae* and *X. campestris* pv. *Punicae* (Gowrish et al. 2016). The leaf saponin fraction showed an inhibition zone around the discs against *X. oryzae* and *X. punicae* at 150 mg/mL concentration. The efficacy of the leaf saponin fraction was compared to a positive control tetracycline (Asati and Yadava 2018). The leaf saponin fraction increased outer membrane permeability in both *Xanthomonas* strains, causing disruption of cytoplasmic membranes and cellular leakage (Sannabommaji et al. 2020).

5.6 Glycosides

The word "glycoside" describes a broad range of substances formed by joining sugars with aglycones (non-sugar molecules). Glycosides can be categorized based on their aglycone moiety, glycosidic bond, or glycone (sugar) group. Glycosides are classified according to the type of sugar they contain, such as glucosides (they include glucose), pentosides (they contain pentose sugars like arabinose), and rhamnosides (they contain rhamnose). Sugars like fructose, galactose, and oligosaccharides are examples of natural glycones. It is difficult to classify complex glycosides since they may contain many glycones or aglycones. Glycosides can be categorized based on the kind of glycosidic bond, such as O-linkages, N-linkages, or C-linkages. α- and β-glycosides differ by bond orientation at the anomeric carbon. The third and most interesting division is based on the aglycone moiety, which divides glycosides into categories, such as anthraquinone glycosides, cardiac glycosides, coumarin glycosides, flavonoid glycosides, alcoholic glycosides, and saponin glycosides, among others (Johnson et al. 2021). Numerous microbial strains are susceptible to the strong antibacterial action of glycosides derived from medicinal plants. The root bark of *Ulmus davidiana* contained phenolic glycosides that showed strong anti-*Helicobacter pylori* action (So et al. 2019). Likewise, the growth of *Escherichia coli, Klebsiella pneumoniae, Staphylococcus aureus,* and *Bacillus subtilis* was suppressed by tiliroside derived from the leaves of *Leucosidea sericea.* These glycosides from plants have potential as natural antimicrobials for bacterial infections, and more research is required to improve their therapeutic effectiveness (Pendota et al. 2018).

5.7 Coumarins

Coumarin is a member of the flavonoid class and may be found in both synthetic and natural products (Vilar et al. 2006). Coumarins are heterocyclic α-benzopyrones derived from *Coumarouna odorata auba* (Dipteryx odorata) (Venugopala et al. 2013). Numerous plants, microbes, and essential oils naturally contain it as a secondary metabolite (Hussain et al. 2018). Because of their immense potential, coumarins and their metabolites have proven beneficial in the pharmaceutical sector and are a common component of the basic nucleus of many compounds with therapeutic uses. Simple coumarins, biscoumarins, dihydrofuranocoumarins, pyranocoumarins, phenylcoumarins, and furanocoumarins are the six primary categories of coumarins (Flores-Morales et al. 2023). One of the primary reasons for the rise of coumarin derivative biosynthesis is drug resistance, which has become a major impediment to the identification and treatment of various cancers. Natural and synthesized drugs derived from coumarins have shown substantial potency in inhibiting multidrug resistance (MDR) in cancer patients (Chen et al. 2012). The hydrocarbon alterations in the basic coumarin nucleus give simple coumarins their antibacterial properties. The plant *Fatoua pilosa* contained umbelliferone and various coumarin derivatives, including phelodenol, psoralen, scopoletin, bergaptene, marmesin rutaretin, and xantholein. These compounds, particularly umbelliferone and scopoletin, were shown to be effective against *Mycobacterium tuberculosis* (Cunha et al. 2015). Coumarins isolated from *Dipteryx odorata* have been proven to be active against *Escherichia coli, Staphylococcus aureus,* and *Bacillus subtilis.* By altering bacterial DNA and membrane structures, coumarin derivatives—such as chalcones and 7-hydroxy coumarins—have shown significant antibacterial activity. The efficiency of coumarin compounds is boosted by chemical changes and complex forms, giving them interesting options for treating multidrug-resistant bacteria (Sahoo et al. 2021). Details about the phytochemicals are summarized in Table 7.1.

Table 7.1 Summary of phytochemicals, their mechanism of action, target pathogens, and key effects

Phytochemical	Type	Mechanism of action	Target pathogen	Key effects	References
Galangin	Flavonoid	Reverse Beta-lactam resistance	Staphylococcus aureus	Inhibits resistance mechanisms	Khodabandehloo and Maghsoudlou (2023)
Resveratrol	Polyphenol	Inhibits efflux pumps	Multidrug resistant Gram-negative bacteria	Increases the efficacy of antibiotics when used in combination	Abedini et al. (2021)
Sophoraflavanone G	Flavonoid	Inhibit peptidoglycan cell wall synthesis	Methicillin resistant Staphylococcus aureus (MRSA)	Strong antimicrobial activity against resistant bacteria	Gutierrez et al. (2022)
Berberine	Alkaloid	Damages Cell Membrane	Escherichia coli, Bacillus subtilis, S. aureus, Salmonella	Leakage of intracellular content	Bao et al. (2020)
Thymol	Terpenoid	Increases membrane permeability	Biofilm-embedded microbes and foodborne pathogens	Disrupts respiration and metabolism	Aljelehawy et al. (2023)
Allicin	Organosulfur	Inhibit DNA replication and mitochondrial translation	Fungal pathogens and Gram-negative bacteria	High potent activity against diverse pathogens	Huo et al. (2025)
Cinnamaldehyde	Phenylpropanoid	Disrupts cell membranes and inhibits ATPase	Multiple Gram-negative and Gram-positive strains	Interference with amino acid biosynthesis	Thirapanmethee et al. (2021)
Catechins	Polyphenol	Destabilizes bacterial membranes	Multiple Gram-negative and Gram-positive strains	Leakage of intracellular content	Veiko et al. (2023)
Aegelinol	Coumarin	Block DNA gyrase	Multiple Gram-negative and Gram-positive strains	Strong antibacterial activity at low dosage	Angelini (2024)

6 Exploring the Antibacterial Effects of Phytochemical: Findings from Pre-clinical and Clinical Studies

In the world of science, phytochemicals have caught significant importance for their considerable contributions to the supply of food, herbal medication, cosmetics, and antibacterial products. Additionally, phytochemicals are renowned for holding antimicrobial properties, with significant efficacy in fighting against MDR strains (Kongkham et al. 2020). The restrictive abilities of phytochemicals have made clinical isolates useful. A study conducted by Dahiya and Purkayastha in 2012 significantly proved that they'd tested solvent-based extracts from plants, including lemongrass, neem, aloe vera, oregano, rosemary, thyme, tulsi, and bryophyllum, which fought against MDR bacteria, including both Gram-positive and Gram-negative types. They came up with the conclusion that the extracts with flavonoids and tannins proved to be useful to defend against methicillin-resistant S. aureus. Furthermore, to combat both Gram-negative *(E. coli, S. typhi, Klebsiella* spp., *S. sonnei)* and Gram-positive *(B. subtilis, S. aureus, S. epidermidis)* bacteria, the extracts from *Lawsonia inermis* were significantly beneficial (Purkayastha and Dahiya 2012).

The antibacterial activity of 28 plant extracts and oils against four Gram-negative bacterial species was assessed in a study: *E. coli* O157:H7, *Yersinia enterocolitica* O9, *Proteus* spp., and *Klebsiella pneumoniae.* Furthermore, the antibacterial activity of many extracts and oils was also found, including *Origanum syriacum L., Thymus syriacus Boiss., Syzygium aromaticum L., Juniperus foetidissima Wild, Allium sativum L., Myristica fragrans* Houtt, and *Cinnamomum zeylanicum L.* The least inhibitory concentration (MIC) values of these extracts and oils were discovered to range from 1.5 to 12.5 μl/ml, with *O. syriacum L., T. syriacus Boiss., C. zeylanicum L.,* and *S. aromaticum L.* proven to be the most useful, showing potential as bactericidal agents against Gram-negative bacteria (Al-Mariri and Safi 2014).

7 Challenges Associated with Plant-Based Drug Discovery and Their Possible Solutions

There are many hurdles in the discovery of phytochemical-based drug discovery. First, the extinction of biochemical compounds can be led by the separation process, harder to recognize the active constituents (Nothias et al. 2018). Additionally, during fractional, synergistic compounds might be ignored because of their inability to show biological activity on their own (Caesar and Cech 2019). The complexity of natural mixtures, with several different compounds, can be challenging in the identification of the active chemicals, and the understanding of their interactive ways. The requirement of modern and upgraded tools for the insights of these bioactive compounds and their catalysts, like mass spectrometry, NMR, and molecular networking, could be affordable and accessible (Chen et al. 2020). Strategies include advanced chromatography techniques, molecular networking, mass spectrometry, NMR spectroscopy, metabolomics and chemometrics, machine learning and artificial intelligence, microfluidics and nanotechnology, green chemistry approaches, collaboration and data sharing, and integrating traditional knowledge can help identify and prioritize plants with potential therapeutic value, optimize the separation and identification process, and minimize the use of harsh chemicals. By addressing these challenges, researchers can unlock the full potential of plant-based molecules for therapeutic discovery and development (Zhang et al. 2023). The most prominent can be phytochemicals which target cell membranes with many solutions. These chemicals could result in the disruption of cell membranes, prevention of enzymes, and interference. Recently, a sudden rise in phytochemicals effectiveness can be seen in combating MDR pathogens (Ghosh et al. 2024a) that contributed to the formation of the prevention and of antibiotics side effects. A strong antimicrobial potential can be observed by Chalcone derivatives to combat (MDR strains. A subcategory of flavonoids, known as catechins, is responsible for the loss of bacterial membranes, leakage and weakness of cell membranes can also be a significant cause of potassium, which demonstrates the effectiveness of phytochemicals as a significant antibacterial agent (Xu et al. 2019). The latest development shows the significant advantages for the treatment of breast cancer by phytomedicine and nanotechnology. The herbal plants that can be produced by existing phytochemicals can be canonized for the making of non-herbal medicine with abundant bioavailability, can increase half-life, and minimize the transportation of tumor areas. The nanoformulation proves itself an effective source of defense against breast cancer cells, including MCF-7 and MDA-MB-231 (Sartaj et al. 2021).

The advancement of nanotechnology and the collaboration among molecular biologists, phytochemists, formulators, and clinicians can be seen. Significant efficient nanoformulations are made for the cancer prevention cure, unlocking a new era of herbal chemical discovery (Siddiqui et al. 2022). The phytochemical nanoformulation of eugenol is renowned for its several characteristics of pharmacology and is responsible for water-soluble nano-eugenol with increased bioavailability. The eugenol nanoparticles (ENPs) passed the size which ranges from 20 to 30 nm, and an 80% fast speed with the provision of sustained release over 5 days. The nanoformulation shows superior anti biofilm activity against *Pseudomonas aeruginosa*, in comparison to the number of eugenols, showing significance in the improvement in the treatment level of phytochemicals (Ghosh et al. 2024b). A flavonoid phytochemical, known as the nanonization of chrysin (CHR), can drastically improve the P-glycoprotein (P-gp) mediated drug efflux. Nanonized chrysin (NCH) can increase the intestinal absorption and oral bioavailability of the P-gp probe substrate fexofenadine in rats. The apparent permeability and effective permeability of fexofenadine can be improved by the NCH pretreatment that results in higher Cmax, AUC0-t, and AUC0-∞, and reduction clearance and volume of distribution. The above-mentioned insights provide many suggestions that there is a chance of increment in bioavailability and therapeutic potential by the nanonization of phytochemicals like chrysin (Sashidhar et al. 2022).

Plant-based chemicals can be used in green nanotechnology to create eco-friendly nanomaterials and nanoproducts. This approach integrates principles of green chemistry and green engineering (Nasrollahzadeh et al. 2019), using sustainable materials and operating at lower temperatures (Patra and Baek 2014). Plants like alkaloids, phenolics, and terpenoids have biochemicals that decrease chemicals (Vijayaraghavan and Ashokkumar 2017). The bio-reduction process involves three stages: activation, development, and termination (Md Ishak et al. 2019). Plant-mediated nanoparticle production is a trustworthy, cheaper, and eco-friendly method that reduces the need for toxic solvents and reagents (Castillo-Henríquez et al. 2020).

8 Conclusion

The future exploration and utilization of phytochemicals as antibacterial agents hold great promise amidst rising antibiotic resistance challenges worldwide. By harnessing nature's diversity through focused research efforts, interdisciplinary collaboration, and innovation in formulation technologies, we could pave the way toward developing safe alternatives that integrate seamlessly into existing healthcare paradigms while addressing global public health concerns related to infectious diseases. Continued investment in research will be essential to unlock their full potential as viable therapeutics against bacterial infections that threaten human health today.

References

Abedini E, Khodadadi E, Zeinalzadeh E, Moaddab SR, Asgharzadeh M, Mehramouz B et al (2021) A comprehensive study on the antimicrobial properties of resveratrol as an alternative therapy. Evid Based Complement Alternat Med 2021(1):8866311. https://doi.org/10.1155/2021/8866311

Aghayan SS, Mogadam HK, Fazli M, Darban-Sarokhalil D, Khoramrooz SS, Jabalameli F et al (2017) The effects of berberine and palmatine on efflux pumps inhibition with different gene patterns in Pseudomonas aeruginosa isolated from burn infections. Avicenna J Med Biotechnol 9(1):2

Aljelehawy QHA, Mohammadi S, Mohamadian E, Raji Mal Allah O, Mirzaei A, Ghahremanlou M (2023) Antimicrobial, anticancer, antidiabetic, antineurodegenerative, and antirheumatic activities of thymol: clarification of mechanisms. Micro Nano Bio Aspects 2(1):1–7. https://doi.org/10.22034/mnba.2023.381107.1019

Al-Mariri A, Safi M (2014) In vitro antibacterial activity of several plant extracts and oils against some gram-negative bacteria. IJMS 39(1)

AlSheikh HMA, Sultan I, Kumar V, Rather IA, Al-Sheikh H, Tasleem Jan A et al (2020) Plant-based phytochemicals as possible alternative to antibiotics in combating bacterial drug resistance. Antibiotics 9(8):480. https://doi.org/10.3390/antibiotics9080480

Amirkia V, Heinrich M (2014) Alkaloids as drug leads–a predictive structural and biodiversity-based analysis. Phytochem Lett 10:xlviii–liii. https://doi.org/10.1016/j.phytol.2014.06.015

Angelini P (2024) Plant-derived antimicrobials and their crucial role in combating antimicrobial resistance. Antibiotics 13(8):746. https://doi.org/10.3390/antibiotics13080746

Asati N, Yadava RN (2018) Antibacterial activity of a triterpenoid saponin from the stems of Caesalpinia pulcherrima Linn. Nat Prod Res 32(5):499–507. https://doi.org/10.1080/14786419.2017.1317772

Bailly C, Vergoten G (2020) Proposed mechanisms for the extracellular release of PD-L1 by the anticancer saponin platycodin D. Int Immunopharmacol 85:106675. https://doi.org/10.1016/j.intimp.2020.106675

Baker RE, Mahmud AS, Miller IF, Rajeev M, Rasambainarivo F, Rice BL et al (2022) Infectious disease in an era of global change. Nat Rev Microbiol 20(4):193–205. https://doi.org/10.1038/s41579-021-00639-z

Bao YH, Zhang JS, Fu Q, Zhang HT (2020) Antibacterial activity and mechanism of berberine from the fruit of Berberis poiretii. https://doi.org/10.7506/spkx1002-6630-20190728-372

Barbieri R, Coppo E, Marchese A, Daglia M, Sobarzo-Sánchez E, Nabavi SF, Nabavi SM (2017) Phytochemicals for human disease: an update on plant-derived compounds antibacterial activity. Microbiol Res 196:44–68. https://doi.org/10.1016/j.micres.2016.12.003

Barbosa F, Pinto E, Kijjoa A, Pinto M, Sousa E (2020) Targeting antimicrobial drug resistance with marine natural products. Int J Antimicrob Agents 56(1):106005. https://doi.org/10.1016/j.ijantimicag.2020.106005

Barrows JM, Goley ED (2021) FtsZ dynamics in bacterial division: what, how, and why? Curr Opin Cell Biol 68:163–172. https://doi.org/10.1016/j.ceb.2020.10.013

Bhalodia NR, Shukla VJ (2011) Antibacterial and antifungal activities from leaf extracts of Cassia fistula l.: an ethnomedicinal plant. J Adv Pharm Technol Res 2(2):104–109. https://doi.org/10.4103/2231-4040.82956

Boyd NK, Teng C, Frei CR (2021) Brief overview of approaches and challenges in new antibiotic development: a focus on drug repurposing. Front Cell Infect Microbiol 11:684515. https://doi.org/10.3389/fcimb.2021.684515

Byrne M, Miellet S, McGlinn A, Fish J, Meedya S, Reynolds N et al (2019) Antibiotic use and misuse: the development and investigation of a theory driven community measure. https://doi.org/10.21203/rs.2.12696/v1

Cabello FC (2006) Heavy use of prophylactic antibiotics in aquaculture: a growing problem for human and animal health and for the environment. Environ Microbiol 8(7):1137–1144. https://doi.org/10.1111/j.1462-2920.2006.01054.x

Caesar LK, Cech NB (2019) Synergy and antagonism in natural product extracts: when 1 + 1 does not equal 2. Nat Prod Rep 36(6):869–888. https://doi.org/10.1039/C9NP00011A

Cappiello F, Loffredo MR, Del Plato C, Cammarone S, Casciaro B, Quaglio D et al (2020) The revaluation of plant-derived terpenes to fight antibiotic-resistant infections. Antibiotics 9(6):325. https://doi.org/10.3390/antibiotics9060325

Castillo-Henríquez L, Alfaro-Aguilar K, Ugalde-Álvarez J, Vega-Fernández L, Montes de Oca-Vásquez G, Vega-Baudrit JR (2020) Green synthesis of gold and silver nanoparticles from plant extracts and their possible applications as antimicrobial agents in the agricultural area. Nano 10(9):1763. https://doi.org/10.3390/nano10091763

Charani E, Mendelson M, Ashiru-Oredope D, Hutchinson E, Kaur M, McKee M et al (2021) Navigating sociocultural disparities in relation to infection and antibiotic resistance—the need for an intersectional approach. JAC Antimicrob Resist 3(4):dlab123. https://doi.org/10.1093/jacamr/dlab123

Chen G, Seukep AJ, Guo M (2020) Recent advances in molecular docking for the research and discovery of potential marine drugs. Mar Drugs 18(11):545. https://doi.org/10.3390/md18110545

Chen Y, Liu HR, Liu HS, Cheng M, Xia P, Qian K et al (2012) Antitumor agents 292. Design, synthesis and pharmacological study of S-and O-substituted 7-mercapto-or hydroxy-coumarins and chromones as potent cytotoxic agents. Eur J Med Chem 49:74–85. https://doi.org/10.1016/j.ejmech.2011.12.025

Church NA, McKillip JL (2021) Antibiotic resistance crisis: challenges and imperatives. Biologia 76(5):1535–1550. https://doi.org/10.1007/s11756-021-00697-x

Cunha S, Iunes CEM, Oliveira CC, Santana LLBD (2015) Síntese de ácidos cumarino-3-carboxílicos e sua aplicação na síntese total da aiapina, cumarina e umbeliferona. Química Nova 38(8):1125–1131. https://doi.org/10.5935/0100-4042.20150096

Darabpour E, Bavi AP, Motamedi H, Nejad SMS (2011) Antibacterial activity of different parts of Peganum harmala L. growing in Iran against multi-drug resistant bacteria. EXCLI J 10:252

Dias MC, Pinto DC, Silva AM (2021) Plant flavonoids: chemical characteristics and biological activity. Molecules 26(17):5377. https://doi.org/10.3390/molecules26175377

Donadio G, Mensitieri F, Santoro V, Parisi V, Bellone ML, De Tommasi N et al (2021) Interactions with microbial proteins driving the antibacterial activity of flavonoids. Pharmaceutics 13(5):660. https://doi.org/10.3390/pharmaceutics13050660

Dougan G, Dowson C, Overington J, Participants NGADS (2019) Meeting the discovery challenge of drug-resistant infections: progress and focusing resources. Drug Discov Today 24(2):452–461. https://doi.org/10.1016/j.drudis.2018.11.015

Evans WC, Evans D (2009) Chapter 26-Alkaloids. In: Evans WC, Evans D (eds) Trease and Evans' pharmacognosy, 16th edn, pp 353–415

Fathi F, Sadrnia M, Arjomandzadegan M, Mohajerani HR (2021) In vitro and in vivo evaluation of antibacterial and anti-biofilm properties of five ethnomedicinal plants against oral bacteria by TEM. Avicenna J Phytomed 11(2):180

Fink G, D'Acremont V, Leslie HH, Cohen J (2020) Antibiotic exposure among children younger than 5 years in low-income and middle-income countries: a cross-sectional study of nationally representative facility-based and household-based surveys. Lancet Infect Dis 20(2):179–187. https://doi.org/10.1016/S1473-3099(19)30572-9

Flores-Morales V, Villasana-Ruíz AP, Garza-Veloz I, González-Delgado S, Martinez-Fierro ML (2023) Therapeutic effects of coumarins with different substitution patterns. Molecules 28(5):2413. https://doi.org/10.3390/molecules28052413

Ghosh S, Basu S, Anbarasu A, Ramaiah S (2024a) A comprehensive review of antimicrobial agents against clinically important bacterial pathogens: prospects for phytochemicals. Phytother Res 39. https://doi.org/10.1002/ptr.8365

Ghosh S, Sett U, Pal A, Nandy S, Nandi S, Chakrabarty S, Das A, Bandopadhyay P, Basu T (2024b) Antibiofilm potential of nanonized eugenol against Pseudomonas aeruginosa. J Appl Microbiol 135(1). https://doi.org/10.1093/jambio/lxad305

González OA, Escamilla C, Danaher RJ, Dai J, Ebersole JL, Mumper RJ, Miller CS (2013) Antibacterial effects of blackberry extract target periodontopathogens. J Periodontal Res 48(1):80–86. https://doi.org/10.1111/j.1600-0765.2012.01506.x

Gopu V, Shetty PH (2016) Cyanidin inhibits quorum signalling pathway of a food borne opportunistic pathogen. J Food Sci Technol 53:968–976. https://doi.org/10.1007/s13197-015-2031-9

Gowrish A, Vagdevi HM, Rajashekar H (2016) Antibacterial and antitubercular activity of Gnidia glauca (fresn gilg) Root extracts. Int J Pharma Chem, DOI, p 10

Griffin SG, Wyllie SG, Markham JL, Leach DN (1999) The role of structure and molecular properties of terpenoids in determining their antimicrobial activity. Flavour Fragr J 14(5):322–332. https://doi.org/10.1002/(SICI)1099-1026(199909/10)14:5<322::AID-FFJ837>3.0.CO;2-4

Gu Y, Dong J, Li J, Luo Q, Dong X, Tang G et al (2023) Antibacterial activity and mechanism of sanguinarine against Staphylococcus aureus by interfering with the permeability of the cell wall and membrane and inducing bacterial ROS production. Front Vet Sci 10:1121082. https://doi.org/10.3389/fvets.2023.1121082

Guimarães AC, Meireles LM, Lemos MF, Guimarães MCC, Endringer DC, Fronza M, Scherer R (2019) Antibacterial activity of terpenes and terpenoids present in essential oils. Molecules 24(13):2471. https://doi.org/10.3390/molecules24132471

Gutierrez J, Bakke A, Vatta M, Merrill AR (2022) Plant natural products as antimicrobials for control of Streptomyces scabies: a causative agent of the common scab disease. Front Microbiol 12:833233. https://doi.org/10.3389/fmicb.2021.833233

Hancock RE (2005) Mechanisms of action of newer antibiotics for gram-positive pathogens. Lancet Infect Dis 5(4):209–218. https://doi.org/10.1016/S1473-3099(05)70051-7

Hegab A (2016) Antibiotic apocalypse; are we ready. J Dent Heal Oral Disord Ther 5(3):260–261

Huo T, Zhu B, Shan X, Song J, Wang Z, Zhou J et al (2025) Allicin amplifies disulfidptosis during GOx catalyzing glucose-starvation for cancer therapy via simultaneous antimicrobial and antitumor intervention. Chem Eng J 159871. https://doi.org/10.1016/j.cej.2025.159871

Huremović D (2019) Brief history of pandemics (pandemics throughout history). In: Psychiatry of pandemics: a mental health response to infection outbreak, pp 7–35

Hussain MI, Qamar Abbas S, Reigosa MJ (2018) Activities and novel applications of secondary metabolite coumarins. Planta Daninha 36:e018174040. https://doi.org/10.1590/S0100-83582018360100016

Javed B, Nawaz K, Munazir M (2020) Phytochemical analysis and antibacterial activity of tannins extracted from Salix alba L. against different gram-positive and gram-negative bacterial strains. Iran J Sci Technol Trans A Sci 44(5):1303–1314. https://doi.org/10.1007/s40995-020-00937-w

Joakim Larsson D (2014) Antibiotics in the environment. Ups J Med Sci 119(2):108–112

Johnson JB, Mani JS, Broszczak D, Prasad SS, Ekanayake CP, Strappe P et al (2021) Hitting the sweet spot: a systematic review of the bioactivity and health benefits of phenolic glycosides from medicinally used plants. Phytother Res 35(7):3484–3508. https://doi.org/10.1002/ptr.7042

Kalia M, Yadav VK, Singh PK, Sharma D, Narvi SS, Agarwal V (2018) Exploring the impact of parthenolide as anti-quorum sensing and anti-biofilm agent against Pseudomonas aeruginosa. Life Sci 199:96–103. https://doi.org/10.1016/j.lfs.2018.03.013

Kawano M (2021) Novel antimicrobial method to tackle drug-resistant bacteria. Impact 2021(5):48–50. https://doi.org/10.21820/23987073.2021.5.48

Khameneh B, Eskin NM, Iranshahy M, Fazly Bazzaz BS (2021) Phytochemicals: a promising weapon in the arsenal against antibiotic-resistant bacteria. Antibiotics 10(9):1044. https://doi.org/10.3390/antibiotics10091044

Khameneh B, Iranshahy M, Ghandadi M, Ghoochi Atashbeyk D, Fazly Bazzaz BS, Iranshahi M (2015) Investigation of the antibacterial activity and efflux pump inhibitory effect of co-loaded piperine and gentamicin nanoliposomes in methicillin-resistant Staphylococcus aureus. Drug Dev Ind Pharm 41(6):989–994. https://doi.org/10.3109/03639045.2014.920025

Khan MI, Ahhmed A, Shin JH, Baek JS, Kim MY, Kim JD (2018) Green tea seed isolated saponins exerts antibacterial effects against various strains of gram positive and gram negative bacteria, a comprehensive study in vitro and in vivo. Evid Based Complement Alternat Med 2018(1):3486106. https://doi.org/10.1155/2018/3486106

Khodabandehloo M, Maghsoudlou S (2023) Formulations of galangin and its derivatives with antimicrobial, anticancer, and anti-neurodegenerative effects: recent advances and challenges. Micro Nano Bio Aspects 2(4):41–46. https://doi.org/10.22034/mnba.2023.423432.1051

Khwaif JM, Hayyawi AH, Yousif TI (2010) Cholera outbreak in Baghdad in 2007: an epidemiological study. East Mediterr Health J 16(6):584

Koh G (2009) The canon of medicine (Al-Qanun fi'l-tibb), vol 339. JSTOR

Kongkham B, Prabakaran D, Puttaswamy H (2020) Opportunities and challenges in managing antibiotic resistance in bacteria using plant secondary metabolites. Fitoterapia 147:104762. https://doi.org/10.1016/j.fitote.2020.104762

Kregiel D, Berlowska J, Witonska I, Antolak H, Proestos C, Babic M et al (2017) Chapter 1: Saponin-based, biological-active surfactants from plants. In: Application and characterization of surfactants, vol 6, pp 184–205. https://doi.org/10.5772/68062

Kumari P, Pillai VV, Benedetto A (2020) Mechanisms of action of ionic liquids on living cells: the state of the art. Biophys Rev 12(5):1187–1215. https://doi.org/10.1007/s12551-020-00754-w

Kzhyshkowska J, De Berardinis P, Bettencourt P (2022) Research and development: the past, present and future, including novel therapeutic strategies. In: Fighting an elusive enemy: Staphylococcus aureus and its antibiotic resistance, immune-evasion and toxic mechanisms

Li D, Liu Z, Yuan Y, Liu Y, Niu F (2015) Green synthesis of gallic acid-coated silver nanoparticles with high antimicrobial activity and low cytotoxicity to normal cells. Process Biochem 50(3):357–366. https://doi.org/10.1016/j.procbio.2015.01.002

Li Y, Sun Y, Tang T, Niu Y, Li X, Xie M et al (2019) Paris saponin VII reverses chemoresistance in breast MCF-7/ADR cells. J Ethnopharmacol 232:47–54. https://doi.org/10.1016/j.jep.2018.12.018

Liu Y, Li JJ, Li HY, Deng SM, Jia AQ (2021) Quorum sensing inhibition of hordenine analogs on Pseudomonas aeruginosa and Serratia marcescens. Synth Syst Biotechnol 6(4):360–368. https://doi.org/10.1016/j.synbio.2021.09.010

Lu L, Hu W, Tian Z, Yuan D, Yi G, Zhou Y et al (2019) Developing natural products as potential anti-biofilm agents. Chin Med 14:1–17. https://doi.org/10.1186/s13020-019-0232-2

Mabhiza D, Chitemerere T, Mukanganyama S (2016) Antibacterial properties of alkaloid extracts from Callistemon citrinus and Vernonia adoensis against Staphylococcus aureus and Pseudomonas aeruginosa. Int J Med Chem 2016(1):6304163. https://doi.org/10.1155/2016/6304163

Marshall E (1988) Sverdlovsk: anthrax capital? Soviet doctors answer questions about an unusual anthrax epidemic once thought to have been triggered by a leak from a weapons lab. Science 240(4851):383–385. https://doi.org/10.1126/science.3358121

Masák J, Čejková A, Schreiberová O, Řezanka T (2014) Pseudomonas biofilms: possibilities of their control. FEMS Microbiol Ecol 89(1):1–14. https://doi.org/10.1111/1574-6941.12344

Md Ishak NAI, Kamarudin SK, Timmiati SN (2019) Green synthesis of metal and metal oxide nanoparticles via plant extracts: an overview. Mater Res Express 6(11):112004. https://doi.org/10.1088/2053-1591/ab4458

Miethke M, Pieroni M, Weber T, Brönstrup M, Hammann P, Halby L et al (2021) Towards the sustainable discovery and development of new antibiotics. Nat Rev Chem 5(10):726–749

Moses T, Papadopoulou KK, Osbourn A (2014) Metabolic and functional diversity of saponins, biosynthetic intermediates and semi-synthetic derivatives. Crit Rev Biochem Mol Biol 49(6):439–462. https://doi.org/10.3109/10409238.2014.953628

Murray CJ, Ikuta KS, Sharara F, Swetschinski L, Aguilar GR, Gray A et al (2022) Global burden of bacterial antimicrobial resistance in 2019: a systematic analysis. Lancet 399(10325):629–655. https://doi.org/10.1016/S0140-6736(21)02724-0

Muteeb G, Rehman MT, Shahwan M, Aatif M (2023) Origin of antibiotics and antibiotic resistance, and their impacts on drug development: a narrative review. Pharmaceuticals 16(11):1615. https://doi.org/10.3390/ph16111615

Nasrollahzadeh M, Sajjadi M, Sajadi SM, Issaabadi Z (2019) Green nanotechnology (pp. 145–198). https://doi.org/10.1016/B978-0-12-813586-0.00005-5

Nazzaro F, Fratianni F, De Martino L, Coppola R, De Feo V (2013) Effect of essential oils on pathogenic bacteria. Pharmaceuticals 6(12):1451–1474. https://doi.org/10.3390/ph6121451

Nothias LF, Nothias-Esposito M, Da Silva R, Wang M, Protsyuk I, Zhang Z, Sarvepalli A, Leyssen P, Touboul D, Costa J, Paolini J, Alexandrov T, Litaudon M, Dorrestein PC (2018) Bioactivity-based molecular networking for the discovery of drug leads in natural product bioassay-guided fractionation. J Nat Prod 81(4):758–767. https://doi.org/10.1021/acs.jnatprod.7b00737

O'Donnell G, Gibbons S (2007) Antibacterial activity of two canthin-6-one alkaloids from Allium neapolitanum. Phytother Res 21(7):653–657. https://doi.org/10.1002/ptr.2136

Othman L, Sleiman A, Abdel-Massih RM (2019) Antimicrobial activity of polyphenols and alkaloids in middle eastern plants. Front Microbiol 10:911. https://doi.org/10.3389/fmicb.2019.00911

Overbye KM, Barrett JF (2005) Antibiotics: where did we go wrong? Drug Discov Today 10(1):45–52. https://doi.org/10.1016/S1359-6446(04)03285-4

Patra JK, Baek K-H (2014) Green nanobiotechnology: factors affecting synthesis and characterization techniques. J Nanomater 2014(1). https://doi.org/10.1155/2014/417305

Pendota SC, Aremu AO, Slavětínská LP, Rárová L, Grúz J, Doležal K, Van Staden J (2018) Identification and characterization of potential bioactive compounds from the leaves of Leucosidea sericea. J Ethnopharmacol 220:169–176. https://doi.org/10.1016/j.jep.2018.03.035

Phillips I, Casewell M, Cox T, De Groot B, Friis C, Jones R et al (2004) Does the use of antibiotics in food animals pose a risk to human health? A critical review of published data. J Antimicrob Chemother 53(1):28–52. https://doi.org/10.1093/jac/dkg483

Phipps RK, Shanmugham MS, Christensen LD, Jakobsen TH, Gennip MV, Alhede M et al (2012) Ajoene, a sulfur-rich molecule from garlic, inhibits genes controlled by quorum sensing. Antimicrob Agents Chemother 56(5):2314–2325. https://doi.org/10.1128/2FAAC.05919-11

Purkayastha S, Dahiya P (2012) Phytochemical screening and antimicrobial activity of some medicinal plants against multi-drug resistant bacteria from clinical isolates. Indian J Pharm Sci 74(5):443. https://doi.org/10.4103/0250-474X.108420

Qasim M, Islam W, Ashraf HJ, Ali I, Wang L (2020) Saponins in insect pest control. In: Co-evolution of secondary metabolites, pp 897–924. https://doi.org/10.1007/978-3-319-96397-6_39

Rather IA, Kim B-C, Bajpai VK, Park Y-H (2017) Self-medication and antibiotic resistance: crisis, current challenges, and prevention. Saudi J Biol Sci 24(4):808–812. https://doi.org/10.1016/j.sjbs.2017.01.004

Riihinen KR, Ou ZM, Gödecke T, Lankin DC, Pauli GF, Wu CD (2014) The antibiofilm activity of lingonberry flavonoids against oral pathogens is a case connected to residual complexity. Fitoterapia 97:78–86. https://doi.org/10.1016/j.fitote.2014.05.012

Robbers JE, Speedie MK, Tyler VE (1996) Pharmacognosy and pharmacobiotechnology

Różalski M, Walencka E, Różalska B, Wysokińska H (2007) Antimicrobial activity of diterpenoids from hairy roots of Salvia sclarea L.: Salvipisone as a potential anti-biofilm agent active against antibiotic resistant Staphylococci. Phytomedicine 14(1):31–35. https://doi.org/10.1016/j.phymed.2005.10.008

Sahoo CR, Sahoo J, Mahapatra M, Lenka D, Sahu PK, Dehury B et al (2021) Coumarin derivatives as promising antibacterial agent (s). Arab J Chem 14(2):102922. https://doi.org/10.1016/j.arabjc.2020.102922

Sannabommaji T, Kumar V, Poornima DV, Basappa G, Gajula H, Rajashekar J (2020) Exploration of antibacterial properties of Gnidia glauca (Fresen.) Gilg. Leaf Saponin fraction. Int J Pharm Phyto Res 12(1):66–71. https://doi.org/10.25258/phyto.12.1.7

Sartaj A, Baboota S, Ali J (2021) Exploring the therapeutic potential of nanostructured lipid carrier approaches to tackling the inherent lacuna of chemotherapeutics and herbal drugs against breast cancer. J Drug Deliv Sci Technol 63:102451. https://doi.org/10.1016/j.jddst.2021.102451

Sashidhar RB, Dhurke R, Alikatte K (2022) Effect of nanonized Chrysin on fexofenadine pharmacokinetics mediated by p-glycoprotein in rats. J Adv Sci Res 13(06):79–88. https://doi.org/10.55218/JASR.202213613

Schopf JW, Barghoorn ES, Maser MD, Gordon RO (1965) Electron microscopy of fossil bacteria two billion years old. Science 149(3690):1365–1367. https://doi.org/10.1126/science.149.3690.1365

Septama AW, Jantan I, Panichayupakaranant P, Aluwi MFF, Rahmi EP (2020) Immunosuppressive and antibacterial activities of dihydromorin and norartocarpetin isolated from Artocarpus heterophyllus heartwoods. Asian Pac J Trop Biomed 10(8):361–368. https://doi.org/10.4103/2221-1691.287162

Serrano J, Puupponen-Pimiä R, Dauer A, Aura AM, Saura-Calixto F (2009) Tannins: current knowledge of food sources, intake, bioavailability and biological effects. Mol Nutr Food Res 53(S2):S310–S329. https://doi.org/10.1002/mnfr.200900039

Shamsudin NF, Ahmed QU, Mahmood S, Ali Shah SA, Khatib A, Mukhtar S et al (2022) Antibacterial effects of flavonoids and their structure-activity relationship study: a comparative interpretation. Molecules 27(4):1149. https://doi.org/10.3390/molecules27041149

Shaw MK, Ingraham JL (1967) Synthesis of macromolecules by Escherichia coli near the minimal temperature for growth. J Bacteriol 94(1):157–164. https://doi.org/10.1128/jb.94.1.157-164.1967

Shriram V, Khare T, Bhagwat R, Shukla R, Kumar V (2018) Inhibiting bacterial drug efflux pumps via phyto-therapeutics to combat threatening antimicrobial resistance. Front Microbiol 9:2990. https://doi.org/10.3389/fmicb.2018.02990

Siddiqui B, Anthony V, Pathak K, Singh S, Chandrakar R, Jain D (2022) Review on herbal drugs nanonization for the treatment of breast cancer. Biol Sci 02(03):10.55006/biolsciences.2022.2302

Skariyachan S, Muddebihalkar AG, Badrinath V, Umashankar B, Eram D, Uttarkar A, Niranjan V (2020) Natural epiestriol-16 act as potential lead molecule against prospective molecular targets of multidrug resistant Acinetobacter baumannii-Insight from in silico modelling and in vitro investigations. Infect Genet Evol 82:104314. https://doi.org/10.1016/j.meegid.2020.104314

Šmejkal K, Chudík S, Kloucek P, Marek R, Cvacka J, Urbanová M et al (2008) Antibacterial C-geranylflavonoids from Paulownia tomentosa fruits. J Nat Prod 71(4):706–709. https://doi.org/10.1021/np070446u

So HM, Yu JS, Khan Z, Subedi L, Ko YJ, Lee IK et al (2019) Chemical constituents of the root bark of Ulmus davidiana var. japonica and their potential biological activities. Bioorg Chem 91:103145. https://doi.org/10.1016/j.bioorg.2019.103145

Subramaniam G, Girish M (2020) Antibiotic resistance—a cause for reemergence of infections. Indian J Pediatr 87(11):937–944. https://doi.org/10.1007/s12098-019-03180-3

Swamy MK, Akhtar MS, Sinniah UR (2016) Antimicrobial properties of plant essential oils against human pathogens and their mode of action: an updated review. Evid Based Complement Alternat Med 2016(1):3012462. https://doi.org/10.1155/2016/3012462

Tacconelli E (2017) Global priority list of antibiotic-resistant bacteria to guide research, discovery, and development. https://coilink.org/20.500.12592/khnnff

Thawabteh A, Juma S, Bader M, Karaman D, Scrano L, Bufo SA, Karaman R (2019) The biological activity of natural alkaloids against herbivores, cancerous cells and pathogens. Toxins 11(11):656. https://doi.org/10.3390/toxins11110656

Thirapanmethee K, Kanathum P, Khuntayaporn P, Huayhongthong S, Surassmo S, Chomnawang MT (2021) Cinnamaldehyde: a plant-derived antimicrobial for overcoming multidrug-resistant Acinetobacter baumannii infection. Eur J Integr Med 48:101376. https://doi.org/10.1016/j.eujim.2021.101376

Tsuchiya H, Sato M, Miyazaki T, Fujiwara S, Tanigaki S, Ohyama M et al (1996) Comparative study on the antibacterial activity of phytochemical flavanones against methicillin-resistant Staphylococcus aureus. J Ethnopharmacol 50(1):27–34. https://doi.org/10.1016/0378-8741(96)85514-0

Upadhyay A, Upadhyaya I, Kollanoor-Johny A, Venkitanarayanan K (2014) Combating pathogenic microorganisms using plant-derived antimicrobials: a minireview of the mechanistic basis. Biomed Res Int 2014(1):761741. https://doi.org/10.1155/2014/761741

Vadhana P, Singh BR, Bharadwaj M, Singh SV (2015) Emergence of herbal antimicrobial drug resistance in clinical bacterial isolates. Pharm Anal Acta 6(10):434. https://doi.org/10.4172/2153-2435.1000434

Veiko AG, Olchowik-Grabarek E, Sekowski S, Roszkowska A, Lapshina EA, Dobrzynska I et al (2023) Antimicrobial activity of quercetin, naringenin and catechin: flavonoids inhibit Staphylococcus aureus-induced hemolysis and modify membranes of bacteria and erythrocytes. Molecules 28(3):1252. https://doi.org/10.3390/molecules28031252

Venugopala KN, Rashmi V, Odhav B (2013) Review on natural coumarin lead compounds for their pharmacological activity. Biomed Res Int 2013(1):963248. https://doi.org/10.1155/2013/963248

Vijayaraghavan K, Ashokkumar T (2017) Plant-mediated biosynthesis of metallic nanoparticles: a review of literature, factors affecting synthesis, characterization techniques and applications. J Environ Chem Eng 5(5):4866–4883. https://doi.org/10.1016/j.jece.2017.09.026

Vilar S, Quezada E, Santana L, Uriarte E, Yánez M, Fraiz N et al (2006) Design, synthesis, and vasorelaxant and platelet antiaggregatory activities of coumarin–resveratrol hybrids. Bioorg Med Chem Lett 16(2):257–261. https://doi.org/10.1016/j.bmcl.2005.10.013

Walencka E, Rozalska S, Wysokinska H, Rozalski M, Kuzma L, Rozalska B (2007) Salvipisone and aethiopinone from Salvia sclarea hairy roots modulate staphylococcal antibiotic resistance and express anti-biofilm activity. Planta Med 73(06):545–551. https://doi.org/10.1055/s-2007-967179

Walker W (1965) The Aberdeen typhoid outbreak of 1964. Scott Med J 10(12):466–479. https://doi.org/10.1177/003693306501001203

Weart RB, Lee AH, Chien AC, Haeusser DP, Hill NS, Levin PA (2007) A metabolic sensor governing cell size in bacteria. Cell 130(2):335–347. https://doi.org/10.1016/j.cell.2007.05.043

Williams OJ, Raghavan GSV, Orsat V, Dai J (2004) Microwave-assisted extraction of Capsaicinoids from capsicum fruit. J Food Biochem 28(2):113–122. https://doi.org/10.1111/j.1745-4514.2004.tb00059.x

Xu M, Wu P, Shen F, Ji J, Rakesh KP (2019) Chalcone derivatives and their antibacterial activities: current development. Bioorg Chem 91:103133. https://doi.org/10.1016/j.bioorg.2019.103133

Young M, Craig J (2020) Urgent global action is needed on multi drug-resistant tuberculosis (MDR-TB)–can small cone moxa contribute to a global response? Eur J Integr Med 37(101072). https://doi.org/10.1016/j.eujim.2020.101072

Zhang M, Otsuki K, Li W (2023) Molecular networking as a natural products discovery strategy. Acta Materia Medica 2(2):126–141

Understanding Fungal Pathogens and Their Challenges in Therapy

8

Muhammad Kamran, Muhammad Bilal Habib, Saba Zafar, Adeel Shahid, Muhammad Ali, and Rehan Sadiq Shaikh

Abstract

The increasing incidence of fungal infections has led to higher mortality rates and greater economic burden in low- and middle-income countries (LMIC). The World Health Organization has included fungal species, especially Candida, in its top priority list of critical pathogens. Antimicrobial resistance is a major problem in treating fungal infections, as fungal species develop resistance to commonly used antifungal drugs through various mechanisms, such as efflux pump activation and biofilm formation. Fungal pathogens possess virulence factors,

M. Kamran (✉)
Department of Medical Laboratory Technology, The University of Chenab, Gujrat, Pakistan

Department of Biosciences, COMSATS University Islamabad, Islamabad, Pakistan
e-mail: kamran@ahs.uchenab.edu.pk

M. B. Habib
Department of Biosciences, COMSATS University Islamabad, Islamabad, Pakistan
e-mail: bilalhabib62@gmail.com

S. Zafar
Department of Biochemistry and Biotechnology, The Women University, Multan, Pakistan
e-mail: saba.6498@wum.edu.pk

A. Shahid
University Institute of Medical Laboratory Technology, The University of Lahore, Lahore, Pakistan
e-mail: adeel.shahid@mlt.uol.edu.pk

M. Ali
Department of Pharmacy, The University of Chenab, Gujrat, Pakistan
e-mail: ali@pharm.uchenab.edu.pk

R. S. Shaikh (✉)
Centre for Applied Molecular Biology, The University of Punjab, Lahore, Pakistan
e-mail: rehan.camb@pu.edu.pk

K. Kuca et al. (eds.), *Medicinal Plants and Their Bioactives in Human Diseases*,
https://doi.org/10.1007/978-3-032-01356-9_8

including secretory products, surface molecules, and dimorphic growth, to evade host immune responses and cause infections. The emergence of antifungal resistance has necessitated the exploration of alternative therapeutic strategies. Plant-based bioactive compounds, with their diverse mechanisms of action and limited side effects, have shown potential as antifungal agents. These phytochemicals can disrupt fungal cell membranes, interfere with metabolic pathways, induce oxidative stress, and modulate immune responses. However, the use of plant-based therapeutics faces challenges related to quality control, standardization, and safety. This chapter provides a comprehensive overview of fungal pathogens, challenges in their treatment, and how plant-based therapeutics can redefine treatment options while promoting sustainable and effective antifungal strategies. Future recommendations include the development of novel antifungal agents with different mechanisms of action, the implementation of antifungal combinations, and the improvement of diagnostic methodologies to address the complex nature of the threats posed by fungal pathogens.

Keywords

Antimicrobial Resistance · Bioactive Compounds · Biofilm Formation · Fungal Infections · Immunocompromised · Phytochemicals

1 Introduction

Fungi are the most common pathogens affecting both animals and humans and are becoming a major public health issue. Over the last 5 years, a significant increase in the incidence of endemic and superficial fungal infections has been observed due to recurrent and uncontrolled chronic infections. Fungal infections are transmitted between humans and animals as their spores are present in the soil and air, causing severe infections in more than 300 million individuals of all ages annually. Fungal infections are classified into three types: superficial, mucous, and invasive infections. Mucous fungal infections (MFIs) occur when a fungus infects the mucus membranes of the human tract. Dermatophytes group of mycoses (pathogenic fungi)which have a high economic burden worldwide. These are predominantly cause superficial infections of the nails, skin, and hair in more than one billion individuals (Gnat et al. 2021). Another type of infection is invasive fungal infection (IFI), which occurs when the fungus is located in an organ or tissue through the systemic circulation. IFIs are deadly to individuals, with increased mortality and morbidity rates compared with superficial fungal infections (SFIs) (Fang et al. 2023). *Candida, Pneumocystis, Aspergillus, Cryptococcus, Coccidioides,* and *Histoplasma* species are the major causative agents of IFIs (Bassetti and Righi 2015). Fungal infections pose a substantial risk to humans, especially immunocompromised, diabetic, cancer, AIDS, burn patients, and organ transplant recipients receiving immunosuppressive therapy. Around 1.9 million individuals suffer from IFIs annually, with 1.6 million fatalities due to fungal diseases. Invasive candidiasis

(IC) accounts for 70% of all IFIs, cryptococcosis (20%), and aspergillosis (10%). According to the CDC, hospitalized invasive aspergillosis (IA) in immunocompromised individuals showed up to a 90% mortality rate (Fang et al. 2023). Immunocompromised individuals have a higher risk of IFIs due to impaired defence mechanisms, such as drug-induced neutropenia, dysfunctional cell-mediated immunity, humoral immunodeficiency, invasive devices, or splenectomy (Gnat et al. 2021). Most SFIs and IFIs are easy to detect and diagnose, based on microscopic and clinical results. Diagnostic imaging tools are not required for such infections. However, for most IFIs and a few MFIs, imaging techniques are important and noteworthy requirements for diagnosis. This need is directly related to the increase in the prevalence and complexity of fungal infections due to an increase in immunocompromised individuals resulting from the intensive use of immunosuppressive and chemotherapeutic drugs, organ transplants, and HIV incidence (Ankrah et al. 2016).

2 Virulence Factors and Mechanism of Pathogenicity

Of the approximately 1.5 million fungal species, around 600 are known to be pathogenic to humans. Fungal infections are generally referred to as mycoses. These pathogenic fungi develop virulence factors that can overcome the host immune system, allowing them to survive in harsh host environments. Some fungal pathogens, such as *Coccidioides immitis, Histoplasma capsulatum, Blastomyces dermatitidis*, and *Paracoccidioides brasiliensis*, are ubiquitous and can cause infections in immunocompetent individuals. *Candida albicans, Cryptococcus neoformans*, and *Aspergillus fumigatus* are opportunistic pathogens that cause infections primarily in immunocompromised individuals (Iyalla 2017). Virulence factors are molecules, such as proteins, enzymes, RNAs, and toxins, that contribute to the infection cycle by adhesion, penetration into cells, spreading of infections, and host immune evasion. Previously, virulence factors were categorized as altered pathogenic phenotypes in the mutants. Thus, genes of adherence, signaling, metabolism, and cell cycle are considered virulence factor-encoding genes (González et al. 2011). The virulence factors of the pathogenic fungus have a wide range that can help prevent infection by bypassing the host's defensive mechanisms. These factors range from secretory products, such as enzymes, toxins, and exopolysaccharides, to surface molecules, such as capsules, lipo−/glycoproteins, and lipopolysaccharides (Leitão 2020).

The host immune system features pattern-recognizing receptors (PRRs), such as dectins, mannose-binding lectins (MBL), and toll-like receptors (TLRs), which are located on monocytes, macrophages, dendritic cells, and natural killer cells. These receptors serve as the first line of defense in innate immunity. These cells can recognize specific fungal pathogen-associated molecular patterns (PAMPs), such as β-1,3-glucans and chitin. The binding of PRRs to PAMPs stimulates the phagocytosis of fungi, a respiratory burst of reactive nitrogen and oxygen species (RNS and ROS), and the production of chemokines and cytokines, which initiate a strong inflammatory response that neutralizes pathogens. Additionally, these immune cells may function as antigen-presenting cells (APCs) and activate cell-mediated immunity against fungal infections (Reddy et al. 2022). However, the fungus has developed several survival strategies to evade the host's immunity.

Pathogenic fungi can penetrate the systemic system into cells, reside in host tissues, and develop infections of multiple severities. The pathogenicity mechanism consists of key steps such as body entrance via ingestion, inhalation, or direct skin contact. Adherence of fungal pathogens to host cell receptors is the first step in the development of infection via the adhesion of surface molecules. Fungal cells modify their surface molecules to protect conserved PAMPs so that they cannot be recognized by PRR. Some pathogenic fungi develop biofilms to evade host immune cells and recognize PAMPs (Rajendhran et al. 2023). For example, *C. albicans* can form biofilms on medical devices to enhance its pathogenicity. *C. albicans has* eight genes that are involved in agglutinin-like sequence (ALS) proteins that bind to various host cells, such as endothelial or epithelial cells and collagen (Iyalla 2017). *Pneumocystis jirovecii* produces specific glycoproteins that cover β-1,3-glucans (PAMP) and cell walls without chitin, which evade the host's innate immune cells (Reddy et al. 2022).

Additionally, some fungal species form polysaccharide capsules as virulence factors that negatively affect T lymphocyte and neutrophil activation. *C. neoformans* secretes capsules to protect themselves from phagocytosis. This polysaccharide capsule has been observed in pathogenic *C. neoformans* strains; however, environmental strains are poorly encapsulated. Unencapsulated strains are not pathogenic because they are easily phagocytosed. CAP 59 and 64 genes of *C. neoformans* are involved in capsule formation, which impairs cytokine production and reduces complement activation, leading to decreased leukocyte recruitment to infected sites (Iyalla 2017). Some pathogenic fungi develop different structures to evade the host immune system; for example, *S. schenckii* develops asteroidal bodies, and *C. neoformans* develops titan cells (Reddy et al. 2022). Another virulence factor is the dimorphic growth of some pathogenic fungi, which can convert from a non-pathogenic form to a pathogenic form. For example, C. albicans grows in the yeast form (non-pathogenic), readily develops hyphae, and causes infection when it finds a suitable environment. Other dimorphic fungi include *C. immitis, B. dermatitidis, P. brasiliensis,* and *H. capsulatum* (Aldejohann et al. 2022). *C. posadasii* and *C. immitis* convert their morphology from arthroconidia to reproductive-form spherules that show tolerance to RNS, and inhibit phagocytic activity through the production of urease and arginase enzymes (Diep and Hoyer 2020).

Some pathogenic fungal species secrete mycotoxins to suppress host defense mechanisms. These mycotoxins can damage host cells and tissues and impair cellular processes, leading to inflammation and immune suppression. Fungal species secrete immunomodulatory molecules that resist the host's immune system (Rajendhran et al. 2023). *A. fumigates* produces gliotoxins and aflatoxins. Aflatoxins are carcinogenic and hepatotoxic, whereas gliotoxins are immunosuppressive agents that inhibit macrophages to phagocytose and activate T lymphocytes. These mycotoxins also reduce the ciliary movement of epithelial cells, thereby interfering with fungal clearance mechanisms. Thermotolerance is another virulence factor of fungal pathogens. Invasive fungal species can grow at body temperature (37 °C) but may be able to cause infections at high temperatures (38–42 °C). For example, *A. fumigatus* is thermophilic and can grow at temperatures of up to 55–77 °C. it was considered in previous studies that protein HSP 70 (heat shock protein 70) is involved in the thermotolerance of fungal species (Chen et al. 2020).

Macrophages and neutrophils produce respiratory bursts of ROS and RNS in response to fungal infections, which damage fungal cells via lipid peroxidation and cause nucleic acid damage. Pathogenic fungi secrete enzymes that protect themselves against oxidative damage. Catalases are secreted to inhibit oxidative damage. *C. albicans* secretes HSP and superoxide dismutase (SOD) to bypass reactive oxygen and nitrogen species. *C. neoformans* uses zinc, copper, and peroxidase to inhibit oxidative damage (Kronstad et al. 2011). Most pathogenic fungi secrete degradative enzymes as virulence factors for the evasion of host immune responses and are involved in pathogenicity. These enzymes damage the host cells and tissues. Pathogenic *Candida* species secrete more phospholipases than commensal strains such as *C. albicans*, which produce extracellular lipases, phospholipases, and proteases. Phospholipases A, B, C, and D break ester bonds and are important for nutrient uptake and iron acquisition (Schaller et al. 2005). *C. albicans* also secretes aspartyl proteinases that hydrolyze the coagulation factors, extracellular matrix, IgA, mucin, and complement proteins. *C. neoformans* invade the brain and spinal cord by producing urease (Naglik et al. 2003) (Fig. 8.1).

3 Antimicrobial Resistance in Fungal Infections

With the emergence of more resistant fungi, the number of antibiotic-resistant fungal infections has increased worldwide. Antimicrobial resistance (AR) occurs when bacteria and fungi can resist the effects of medications intended to eradicate them. Antifungals are a class of antimicrobials used to treat fungal infections (Centers for Disease Control and Prevention 2023). Among the most significant discoveries in the twentieth century, antibiotics and antifungals have prevented infectious diseases from killing millions of people. As their usage and abuse have increased over time, microbes have evolved to acquire antimicrobial resistance (AMR) to numerous medications due to strong selection pressure. Both inside and outside of healthcare facilities, human–human interfaces are the main way that AMR is acquired and transmitted (Salam et al. 2023). The majority of fatalities are caused by organisms belonging to the genera Candida, Aspergillus, and Cryptococcus, and fewer than 100 of the estimated five million fungal species are common causes of human illness (Brown 2019). Stable and heritable point mutations or insertions/deletions that directly affect the interaction of a drug with its target are typically the cause of antifungal drug resistance (Lee et al. 2023). Genetic mutations and triggered defense systems are the causes of resistance. In contrast to bacteria, fungi do not show to exhibit rapid dissemination of plasmid-mediated resistance. Nevertheless, the development of aneuploidy or other types of copy number variation, as well as the

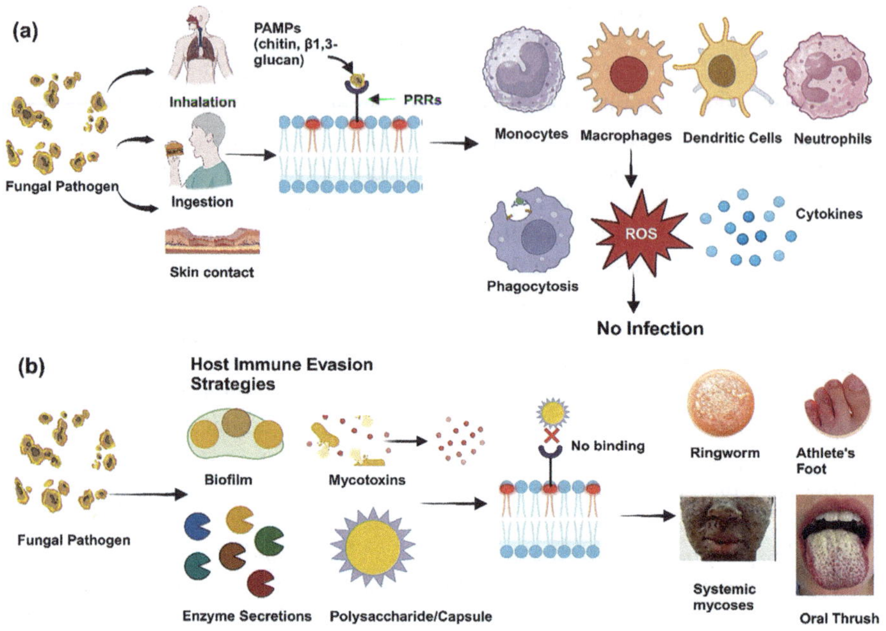

Fig. 8.1 (**a**) Fungal pathogens enter the human body and are recognized by PRRs on the surfaces of innate immune cells that activate the phagocytosis of fungal pathogens and the production of cytokines and ROS results in fungal cell death. (**b**) Fungi develop evasion strategies, such as biofilm formation, secretion of enzymes and mycotoxins, and formation of capsules that cause infections. (Illustration was generated using BioRender)

stimulation of protective stress response pathways, can result in rapid acquisition of antifungal resistance and tolerance. Therefore, treating medical, veterinary, and agricultural hosts becomes extremely difficult when fungal strains and species with one or more medication resistance profiles appear (Fisher et al. 2022) (Table 8.1).

According to the US Centers for Disease Control and Prevention (CDC), fungi and other microbial diseases have flexible genomes, short generation times, and the capacity to adapt to their ecological surroundings, which are full of potentially harmful substances that apply intense selection pressures. When antibiotic and antifungal resistance are presented together, many policymakers may be more inclined to consider these two issues (United States Centers for Disease Control 2022). Aspergillus fumigatus was added to the US CDC's "watch list" of antibiotic-resistant dangers, while C. auris was designated as an "urgent threat." Therefore, despite being fungi, these pathogens were added to policy bodies such as the Presidential Advisory Council for Combating Antibiotic-Resistant Bacteria. Owing to their classification, both bacteria were subsequently granted special funding and added to the Antibiotic-Resistant Laboratory Network. As they were exposed to other significant dangers, these diseases received greater attention and resources, although they might not have received as much attention when communicating alone. Similar instances worldwide can be observed in the inclusion of fungal

Table 8.1 Antifungal classes, mechanism of action, and therapeutic activity

Antifungal	Year	Action	Potential therapeutic activity against	Reference
Polyenes	1950	They alter the permeability of the membrane and disturb the membrane potential by targeting the special fungal membrane component ergosterol.	*Aspergillus* spp., *Candida* spp., *Cryptococcus* spp., *Rhizopus* spp., *Fusarium* spp	Carolus et al. (2021)
Echinocandins	1980s and 1990s	In order to damage the fungal cell wall and cause cell death, they target β(1,3)-glucan, a crucial component of the fungal cell wall.	Candida and Aspergillus species	Hector (1993)
Pyrimidine analogs	1982	Affects the fungal cell by interfering with pyrimidine metabolism, RNA, DNA, and protein synthesis.	*Candida, Cryptococcus neoformans*	Defever et al. (1982)
Olorofim	2000–2010	Inhibits the fungal dihydroorotate dehydrogenase (DHODH) enzyme, which is found in mitochondria and is involved in the synthesis of pyrimidines, hence preventing fungal development.	Histoplasma and Coccidioides spp. or hyaline hyphomycetes such as *Aspergillus* spp	Wiederhold (2021)
Fosfomanogepix	2014–2020	Inhibits the fungal enzyme Gwt1 at the endoplasmic reticulum, which stops glycosylphosphatidylinositol (GPI)-anchored proteins from maturing.	*Candida* and *Aspergillus* spp	Shaw and Ibrahim (2020)

diseases in programs such as the Global Antimicrobial Resistance Surveillance System (GLASS) of the World Health Organization (World Health Organization 2022). Certain novel drugs belong to well-established classes that may have benefits over those currently available in the market. These include (1) encoded amphotericin B, which is taken orally; (2) rezafungin, an echinocandin with a long half-life that may allow for less frequent intravenous administration; (3) oteseconazole, a tetrazole whose structure may be more specific for fungal lanosterol 14α-demethylase than human cytochrome P450 enzymes, resulting in fewer drug–drug interactions; and (4) opelconazole, a triazole made especially for inhaled administration. (Hoenigl et al. 2021). The new antifungal agents used in the clinical trial are shown in Fig. 8.2.

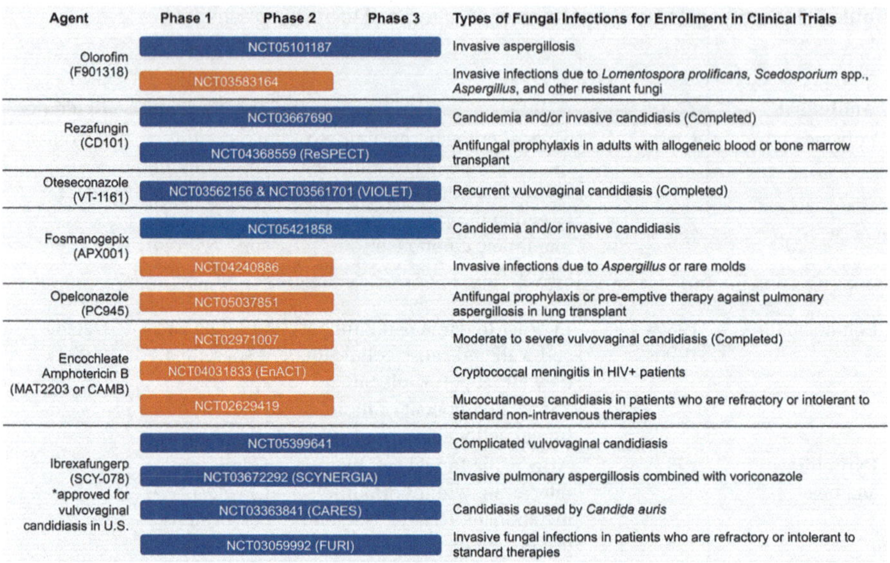

Fig. 8.2 New antifungal drugs in the clinical pipeline (Gow et al. 2022)

4 Overview of Plant-Based Therapeutics

Owing to the ongoing global emergence of fungal infections and the insufficiency of current conventional antifungal medications, there has been increased interest in plant-mediated therapeutics against fungal pathogens in recent years. Candida spp. are fungal pathogens that can have major negative effects, especially in patients with compromised immune systems. The rise in cases of antifungal resistance has highlighted the need to investigate alternative therapeutic approaches, such as plant-based remedies. Recent studies have shown that plant extracts can be effective against fungi. For instance, a study conducted by Kumar examined the efficacy of *Eichhornia crassipes* and *Boerhavia diffusa* against *Candida tropicalis* and *C. albicans*, suggesting that these plants could be employed as effective natural antifungal agents in the future. Similarly, Srikishen and Subramanian described alcoholic extracts of *Batis maritima*, which have been used traditionally for various illnesses in Puerto Rico and Mexico, as having antifungal properties (Srikishen and Subramanian 2022). These findings indicate that traditional knowledge systems are effective tools for selecting plants that are useful for managing fungal infections. The production of these plant extracts and their antifungal activities occurs via several pathways. Flavonoids, terpenoids, and alkaloids found in these plants contribute to the integrity of fungal cell membranes, inhibit enzymatic activity, and hinder the production of essential biomolecules. An additional advantage of synergistic interactions between these compounds is their potential to address the problem of antifungal resistance. However, it is important to note that the use of natural products to manage disorders has both advantages and disadvantages. Contamination and adulteration of herbal products have become significant concerns, primarily owing to issues of quality and safety. Alfaifi et al. (2020) identified several risks associated with such products; some products contain heavy metals and microbial contaminants that may reduce their effectiveness (Alfaifi et al. 2020). Therefore, it is crucial to establish stringent quality control measures and standardization protocols for herbal formulations to ensure their safety and effectiveness in clinical applications.

5 Mechanism of Action of Phytochemicals Against Fungal Pathogens

Interference with the fungal cell membrane is likely the most common pathway through which phytochemicals exhibit antifungal activity. For example, researchers have demonstrated that quercetin and rutin interact with key enzymes in fungal cells, resulting in cell membrane disruption and increased porosity (Khanzada et al. 2021). This disruption causes intracellular contents to be released into the extracellular environment, ultimately leading to cell death. Furthermore, phytochemicals from *Piper crocatum* reduce the activity of the ergosterol biosynthesis pathway enzyme lanosterol 14α-demethylase (Siswina et al. 2022). Inhibition of this enzyme affects membrane formation and induces morphological alterations in fungal cells, rendering them more vulnerable to antifungal agents. Phytochemicals can inhibit fungal metabolic processes by blocking or altering fungal enzymes. For example, flavonoids and phenolic compounds have been shown to suppress the activity of several enzymes involved in fungal energy metabolism and macromolecule biosynthesis (Mondal et al. 2022). This inhibition may lead to structural pathological changes in fungal mitochondria, significantly reducing the viability of the organism. Mondal et al. (2022) revealed that phytochemicals have the potential to suppress fungal growth and invasiveness by targeting specific metabolic pathways (Mondal et al. 2022).

Oxidative stress is another apoptosis-related process that is involved in the antifungal effects of phytochemicals. Many phytochemicals, including those isolated from *Allium sativum* (garlic), promote the generation of reactive oxygen species (ROS) within fungal cells, leading to oxidative stress and cell death (Sharma and Kadosh 2023). Oxidative stress can interfere with various cellular processes, such as DNA and protein synthesis, thereby enhancing the antifungal efficacy of these substances. In addition to their role in mobilizing processes to eradicate fungi, phytochemicals stimulate the immune response to fungal infections. For instance, some phytochemicals have demonstrated immunomodulatory properties that enhance the activity of immune cells against fungal pathogens that need to be eliminated (Adnan 2023). Together, these findings the direct antifungal effects alongside immune activation suggest that phytochemicals play a valuable role in the formulation of new antifungal drugs. The synergistic action of phytochemicals against fungal pathogens is promising for the development of effective antifungal agents. These compounds act as antifungal agents by disrupting cell membranes, interfering with metabolic

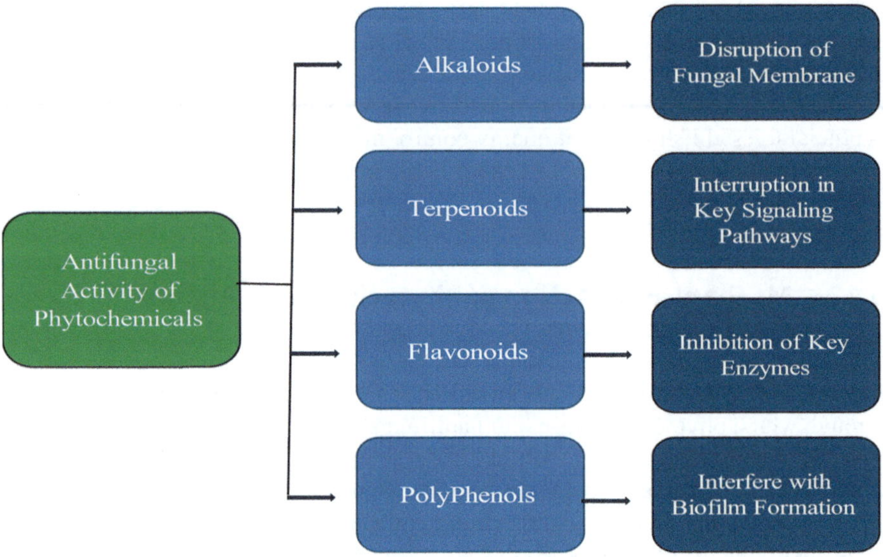

Fig. 8.3 Antifungal activity of different components of phytochemicals

pathways, inducing oxidative stress, and modulating immune responses. Therefore, they represent a potential source of new antifungal drugs, particularly given the increasing emergence of antifungal resistance. The overall effect of phytochemicals on fungal cells is illustrated below (Fig. 8.3).

6 Limitations and Future Recommendations

Antifungal agents have often become ineffective against fungi and fungicides due to the emergence of antifungal resistance, as exemplified by strains such as *C. auris*. Resistance may arise from alterations in target enzymes, increased expression and activity of efflux pumps, and development of biofilms that protect fungal cells from medications. Although standard antifungal agents exhibit broad-spectrum activity, they are ineffective against certain fungal species. Conventional chemotherapy presents challenges due to its toxicity and associated side effects, which pose greater risks for the elderly or immunocompromised individuals. Variations in pharmacokinetics among patients can lead to adverse effects and suboptimal treatment outcomes. The absence of rapid diagnostic tests, combined with normal tissue colonization through biofilm formation on medical devices and tissues, further complicates the treatment.

Currently available antifungal drugs are effective in various clinical situations; however, issues persist regarding their narrow-spectrum activity, toxicity, and capacity to develop resistance against fungal pathogens. Therefore, effective strategies for the prevention and control of these infections are critical. This chapter elaborates on the distinct profiles of phytochemicals derived from plants, which possess significant potential as additional or alternative treatments for fungal diseases. These natural products can influence fungal cell membrane function, inhibit critical metabolic pathways, and increase the production of reactive oxygen species, thereby enhancing the antifungal effects. Phytochemicals have potential applications in clinical settings; however, further research is required to elucidate the mechanisms underlying their effects and assess their safety and efficacy. Recent developments in antifungal management have necessitated the exploration of novel targets for antifungal agents with distinct mechanisms of action, implementation of antifungal combinations, and improvement of diagnostic methodologies. Furthermore, the application of pharmacogenomics in genomic medicine can enhance antifungal therapy for affected patients. Additional prevention strategies include improving community awareness of basic infection prevention practices and monitoring resistance to agents used in the treatment of fungal infections. Current analyses indicate that the intricate issue of fungal pathogens cannot be addressed solely through traditional methods; thus, new approaches and techniques are required. To maximize the effectiveness of new antifungal drugs, it is essential to utilize phytochemicals for drug development, advance academic research on antifungal agents, and implement optimal public health management strategies to address the threat of antifungal resistance.

7 Conclusion

Most antifungal drugs are effective, but their administration requires ongoing consideration of several limitations, including a restricted spectrum of action, potential adverse effects, and susceptibility of fungal strains to resistance. Therefore, there is an urgent need for effective methods to treat and prevent these diseases. This chapter focuses on supplementary or alternative treatments and provides information on the profiles of phytochemicals that are likely to be utilized in the fight against fungal illnesses. One important finding is that these natural substances affect the function of the fungal cell membrane, interfere with essential metabolic processes, increase oxidative stress, and thus, enhance antifungal potential. However, in clinical practice, phytochemicals require well-defined mechanisms of action, effectiveness, and safety, which necessitates further research. New conceptual advancements in antifungal management plans should include identifying novel molecules with different mechanisms of action, administering antifungals in combination, and improving diagnostic tests. Additionally, concepts such as pharmacogenomics in genomic medicine can enhance antifungal therapy. It is also crucial to search for microbes resistant to azoles and raise public awareness of standard infection prevention procedures. In summary, this study aimed to demonstrate that the complex nature of the dangers posed by fungal pathogens indicates that a diverse strategy, encompassing both conventional and novel techniques, is the only viable approach to addressing these threats. Incorporating phytochemicals into medicinal formulations, advancing scholarly research on antifungal compounds, and maximizing public health efforts in the fight against fungal resistance are essential for fully realizing the benefits of novel antifungal medications.

References

Adnan M, Siddiqui AJ, Ashraf SA, Bardakci F, Alreshidi M, Badraoui R et al (2023) Network pharmacology, molecular docking, and molecular dynamics simulation to elucidate the molecular targets and potential mechanism of *Phoenix dactylifera* (ajwa dates) against candidiasis. Pathogens 12(11):1369

Aldejohann AM, Wiese-Posselt M, Gastmeier P, Kurzai O (2022) Expert recommendations for prevention and management of *Candida auris* transmission. Mycoses 65(6):590–598

Alfaifi FF, Alghamdi AM, Al-omarani B, Dawaji EH, Aljabry H, Al-Gayyar MMH (2020) Comprehensive study of scientific evidence and potential risk of herbal medicine use for body weight reduction in North West Saudi Arabia. Cureus 12(10):e11028

Ankrah AO, Sathekge MM, Dierckx RA, Glaudemans AW (2016) Imaging fungal infections in children. Clin Translat Imaging 4:57–72

Bassetti M, Righi E (2015) Overview of fungal infections—the Italian experience. Semin Respir Crit Care Med 36(5):796–806

Brown O (2019) The feasibility of a text-delivered intervention to improve dietary habits, stress management behaviors and create awareness of food assistance resources among college students. Ann Rev Res 51(7):S4–S5

Carolus H, Pierson S, Muñoz JF, Subotić A, Cruz RB, Cuomo CA, Van Dijck P (2021) Genome-wide analysis of experimentally evolved *Candida auris* reveals multiple novel mechanisms of multidrug resistance. MBio 12(2):e03333–e03320

Centers for Disease Control and Prevention (2023) Antimicrobial-resistant fungi. CDC. https://www.cdc.gov/fungal/antimicrobial-resistant-fungi/index.html

Chen J, Tian S, Han X, Chu Y, Wang Q, Zhou B, Shang H (2020) Is the superbug fungus really so scary? A systematic review and meta-analysis of global epidemiology and mortality of *Candida auris*. BMC Infect Dis 20:1–10

Defever KS, Whelan WL, Rogers AL, Beneke ES, Veselenak JM, Soll DR (1982) *Candida albicans* resistance to 5-fluorocytosine: frequency of partially resistant strains among clinical isolates. Antimicrob Agents Chemother 22:810–815

Diep AL, Hoyer KK (2020) Host response to *Coccidioides* infection: fungal immunity. Front Cell Infect Microbiol 10:581101

Fang W, Wu J, Cheng M, Zhu X, Du M, Chen C et al (2023) Diagnosis of invasive fungal infections: challenges and recent developments. J Biomed Sci 30(1):42

Fisher MC, Alastruey-Izquierdo A, Berman J, Bicanic T, Bignell EM, Bowyer P et al (2022) Tackling the emerging threat of antifungal resistance to human health. Nat Rev Microbiol 20(9):557–571

Gnat S, Łagowski D, Nowakiewicz A, Dyląg M (2021) A global view on fungal infections in humans and animals: infections caused by dimorphic fungi and dermatophytoses. J Appl Microbiol 131(6):2688–2704

González R, Ballester I, López-Posadas R, Suárez MD, Zarzuelo A, Martínez-Augustin O, Medina FSD (2011) Effects of flavonoids and other polyphenols on inflammation. Crit Rev Food Sci Nutr 51(4):331–362

Gow NA, Johnson C, Berman J, Coste AT, Cuomo CA, Perlin DS et al (2022) The importance of antimicrobial resistance in medical mycology. Nat Commun 13(1):5352

Hector RF (1993) Compounds active against cell walls of medically important fungi. Clin Microbiol Rev 6:1–21

Hoenigl M, Sprute R, Egger M, Arastehfar A, Cornely OA, Krause R et al (2021) The antifungal pipeline: Fosmanogepix, Ibrexafungerp, Olorofim, Opelconazole, and Rezafungin. Drugs 81:1703–1729

Iyalla C (2017) A review of the virulence factors of pathogenic fungi. Afr J Clin Exp Microbiol 18(1):53–58

Khanzada B, Akhtar N, Okla MK, Alamri SA, Al-Hashimi A, Baig MW et al (2021) Profiling of antifungal activities and in silico studies of natural polyphenols from some plants. Molecules 26(23):7164

Kronstad JW, Attarian R, Cadieux B, Choi J, D'Souza CA, Griffiths EJ, Geddes JMH, Hu G, Jung WH, Kretschmer M, Saikia S, & Wang J (2011) Expanding fungal pathogenesis: Cryptococcus breaks out of the opportunistic box. Nature Reviews Microbiology 9(3): 193–203.

Lee SH, Park SY, Lee JS, Oh J (2023) Infection of the cavernous sinus caused by *Candida auris*. Korean J Healthc Assoc Infect Control Prev 28(1):167–171

Leitão JH (2020) Microbial virulence factors. Int J Mol Sci 21(15):5320

Mondal S, Santra S, Uddin H, Pal K, Halder SK, Chattopadhyay S, Mondal KC (2022) Application of phytochemicals to combat fungal pathogens of pulses: an approach toward inhibition of fungal propagation and invasin activity. J Agric Food Chem 70(25):7662–7673

Naglik JR, Challacombe SJ, Hube B (2003) *Candida albicans* secreted aspartyl proteinases in virulence and pathogenesis. Microbiol Mol Biol Rev 67(3):400–428

Rajendhran J, Yun Y, Tang W, Li Y (2023) Molecular mechanism in the development and pathogenesis of fungi. Front Fungal Biol 4:1231925

Reddy GKK, Padmavathi AR, Nancharaiah Y (2022) Fungal infections: pathogenesis, antifungals and alternate treatment approaches. Curr Res Microb Sci 3:100137

Salam MA, Al-Amin MY, Salam MT, Pawar JS, Akhter N, Rabaan AA, Alqumber MA (2023) Antimicrobial resistance: a growing serious threat for global public health. Healthcare 11(13):1946

Sharma C, Kadosh D (2023) Perspective on the origin, resistance, and spread of the emerging human fungal pathogen *Candida auris*. PLoS Pathog 19(3):e1011190

Shaw KJ, Ibrahim AS (2020) Fosmanogepix: a review of the first-in-class broad spectrum agent for the treatment of invasive fungal infections. J Fungi 6:239

Siswina T, Miranti Rustama M, Sumiarsa D, Kurnia D (2022) Phytochemical profiling of *Piper crocatum* and its antifungal mechanism action as lanosterol 14 alpha demethylase CYP51 inhibitor: a review. F1000Res 11:1115

Srikishen A, Subramanian NG (2022) Antifungal activity of alcoholic extracts of *Batis maritima* against animal pathogens. Int J Biol Pharm Sci Arch 3(2):121–126

Schaller M, Borelli C, Korting HC & Hube B (2005) Hydrolytic enzymes as virulence factors of Candida albicans. Mycoses 48(6): 365–377.

United States Centre for Disease Control (2022) 2019 AR threats report. https://www.cdc.gov/drugresistance/pdf/threats-report/2019-ar-threats-report-508.pdf

Wiederhold NP (2021) Review of T-2307, an investigational agent that causes collapse of fungal mitochondrial membrane potential. J Fungi 7(2):2

World Health Organization (2022) Global Antimicrobial Resistance Use and Surveillance System (GLASS). https://www.who.int/initiatives/glass

Synergistic Effects of Medicinal Plants with Conventional Antibiotics Against Bacterial and Fungal Infections

9

Vikas Kumar, Nidhi Gupta, Nitin Sharma, and Kamal Dev

Abstract

Over the years, antimicrobial resistance (AMR) and multidrug resistance (MDR) have become a global threat to universal health. Emergence of AMR and MDR has led to a decline in the efficacy of conventional antibiotics against bacterial and fungal infections. The current chapter investigates the prospective of medicinal plants and their phytocompounds as synergistic agents that enhance the activities of the currently employed antibiotics. Medicinal plants produce secondary metabolites, including flavonoids, alkaloids, and terpenoids, and exhibit multiple mechanisms of synergy, including disruption of microbial membranes, inhibition of resistance-related enzymes and efflux pumps, and suppression of biofilm formation. Combining these natural agents with antibiotics not only enhances antimicrobial efficacy but also lowers the minimum inhibitory concentrations (MICs), delays resistance development, and minimizes adverse drug effects. Evidence from numerous studies demonstrates the effectiveness of such combinations against various MDR pathogens, including *Acinetobacter baumannii, Pseudomonas aeruginosa*, *Staphylococcus aureus*, *Escherichia coli*, *Neisseria gonorrhoeae*, and *Candida* species. This chapter highlights molecular mechanisms, examples of successful plant–antibiotic combinations, and therapeutic implications. Despite promising results, challenges such as standardization, clinical validation, and safety concerns remain. Future perspectives emphasize the

V. Kumar (✉)
University Institute of Biotechnology, Chandigarh University, Gharuan, Mohali, Punjab, India

N. Gupta · N. Sharma
Department of Biotechnology, Chandigarh Group of Colleges, Landran, Mohali, Punjab, India

K. Dev
Department of Pharmacology and Toxicology, Wright State University, Dayton, OH, USA

K. Kuca et al. (eds.), *Medicinal Plants and Their Bioactives in Human Diseases*,
https://doi.org/10.1007/978-3-032-01356-9_9

need for in vivo studies, clinical trials, and regulatory frameworks to ensure the safe integration of plant-based adjuvants in antimicrobial therapy. Overall, synergistic use of medicinal plants and antibiotics offers a sustainable and innovative approach to combat resistant infections.

Keywords

Antimicrobial resistance · Multidrug resistance · Medicinal plants · Synergy · Antibiotics

1 Introduction

Antibiotics have been vital in lowering morbidity and death from many different microbial illnesses since the first successful manufacture of the first antibiotic penicillin in 1943 (Gaynes 2017; Keenan et al. 2018). In the year 2023, global human consumption for antibiotics was estimated to be 49.3 billion defined daily doses (DDDs), and it is estimated to increase to 75.1 billion DDDs by 2030 (Klein et al. 2024). AMR has become a danger to the world health system in the last two decades, primarily due to the improper utilization or excessive consumption of antibiotics by individuals, thus reducing the effectiveness of conventional antibiotics in combating bacterial and fungal infections (Ferri et al. 2017). Antibiotic overuse and misuse have accelerated the development of antibiotic-resistant strains, requiring advanced approaches to restore the efficacy of these treatments. According to a WHO report (2023), annually, 4.95 million deaths occur linked with AMR globally. In 2021, bacterial AMR caused 1.14 million deaths and 4.71 million deaths were associated with bacterial AMR globally and by 2050, globally bacterial AMR deaths and associated with AMR deaths could reach 1.91 million and 8.22 million annually respectively, with the highest burden in Caribbean, South Asia and Latin America (Naghavi et al. 2024). It is predicted that by the year 2050, drug-resistant infections could cause two million annual fatalities, mostly to people over 70 years old, marking a 70% increase in mortality (Naddaf 2024). The incidence of multidrug resistance (MDR) is also increasing all over the world. MDR pathogens delay effective treatment, increasing mortality, hospitalization period, and medical expenses (Kalın et al. 2023).

The use of natural compounds alongside antibiotics can be one of the major approaches to combat increased levels of antibiotic resistance. This approach increases or preserves the performance of those antibiotics that are currently non-effective because of the existence of bacterial or fungal resistance, thus offering ways by which resistance mechanisms can be overcome and treatment efficacy might be better (Sanhueza et al. 2017). In this context, medicinal plants, known for their diverse bioactive compounds, have garnered significant attention as a complementary strategy in infectious disease management. Plants have been inherent to ancient medicinal practices because of the antimicrobial, antidiabetic, anti-inflammatory, antioxidant, and immunomodulatory properties possessed by them

(Shahrajabian and Sun 2023). Plants undergo abiotic stress such as drought stress, salinity stress, temperature stress, and pollutant stress as well as biotic stress from pests, herbivores, nematodes, or weeds (Nawaz et al. 2023). Plants initiate immune mechanisms, enhance cell walls and synthesize lytic enzymes and pathogenesis-related proteins to protect itself (Nishad et al. 2020; Kaur et al. 2022). Secondary metabolites produced by plants, such as polyphenols, alkaloids, terpenoids, and flavonoids, are useful for plants defence mechanisms and serve as a potential source of therapeutic compounds with antimicrobial and other medicinal values for humans (Dehghan et al. 2016; Allemailem 2021). Several studies have shown that phytocompounds from plants can improve the efficacy of conventional antibiotics by showing synergistic interactions (Angelini 2024; Lorca et al. 2024). These benefits not only increase the antibacterial and antifungal efficacy of antibiotics but also help to lower the necessary doses, hence lowering the probability of the development of microbial resistance and adverse side effects. Focusing on their combination modes of action against bacterial and fungal infections, this chapter investigates the possibility of medicinal plants working in concert with conventional antibiotics. It explores the phytochemicals responsible for these effects, mechanism of synergy, therapeutic implications, advantages, challenges, and limitations.

2 Mechanism of Synergy

Synergistic interactions of phytocompounds with conventional antibiotics can control AMR and MDR. Numerous studies have found that when combined with antibiotics, crude plant extracts, essential oils from plants, and phytocompounds derived from plants show synergy (Dassanayake et al. 2021; Bonincontro et al. 2023). By changing the drug-resistance mechanisms of infection-causing bacteria, plant-derived phytocompounds can significantly lower the minimum inhibitory concentration (MIC) of conventional antibiotics (Ayaz et al. 2019). There are several ways by which bacteria can develop antibiotics resistance, including modifying antibiotic targets, production of β-lactamases, a decrease in permeability of cell wall, and increase in energy-dependent antibiotic efflux (Alekshun and Levy 2007).

2.1 Disruption of Cell Membrane/Altering the Permeability of Cell Membrane/Organelles

Phytocompounds thymol and carvacrol extracted from *Thymus capitatus* show antibacterial action against *Pseudomonas aeruginosa* by damaging its cell membrane (Althunibat et al. 2016). Peppermint essential oil (PEO) and antibiotic meropenem work against multidrug-resistant *Escherichia coli* by disrupting bacterial membranes, increasing permeability, and inhibiting quorum sensing, leading to effective bacterial killing and aiding in reversing antibiotic resistance (Yang et al. 2018). Studies on medicinal plants have also revealed their targeting of methicillin-resistant *Staphylococcus aureus* (MRSA) by various mechanisms, for instance, cell wall and membrane disruption, interfering with its genetic material (nucleoids), and altering the structure of bacteria to appearances of pseudo-multicellularity (He et al. 2024). Thymoquinone, isolated from the essential oil of *Nigella sativa* seed, produces extensive damage to the fungal cell wall along with alteration in the cytoplasmic membranes, disruption, and distortion in the nucleus, suggesting considerable structural and functional impairment (İşcan et al. 2016). Phytocompounds like curcumin derivatives used in combination with antibiotic fluconazole, can synergistically suppress cellular ATP synthesis and disrupt membrane fluidity and transport in pathogenic fungi *C. albicans* (Dong et al. 2021).

2.2 Inhibition of Enzymes/Gene Expression

Medicinal plant extracts can also inhibit those enzymes, which interfere with antibiotic activity. Isothiocyanate (ITC) sulforaphane isolated from edible crucifers showed antimicrobial properties against *Helicobacter pylori* by inactivating urease expressed by bacteria, which aggravates infection (Fahey et al. 2013). Another mechanism of action, as shown by Tomatidine derived from solanaceous plants, targets bacterial ATP synthase subunit C in *S. aureus* small-colony variants (SCVs), disrupting the energy production necessary for bacterial survival (Lamontagne et al. 2018). Allicin and ajoene are sulfur-containing compounds derived from garlic, which exhibit antimicrobial activity by having multiple inhibitory effects on thiol-dependent enzymatic systems. Their antifungal mechanism involves the inhibition of phosphatidyl choline biosynthesis, a critical component of fungal cell membrane formation and integrity (Rehman and Mairaj 2013; Reiter et al. 2017).

2.3 Inhibition of Permeability Enhancers/Efflux Pumps

One of the main bacterial mechanisms whereby toxic chemicals, including antibiotics, are actively expelled from the cell is active efflux. It significantly contributes to the emergence of MDR, acting upon the fact that antibiotics cannot reach a therapeutic concentration within the bacteria (Wright 2011). Plant-derived essential oils have been demonstrated to block efflux pump mechanisms in gram-positive as well as gram-negative bacteria. Leaves and bark methanolic extract from *Nauclea pobeguiinii*, as well as six compounds from the bark extract, can independently and synergistically enhance with antibiotics can overcome bacterial resistance by inhibiting active efflux pumps (Seukep et al. 2016). Purified phytocompounds like piperine alone as well as in combination with gentamicin have been found to be useful targeting MRSA through efflux pump suppression (Khameneh et al. 2015). Essential oil from the plant *Chenopodium ambrosioides* L. and its component α-terpinene, along with antibiotics, is reported to have increased inhibitory activity on TetK efflux-pump against *S. aureus* (Limaverde et al. 2017). Fluconazole antibiotic resistance in *Candida albicans* arises from the overexpression of efflux-pump genes CDR1, CDR2, MDR1, or mutations alongside the overexpression of the target gene

ERG11. Two main components of thyme oil, Thymol and carvacrol, exhibit a synergistic antifungal effect along with fluconazole, which can be attributed to their ability to inhibit drug efflux pumps (Ahmad et al. 2013).

2.4 Inhibition of Biofilm Formation

Essential oil derived from *Thymus vulgaris* and its principal active component, thymol, have demonstrated the ability to inhibit and eliminate biofilms of *Candida* species. Synergistic interactions between antifungal drugs and the essential oil of *Thymus vulgaris* disrupt *Candida* species biofilm cells, leading to biofilm disaggregation, deformation of cell shape, and reduced hyphae formation, which are critical for biofilm development and fungal virulence (Jafri and Ahmad 2020). Epigallocatechin gallate (EGCG), a polyphenol found in green tea, has the potential to disrupt bacterial biofilms by 95–99%, particularly in *S. aureus*, *E. coli*, *P. aeruginosa*, *S. epidermidis*, and *Mycobacterium smegmatis*, both independently and in conjunction with antibiotics (Shinde et al. 2021). Treatment with phytocompounds and antibiotic combination also causes a synergistic inhibitory effect on gene expression in pathogenic fungi, specifically ALS2, ALS1, and ALS4, which are involved in adhesion, and ECE1, PRA1, and TEC1, which regulate the critical yeast-to-hypha transition affecting biofilm formation (Zhu et al. 2022). The synergy between essential oils from plants and antibiotics can cause defects in hyphal development and biofilm formation by disrupting the localization of the Kar3/Cik1 kinesin motor protein complex and interfering with microtubules, causing their dislocation (Shahina et al. 2023).

3 Synergistic Effect of Plants Against Bacteria

Numerous research have reported that the phytocompounds isolated from the plants can reduce the MIC of the conventional antibiotics employed for bacteria, thus proving to have effective synergistic mechanisms. In a recent study, ethanolic extract of chamomile flower, lavender flower, and cempasuchitl leaves and flower, when combined with vancomycin and ethanolic extract of lavender flower, roselle flower, marigold flower, chamomile flower, and cempasuchil flower when combined with metronidazole, have synergic effects (Ruiz-Pérez et al. 2025). Extracts from the plant *Carthamus tinctorius* (Safflower) are found to be in synergy with various antibiotics such as ampicillin, vancomycin, tetracyclines, cefotaxime, chloramphenicol, penicillin, erythromycin, and neomycin against the tested organisms such as *Acinetobacter baumannii, Klebsiella pneumoniae, S. aureus, P. aeruginosa, E. coli,* and *Salmonella* spp. (Abdel Moneim et al. 2018). Ethyl acetate fraction (S20b) of *Pithecellobium clypearia* branches and leaves have been found to increase the antibacterial activity of amikacin, imipenem, cefoperazone, polymyxin B sulfate, levofloxacin, tetracycline, and ceftazidime antibiotics against MDR bacteria *A. baumannii* and *P. aeruginosa* by causing cell wall damage (Liu et al. 2019).

Isolated plant compound berberine is also found to be in synergy with antibiotics when used against the bacteria *A. baumannii* by decreasing MIC by 0-1-fold and 0-2-fold for Ciprofloxacin and Imipenem, respectively (Mahmoudi et al. 2020). Another plant compound, Galangin enhances the effectiveness of vancomycin against AMR *Enterococcus faecium*, achieving a fractional inhibitory concentration index (FICI) of 0.26–0.28 by damaging the cell envelope and avoiding biofilm formation (Ayamuang et al. 2025). Table 9.1 contains lists of the medicinal plants having synergistic activity with various antibiotics against pathogenic bacteria.

4 Synergistic Effect of Plants Against Fungi

Although less frequent than bacterial ones, fungal diseases are growing in concern due to the increased number of people with a weakened immune system such as diabetic patients, aged patients, those on broad spectrum antibiotics, patients on chemotherapy, patients on corticosteroids and psychiatric drugs, and the aged. Among Candida species, *C. albicans*, *C. tropicalis, C. glabrata, C. parapsilosis*, and *C. krusei* are the most common pathogen of humans, causing 90% of Candida infections (Turner and Butler 2014). According to several studies, the emergence of resistance in fungal pathogens to antifungal drugs makes it very difficult to treat these infections, common causes of diseases and death (Berman and Krysan 2020). Combination of medicinal plants and antibiotics is found to be effective in treating fungal infections. Ethanolic and aqueous leaf extracts of *B. holophylla*, when combined with fluconazole and nystatin, exhibit a synergistic effect, reducing the required concentration to inhibit *C. albicans* growth by 50% (de Oliveira et al. 2025). Badrashboo essential oil from *Dracocephalum moldavica* in combination with Nystatin has been found to have a synergistic effect against fungal pathogens *C. albicans*, *Saccharomyces cerevisiae*, *Fusarium solani*, *Aspergillus niger,* and *Penicillium expansum* (Sheykhjan et al. 2025). Table 9.2 contains the lists of medicinal plants possessing synergistic activity with various antibiotics against pathogenic fungi.

5 Advantages of Synergistic Combinations

Phytocompounds can complement antibiotics by providing a broader range of action against diverse pathogens. Synergistic combinations between medicinal plants or their phytocompounds and antibiotics enhance the effectiveness of antibiotics by lowering the minimum inhibitory concentration (Dhaouadi et al. 2025). Synergistic combinations of cinnamic acid compounds and colistin significantly reduces MIC values by 16- to 128-fold (Korkut et al. 2025). Extracts from plants can help cure fungal and bacterial diseases resistant to antibiotics by acting as adjuvants to those medications. Plant extracts can increase the strength of antibiotics (Haq et al. 2019). These combinations attack several routes in bacteria and fungi, hence lowering antibiotic resistance (AMR), as well as multidrug resistance (MDR)

Table 9.1 Medicinal plants having synergistic activity with various antibiotics against pathogenic bacteria

Plant	Family	Plant part	Extract/essential oil	Antibiotic used	Types of bacteria	Bioactivity enhancement	Reference
Phyllanthus emblica	Phyllanthaceae	Fruits	Ethanolic	CHL	*P. aeruginosa* (MTCC424)	8.01–16.02 Folds	Saini et al. (2024b)
					E. coli (ATCC25922)	16.02–32.05 Folds	
					S. aureus (ATCC29213)	8.01–16.02 Folds	
					K. pneumoniae (MTCC39)	4–17.85 Folds	
					B. subtilis (MTCC441)	8–142.85 Folds	
Ptychotis verticillata	Apiaceae	Leaves	Essential oil	AMX, ERY, AMP	*S. aureus*, *Micrococcus luteus*, *B. subtilis*, *P. aeruginosa*, *E. coli*, and *K. pneumoniae*	*K. pneumoniae*, *S. aureus*, *B. subtilis*: 2–7 folds (AMX) *S. aureus*, *M. luteus:* 3–10 folds (ERY) *S. aureus*, *B. subtilis*, *K. pneumoniae*, *M. luteus*, *P. aeruginosa:* 2–4 folds (AMP)	Taibi et al. (2024)
Phyllanthus emblica	Phyllanthaceae	Fruits	Ethanolic	AMP	*P. aeruginosa* (MTCC424)	4–277 Folds	Saini et al. (2024a)
					E. coli (ATCC25922)	4–32 Folds	
					S. aureus (ATCC29213)	4–128 Folds	
					K. pneumoniae (MTCC39)	4–64 Folds	
					B. subtilis (MTCC441)	4–64 Folds	
Fagonia indica	Zygophyllaceae	Aerial part	Ethyl acetate, *n*-hexane, methanol, and water extract	CEF	Resistant *P. aeruginosa*	4-Folds (*n*-hexane extract)	Abrar et al. (2024)

(continued)

Table 9.1 (continued)

Plant	Family	Plant part	Extract/essential oil	Antibiotic used	Types of bacteria	Bioactivity enhancement	Reference
Momordica charantia, Nigella sativa	Cucurbitaceae Ranunculaceae	Fruits	Ethyl acetate	CIP, CLA, LIN, CEFI, DOX	*E. coli*	*E. coli*– 4–8 folds (*M. charantia)* *E. coli*– 4 folds (*N. sativa*)	Atta et al. (2023)
Cymbopogon citratus	Gramineae	Leaves	Essential oil	CHL, AMP, and ERY	*E. coli*	3–4 Folds	Sharma et al. (2023)
Arum italicum	Araceae	Aerial	Acetone: water (1:1, v:v)	CIP	*P. aeruginosa* (ATCC 27853)	8 folds	Ağalar et al. (2022)
Thymus serpyllum	Lamiaceae	Whole plant	Essential oil	TET, VAN	*K. pneumonia* (MTCC39)	65 Folds (TET) 128 Folds (VAN)	Salaria et al. (2022)
					E. coli (ATCC29213)	32 Folds (TET) 65 Folds (VAN)	
					S. aureus (ATCC25922)	32 Folds (TET) 16 Folds (VAN)	
Bergenia ligulata	Saxifragaceae	Rhizome	Ethyl acetate fraction	TET, KAN, CHL	*E. coli* (ATCC29213)	32 Folds (TET); 128 folds (KAN, CHL)	Rolta et al. (2022)
					S. aureus (ATCC25922),	2 Folds (TET, KAN); 4 folds (CHL)	
Artemisia anuua L.	Rutaceae	Flowers and Leaves	Methanol, and petroleum ether	VAN, ERY, CHL, TET, KAN	*E. coli* (ATCC25922),	ME-8 folds (VAN, ERY, CHL, TET, KAN) PE-8 folds (VAN, ERY, CHL, TET), and 32 folds (KAN)	Rolta et al. (2021)
					S. aureus (ATCC29213)	ME-8 folds (VAN, ERY, CHL, KAN); 15-folds (TET) PE-4 folds (VAN, ERY); 129 folds (CHL); 16 folds (TET); and 8 folds (KAN)	

Hypericum roeperianum	Hypericaceae	Bark	Crude methanol extract	CLO, DOX, TET	MDR *E. aerogenes* EA-CM64	2–128 Folds	Demgne et al. (2021)
Rheum emodi Wall.	Polygonaceae	rhizome	Methanol, chloroform, ethyl acetate, n-hexane	TET, CHL, ERY	*S. aureus* (ATCC 29213),	8 Folds (TET, CHL, ERY)	Rolta et al. (2020a)
					E. coli (ATCC25922)	8 Folds (TET, CHL); 16 folds (ERY)	
					K. pneumoniae (MTCC39)	4 Folds (CHL); 16 folds (ERY, TET)	
Juniperus communis, Urtica dioica, and *Coleus forskohlii*	Cupressaceae Urticaceae Lamiaceae	Leaves	Methanol	TET, CHL, ERY	*S. aureus* (ATCC 29213)	8 Folds (TET)- *J. communis*	Rolta et al. (2020b)
						2 Folds (TET); 4 folds (CHL); 16 folds (ERY)- *U. dioica*	
						4 Folds (CHL); 8 folds (ERY)- *C. forskohlii*	
					E. coli (ATCC25922)	16 Folds (TET); 4 folds (CHL, ERY)- *J. communis*	
						2 Folds (TET, CHL); 67 folds (ERY)- *U. dioica*	
						4 Folds (ERY)- *C. forskohlii*	
					K. pneumoniae (MTCC39)	4 Folds (TET, CHL, ERY)- *J. communis*	
						8 Folds (TET, Chl); 16 folds (ERY)- *U. dioica*	
						515 Folds (TET); 8 folds (CHL, ERY)- *C. forskohlii*	

(continued)

Table 9.1 (continued)

Plant	Family	Plant part	Extract/essential oil	Antibiotic used	Types of bacteria	Bioactivity enhancement	Reference
Zanthoxylum armatum	Rutaceae	Leaves	Methanol and butanol extracts	TET, KAN	*S. aureus* (ATCC 29213)	16–128 Folds (KAN, TET)	Gautam et al. (2020)
					E. coli (ATCC25922)	16–515 Folds (KAN, TET)	
Punica granatum *Rosmarinus officinalis*	Lythraceae Lamiaceae	Peels Leaves	Ethanolic	PRL, CDZ, IPM, GTM, and LEV	*P. aeruginosa*	PRL-32 folds CDZ-256 folds IPM-256 folds CN-512 folds LEV-256 folds	Abu El-Wafa et al. (2020)
Senecio anteuphorbium	Asteraceae	Aerial parts	Essential oil	CIP	*S. aureus* (CCMMB3)	8-Folds	Elhidar et al. (2019)
					Micrococcus luteus (ATCC 10240)	2-Folds	
					B. subtilis (ATCC 9524)	2-Folds	
					P. aeruginosa (DSM 5090)	16-Folds	
					K. pneumoniae	4-Folds	
					E. coli (ATCC 8739)	2-Folds	

AMX Amoxicillin, *AMP* Ampicillin, *AMK* Amikacin, *CDZ* ceftazidime, *CHL* Chloramphenicol, *CEFI* cefixime, *CIP* ciprofloxacin, *CLO* cloxacillin, *DOX* doxycycline, *ERY* Erythromycin, *GTM* Gentamycin, *IPM* imipenem, *KAN* kanamycin, *STM* Streptomycin, *TET* Tetracycline, *PEN* Penicillin, *OFL* Ofloxacin, *TOB* Tobramycin, *LEV* Levofloxacin, *RIF* Rifampicin, *CLA* clarithromycin, *LIN* lincomycin, *PRL* piperacillin, *LEV* levofloxacin, *VAN* Vancomycin, *ND* Not determined

Table 9.2 Medicinal plants having synergistic activity with various antibiotics against pathogenic fungi

Plant	Family	Plant part	Extract/essential oil	Antibiotic used	Types of fungi	Bioactivity enhancement	Reference
Galenia africana	Aizoaceae	Leaves and shoots	Ethanolic	FLC	*C. albicans (ATCC 90028)*	FICI- 1.002	Ng'uni et al. (2022)
					C. glabrata (ATCC 26512)	FICI- 0.36	
Syzygium cumini	Myrtaceae	Leaves	Aqueous and ethanolic	FLC	*C. utilis, C. albicans (ATCC 76615)*	3-Fold	Adelakun et al. (2024)
					C. parapsilosis (ATCC 22019)	3-Fold	
					C. glabrata (ATCC 66032)	3-Fold	
					C. krusei (ATCC 14243)	3-Fold	
Moringa oleifera	Moringaceae			FLC	*C. krusei* (ATCC 14243)	3-Fold	
					C. glabrata (ATCC 66032)	3-Fold	
					C. utilis, C. albicans (ATCC 76615)	3-Fold	
					C. parapsilosis (ATCC 22019)	3-Fold	
Tinospora cordifolia	Menispermaceae			FLC	*C. krusei* (ATCC 14243)	3-Fold	
					C. parapsilosis (ATCC 22019)	3-Fold	
					C. utilis, C. albicans (ATCC 76615)	3-Fold	
					C. glabrata (ATCC 66032)	3-Fold	
Punica granatum	Punicaceae	Fruit peel	Ethanolic	FLC	*C. albicans* (ATCC 10231), *C. parapsilosis* (ATCC 22019)	2-Fold (compound Punicalagin)	Endo et al. (2010)
Cinnamomum zeylanicum, Satureja montana, Cymbopogon martini, Eucalyptus Citriodora, Melaleuca teretifolia	Lauracae, Lamiaceae, Poaceae, Myrtaceae, Myrtaceae	Leaves, stalks	Essential oils (Palmarosa, Satureja montana, Lemon eucalyptus, Cinnamon, and Honey myrtle)	AmB	*C. albicans (CMCC 98001)*	Palmarosa FICI = 0.375, Cinnamon FICI = 0.750, Honey myrtle FICI = 0.516, Satureja montana FICI = 0.625, Lemon eucalyptus FICI = 0.750	Yuan et al. (2024)

(cotinued)

Table 9.2 (continued)

Plant	Family	Plant part	Extract/essential oil	Antibiotic used	Types of fungi	Bioactivity enhancement	Reference
Acorus calamus	Acoraceae	Rhizomes	Methanolic	Azoles and AmB	*C. tropicalis* (MTCC 184) *C. albicans* (MTCC 277)	8- Folds (α- and β-Asarones) for all combinations	Kumar et al. (2015)
6,7,4′-O-triacetylscutellarein (TA)				FLC	*C. albicans strains* *CA602* *CA550* *CA4508* *CA381* *CA187* *CA800* *CA799* *CA3816* *CA808* *CA23*	*CA602*-FICI: 0.0313 *CA550*-FICI:0.0609 *CA4508*-FICI:0.0738 *CA381*-FICI:0.0301 *CA187*-FICI: 0.0188 *CA800*-FICI:0.0364 *CA799*-FICI:0.0447 *CA3816*-FICI:0.0395 *CA808*-FICI:0.0234 *CA23*-FICI:0.0738	Su et al. (2021)
Camellia sinensis var. assamica, Camellia sinensis var. sinensis	Theaceae	Leaves	(EGCG: epigallocatechin gallate or epigallocatechin 3-gallate or epigallocatechin 3-O-gallate)	KCZ FLC	*C. dubliniensis (CDC 27963)* *C. glabrata (CDC 27845)* *C. glabrata (CDC 28398)* *C. dubliniensis (CDC 28551)* *C. albicans (ATCC 24433)* *C. glabrata (ATCC 15126)* *C. albicans (CDC 27974)* *C. albicans (CDC 28304)*	3 to 5-fold	Behbehani et al. (2019)
Pseudolarix kaempferi	Pinaceae	Root bark	Pseudolaric acid A (PAA)	FLC	*C. albicans (ATCC 90028)*	56% MIC reduction	Zhu et al. (2022)

Piper nigrum, Piper longum	Piperaceae	–	Piperine	FLC CLO NYS AmB	*C. albicans Ca01, Ca02, Ca03, Ca04, Ca05, Ca06, Ca07, Ca08, Ca09, Ca10,Ca11, Ca12, Ca13, Ca14, Ca15, Ca16, Ca17, Ca18, Ca19, Ca20, ATCC 9002*	*FLC* 32-fold (*Ca03*) 4-fold (*Ca02*) 8-fold (*Ca14*) 16-fold (ATCC 9002) *CLO* FICI: 0.28 (*Ca03*) FICI: 0.37 (*Ca02*) FICI: 0.37 (*Ca14*) FICI: 0.31 (ATCC 9002) *NYS* 8 to 64-fold (*Ca16*, *Ca14*, and ATCC 9002) *AmB* 8-fold (*Ca02*) 4-fold (*Ca14*)	Tsopmene et al. (2024)
Curcuma longa	Zingiberaceae	–	Curcumin			*NYS* 4-fold (ATCC 9002) 64-fold (*Ca13*) 4-fold (*Ca14*) *AmB* 4-fold (*Ca14*) *FLC* 8-fold (*Ca02*) 32-fold (*Ca03*)	
Rubus chingii	Rosaceae	Fruit	Ethanolic	FLC	*C. albicans 100*	FICI 4.88–312.5	Han et al. (2016)

(cotinued)

Table 9.2 (continued)

Plant	Family	Plant part	Extract/essential oil	Antibiotic used	Types of fungi	Bioactivity enhancement	Reference
Eugenia uniflora	Myrtaceae	Leaves	Acetone	FLC AmB MCF	*C. parapsilosis (ATCC 22019)* *C. tropicalis (ATCC 13803)* *C. dubliniensis (CBS 7987)* *C. albicans (ATCC 90028)* *C. glabrata (ATCC 2001)*	2 to 8-fold	Souza et al. (2024)
Acantholippia seriphioides, Azorella cryptantha, Clinopodium gilliesii, Gymnophyton polycephalum, Lippia integrifolia, Seriphidium mendozanum, Tagetes mendocina	Verbenaceae, Apiaceae, Asteraceae, Apiaceae, *Verbenacea, Asteraceae, Asteraceae*	Aerial parts	Essential oil	FLC ITC	*Microsporum gypseum (CCC115)* *C. albicans (ATCC 10231)* *Cryptococcus neoformans (ATCC 32264)* *Trichophyton rubrum (CCC110)*	FIC = 0.31 (*A. cryptantha*-B EO with FLC *C. Albicans)*	Lima et al. (2020)

FLC Fluconazole, *AmB* Amphotericin B, *KCZ* Ketoconazole, *NYS* Nystatin, *CLO* Clotrimazole, *MCF* Micafungin, *ITC* Itraconazole, *ND* Not determined

(Jeong et al. 2023). These synergistic combinations can also prevent biofilm development in bacteria and fungi, hence enhancing antibiotic action. This could lead to improved effectiveness and fewer antibiotic doses as well as less medical costs. Synergistic combinations can also create possibilities for creating creative treatments for resistant infections. Another significant point of view is that using bioactive chemicals from plants not only helps to fight antibiotic resistance but also guarantees the sustainable use of natural resources, therefore stressing the function of green chemistry in producing efficient and environmentally friendly antimicrobial therapies (Schiavone et al. 2025).

6 Challenges and Limitations

Although medicinal plants and antibiotics can be a good synergistic combination to treat bacterial and fungi infections and lower AMR and MDR incidences but there are certain challenges and limitations associated with their use. Understanding the precise mechanisms of synergy between plant compounds and antibiotics can be challenging due to the diverse phytochemical profiles of plants. One of the major challenges is that most of the synergistic effects are researched in vitro with limited clinical trials to confirm their efficacy and safety in humans (Mundy et al. 2016). Factors like cultivation practices, environmental conditions, geographic location, method of extraction, and storage conditions also impact the phytochemical constituents of plant extracts, thus inhibiting reproducibility and standardization of plants (Njume and Goduka 2012; Wang et al. 2023). Another major challenge is that the plant-based compounds are often considered safe, but some extracts or combinations may lead to unexpected toxicity or adverse reactions in humans. Few plant formulations, for example, a formulation used in treating urinary tract infections, show limited effectiveness in inhibiting the urease activity of *Staphylococcus saprophyticus,* thus indicating further research is required to ensuring better efficacy and reliability in clinical applications (Deutch 2017). Various medicinal plants and herbal medicinal products have been found to possess nephrotoxicity (Xu et al. 2020), cardiotoxicity (Gavanji 2023), genotoxicity (Bardoloi and Soren 2022), and hepatotoxicity (Quan et al. 2020). Also, natural sources and anthropogenic activities like agriculture and industry can cause contamination of heavy metals (HMs) in herbal preparations, thus implying the importance of stringent regulation and monitoring systems (Vinogradova et al. 2023; Opuni et al. 2023). Ensuring the safety and efficacy of natural supplements requires proactive measures to prevent contamination and protect public health.

7 Future Perspectives and Conclusions

The future of synergistic combinations of medicinal plants and conventional antibiotics offers immense promise in fighting bacterial and fungal diseases as the incidence of AMR is rising. Examining the molecular mechanisms behind these synergies calls for more advanced study. A change from in vitro research to clinical trials of these combinations is also required. Furthermore, plant–antibiotic combinations can target biofilm-related diseases and MDR organisms, hence providing a unique way to manage such infections. While maintaining biodiversity, ethnobotanical research, and sustainable farming techniques can reveal fresh bioactive chemicals. Strong regulatory systems and thorough clinical studies are necessary to guarantee safe and effective usage of mixtures of medicinal plants and conventional antibiotics. These initiatives might open the path for creative, environmentally friendly treatments that tackle world health issues and support sustainability.

Using herbal medicines with traditional antibiotics to treat and combat AMR and MDR is a very good synergistic mix. The combinations reduce the MIC values relative to their individual use. Medicinal plants have phytocompounds that can interact synergistically with antibiotics by affecting altering microbial resistance processes, preventing biofilm formation, and enhancing drug penetration. To maximize their full potential, nevertheless, concerns including heterogeneity in plant extracts, few clinical investigations, and standardization must be tackled. These synergistic strategies could greatly help to provide efficient and sustainable treatments for resistant infections as phytochemistry, nanotechnology, and clinical research progress.

References

Abdel Moneim ES, Sherif MS, Ahmed AE, Mohanad A, Vajid NV (2018) Evaluation of antimicrobial activity of safflower (*Carthamus tinctorius*) and its synergistic effect with antibiotic. EC Microbiol 14(3):160–166

Abrar A, Zafar A, Fatima M, Muntaqua D, Naz I, Fatima H, Haq IU (2024) Mechanistic insight into the synergistic antimicrobial potential of *Fagonia indica* Burm. f. extracts with cefixime. Saudi Pharm J 32(1):101893

Abu El-Wafa WM, Ahmed RH, Ramadan MA (2020) Synergistic effects of pomegranate and rosemary extracts in combination with antibiotics against antibiotic resistance and biofilm formation of *Pseudomonas aeruginosa*. Braz J Microbiol 51(3):1079–1092

Adelakun AO, Awosika A, Adabanya U, Omole AE, Olopoda AI, Bello ET (2024) Antimicrobial and synergistic effects of *Syzygium cumini*, *Moringa oleifera*, and *Tinospora cordifolia* against different *Candida* infections. Cureus 16(1)

Ağalar HG, Öztürk G, Kırımer N (2022) *Arum italicum* Miller tuber extracts: evaluation of synergistic activities with ciprofloxacin against some pathogens. Eur J Life Sci 1(3):97–106

Ahmad A, Khan A, Manzoor N (2013) Reversal of efflux mediated antifungal resistance underlies synergistic activity of two monoterpenes with fluconazole. Eur J Pharm Sci 48(1–2):80–86

Alekshun MN, Levy SB (2007) Molecular mechanisms of antibacterial multidrug resistance. Cell 128(6):1037–1050

Allemailem KS (2021) Antimicrobial potential of naturally occurring bioactive secondary metabolites. J Pharm Bioallied Sci 13(2):155–162

Althunibat OY, Qaralleh H, Ahmed Al-Dalin SY, Abboud M, Khleifat K, Majali IS, Aldal'in HK, Rayyan WA, Jaafraa A (2016) Effect of thymol and carvacrol, the major components of *Thymus capitatus*, on the growth of *Pseudomonas aeruginosa*. J Pure Appl Microbiol 10(1)

Angelini P (2024) Plant-derived antimicrobials and their crucial role in combating antimicrobial resistance. Antibiotics 13(8):746

Atta S, Waseem D, Fatima H, Naz I, Rasheed F, Kanwal N (2023) Antibacterial potential and synergistic interaction between natural polyphenolic extracts and synthetic antibiotic on clinical isolates. Saudi J Biol Sci 30(3):103576

Ayamuang IO, Teethaisong Y, Sirichaiwetchakoon K, Suknasang S, Wattana S, Chaiseha Y, Eumkeb G (2025) Galangin synergistically revives the antibacterial activity of vancomycin against vancomycin-resistant *Enterococcus faecium*. J Appl Microbiol 136(1):lxaf011

Ayaz M, Ullah F, Sadiq A, Ullah F, Ovais M, Ahmed J, Devkota HP (2019) Synergistic interactions of phytochemicals with antimicrobial agents: potential strategy to counteract drug resistance. Chem Biol Interact 308:294–303

Bardoloi A, Soren AD (2022) Genotoxicity induced by medicinal plants. Bull Natl Res Centre 46(1):119

Behbehani JM, Irshad M, Shreaz S, Karched M (2019) Synergistic effects of tea polyphenol epigallocatechin 3-O-gallate and azole drugs against oral *Candida* isolates. J Mycol Med 29(2):158–167

Berman J, Krysan DJ (2020) Drug resistance and tolerance in fungi. Nat Rev Microbiol 18(6):319–331

Bonincontro G, Scuderi SA, Marino A, Simonetti G (2023) Synergistic effect of plant compounds in combination with conventional antimicrobials against biofilm of *Staphylococcus aureus*, *Pseudomonas aeruginosa*, and *Candida* spp. Pharmaceutics 16(11):1531

Dassanayake MK, Khoo TJ, An J (2021) Antibiotic resistance modifying ability of phytoextracts in anthrax biological agent *Bacillus anthracis* and emerging superbugs: a review of synergistic mechanisms. Ann Clin Microbiol Antimicrob 20:1–36

Dehghan H, Sarrafi Y, Salehi P (2016) Antioxidant and antidiabetic activities of 11 herbal plants from Hyrcania region. Iran J Food Drug Anal 24(1):179–188

Demgne OM, Damen F, Fankam AG, Guefack MG, Wamba BE, Nayim P, Mbaveng AT, Bitchagno GT, Tapondjou LA, Penlap VB, Tane P (2021) Botanicals and phytochemicals from the bark of *Hypericum roeperianum* (Hypericaceae) had strong antibacterial activity and showed synergistic effects with antibiotics against multidrug-resistant bacteria expressing active efflux pumps. J Ethnopharmacol 277:114257

Deutch CE (2017) Limited effectiveness of over-the-counter plant preparations used for the treatment of urinary tract infections as inhibitors of the urease activity from *staphylococcus saprophyticus*. J Appl Microbiol 122(5):1380–1388

Dhaouadi S, Ghorbel SK, Bouglita W, Chaari S, Dhifi W, Khrouf R, Cherif A, Elandoulsi RB (2025) The potency of *Cupressus sempervirens* and *Eucalyptus globulus* essential oils against antibiotic-resistant *Escherichia coli* and *Mammaliicoccus sciuri* from diseased animals in Tunisia. Curr Microbiol 82(1):14

Dong HH, Wang YH, Peng XM, Zhou HY, Zhao F, Jiang YY, Zhang DZ, Jin YS (2021) Synergistic antifungal effects of curcumin derivatives as fungal biofilm inhibitors with fluconazole. Chem Biol Drug Des 97(5):1079–1088

Elhidar N, Nafis A, Kasrati A, Goehler A, Bohnert JA, Abbad A, Hassani L, Mezrioui NE (2019) Chemical composition, antimicrobial activities and synergistic effects of essential oil from *Senecio anteuphorbium*, a Moroccan endemic plant. Ind Crop Prod 130:310–315

Endo EH, Cortez DA, Ueda-Nakamura T, Nakamura CV, Dias Filho BP (2010) Potent antifungal activity of extracts and pure compound isolated from pomegranate peels and synergism with fluconazole against *Candida albicans*. Res Microbiol 161(7):534–540

Fahey JW, Stephenson KK, Wade KL, Talalay P (2013) Urease from *Helicobacter pylori* is inactivated by sulforaphane and other isothiocyanates. Biochem Biophys Res Commun 435(1):1–7

Ferri M, Ranucci E, Romagnoli P, Giaccone V (2017) Antimicrobial resistance: a global emerging threat to public health systems. Crit Rev Food Sci Nutr 57(13):2857–2876

Gautam M, Rolta R, Kumar V, Sourirajan A, Sharma DR, Dev K (2020) Comparative antimicrobial and synergistic potential of extracts of leaves of *Zanthoxylum armatum* DC. from different geographical regions of Himachal Pradesh, India. Plant Arch 20(2):8127–8135

Gavanji S (2023) Cardiotoxicity effects of herbal medicine, a review article. Biol Med Nat Prod Chem 12(1):89–96

Gaynes R (2017) The discovery of penicillin—new insights after more than 75 years of clinical use. Emerg Infect Dis 23(5):849

Han B, Chen J, Yu YQ, Cao YB, Jiang YY (2016) Antifungal activity of *Rubus chingii* extract combined with fluconazole against fluconazole-resistant *Candida albicans*. Microbiol Immunol 60(2):82–92

Haq A, Siddiqi M, Batool SZ, Islam A, Khan A, Khan D, Khan S, Khan H, Shah AA, Hasan F, Ahmed S (2019) Comprehensive investigation on the synergistic antibacterial activities of *Jatropha curcas* pressed cake and seed oil in combination with antibiotics. AMB Express 9:1–21

He Q, Meneely J, Grant IR, Chin J, Fanning S, Situ C (2024) Phytotherapeutic potential against MRSA: mechanisms, synergy, and therapeutic prospects. Chin Med 19(1):1–6

İşcan G, İşcan A, Demirci F (2016) Anticandidal effects of thymoquinone: mode of action determined by transmission electron microscopy (TEM). Nat Prod Commun 11(7):1934578X1601100726

Jafri H, Ahmad I (2020) *Thymus vulgaris* essential oil and thymol inhibit biofilms and interact synergistically with antifungal drugs against drug-resistant strains of *Candida albicans* and *Candida tropicalis*. J Mycol Med 30(1):100911

Jeong JY, Jung IG, Yum SH, Hwang YJ (2023) In vitro synergistic inhibitory effects of plant extract combinations on bacterial growth of methicillin-resistant *Staphylococcus aureus*. Pharmaceuticals 16(10):1491

Kalın G, Alp E, Chouaikhi A, Roger C (2023) Antimicrobial multidrug resistance: clinical implications for infection management in critically ill patients. Microorganisms 11(10):2575

Kaur S, Samota MK, Choudhary M, Choudhary M, Pandey AK, Sharma A, Thakur J (2022) How do plants defend themselves against pathogens—biochemical mechanisms and genetic interventions. Physiol Mol Biol Plants 28(2):485–504

Keenan JD, Bailey RL, West SK, Arzika AM, Hart J, Weaver J, Kalua K, Mrango Z, Ray KJ, Cook C, Lebas E (2018) Azithromycin to reduce childhood mortality in sub-Saharan Africa. N Engl J Med 378(17):1583–1592

Khameneh B, Iranshahy M, Ghandadi M, Ghoochi Atashbeyk D, Fazly Bazzaz BS, Iranshahi M (2015) Investigation of the antibacterial activity and efflux pump inhibitory effect of co-loaded piperine and gentamicin nanoliposomes in methicillin-resistant *Staphylococcus aureus*. Drug Dev Ind Pharm 41(6):989–994

Klein EY, Impalli I, Poleon S, Denoel P, Cipriano M, Van Boeckel TP, Pecetta S, Bloom DE, Nandi A (2024) Global trends in antibiotic consumption during 2016–2023 and future projections through 2030. Proc Natl Acad Sci 121(49):e2411919121

Korkut A, Özkaya Gül S, Aydemir E, Er H, Odabaş Köse E (2025) Cinnamic acid compounds (p-Coumaric, ferulic, and p-methoxycinnamic acid) as effective antibacterial agents against colistin-resistant *Acinetobacter baumannii*. Antibiotics 14(1):71

Kumar SN, Aravind SR, Sreelekha TT, Jacob J, Kumar BD (2015) Asarones from *Acorus calamus* in combination with azoles and amphotericin B: a novel synergistic combination to compete against human pathogenic *Candida* species *in vitro*. Appl Biochem Biotechnol 175:3683–3695

Lamontagne Boulet M, Isabelle C, Guay I, Brouillette E, Langlois JP, Jacques PÉ, Rodrigue S, Brzezinski R, Beauregard PB, Bouarab K, Boyapelly K (2018) Tomatidine is a lead antibiotic molecule that targets *Staphylococcus aureus* ATP synthase subunit C. Antimicrob Agents Chemother 62(6):10128

Lima BV, Sortino MA, Tapia A, Feresin GE (2020) Synergistic antifungal effectiveness of essential oils from Andean plants combined with commercial drugs. Int J Pharm Sci Dev Res 8(1):23–31

Limaverde PW, Campina FF, da Cunha FA, Crispim FD, Figueredo FG, Lima LF, Oliveira-Tintino CD, de Matos YM, Morais-Braga MF, Menezes IR, Balbino VQ (2017) Inhibition of the TetK efflux-pump by the essential oil of *Chenopodium ambrosioides* L. and α-terpinene against *Staphylococcus aureus* IS-58. Food Chem Toxicol 109:957–961

Liu C, Huang H, Zhou Q, Liu B, Wang Y, Li P, Liao K, Su W (2019) Antibacterial and antibiotic synergistic activities of the extract from *Pithecellobium clypearia* against clinically important multidrug-resistant gram-negative bacteria. Eur J Integr Med 32:100999

Lorca G, Ballestero D, Langa E, Pino-Otín MR (2024) Enhancing antibiotic efficacy with natural compounds: synergistic activity of tannic acid and nerol with commercial antibiotics against pathogenic bacteria. Plants 13(19):2717

Mahmoudi H, Zare Fahim N, Alikhani MY, Shokoohizadeh L (2020) Investigation of antimicrobial effect of berberine on ciprofloxacin and imipenem resistance *Acinetobacter baumannii* isolated from Hamadan hospitals. Iran J Med Microbiol 14(1):44–54

Mundy L, Pendry B, Rahman M (2016) Antimicrobial resistance and synergy in herbal medicine. J Herbal Med 6(2):53–58

Naddaf M (2024) 40 million deaths by 2050: toll of drug-resistant infections to rise by 70%. Nature 633(8031):747–748

Naghavi M, Vollset SE, Ikuta KS, Swetschinski LR, Gray AP, Wool EE, Aguilar GR, Mestrovic T, Smith G, Han C, Hsu RL (2024) Global burden of bacterial antimicrobial resistance 1990–2021: a systematic analysis with forecasts to 2050. Lancet 404(10459):1199–1226

Nawaz M, Sun J, Shabbir S, Khattak WA, Ren G, Nie X, Bo Y, Javed Q, Du D, Sonne C (2023) A review of plants strategies to resist biotic and abiotic environmental stressors. Sci Total Environ 900:165832

Ng'uni TL, dos Santos Abrantes PM, McArthur C, Klaasen JA, Fielding BC (2022) Evaluation of synergistic anticandidal activity of *Galenia africana* extract and fluconazole against *Candida albicans* and *Candida glabrata*. J Herbal Med 32:100503

Nishad R, Ahmed T, Rahman VJ, Kareem A (2020) Modulation of plant defense system in response to microbial interactions. Front Microbiol 11:1298

Njume C, Goduka NI (2012) Treatment of diarrhoea in rural African communities: an overview of measures to maximise the medicinal potentials of indigenous plants. Int J Environ Res Public Health 9(11):3911–3933

de Oliveira RS, de Almeida-Apolonio AA, Dos Santos JV, da Rocha LC, Cupozak-Pinheiro WJ, Cardoso CA, de Castro TL, Ferreira DG, Negri M, de Oliveira KM (2025) Synergistic potential of *Bauhinia holophylla* leaf extracts with conventional antifungals in the inhibition of *Candida albicans*: a new approach for the treatment of oral candidiasis. Arch Oral Biol 170:106133

Opuni KF, Kretchy JP, Agyabeng K, Boadu JA, Adanu T, Ankamah S, Appiah A, Amoah GB, Baidoo M, Kretchy IA (2023) Contamination of herbal medicinal products in low-and-middle-income countries: a systematic review. Heliyon 9(8):e19370

Quan NV, Dang Xuan T, Teschke R (2020) Potential hepatotoxins found in herbal medicinal products: a systematic review. Int J Mol Sci 21(14):5011

Rehman F, Mairaj SA (2013) Antimicrobial studies of allicin and ajoene. Int J Pharm Bio Sci 4(2):1095–1105

Reiter J, Levina N, Van der Linden M, Gruhlke M, Martin C, Slusarenko AJ (2017) Diallylthiosulfinate (Allicin), a volatile antimicrobial from garlic (*Allium sativum*), kills human lung pathogenic bacteria, including MDR strains, as a vapor. Molecules 22(10):1711

Rolta R, Kumar V, Sourirajan A, Upadhyay NK, Dev K (2020a) Bioassay guided fractionation of rhizome extract of *Rheum emodi* wall as bio-availability enhancer of antibiotics against bacterial and fungal pathogens. J Ethnopharmacol 257:112867

Rolta R, Kumar V, Sourirajan A, Upadhyay NK, Dev K (2020b) Phytocompounds of three medicinal plants (*Juniperus communis*, *Urtica dioica*, and *Coleus forskohlii*) of North West Himalayas increases the potency of antibacterial and antifungal antibiotics. Plant Arch 20(2):481–489

Rolta R, Sharma A, Sourirajan A, Mallikarjunan PK, Dev K (2021) Combination between antibacterial and antifungal antibiotics with phytocompounds of *Artemisia annua* L: a strategy to control drug resistance pathogens. J Ethnopharmacol 266:113420

Rolta R, Goyal M, Sharma S, Bharaj D, Salaria D, Upadhyay NK, Lal UR, Dev K, Sourirajan A (2022) Bioassay guided fractionation of phytocompounds from *Bergenia ligulata*: a synergistic approach to treat drug-resistant bacterial and fungal pathogens. Pharmacol Res-Modern Chinese Med 3:100076

Ruiz-Pérez NC, Bayona-Pérez Y, Guzmán-Gutiérrez SL, Reyes-Chilpa R, Luna-Pineda VM, Torres J, Romo-Castillo M (2025) Metronidazole and vancomycin have a synergic effect, with plant extracts as helpful tools to combat *Clostridioides difficile* infections. Antibiotics 14(1):54

Saini R, Kumar V, Patel CN, Sourirajan A, Dev K (2024a) Synergistic antibacterial activity of *Phyllanthus emblica* fruits and its phytocompounds with ampicillin: a computational and experimental study. Naunyn Schmiedeberg's Arch Pharmacol 397(2):857–871

Saini R, Kumar V, Sourirajan A, Dev K (2024b) Fruit extract and phenolic compounds of *Phyllanthus emblica* fruits as bioactivity enhancer of chloramphenicol against bacterial species. Plant Foods Hum Nutr 79(3):656–661

Salaria D, Rolta R, Patel CN, Dev K, Sourirajan A, Kumar V (2022) In vitro and in silico analysis of *Thymus serpyllum* essential oil as bioactivity enhancer of antibacterial and antifungal agents. J Biomol Struct Dyn 40(20):10383–10402

Sanhueza L, Melo R, Montero R, Maisey K, Mendoza L, Wilkens M (2017) Synergistic interactions between phenolic compounds identified in grape pomace extract with antibiotics of different classes against *Staphylococcus aureus* and *Escherichia coli*. PLoS One 12(2):e0172273

Schiavone ML, Barletta R, Trezza A, Geminiani M, Millucci L, Figura N, Santucci A (2025) Green chemistry within the circular bioeconomy to harness chestnut burr extract's synergistic antimicrobial activity against *Helicobacter pylori*. Molecules 30(2):324

Seukep JA, Sandjo LP, Ngadjui BT, Kuete V (2016) Antibacterial and antibiotic-resistance modifying activity of the extracts and compounds from *Nauclea pobeguinii* against Gram-negative multi-drug-resistant phenotypes. BMC Complement Altern Med 16:1–8

Shahina Z, Yennamalli RM, Dahms TE (2023) Key essential oil components delocalize *Candida albicans* Kar3p and impact microtubule structure. Microbiol Res 272:127373

Shahrajabian MH, Sun W (2023) Survey on medicinal plants and herbs in traditional Iranian medicine with antioxidant, anti-viral, anti-microbial, and anti-inflammation properties. Lett Drug Des Discov 20(11):1707–1743

Sharma HK, Gupta P, Nagpal D, Mukherjee M, Parmar VS, Lather V (2023) Virtual screening and antimicrobial evaluation for identification of natural compounds as the prospective inhibitors of antibacterial drug resistance targets in *Staphylococcus aureus*. Fitoterapia 168:105554

Sheykhjan MG, Fazlara A, Hojjati M, Behbahani BA (2025) Evaluation of the antifungal effect of Badrashboo (*Dracocephalum moldavica*) essential oil and its interaction with Nystatin on some fungal strains. J Food Sci Technol 21(156):226–240

Shinde S, Lee LH, Chu T (2021) Inhibition of biofilm formation by the synergistic action of EGCG-S and antibiotics. Antibiotics 10(2):102

Souza LB, de Oliveira BA, Lourenço EM, Ferreira MR, Oliveira WN, Soares LA, Barbosa EG, Rocha HA, Chaves GM (2024) Mechanism of action and synergistic effect of *Eugenia uniflora* extract in *Candida* spp. PLoS One 19(8):e0303878

Su LY, Ni GH, Liao YC, Su LQ, Li J, Li JS, Rao GX, Wang RR (2021) Antifungal activity and potential mechanism of 6, 7, 4′-O-triacetylscutellarein combined with fluconazole against drug-resistant *C. Albicans*. Front Microbiol 12:692693

Taibi M, Elbouzidi A, Haddou M, Loukili EH, Bellaouchi R, Asehraou A, Douzi Y, Addi M, Salamatullah AM, Nafidi HA, Bourhia M (2024) Chemical profiling, antibacterial efficacy, and synergistic actions of *Ptychotis verticillata* Duby essential oil in combination with conventional antibiotics. Nat Prod Commun 19(1):1934578X231222785

Tsopmene UJ, Tokam Kuaté CR, Kayoka-Kabongo PN, Bisso BN, Metopa A, Mofor CT, Dzoyem JP (2024) Antibiofilm activity of curcumin and piperine and their synergistic effects with antifungals against *Candida albicans* clinical isolates. Scientifica 1:2025557

Turner SA, Butler G (2014) The *Candida* pathogenic species complex. Cold Spring Harb Perspect Med 4(9):a019778

Vinogradova N, Glukhov A, Chaplygin V, Kumar P, Mandzhieva S, Minkina T, Rajput VD (2023) The content of heavy metals in medicinal plants in various environmental conditions: a review. Horticulturae 9(2):239

Wang H, Chen Y, Wang L, Liu Q, Yang S, Wang C (2023) Advancing herbal medicine: enhancing product quality and safety through robust quality control practices. Front Pharmacol 14:1265178

World Health Organization (WHO) (2023) Vaccines could avert half a million deaths associated with antimicrobial resistance a year. WHO, Geneva

Wright GD (2011) Molecular mechanisms of antibiotic resistance. Chem Commun 47(14):4055–4061

Xu X, Zhu R, Ying J, Zhao M, Wu X, Cao G, Wang K (2020) Nephrotoxicity of herbal medicine and its prevention. Front Pharmacol 11:569551

Yang SK, Yap PS, Krishnan T, Yusoff K, Chan KG, Yap WS, Lai KS, Lim SH (2018) Mode of action: synergistic interaction of peppermint (*Mentha x piperita* L. Carl) essential oil and meropenem against plasmid-mediated resistant *E. coli*. Rec Nat Prod 12(6):582–594

Yuan X, Cao D, Xiang Y, Jiang X, Liu J, Bi K, Dong X, Wu T, Zhang Y (2024) Antifungal activity of essential oils and their potential synergistic effect with amphotericin B. Sci Rep 14(1):31125

Zhu B, Li Z, Yin H, Hu J, Xue Y, Zhang G, Zheng X, Chen W, Hu X (2022) Synergistic antibiofilm effects of pseudolaric acid a combined with fluconazole against *Candida albicans* via inhibition of adhesion and yeast-to-hypha transition. Microbiol Spectr 10(2):e01478–e01421

Safety, Dosage, and Regulatory Aspects in the Use of Phytochemicals

10

Manivannan Karthikeyan, Anisha Mehta,
Sahasra Suresh Kumar, Harikrishnan Mohan,
and Balasundaram Usha

Abstract

Phytochemicals are naturally occurring bioactive compounds that are vital to human health, nutrition, and medicine. The medicinal potential of these compounds, which include polyphenols, alkaloids, glycosides, tannins, terpenes, and saponins, in disease prevention and treatment has been extensively studied. Modern drug discovery began with phytochemicals, which have been fundamental to traditional medicinal systems such as Ayurveda, Siddha, Traditional Chinese Medicine, and ancient Egyptian and Greek pharmacology. Despite their well-documented therapeutic benefits, phytochemicals present safety concerns about toxicity, bioavailability, metabolic interactions, and herb–drug interactions. Artificial intelligence and nanotechnology, together with other developing technologies, are enhancing phytochemical research and facilitating its application in individualized therapy. Regulatory frameworks established by organizations such as the FDA, EMA, WHO, and AYUSH ensure product safety, quality, and commercialization; however, appropriate dosing and rigorous standardization remain essential to maximize efficacy and minimize adverse effects. To ensure their safe and effective use, more stringent clinical trials and improved regulatory frameworks are necessary. This chapter focuses on a thorough examination of the relevance of phytochemicals, their historical background, safety issues, dosage difficulties, regulatory factors, and future viewpoints in the field of medicine and pharmacology.

M. Karthikeyan · A. Mehta · S. S. Kumar · B. Usha (✉)
Medicinal Plants and PCOS Research Laboratory, Department of Genetic Engineering, SRM Institute of Science and Technology, Kattankulathur, Chengalpattu, Tamil Nadu, India
e-mail: km4437@srmist.edu.in; ushab@srmist.edu.in

H. Mohan
Department of Pathobiological Sciences, School of Veterinary Medicine, Louisiana State University, Baton Rouge, LA, USA

K. Kuca et al. (eds.), *Medicinal Plants and Their Bioactives in Human Diseases*,
https://doi.org/10.1007/978-3-032-01356-9_10

Keywords

Bioavailability · Herb-drug interactions · Hormonal imbalance · Pharmacokinetics · Therapeutic possibilities

1 Introduction

For centuries, the human race has relied on nature for healing. Traditional medicine systems, such as Ayurveda, Siddha, the ancient Egyptian papyri, Traditional Chinese Medicine, and African ethnobotany, have long relied on plant-based compounds for treatments. The idea that plants harbor medicinal properties is as old as civilization itself. Plants produce both primary and secondary metabolites, each serving distinct functions. The primary metabolites focus on basic physiological and biochemical processes, including protein synthesis, energy production, membrane biosynthesis, building genetic material, and the growth and development of plants. In contrast, secondary metabolites, also known as phytocompounds, present in plants are not directly involved in growth and reproduction but play a crucial role in plant–environmental interactions (Wani et al. 2022; Chandi Charan Kandar 2020). Their primary role includes defense mechanisms, attracting pollinators, seed dispersion, stress tolerance, and symbiotic interaction. Apart from these ecological roles, these phytochemicals have an important role in medicine and nutrition (Thacker and Ram 2024).

Phytochemicals are categorized into alkaloids, polyphenols, terpenoids, tannins, glycosides, sulphur-containing compounds, coumarins, and saponins, each exhibiting diverse pharmacological activities (Newman et al. 2003). Within these major classes, further subdivisions exist based on structural characteristics and biosynthetic pathways. Among these, alkaloids represent a highly diverse group of nitrogen-containing organic compounds known for their potent biological effects: purine, pyridine/piperidine, tropane, indole, quinoline, isoquinoline, steroidal, and pyrrolizidine. These groups demonstrate diverse biological activities like analgesic, antimalarial, cardiovascular benefits, anticancer, and antimicrobial properties (Bhambhani et al. 2021). Whereas, polyphenols, which are widely classified into four sub-groups: flavonoids, phenolic acids, stilbenes, and lignans, exhibit various biological activities, such as antioxidant, anti-inflammatory, anti-cancer, neuroprotective, and cardiovascular protection, making them crucial for human health (Awuchi 2019).

Owing to their importance, these phytochemicals have been pivotal in drug discovery, offering a natural arsenal against diseases ranging from microbial infections to cancer treatments (Ahmed et al. 2024). Their undeniable significance in disease prevention and health enhancement is seen in nutraceuticals, functional foods, herbal medicine, and pharmaceutical development (Tahir 2024). Researchers have emphasized that regular consumption of a plant rich diet that comprises phytochemicals contributes to the prevention of chronic diseases, such as type 2 diabetes, cardiovascular diseases, cancer, and neurodegenerative disorders, by modulating inflammatory pathways, decreasing oxidative damage, enhancing metabolic functions, and promoting cellular homeostasis. Despite their recognized health benefits, phytochemicals must be meticulously assessed for safety, as their natural origin does not inherently imply the absence of toxicity (Shuaib et al. 2021). Their effects on the human body are shaped by multiple factors, including dose, bioavailability, metabolic interactions, and the risks associated with long-term exposure. The relationship between dose and response is particularly complex, as many phytochemicals exhibit biphasic effects, providing health benefits at low doses while becoming harmful at higher concentrations. Additionally, bioavailability plays a crucial role in determining both efficacy and safety, as factors such as absorption, metabolism, and interactions with other dietary components can influence their biological activity and toxicity. Metabolic interactions further complicate their impact, as phytochemicals can modulate drug-metabolizing enzymes, potentially altering the pharmacokinetics of concurrently administered medications. Furthermore, prolonged exposure to certain phytochemicals may yield unforeseen outcomes such as cumulative toxicity, endocrine disruption, or genotoxic effects, necessitating thorough toxicological investigations (Jodynis-Liebert and Kujawska 2020).

In light of these concerns, regulatory agencies such as the United States Food and Drug Administration (FDA), the European Medicines Agency (EMA), the World Health Organization (WHO), and the Food Safety and Standards Authority of India (FSSAI) have developed guidelines to ensure that phytochemicals used in pharmaceuticals, functional foods, and dietary supplements to meet safety and efficacy standards (Thakur et al. 2020). These recommendations highlight the need for clinical studies, standardized dosage, and toxicological reviews to ensure consumer safety and reduce adverse effects. Standardized processes for testing the purity, stability, and concentration of active chemicals are required to provide uniform therapeutic benefits and reduce dangers associated with phytochemical usage (Prajapati 2024). Despite phytochemicals' enormous therapeutic potential, their usage must be guided by clinical evidence, safety regulations, and specific considerations to maximize benefits and reduce potential risks. While many phytochemicals have been used in traditional medicine for centuries, their transition into mainstream healthcare requires a deeper understanding of their pharmacological properties, interactions, and long-term effects. The lack of standardized dosage and potential variability in bioactive compound concentrations further complicate their safe use.

This chapter explores the safety, dosage, and regulatory landscapes surrounding phytochemicals, shedding light on both their potential and challenges. While modern pharmaceuticals undergo rigorous testing, phytochemicals often straddle the line between food supplements and therapeutic agents, making regulation a complex issue. Their classification as either dietary components or medicinal substances varies across regulatory frameworks, further complicating their approval and standardization. Unlike synthetic drugs, whose safety and efficacy are evaluated through extensive clinical trials, many phytochemicals are consumed based on traditional knowledge, often without stringent oversight. This gap in regulatory scrutiny raises concerns regarding purity, consistency, and potential interactions with other compounds. Understanding their safety profiles, standardization processes, and legal frameworks is crucial to harnessing their benefits while minimizing risks.

2 Safety Concerns in Phytochemical Use

Despite their numerous health benefits, phytochemicals require careful safety assessment to prevent potential adverse effects, such as hormonal imbalances or neurotoxicity. While they are often perceived as natural and harmless, they can still pose risks such as toxicity, drug interactions, and inconsistent bioavailability. Their safety depends on several factors, including dosage, duration of exposure, individual metabolism, and interactions with medications or other dietary components (Gupta et al. 2022). For instance, excessive consumption of certain phytochemicals can lead to liver toxicity, gastrointestinal issues, or even interfere with essential nutrient absorption (Shahnawaz et al. 2024). Phytochemicals can also exhibit synergistic effects, where their combined action is greater than the sum of their individual effects, potentially enhancing therapeutic benefits. However, they may also produce antagonistic effects, where their combined impact is weaker or counteracts expected outcomes, reducing their efficacy or leading to impulsive side effects (Boik et al. 2009). Given these complexities, regulatory oversight, rigorous scientific evaluation, and controlled consumption are essential to ensure their safe and effective use in both dietary and medicinal applications (Marakis et al. 2018; Rathaur 2019). Some phytochemicals can enhance or weaken the effects of pharmaceutical drugs, leading to unintended health consequences.

2.1 Bioavailability and Metabolic Pathways of Phytochemicals

The efficacy and safety of phytochemicals are highly dependent on their bioavailability, which determines their absorption, distribution, metabolism, and excretion (ADME) within the body (Bohn et al. 2015). Several factors affect bioavailability, including poor absorption, which is particularly common among polyphenols, terpenoids, and certain alkaloids. Phytochemicals have low water solubility, leading to reduced gastrointestinal absorption and limited systemic availability; rapid metabolism further affects the bioavailability of many phytochemicals, particularly flavonoids, alkaloids, and isothiocyanates. These compounds are quickly metabolized by the liver and excreted, diminishing their bioactivity before exerting therapeutic effects, synergistic and antagonistic interactions. Flavonoids and organosulfur compounds can enhance the absorption of other phytochemicals by modulating enzyme activity or altering membrane permeability. Conversely, tannins and dietary fibers may bind to certain phytochemicals, reducing their bioavailability by limiting intestinal uptake. For instance, piperine from black pepper significantly increases curcumin absorption, whereas dietary fiber may reduce the absorption of carotenoids (Bohn et al. 2015).

2.2 Interaction Between Drugs and Phytochemicals

Understanding the body's metabolic and regulatory processes is crucial when assessing the safety of phytochemicals. Key hormones such as insulin, glycogen, cortisol, and melatonin play interconnected roles in maintaining physiological balance because they regulate energy metabolism, stress response, and circadian rhythm, all of which can be influenced by phytochemical interactions (Baxter and Ray 2020). Insulin, produced by the pancreas, facilitates glucose uptake into cells, lowering blood sugar levels and promoting energy storage. Glycogen, though not a hormone, serves as a stored form of glucose in the liver and muscles, which is converted back to glucose when energy is needed. Cortisol, a stress hormone released by the adrenal glands, helps regulate metabolism, manage stress, and maintain blood sugar levels through gluconeogenesis (Poggiogalle et al. 2018). Meanwhile, melatonin, secreted by the pineal gland, governs the sleep–wake cycle by signaling the body to prepare for rest, playing a crucial role in circadian rhythm regulation. Together, these hormones contribute to the complex interplay of energy balance, stress response, and sleep regulation. Given that phytochemicals can influence these processes through various mechanisms, regulatory oversight and controlled consumption are essential to ensure their safe use. Below are some key factors that affect phytochemical interactions (Fasinu et al. 2012).

2.3 Phytochemicals and CPY450 Enzyme Modulation

Phytochemicals can significantly impact drug metabolism by modulating the activity of cytochrome P450 (CYP) enzymes, which are responsible for the breakdown and clearance of many pharmaceutical compounds. One well-documented example is *Hypericum perforatum,* commonly known as St. John's Wort, whose phytocompounds are responsible for inducing CYP3A4, an enzyme that plays a key role in metabolizing a wide range of drugs. By increasing CYP3A4 activity, the phytochemicals accelerate the breakdown of medications such as immunosuppressants, anticoagulants, and contraceptives, ultimately reducing their efficacy and potentially leading to treatment failure. A daily dose of 900–1800 mg of St. John's Wort extract has been shown to significantly alter drug metabolism, making it unsafe for individuals on critical medications (Hall et al. 2003). Flavonoids present in grapefruit are potent inhibitors of CYP3A4; this inhibition prevents the normal breakdown of drugs, leading to elevated drug levels in the bloodstream. Consuming just 200–300 mL of grapefruit juice can inhibit CYP3A4 for up to 24 h, which may dangerously increase drug concentrations in individuals taking medications metabolized by this enzyme (Bailey et al. 2013).

2.4 Phytochemical Interactions with Anticoagulants

The interaction between phytochemicals and anticoagulant medications is another critical safety concern. Flavonoids, which are widely found in fruits, vegetables, and herbal supplements, have been shown to influence blood coagulation. Quercetin, a well-known flavonoid present in onions, apples, and tea, has been found to potentiate the effects of blood thinners like warfarin. This can lead to an excessive reduction in blood clotting, significantly increasing the risk of spontaneous bleeding, hemorrhages, and prolonged wound healing. In a typical diet, quercetin intake varies based on food choices, with most individuals consuming between 10 and 100 mg per day from natural sources. However, daily quercetin intake above 1000 mg has been associated with increased bleeding risk in individuals taking anticoagulants, making it crucial for those on blood thinners to limit their intake (Patel et al. 2022).

2.5 Hypoglycemic Risks of Phytochemicals

Certain phytochemicals have glucose-lowering properties, which, while beneficial for managing diabetes, can also pose risks when combined with pharmaceutical hypoglycemic agents. Berberine, a plant-derived alkaloid found in herbs like *Berberis vulgaris* (barberry), has been extensively studied for its role in enhancing glucose metabolism and improving insulin sensitivity. However, when taken alongside diabetic medications such as metformin or insulin, berberine can further lower blood sugar levels, increasing the risk of hypoglycemia. Doses of 900–1500 mg per day have been linked to dangerous drops in blood sugar, particularly when combined with diabetic medications. Severe hypoglycemia can cause dizziness, confusion, loss of consciousness, and, in extreme cases, coma. Thus, patients using blood sugar-regulating medications should be closely monitored when incorporating berberine into their regimen (Och et al. 2022).

2.6 Neurological Effects of Phytochemicals

Phytochemicals can also influence neurological function by interacting with the central nervous system. Alkaloids, a diverse class of plant-derived compounds, are particularly notable for their effects on neurotransmitter signaling. Caffeine, one of the most widely consumed alkaloids, acts as a central nervous system stimulant by blocking adenosine receptors, leading to increased alertness and reduced fatigue. However, when consumed in high quantities or combined with other stimulants, caffeine can cause insomnia, anxiety, and palpitations. A normal diet consists of 250 ml of coffee/tea containing 65 mg of caffeine; however, a daily caffeine intake exceeding 400 mg has been associated with nervousness, heart palpitations, and sleep disturbances, while doses above 1000 mg can cause serious cardiac and neurological effects (Rathod 2023). On the other hand, phytochemicals may also interfere with sedatives and depressants, leading to unpredictable neurological responses

(Shaikh et al. 2018). For instance, caffeine can reduce the efficacy of benzodiazepines, which are commonly prescribed for anxiety and sleep disorders, by counteracting their sedative effects (Ye et al. 2023). This can result in inadequate symptom control and increased nervous system excitability. Given the potential for such interactions, individuals using central nervous system medications should carefully regulate their intake of neuroactive phytochemicals to avoid unintended effects on brain function.

3 Long-Term Safety and Chronic Exposure Risks

Prolonged consumption of certain phytochemicals may pose chronic health risks, including genotoxicity, carcinogenicity, endocrine disruption, and organ toxicity (Chen et al. 2011; Süntar and Yakıncı 2020). Understanding these potential adverse effects is crucial for informed dietary and therapeutic choices.

3.1 Genotoxicity and Carcinogenicity

Aristolochic acids (AAs) are a group of naturally occurring nitrophenanthrene carboxylic acids found in plants of the Aristolochia and Asarum genera, commonly used in traditional medicine across various cultures, particularly in Chinese, Ayurvedic, and European herbal practices. These compounds have historically been incorporated into herbal formulations for treating ailments such as arthritis, edema, menstrual disorders, gastrointestinal issues, and respiratory conditions due to their purported anti-inflammatory and analgesic properties. Additionally, they have been included in weight-loss and detoxification regimens, further increasing their prevalence in herbal medicine. However, despite their therapeutic applications, AAs have been implicated in severe nephrotoxicity and urothelial malignancies, raising significant safety concerns. Studies have demonstrated that rats administered with AAs at doses of 0.1, 1, and 10 mg/kg body weight over three months developed tumors in the forestomach, kidney, renal pelvis, and urinary bladder, with tumor incidences of approximately 25%, 85%, and 100%, respectively. These findings underscore the potent carcinogenic potential of AAs, even at relatively low exposure levels (Han et al. 2019; Li et al. 2020).

3.2 Endocrine Disruption

Phytoestrogens, such as isoflavones found in soy products, have a chemical structure similar to endogenous estrogens, allowing them to interact with estrogen receptors and potentially influence hormonal balance. While moderate consumption is generally considered safe, excessive intake has raised concerns about potential endocrine-disruption effects. Some studies suggest that high levels of phytoestrogens may interfere with reproductive health and hormone-regulated processes, but the overall impact remains a subject of ongoing research. Further investigation is needed to clarify their long-term effects on hormonal regulation and overall health (Rietjens et al. 2016).

3.3 Hepatic and Renal Toxicity

Certain polyphenols and herbal extracts, when consumed in excessive amounts, may overwhelm liver detoxification pathways, leading to hepatotoxicity. For example, pyrrolizidine alkaloids, present in various herbal remedies, have been associated with liver damage, including hepato-sinusoidal obstruction syndrome (Yang et al. 2017). Additionally, some alkaloids and tannins have demonstrated nephrotoxic effects with prolonged exposure. Research indicates that high concentrations of tannins can adversely affect kidney tissues in rats (Church et al. 2015). Tannins are commonly found in foods such as tea, coffee, wine, nuts, and certain fruits like pomegranates and berries. While moderate consumption of tannin-rich foods is generally considered beneficial due to their antioxidant and anti-inflammatory properties, excessive intake or prolonged consumption of highly tannic herbal teas or unprocessed plant extracts has been linked to potential nephrotoxicity. Studies suggest that high tannin exposure may interfere with iron absorption, contribute to renal oxidative stress, and promote kidney damage over time. These findings highlight the importance of moderating intake of tannin-rich foods and herbal supplements to ensure safety while maximizing their health benefits (Patel-Rodrigues et al. 2024).

4 Safety Strategies for Phytochemical Use

To mitigate the risks associated with phytochemicals while maximizing their health benefits, the following safety strategies should be adopted (Bode and Dong 2015).

4.1 Clinical Trials and Toxicology Studies

Extensive preclinical and clinical studies are essential to establish the safety, efficacy, and optimal dosage of phytochemicals. These studies systematically evaluate potential adverse effects, safe consumption thresholds, and long-term health implications. Research typically progresses from in vitro and animal models to human clinical trials, where factors such as bioavailability, metabolism, and drug interactions are closely monitored. Understanding the pharmacokinetics and pharmacodynamics of phytochemicals through such investigations ensures their safe and effective incorporation into therapeutic and dietary applications (Scott et al. 2009).

4.2 Standardized Extracts

Ensuring the reliability of phytochemical-based interventions requires the use of standardized extracts. Standardization involves precisely identifying and quantifying bioactive compounds within formulations to maintain consistency in potency, purity, and therapeutic effects across different batches. Since plant-derived compounds can vary significantly due to environmental factors, cultivation methods, and extraction processes, standardization minimizes these inconsistencies. This approach allows researchers and healthcare providers to offer more predictable and reproducible health benefits from phytochemicals (Pathrikar et al. 2025).

4.3 Personalized Nutrition and Medicine

Individual genetic makeup, metabolic efficiency, and pre-existing health conditions play a crucial role in how the body processes phytochemicals. Some individuals metabolize these compounds more efficiently, while others may experience heightened sensitivity or adverse reactions due to genetic variations in metabolic enzymes, such as those in the cytochrome P450 family. Additionally, underlying health conditions, medication use, and dietary patterns must be considered to avoid harmful interactions and optimize therapeutic outcomes. Emerging fields like nutrigenomics and precision medicine enable tailored phytochemical interventions, enhancing both safety and effectiveness (Chen 2020).

4.4 Monitoring and Regulation

Ongoing pharmacovigilance and strict regulatory oversight are critical to tracking potential adverse effects and ensuring consumer safety. Organizations such as the FDA, EMA, and WHO establish safety standards and guidelines for phytochemical-based products, requiring evidence of efficacy and safety before approval. Post-market surveillance helps identify unforeseen side effects, ensuring that recommendations remain current and scientifically supported. Through a combination of rigorous clinical research, standardization, personalized approaches, and regulatory oversight, phytochemicals can be safely integrated into medicine and nutrition, maximizing their benefits while minimizing potential risks. Phytochemicals, while offering significant health benefits, require careful evaluation to prevent potential toxicity, drug interactions, and long-term adverse effects (Dillard and German 2000). Regulatory frameworks, standardized dosing, and rigorous scientific research play vital roles in ensuring their safe and effective utilization. A balanced approach, incorporating evidence-based guidelines and personalized recommendations, is essential to harness the therapeutic potential of phytochemicals while minimizing risks to human health (Mtewa et al. 2021).

5 Dosage and Standardization of Phytochemicals

Phytochemicals are gaining prominence for their role in disease prevention and treatment, but their true potential lies in achieving the right balance of efficacy and safety. While improper dosing can lead to reduced effectiveness or toxic effects, the absence of standardization introduces variability that undermines the consistency and reliability of herbal formulations. Ensuring precise dosage and rigorous standardization is key to harnessing their therapeutic benefits while minimizing risks. For instance, curcumin from turmeric is widely known for its anti-inflammatory properties, but its poor bioavailability necessitates careful dosage optimization and bioenhancers like piperine (Rajput et al. 2022). Similarly, excessive consumption of green tea catechins that contain epigallocatechin gallate (EGCG) has been linked to liver toxicity, highlighting the need for precise dosing guidelines (Younes et al. 2018). Standardization through advanced analytical techniques such as high-performance liquid chromatography (HPLC), gas chromatography-mass spectrometry (GC-MS), and nuclear magnetic resonance (NMR) spectroscopy ensures consistent potency and safety, making phytochemical-based therapies more reliable.

5.1 Importance of Proper Dosage in Phytochemical Efficacy and Safety

As synthetic drug use has increased, concerns about its side effects have led more and more people to rely on herbal medicines extracted from different parts of plants. Nevertheless, the efficacy and safety of these natural remedies can vary due to huge fluctuations in the concentration of bioactive phytochemicals. Various factors such as plant species, climatic change, and geographical conditions of plant harvesting time and extraction methods lead to inconsistencies, making it essential to regulate their dosage and standardize the phytochemical drug (Sachan et al. 2016). An optimal dose ensures maximum therapeutic effect and minimum side effects, and this balance becomes important in both clinical and nutritional settings. The right amount of dose is the key because many phytochemicals show biological activity in a dose-dependent curve. At subtherapeutic levels, these compounds will show lower or undesired biological activity, hence becoming ineffective. However, an overdose may, in turn, become toxic or counterproductive. Some suggest that antioxidants may be helpful in moderate doses, but overdosing is strongly linked with increased mortality in clinical studies (Demonty et al. 2008). An important factor to be considered for determining the clinical dosage is the Therapeutic Index (TI), which reflects a drug's safety by showing the gap between its effective and toxic doses, influenced by drug properties (Muller and Milton 2012). It is calculated by the formula,

$$TI = LD_{50} / ED_{50}$$

where LD_{50} (Lethal Dose 50%) is the dose at which 50% of test subjects experience fatal toxicity, and ED_{50} (Effective Dose 50%) is the dose at which 50% of subjects exhibit the intended therapeutic effect. A higher TI indicates a greater safety margin, meaning the drug can be administered with a lower risk of toxicity, while a lower TI suggests a narrow therapeutic window, requiring precise dosage control to avoid adverse effects (Tamargo et al. 2015). This precision is particularly crucial in herbal medicine, where standardization is often lacking, leading to variability in the concentration of active compounds (Panossian 2023). Unlike synthetic drugs, where dosage is carefully regulated, the potency of phytochemicals can vary due to differences in plant species, cultivation methods, and extraction processes. Standardizing the dosage of phytochemicals is further complicated by biological and pharmacokinetic factors such as bioavailability, metabolism, and individual patient variability, which influence how effectively the body absorbs, distributes, and eliminates these compounds. Understanding and optimizing the TI of phytochemicals is, therefore, essential to ensure both their efficacy and safety in clinical applications. Bioavailability refers to the number of phytochemicals that escape radical clearance and make it to systemic circulation in its active form. This variation reinforces the importance of rigorous clinical trials and the toxicological assessments needed for setting up safe and effective dosing guidelines.

5.2 Challenges in Determining Therapeutic Doses

The complexity and variability of natural compounds make it difficult to determine the therapeutic dosages for phytochemicals. Phytochemicals are frequently found in complex mixtures within plant extracts, in contrast to synthetic drugs, which usually contain a single active ingredient. The identification and measurement of the precise compounds causing therapeutic effects are made more difficult by this complexity. The absence of standardization in herbal products is a major obstacle. Inconsistencies in the concentration of active constituents can result from differences in plant cultivation, harvesting, and processing. As a result, one cannot simply extrapolate from one region's experience with a specific plant's safety and effectiveness at a given dose to another (Kofi Busia 2024).

Furthermore, the bioavailability can be affected by the chemical structure of phytochemicals, the presence of other dietary components, and individual metabolic variations. For example, some substances may be quickly metabolized or have poor absorption rates, which limits their potential as medicines. Herbal products can have negative side effects or reduced therapeutic efficacy because they frequently lack standardized dosages and can interact with other medications (Bhusnure et al. 2019). To evaluate the safety and effectiveness of phytochemicals, both by themselves and in conjunction with other treatments, extensive clinical trials and toxicological evaluations are required (Balkrishna et al. 2024). In addition, inadequate reporting, problems with methodological quality, and an unclear risk of bias frequently impede the reproducibility of clinical studies involving phytochemicals. A steady-state concentration of the active ingredient must be reached during the course of treatment in order to achieve repeatable therapeutic efficacy, but this is frequently not clinically demonstrated (Panossian 2023).

5.3 Methods for Standardization and Quality Control

Standardizing herbal medicines is to define the permeable limits of qualitative and quantitative attributes in order to ensure uniformity, efficacy, safety, and repeatability (Sachan et al. 2016). To confirm and quantify active compounds, advanced analytics including HPLC, GC- MS, and NMR spectroscopy have been used to minimize the chance of toxicity. Standardization across quality control guidelines thus helps in assuring predictable outcomes of the phytochemical drug. The safety and effectiveness of herbal products and medicinal plants depend on their quality assessment. It includes several criteria grouped into botanical, physical, chemical, and biological evaluations. By means of these studies, herbal formulations' identity, purity, and strength are known, therefore guaranteeing quality standard compliance. The five-stage drug development and approval process is illustrated in Fig. 10.1.

To achieve standardization, several key parameters are assessed to ensure the quality, purity, and efficacy of herbal products. One of the fundamental steps in this process is macro- and microscopic examination, which plays a crucial role in accurately identifying plant varieties and detecting adulterants that may compromise the authenticity of the material. The assessment of foreign organic matter further enhances purity by eliminating extraneous substances that do not originate from the source plant, thereby preventing contamination (Wang et al. 2023). Ash value determination is another essential parameter, particularly for crude drugs in powdered form, as it helps in assessing quality by quantifying inorganic residues left after incineration. This is complemented by moisture content analysis, which is critical

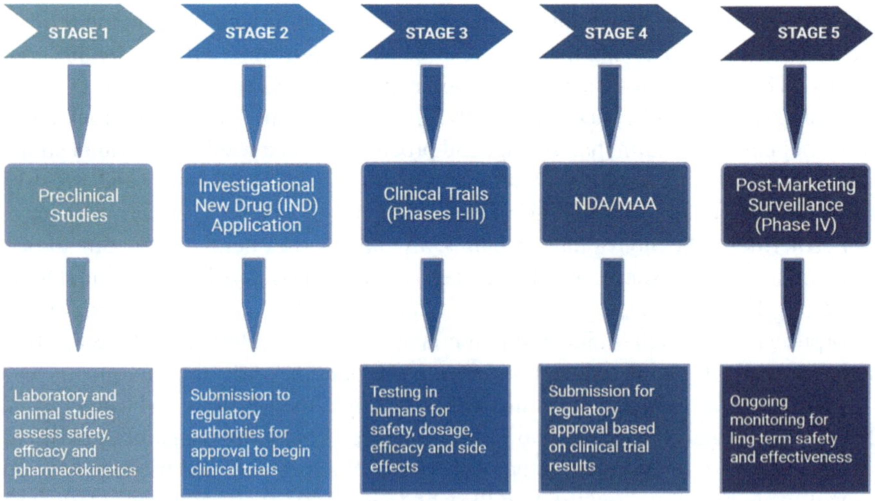

Fig. 10.1 Overview of the five-stage drug development and approval process from preclinical studies to post-marketing surveillance. (Dinesh 2024)

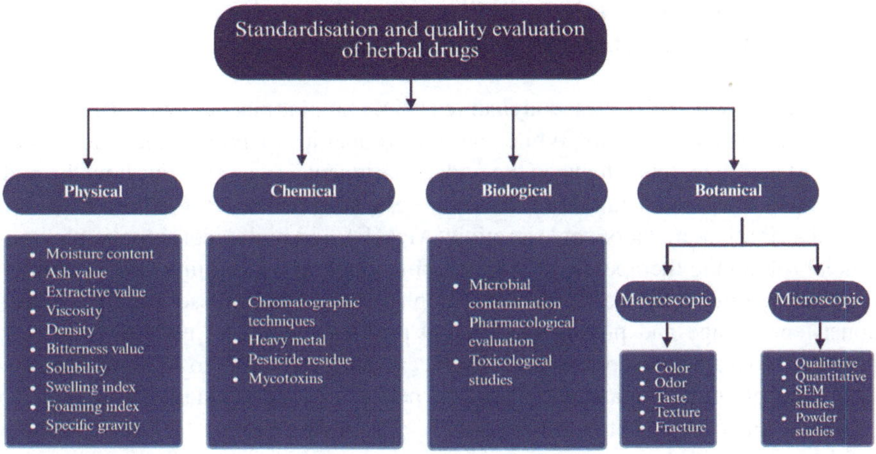

Fig. 10.2 Parameters for the standardization and quality evaluation of herbal drugs. (Hussain et al. 2024)

for evaluating the stability of herbal materials. Lower moisture content is desirable as it reduces the risk of microbial growth, enzymatic degradation, and chemical deterioration, thereby prolonging shelf life. Figure 10.2 illustrates the various parameters involved in the standardization of herbal medicine.

Extractive values provide insight into the presence and quantity of bioactive constituents that can be extracted under different solvent conditions, which is essential for understanding the drug's therapeutic potential. Similarly, crude fiber analysis is conducted to measure the proportion of indigestible plant material, particularly woody components, which serves as an additional criterion for assessing purity (Guo et al. 2019). Chemical evaluation is performed through both qualitative and quantitative methods. Qualitative chemical evaluation involves identifying and characterizing phytochemical constituents using techniques such as phytochemical screening, solvent extraction, purification, and compound characterization. Quantitative chemical evaluation, on the other hand, focuses on estimating the concentration of key bioactive compounds, ensuring that formulations contain the necessary therapeutic components in appropriate amounts. Chromatographic analysis further aids in crude drug identification by using major chemical constituents as markers, allowing for precise characterization and standardization. Additionally, toxicological studies are conducted to evaluate the safety of herbal products, assessing the presence of pesticide residues, heavy metals, and microbial contamination. These studies often include safety assessments in animal models, such as LD_{50} studies, to determine potential toxicity levels. By integrating these assessments, standardization ensures consistency, quality, and therapeutic efficacy of herbal products, making them safer and more reliable for medicinal use.

5.4 Pharmacokinetics and Pharmacodynamics of Key Bioactive Compounds

In drug development, pharmacodynamics (PD) and pharmacokinetics (PK) are two related but distinct concepts. While pharmacodynamics focuses on the drug's mechanism of action and its effects on the body, pharmacokinetics explains how the body absorbs, distributes, metabolizes, and excretes a drug. PD and PK influence absorption, distribution, metabolism, excretion (ADME), and biological activity, playing a crucial role in the therapeutic efficacy of phytochemicals. To minimize variation in therapeutic outcomes, standardization of phytochemicals is essential for ensuring consistent dosage and predictable PK/PD profiles. For many plant-derived compounds, bioavailability remains a significant challenge due to factors like rapid clearance, gut microbiota metabolism, and poor solubility. Turmeric's curcumin, for example, has poor systemic absorption, requiring formulation techniques such as lipid-based delivery or nanoparticle encapsulation (Rajput et al. 2022). The therapeutic effects of phytochemicals are influenced by their pharmacodynamic interactions with cellular targets, such as enzymes and receptors. For instance, polyphenols like resveratrol exhibit anti-inflammatory and antioxidant properties through NF-κB inhibition, while alkaloids like berberine regulate glucose metabolism by modulating AMPK signaling (McCubrey et al. 2017). Quantifying active ingredients, refining extraction techniques, and ensuring batch-to-batch consistency are key aspects of standardization. Reliable dosage forms are developed using sophisticated analytical techniques such as mass spectrometry and HPLC. Understanding PK/PD relationships is essential for determining the right therapeutic dosages while minimizing toxicity risks, ultimately enhancing the efficacy and safety of phytochemical-based interventions. One widely used framework for assessing the drug-likeness of phytochemicals is Lipinski's rule of five, which provides guidelines for evaluating oral bioavailability. According to this rule, a compound is more likely to be orally active if it meets the following criteria: (1) no more than five hydrogen bond donors, (2) no more than ten hydrogen bond acceptors, (3) a molecular weight below 500 Da, (4) a partition coefficient (LogP) of less than five, and (5) fewer than five rotatable bonds. Many phytochemicals, particularly flavonoids and polyphenols, violate one or more of these rules, leading to challenges in their bioavailability and necessitating formulation strategies to enhance their therapeutic potential (Kar et al. 2012). The Absorption, Distribution, Metabolism, Excretion, and Toxicity (ADMET) profile of a compound is another crucial factor in drug development. Computational models and in vitro assays help predict ADMET properties, enabling researchers to optimize lead compounds before clinical trials. Poor ADMET characteristics can lead to rapid metabolism, poor systemic exposure, or toxicity concerns, making early evaluation essential. Several software tools aid in ADMET predictions, including SwissADME, ADMET Predictor, GastroPlus, pkCSM, and QikProp. These tools evaluate solubility, permeability, metabolism by cytochrome P450 enzymes, and potential toxicity risks, providing insights that guide the development of optimized phytochemical formulations (Pires et al. 2018). To overcome bioavailability and pharmacokinetic limitations, advanced formulation approaches such as nanoencapsulation, liposomes, solid lipid nanoparticles, and self-emulsifying drug delivery systems are increasingly employed. These technologies enhance solubility, prolong circulation time, and improve cellular uptake, thereby increasing the clinical efficacy of phytochemical-based therapies. By integrating PK/PD studies with computational modeling and advanced drug delivery systems, the therapeutic potential of bioactive phytochemicals can be maximized, paving the way for their successful translation into clinically effective treatments.

5.5 Case Study of Safe Dosages

Clinical studies have helped establish safe dosage ranges for many herbs, such as *Bougainvillea glabra* and *Ficus religiosa*, particularly in the context of their traditional medicinal uses in India for conditions including heartburn, sore throat, and hepatitis, in Nigeria for inflammation and pain, and in Thailand for stomach problems. It is also used in Africa to enhance intestinal problems (Ornelas García et al. 2023). *F. religiosa* has also been extensively used in Indian traditional medicine; its bark acts as an antibacterial, antiviral, and antidiarrheal agent; its leaves are recognized for their antivenom qualities and function in controlling the menstrual cycle. It has been used in Bangladesh to treat cancer, infections, and inflammation. Its tender branches are also used to treat fever by using it as a toothbrush, its fruits function as laxatives, and its latex is consumed as a tonic (Chandrasekar et al. 2010).

Traditionally, *Saraca asoca* (Ashoka tree) has also been used in Ayurveda to control gynecological problems, especially in the treatment of uterine disorders, menstrual irregularities, and infertility issues. A natural alternative for managing reproductive health, it shows estrogenic qualities that help control the menstrual cycle and enhance endometrial health. *Curcuma longa* (Turmeric) is well known for its anti-inflammatory, antioxidant, and antibacterial effects. Its active ingredient, curcumin, has been investigated for its effects on cancer prevention, wound healing, gastrointestinal diseases, and arthritis. *Ocimum sanctum* (Tulsi) has been used in traditional Indian medicine to treat respiratory disorders, digestive issues, and inflammatory conditions. It is also known for its adaptogenic properties, helping to reduce stress and enhance immunity. Key physicochemical properties, including ash values, extractive yields, and the chromatography retention factor (Rf) values, are shown in Table 10.1, which provides information on their composition and therapeutic potential (Singh et al. 2015; Rakam and Raja 2017; Bisht et al. 2017; Burman and Kanaujia 2019; Kukde et al. 2021; Pooja et al. 2025).

6 Regulatory Framework for Phytochemicals

6.1 Global Regulations

An organized collection of laws, rules, and guidelines created by governmental bodies and global organizations to monitor the efficacy, safety, quality, and marketing

Table 10.1 Comparative standardization and chromatography profiling for safe dosage assessment

Herbs	Physicochemical parameters						Chromatograph—R_f values	Reference
	Total ash (% w/w)	Acid insoluble ash (% w/w)	Sulphated ash (% w/w)	Water soluble ash (% w/w)	Water soluble extractive (% w/w)	Alcohol soluble extractive (% w/w)		
Bougainvillea glabra (whole plant)	6.26 ± 0.05	1.35 ± 0.03	1.10 ± 0.07	1.34 ± 0.04	23.45 ± 1.13	12.32 ± 0.41	TLC for steroids R_f: 0.66	Rakam and Raja (2017)
Ficus religiosa (leaf)	7.8	5.8	1.7	2.5	3.9	2.8	TLC for leaf extract R_f: 0.48	Kukde et al. (2021)
Saraca asoca (flower)	4.73 ± 0.20	0.23 ± 0.03	–	2.5 ± 0.25	24.33 ± 0.76	7.0 ± 1.50	HPTLC for β-sitosterol R_f: 0.80	Bisht et al. (2017)
Curcuma longa (whole plant)	10.03	1.48	–	5.87	23.46	20.65	TLC for variety JHANSI 10 R_f: 0.817	Burman and Kanaujia (2019), Pooja et al. (2025)
Ocimum sanctum (whole plant)	15.6	2.5	16.7	–	20.4	9.6	TLC for eugenol R_f: 0.7	Singh et al. (2015)

of goods is known as a regulatory framework. In the context of phytochemicals, it ensures that before being sold to customers, herbal medications, dietary supplements, and other substances derived from plants fulfill certain safety and health requirements. The need for a global regulatory framework for phytochemical drugs is critical to ensure that all herbal medications, dietary supplements, and other plant-derived substances meet stringent safety and health standards: (i) Safety assurance—nature does not always equate to safety; certain phytochemicals may interact with medications or have harmful effects. Laws guarantee that products are examined for possible adverse effects; (ii) Validation of efficacy—manufacturers might make inflated or fraudulent health claims in the absence of regulations. Therapeutic benefits are validated by scientific evidence and clinical trials; (iii) Quality control—purity and consistency in active ingredients are guaranteed by standardization stops batch-to-batch variations, contamination, and adulteration; (iv) Consumer protection—stops deceptive advertising and unsubstantiated claims guarantees accurate labeling that includes instructions, warnings, and dosage; (v) Market authorization and compliance—laws governing product approval and marketing must be followed by businesses keeps dangerous or unlawful products off the market; (vi) International harmonization—although each nation has its own set of rules, frameworks such as WHO guidelines aid in standardization promotes the international trade in herbal supplements and medications (Table 10.2).

Table 10.2 Global regulatory body for phytochemical drugs

Regulatory body	Country/ region	Key regulations & guidelines
WHO (World Health Organization)	Global	Provides guidelines for assessing the safety, efficacy, and quality of herbal medicines and phytochemicals
FDA (Food and Drug Administration)	United States	Regulates botanical drugs under the Dietary Supplement Health and Education Act (DSHEA) and the FDA's botanical drug development guidelines
EMA (European Medicines Agency)	European Union	Governs herbal medicines under the Committee on Herbal Medicinal Products (HMPC) and Directive 2004/24/EC for Traditional Herbal Medicinal Products
AYUSH (Ministry of Ayurveda, Yoga & Naturopathy, Unani, Siddha, and Homeopathy)	India	Regulates traditional and herbal medicines, ensuring compliance with the Drugs and Cosmetics Act, 1940, and related pharmacopeial standards
TGA (Therapeutic Goods Administration)	Australia	Regulates complementary medicines, including herbal and phytochemical-based products, under the Therapeutic Goods Act 1989
NMPA (National Medical Products Administration, formerly CFDA)	China	Governs Traditional Chinese Medicines (TCM) and phytochemicals under the Pharmacopoeia of the People's Republic of China

Source: Sahane and Basarkar (2023)

6.2 Approval Process for Phytochemical-Based Drugs and Supplements

The approval process for phytochemical-based drugs and supplements varies based on their intended use, classification, and regulatory jurisdiction. Regulatory agencies such as the United States FDA, the EMA, the Central Drugs Standard Control Organization (CDSCO) in India, and other national regulatory bodies oversee the evaluation and approval of these compounds (Aladysheva et al. 2024). While phytochemical-based drugs undergo rigorous testing similar to synthetic drugs, the approval of herbal supplements is generally more lenient. Figure 10.3 presents several countries along with their respective regulatory authorities.

In the United States, phytochemical-based drugs must undergo a rigorous Investigational New Drug application process before clinical trials begin, followed by preclinical studies, Phase I, II, and III clinical trials, and the submission of a New Drug Application (NDA) to the FDA for approval. Herbal supplements, on the other hand, are regulated under the Dietary Supplement Health and Education Act and do not require pre-market approval, but must be Generally Recognized as Safe (GRAS).

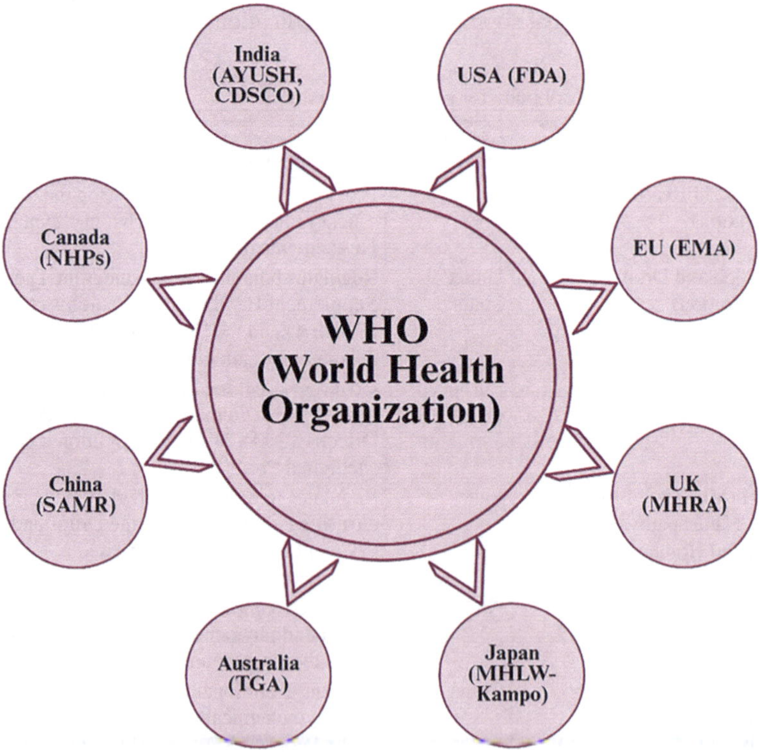

Fig. 10.3 Global regulatory authorities involved in drug approval under the coordination of the World Health Organization (WHO). (Sharma 2015)

Manufacturers must ensure product safety and label accuracy, and the FDA monitors post-market safety. In the European Union (EU), the EMA oversees the approval of phytochemical-based drugs through the Marketing Authorization Application process, evaluated by the Committee for Medicinal Products for Human Use (Lapenna et al. 2013). Herbal medicinal products may be classified under the Traditional Herbal Medicinal Products Directive, which allows market access based on traditional use and safety evidence without full clinical trials, provided they have been used safely for at least 30 years (15 years in the EU).

In India, phytochemical-based drugs are subject to the Drugs and Cosmetics Act, 1940, and require preclinical and clinical evaluations similar to synthetic drugs before obtaining approval from the CDSCO. Ayurvedic, Siddha, and Unani drugs and herbal supplements are regulated by the Ministry of Ayurveda, Yoga & Naturopathy, Unani, Siddha, and Homoeopathy (AYUSH) under the Drugs and Cosmetics Act (Rule 158-B). Traditional formulations may be exempt from clinical trials if referenced in classical texts, whereas proprietary ASU medicines require scientific validation and licensing. In China, phytochemical-based drugs must undergo preclinical research, clinical trials, and NDA to the National Medical Products Administration. Traditional Chinese Medicine products with well-documented use and safety records can undergo a simplified registration process under the Traditional Chinese Medicine Law (2017). In Japan, the Pharmaceuticals and Medical Devices Agency regulates phytochemical-based drugs, requiring rigorous clinical trials for approval under the Pharmaceutical and Medical Device Act. Kampo medicines, or traditional Japanese herbal medicines, are recognized under Japan's National Health Insurance and must comply with Good Manufacturing Practices but have a less stringent approval pathway compared to synthetic drugs. In Australia, the Therapeutic Goods Administration regulates phytochemical-based drugs, requiring full clinical trial evaluations and listing under the Australian Register of Therapeutic Goods. Complementary medicines, including herbal supplements, are subject to Listing (low-risk) or Registration (high-risk) categories and must meet safety, quality, and efficacy standards, though clinical trials are not always required. The approval process for phytochemical-based drugs and supplements varies significantly between countries, reflecting differences in regulatory standards and traditional medicine acceptance. While Western countries generally require extensive clinical trials, regions with strong traditional medicine systems (such as India, China, and Japan) often provide alternative pathways for herbal formulations. Understanding these regulatory frameworks is essential for researchers and companies developing phytochemical-based therapeutics for global markets.

6.3 Challenges in Phytochemical Regulation and Commercialization

6.3.1 Regulatory Disparities and Classification Issues

Different nations categorize phytochemicals under multiple categories (pharmaceuticals, functional foods, herbal remedies, or supplements), which results in uneven approval procedures (Ekor 2014). Trade and commercialization are made more difficult by the absence of unified international regulations. For instance, the FDA and EMA have distinct regulations for botanical medications.

6.3.2 Scientific Validation and Standardization

Due to the intricate and heterogeneous nature of phytochemical compositions, establishing standardized dosage and quality control parameters remains a significant challenge. There are differences in clinical efficacy since there are no widely recognized pharmacokinetics and bioavailability testing procedures (Umamaheswari et al. 2021). Although consistent approaches are missing, regulatory agencies such as the FDA, EMA, and CFDA demand substantial scientific proof (Leufkens 2019).

6.3.3 Intellectual Property Rights and Patenting Issues

Without the development of innovative extraction or formulation methods, natural compounds typically lack strong patentability. Companies are deterred from funding extensive clinical trials by inadequate patent protection (Nath et al. 2020).

6.3.4 Toxicity and Safety Concerns

Although phytochemicals are natural, they may have unstudied harmful consequences when used in large quantities or for an extended period of time (Ekor 2014). Pharmacological interactions necessitate lengthy and expensive clinical trials.

6.3.5 Raw Material Quality and Variability

Geographical and environmental factors influence the phytochemical composition, which results in inconsistent medicinal benefits (Chandra et al. 2019). For botanical medications, the absence of defined good manufacturing practices (GMPs) leads to uneven quality control.

6.3.6 Regulatory Hurdles for Market Approval

Phytochemicals find it challenging to compete with synthetic pharmaceuticals due to strict safety and efficacy standards that delay market access. Randomized controlled trials, which are costly and time-consuming for plant-based substances, are mandated by regulatory bodies (Hossain et al. 2022).

6.3.7 Consumer Perception and Marketing Barriers

Regulatory agencies limit the commercial attraction of products based on phytochemicals by limiting their ability to make health claims (Domínguez Díaz et al. 2020). Labeling and advertising laws limit commercialization even when consumers want natural products (Petty 2015).

6.4 Case Studies of Approved Phytochemicals

Phytochemicals like curcumin, epigallocatechin gallate, and resveratrol have gained significant attention for their therapeutic potential and regulatory acceptance.

Table 10.3 Comparative regulatory approval, dosage, bioavailability, and toxicity aspects of various approved phytochemicals

Aspects	Curcumin	EGCG	Resveratrol
Regulatory status	FDA: GRAS, EU: E100, India: Ayurveda	FDA: GRAS, EFSA: Intake limits	FDA: Allowed in supplements, EU: Approved
Uses	Anti-inflammatory, antioxidant, neuroprotective	Antioxidant, cardioprotective, anticancer	Cardioprotective, anti-aging, anti-inflammatory
Dosage	500–2000 mg/day	300–800 mg/day	150–500 mg/day
Bioavailability	Low, improved with piperine & nano-formulations	Low, improved with lipid-based carriers	Rapid metabolism, improved with micronized forms
Toxicity	High doses (>8000 mg/day) may cause GI issues	High doses (>800 mg/day) linked to liver toxicity	High doses (>5000 mg/day) may cause GI & liver issues

Table 10.3 summarizes key aspects of these compounds, including their approved status, primary health benefits, recommended dosages, bioavailability challenges, and toxicity profiles. Understanding these factors is crucial for optimizing their use in clinical and dietary applications (Smoliga et al. 2011; Amalraj et al. 2017; Hu et al. 2018).

7 Future Perspectives and Challenges

Modern technology and multidisciplinary techniques progressively affect the area of phytochemical research as it develops, presenting both great possibilities and major difficulties. By allowing high-throughput virtual screening of large chemical environments, bioactivity prediction, and unmatched efficiency in molecular structure optimization, emerging technologies such as artificial intelligence and cheminformatics are revolutionizing the discovery process. To accelerate the drug development process, artificial intelligence-driven methods such as deep learning models and generative adversarial networks are being used to predict the pharmacokinetics, toxicity, and therapeutic potential of phytochemicals (Misra et al. 2024). Likewise, developments in nanotechnology are tackling important phytochemical limits, including limited bioavailability and fast metabolism. Nanocarriers include liposomes, dendrimers, and micelles that improve the solubility, stability, and targeted distribution of bioactive chemicals, therefore opening the path for their therapeutic uses in fields like cancer therapy (Fernie 2007; Alharbi et al. 2021). These analytical approaches are essential for ensuring the consistency, efficacy, and safety of nanotechnology-enhanced phytochemical therapies. The intricacy of phytochemical structures and their interactions with biological systems requires strong analytical instruments for characterization. Although techniques such as mass spectrometry and nuclear magnetic resonance spectroscopy have enhanced accuracy in spotting new compounds, scalability calls for more work even in these areas (Chihomvu et al. 2024). Another important challenge is making sure that phytochemicals are safe and effective. Many drugs fail to convert from promising preclinical evidence into successful medicines because of inadequate clinical validation. Establishing uniform doses and reducing side effects depend on rigorous clinical studies and standardized extraction techniques (Kim et al. 2021).

Including phytochemicals into individualized therapy marks a revolutionary front. Genomics and metabolomics have advanced to provide the means to personalize treatments depending on a person's genetic composition and metabolic profile. While lowering dangers, this strategy may maximize therapeutic results. Still, it requires a lot more investigation on exact clinical validation, biomarker identification, and gene-phytochemical interactions. Moreover, regulatory systems have to change to allow these developments while guaranteeing efficiency and safety. Addressing these issues calls for a multidisciplinary team of chemists, biologists, computer scientists, and doctors. To close the distance between conventional botanical knowledge and contemporary medicinal uses, translational research must be more heavily funded. Future phytochemical research has great potential to transform healthcare by means of multidisciplinary collaboration and the use of technology developments.

8 Conclusions

Phytochemicals have long been recognized for their bioactive properties, contributing to both traditional and modern medicine. Increasing research validating their medicinal potential has paved the way for their integration into pharmaceuticals, functional foods, and nutraceuticals, offering promising alternatives to synthetic drugs. However, despite their natural origin, phytochemicals are not inherently free of risks. Concerns regarding safety, bioavailability, pharmacokinetics, and potential toxicities necessitate rigorous scientific scrutiny and evidence-based regulatory frameworks to ensure their efficacy without adverse consequences. The pharmacokinetics of phytochemicals play a crucial role in their therapeutic effects. Many phytochemicals suffer from limited bioavailability, requiring innovative strategies such as nano formulation and co-administration with bioenhancers to optimize their clinical utility. Phytochemical–drug interactions raise critical safety considerations as the use of phytochemicals expands into functional foods and dietary supplements; standardized dosing guidelines and stringent safety assessments are imperative to prevent adverse effects. Moreover, long-term safety concerns such as genotoxicity, endocrine disruption, and organ toxicity necessitate detailed toxicological studies. While polyphenols and flavonoids offer antioxidant and anti-inflammatory benefits, excessive intake of certain alkaloids and terpenoids can lead to systemic toxicity. Therefore, pharmacovigilance systems are essential to monitor and mitigate potential adverse effects.

Regulatory inconsistencies further complicate the global adoption of phytochemicals. Unlike conventional drugs, many phytochemicals lack standardized purity criteria and dosage regulations across different governing bodies. Harmonizing international standards, enforcing GMPs, and establishing post-market surveillance are essential to maintaining product quality, consumer safety, and medical reliability. Furthermore, systems biology approaches, including metabolomics and high-throughput screening, will enable more precise safety assessments and therapeutic applications. The future of phytochemicals in medicine depends on a multidisciplinary approach that integrates scientific innovation, personalized medicine, and robust regulatory frameworks. A deeper understanding of their mechanisms, combined with emerging technologies, will enable their safe and effective utilization. By addressing current challenges related to safety, standardization, and therapeutic optimization, phytochemicals can be harnessed as pivotal components of modern medicine, bridging the gap between traditional botanical knowledge and contemporary pharmacological advancements.

References

Ahmed MH, Karkush SI, Ali SA, Mohammed AA (2024) Phytochemicals: a new arsenal in drug discovery. Int J Med Sci Dent Health 10:29–44. https://doi.org/10.55640/ijmsdh-10-01-03

Aladysheva ZA, Beregovykh VV, Brkich GE, Meshkovskiy AP, Belyaev VV, Pyatigorskaya NV (2024) Regulatory quality in the international circulation of medicinal products. Federalism 29:106–135. https://doi.org/10.21686/2073-1051-2024-3-106-135

Alharbi WS, Almughem FA, Almehmady AM, Jarallah SJ, Alsharif WK, Alzahrani NM, Alshehri AA (2021) Phytosomes as an emerging nanotechnology platform for the topical delivery of bioactive phytochemicals. Pharmaceutics 13:1475. https://doi.org/10.3390/pharmaceutics13091475

Amalraj A, Pius A, Gopi S, Gopi S (2017) Biological activities of curcuminoids, other biomolecules from turmeric and their derivatives—a review. J Tradit Complement Med 7:205–233. https://doi.org/10.1016/j.jtcme.2016.05.005

Awuchi CG (2019) The biochemistry, toxicology, and uses of the pharmacologically active phytochemicals: alkaloids, terpenes, polyphenols, and glycosides. J Food Pharm Sci 7:2. https://doi.org/10.22146/jfps.666

Bailey DG, Dresser G, Arnold JMO (2013) Grapefruit-medication interactions: forbidden fruit or avoidable consequences? CMAJ 185:309–316. https://doi.org/10.1503/cmaj.120951

Balkrishna A, Sharma N, Srivastava D et al (2024) Exploring the safety, efficacy, and bioactivity of herbal medicines: bridging traditional wisdom and modern science in healthcare. Future Integr Med 3:35–49. https://doi.org/10.14218/fim.2023.00086

Baxter M, Ray DW (2020) Circadian rhythms in innate immunity and stress responses. Immunology 161:261–267. https://doi.org/10.1111/imm.13166

Bhambhani S, Kondhare KR, Giri AP (2021) Diversity in chemical structures and biological properties of plant alkaloids. Molecules 26:3374. https://doi.org/10.3390/molecules26113374

Bhusnure OG, Suryawanshi S, Swamy SMV et al (2019) Standardization and quality evaluation of herbal drugs. J Drug Deliv Therap 9:1058–1063. https://doi.org/10.22270/jddt.v9i3-s.2941

Bisht A, Irshad S et al (2017) Pharmacognostical studies on *Saraca asoca* (Roxb.) Willd. Flower. Trop Plant Res 4:153–160. https://doi.org/10.22271/tpr.2017.v4.i1.022

Bode AM, Dong Z (2015) Toxic phytochemicals and their potential risks for human cancer. Cancer Prev Res (Phila) 8:1–8. https://doi.org/10.1158/1940-6207.CAPR-14-0160

Bohn T, McDougall GJ, Alegría A et al (2015) Mind the gap-deficits in our knowledge of aspects impacting the bioavailability of phytochemicals and their metabolites—a position paper focusing on carotenoids and polyphenols. Mol Nutr Food Res 59:1307–1323. https://doi.org/10.1002/mnfr.201400745

Boik J, Kirakosyan A, Kaufman PB, Seymour EM, Spelman K (2009) Interactions of bioactive plant metabolites: synergism, antagonism, and additivity. Springer EBooks:213–230. https://doi.org/10.1007/978-1-4419-0194-1_10

Burman V, Kanaujia H (2019) Characterization of phenolic compounds of turmeric using TLC. J Pharmacogn Phytochem 2:994–998

Chandi Charan Kandar (2020) Secondary metabolites from plant sources. Adv Struct Mater:329–377. https://doi.org/10.1007/978-3-030-54027-2_10

Chandra S, Stanford DA, Fletcher EJ, Walker L (2019) Raw materials production and manufacturing process control strategies. AAPS Adv Pharm Sci Ser. https://doi.org/10.1007/978-3-030-11751-1_10

Chandrasekar SB, Bhanumathy M, Pawar AT, Somasundaram T (2010) Phytopharmacology of ficus religiosa. Pharmacogn Rev 4:195–199. https://doi.org/10.4103/0973-7847.70918

Chen CH (2020) Genetic variations and polymorphisms of metabolic enzymes. Springer EBooks:155–168. https://doi.org/10.1007/978-3-030-41679-9_14

Chen X-W, Serag ES, Sneed KB, Zhou S-F (2011) Herbal bioactivation, molecular targets and the toxicity relevance. Chem Biol Interact 192:161–176. https://doi.org/10.1016/j.cbi.2011.03.016

Chihomvu P, Ganesan A, Gibbons S et al (2024) Phytochemicals in drug discovery—a confluence of tradition and innovation. Int J Mol Sci 25:8792. https://doi.org/10.3390/ijms25168792

Church RJ, Gatti DM, Urban TJ et al (2015) Sensitivity to hepatotoxicity due to epigallocatechin gallate is affected by genetic background in diversity outbred mice. Food Chem Toxicol 76:19–26. https://doi.org/10.1016/j.fct.2014.11.008

Demonty I, Ras RT, van der Knaap HCM et al (2008) Continuous dose-response relationship of the LDL-cholesterol–lowering effect of phytosterol intake. J Nutr 139:271–284. https://doi.org/10.3945/jn.108.095125

Dillard CJ, German JB (2000) Phytochemicals: nutraceuticals and human health. J Sci Food Agric 80:1744–1756. https://doi.org/10.1002/1097-0010(20000915)80:12<1744::aid-jsfa725>3.0.co;2-w

Dinesh KR (2024) The phases of clinical trials: from preclinical studies to post-market surveillance. Glob J Med Clin Case Rep 11:055–065. https://doi.org/10.17352/2455-5282.000189

Domínguez Díaz L, Fernández-Ruiz V, Cámara M (2020) An international regulatory review of food health-related claims in functional food products labeling. J Funct Foods 68:103896. https://doi.org/10.1016/j.jff.2020.103896

Ekor M (2014) The growing use of herbal medicines: issues relating to adverse reactions and challenges in monitoring safety. Front Pharmacol 4:177. https://doi.org/10.3389/fphar.2013.00177

Fasinu PS, Bouic PJ, Rosenkranz B (2012) An overview of the evidence and mechanisms of herb-drug interactions. Front Pharmacol 3:69. https://doi.org/10.3389/fphar.2012.00069

Fernie AR (2007) The future of metabolic phytochemistry: larger numbers of metabolites, higher resolution, greater understanding. Phytochemistry 68:2861–2880. https://doi.org/10.1016/j.phytochem.2007.07.010

Guo Y, Byambasuren K, Liu X et al (2019) Extraction, purification, and characterization of insoluble dietary fiber from oat bran. Trans Tianjin Univ 27. https://doi.org/10.1007/s12209-019-00224-9

Gupta R, Rajpoot K, Tekade M et al (2022) Factors influencing drug toxicity. Elsevier EBooks, pp 27–50. https://doi.org/10.1016/b978-0-323-98367-9.000147

Hall SD, Wang Z, Huang SM et al (2003) The interaction between St John's wort and an oral contraceptive. Clin Pharmacol Ther 74:525–535. https://doi.org/10.1016/j.clpt.2003.08.009

Han J, Xian Z, Zhang Y et al (2019) Systematic overview of aristolochic acids: nephrotoxicity, carcinogenicity, and underlying mechanisms. Front Pharmacol 10:648. https://doi.org/10.3389/fphar.2019.00648

Hossain CM, Gera M, Ali KA (2022) Current status and challenges of herbal drug development and regulatory aspect: a global perspective. Asian J Pharm Clin Res 15:31–41. https://doi.org/10.22159/ajpcr.2022.v15i12.46134

Hu J, Webster D, Cao J, Shao A (2018) The safety of green tea and green tea extract consumption in adults: results of a systematic review. Regul Toxicol Pharmacol 95:412–433. https://doi.org/10.1016/j.yrtph.2018.03.019

Hussain S, Nashid CM, Jamal A et al (2024) A review on standardisation of traditional herbal formulations. Int J Pharm Res Appl 9:116–123. https://doi.org/10.35629/4494-0905116123

Jodynis-Liebert J, Kujawska M (2020) Biphasic dose-response induced by phytochemicals: experimental evidence. J Clin Med 9:718. https://doi.org/10.3390/jcm9030718

Kar K, Sethi A, Archana D et al (2012) Quantitative measurement of some physico-chemical parameters for the medicinally useful natural products. Lett Drug Des Discov 9:706–730. https://doi.org/10.2174/157018012801319427

Kim B, Park JE, Im E et al (2021) Recent advances in nanotechnology with nano-phytochemicals: molecular mechanisms and clinical implications in cancer progression. Int J Mol Sci 22:3571. https://doi.org/10.3390/ijms22073571

Kofi Busia (2024) Herbal medicine dosage standardisation. J Herb Med:100889–100889. https://doi.org/10.1016/j.hermed.2024.100889

Kukde VS, Bhongade MA, Kamde SS (2021) Standardization and phytochemical screening of Ficus religiosa. Res J Pharm Technol 14:971–976. https://doi.org/10.5958/0974-360x.2021.00173.6

Lapenna S, Gemen R, Wollgast J et al (2013) Assessing herbal products with health claims. Crit Rev Food Sci Nutr 55:1918–1928. https://doi.org/10.1080/10408398.2012.726661

Leufkens HG (2019) Regulatory science: regulation is too important to leave it to the regulators. Brit J Clin Pharmacol 86:2333–2334. https://doi.org/10.1111/bcp.13917

Li XL, Guo XQ, Wang HR et al (2020) Aristolochic acid-induced genotoxicity and toxicogenomic changes in rodents. World J Tradit Chin Med 6:12–25. https://doi.org/10.4103/wjtcm.wjtcm_33_19

Marakis G, Ziegenhagen R, Lampen A et al (2018) Risk assessment of substances used in food supplements: the example of the botanical Gymnema sylvestre. EFSA J 16:e16083. https://doi.org/10.2903/j.efsa.2018.e16083

McCubrey JA, Lertpiriyapong K, Steelman LS et al (2017) Effects of resveratrol, curcumin, berberine and other nutraceuticals on aging, cancer development, cancer stem cells and microRNAs. Aging (Albany NY) 9:1477–1536. https://doi.org/10.18632/aging.101250

Misra RC, Thimmappa R, Bonfill M (2024) Editorial: advances in discoveries of plant phytochemicals. Front Plant Sci 15:1414150. https://doi.org/10.3389/fpls.2024.1414150

Mtewa AG, Egbuna C, Beressa TB et al (2021) Phytopharmaceuticals: efficacy, safety, and regulation. Elsevier EBooks, pp 25–38. https://doi.org/10.1016/b978-0-12-820284-5.00010-1

Muller PY, Milton MN (2012) The determination and interpretation of the therapeutic index in drug development. Nat Rev Drug Discov 11:751–761. https://doi.org/10.1038/nrd3801

Nath R, Talukdar AD, Saha P et al (2020) Intellectual property rights and its role in natural product research. Springer Ebooks, pp 449–459. https://doi.org/10.1007/978-981-15-2195-9_29

Newman DJ, Cragg GM, Snader KM (2003) Natural products as sources of new drugs over the period 1981–2002. J Nat Prod 66:1022–1037. https://doi.org/10.1021/np030096l

Och A, Och M, Nowak R et al (2022) Berberine, a herbal metabolite in the metabolic syndrome: the risk factors, course, and consequences of the disease. Molecules 27:1351. https://doi.org/10.3390/molecules27041351

Ornelas García IG, Guerrero Barrera AL, Avelar González FJ et al (2023) Bougainvillea glabra Choisy (Nyctinaginacea): review of phytochemistry and antimicrobial potential. Front Chem 11:1276514. https://doi.org/10.3389/fchem.2023.1276514

Panossian A (2023) Challenges in phytotherapy research. Front Pharmacol 14:1199516. https://doi.org/10.3389/fphar.2023.1199516

Patel R, Stine A, Zitko K (2022) Enhanced anticoagulant effect of warfarin when co-administered with quercetin. J Pharm Technol 38:374–375. https://doi.org/10.1177/87551225221125667

Patel-Rodrigues PA, Cundra L, Alhaqqan D et al (2024) Herbal- and dietary-supplement-induced liver injury: a review of the recent literature. Liver 4:94–118. https://doi.org/10.3390/livers4010008

Pathrikar SY, Mukhmale VS, Deshmukh NB et al (2025) Standardization extraction and phytochemical screening for herbal drugs. GSC Adv Res Rev 22:084–091. https://doi.org/10.30574/gscarr.2025.22.1.0518

Petty RD (2015) "Natural" claims in food advertising: policy implications of filling the regulatory void with consumer class action lawsuits. J Pub Poli Mark 34:131–141. https://doi.org/10.1509/jppm.14.147

Pires DEV, Kaminskas LM, Ascher DB (2018) Prediction and optimization of pharmacokinetic and toxicity properties of the ligand. Methods Mol Biol 1762:271–284. https://doi.org/10.1007/978-1-4939-7756-7_14

Poggiogalle E, Jamshed H, Peterson CM (2018) Circadian regulation of glucose, lipid, and energy metabolism in humans. Metabolism 84:11–27. https://doi.org/10.1016/j.metabol.2017.11.017

Pooja P, Shailja S, Jalwal P (2025) Formulation and efficacy of neem, turmeric, and onion extracts for hyperpigmentation treatment. J Neonatal Surg 14:162–170. https://doi.org/10.52783/jns.v14.1855

Prajapati AK (2024) Profiling the toxicological landscape of herbal drugs. J Prev Med Holist Health 10:6–9. https://doi.org/10.18231/j.jpmhh.2024.003

Rajput A, Mandlik S, Dawre S et al (2022) Chapter 3 herbal bioenhancers in nanoparticulate drug delivery system. Drug Delivery Technology. De Gruyter, pp 45–86

Rakam GK, Raja S (2017) Standardization and phytochemical screening of Bougainvillea glabra. Int J Curr Pharm Res 9:75–80

Rathaur P, JK SR (2019) Metabolism and pharmacokinetics of phytochemicals in the human body. Curr Drug Metab 20:1085–1102. https://doi.org/10.2174/1389200221666200103090757

Rathod H (2023) Caffeine—an enigma. Healthline 14:95–98. https://doi.org/10.51957/healthline_533_2023

Rietjens IMCM, Louisse J, Beekmann K (2016) The potential health effects of dietary phytoestrogens. Br J Pharmacol 174:1263–1280. https://doi.org/10.1111/bph.13622

Sachan AK, Vishnoi G, Kumar R (2016) Need of standardization of herbal medicines in modern era. Int J Phytomedicine 8:300. https://doi.org/10.5138/09750185.1847

Sahane MB, Basarkar GD (2023) A review of herbal regulations and approval process in India and Europe. Int J Drug Regul Aff 11:25–33. https://doi.org/10.22270/ijdra.v11i4.629

Scott EN, Gescher AJ, Steward WP et al (2009) Development of dietary phytochemical chemopreventive agents: biomarkers and choice of dose for early clinical trials. Canc Prev Res 2:525–530. https://doi.org/10.1158/1940-6207.capr-08-0223

Shahnawaz S, Khalid MM, Singh SPS et al (2024) A comprehensive review of phytochemicals: health benefits and therapeutic applications. Int J Pharm Res Appl 9:1373–1379. https://doi.org/10.35629/4494-090613731379

Shaikh MF, Arulsamy A et al (2018) Neuroactive drugs–a perspective on drugs of synthetic and medicinal plants origin. Pharm Pharmaco Int J 6. https://doi.org/10.15406/ppij.2018.06.00212

Sharma S (2015) Current status of herbal product: regulatory overview. J Pharm Bioallied Sci 7:293–296. https://doi.org/10.4103/0975-7406.168030

Shuaib M, Kushwaha PP, Prajapati KS et al (2021) Effect of dietary phytochemicals in obesity and cancer. Obesity Cancer 163–184. https://doi.org/10.1007/978-981-16-1846-8_9

Singh JI, Gupta SJ, Singh AK (2015) Standardization of Tulsi (Ocimum sanctum Linn.). Int J Ayu Pharm Chem 4:165–175

Smoliga JM, Baur JA, Hausenblas HA (2011) Resveratrol and health: a comprehensive review of human clinical trials. Mol Nutr Food Res 55:1129–1141. https://doi.org/10.1002/mnfr.201100143

Süntar I, Yakıncı ÖF (2020) Potential risks of phytonutrients associated with high-dose or long-term use. In: Phytonutrients in food. Elsevier, pp 137–155

Tahir MA (2024) Phytochemicals as nutraceuticals-III. Edited book of dietary supplements and nutraceuticals 1:61–69. https://doi.org/10.58532/nbennurdsch7

Tamargo J, Le Heuzey JY, Mabo P (2015) Narrow therapeutic index drugs: a clinical pharmacological consideration to flecainide. Eur J Clin Pharmacol 71:549–567. https://doi.org/10.1007/s00228-015-1832-0

Thacker H, Ram V (2024) Medicinal properties of phytochemicals: a review. J Pharmacogn Phytochem 13:78–82. https://doi.org/10.22271/phyto.2024.v13.i2a.14873

Thakur M, Singh K, Khedkar R (2020) Phytochemicals. In: Functional and Preservative Properties of Phytochemicals, pp 341–361. https://doi.org/10.1016/b978-0-12-818593-3.00011-7

Umamaheswari D, Muthuraja R, Kumar M, Venkateswarlu BS (2021) Standardization of herbal drugs—an overview. Int J Pharm Sci Rev Res 68. https://doi.org/10.47583/ijpsrr.2021.v68i01.033

Wang H, Chen Y, Wang L et al (2023) Advancing herbal medicine: enhancing product quality and safety through robust quality control practices. Front Pharmacol 14. https://doi.org/10.3389/fphar.2023.1265178

Wani TA, Bhat IA, Guleria K et al (2022) Phytochemicals: diversity, sources and their roles. In: Phytochemical Genetics, pp 3–33. https://doi.org/10.1007/978-981-19-5779-6_1

Yang X, Li W, Sun Y et al (2017) Comparative study of hepatotoxicity of pyrrolizidine alkaloids retrorsine and monocrotaline. Chem Res Toxicol 30:532–539. https://doi.org/10.1021/acs.chemrestox.6b00260

Ye M, Lee S, Yu HJ et al (2023) Sedative-hypnotic effects of Glycine max Merr. Extract and its active ingredient genistein on electric-shock-induced sleep disturbances in rats. Int J Mol Sci 24:7043. https://doi.org/10.3390/ijms24087043

Younes M, Aggett P et al (2018) Scientific opinion on the safety of green tea catechins. EFSA J 16:e05239

Medicinal Plants and Phytochemicals Targeting Specific Cancer Pathways and Mechanisms

11

Akingbolabo Daniel Ogunlakin, Idayat Adeola Akinwumi, Great Oluwamayokun Adebodun, Ajibola David Adelakun, Amaka Agbara, Omolola Oluwadara, Moyosoluwa Dada, Ahamefula Anslem Ahuchaogu, Amel Elbasyouni, Abdullahi Adeyemi Adegoke, Owoola Azeezat Ambali, Tomi Lois Adetunji, and Oyindamola Esther Awosola

Abstract

As one of the biggest causes of death globally, cancer calls for the investigation of innovative treatment approaches. Medicinal plants and their phytochemicals have emerged as promising cancer therapeutic options because they interfere with the pathways and processes that produce cancer. Numerous biological

A. D. Ogunlakin (✉)
Phytomedicine and Drug Discovery Research Laboratory (PDD-RL), Department of Biochemistry, Bowen University, Iwo, Nigeria

I. A. Akinwumi
Department of Pharmacognosy, Faculty of Pharmacy, Lead City University, Ibadan, Nigeria

G. O. Adebodun
Department of Physiology, School of Basic Medical Sciences, Babcock University, Ilisan-Remo, Nigeria
e-mail: alabig@babcock.edu.ng

A. D. Adelakun
Department of Pharmaceutical Sciences, University of Arizona, Tucson, AZ, USA

A. Agbara
Department of Biochemistry, Godfrey Okoye University, Enugu, Nigeria

O. Oluwadara
Project Development and Design Department, Federal Institute of Industrial Research, Oshodi, Lagos State, Nigeria

M. Dada
Department of Anatomy, Babcock University, Ilisan-Remo, Nigeria

K. Kuca et al. (eds.), *Medicinal Plants and Their Bioactives in Human Diseases*,
https://doi.org/10.1007/978-3-032-01356-9_11

activities, such as pro-apoptotic, anti-inflammatory, antioxidant, and antiproliferative effects, have been reported to support botanicals' anticancer potential. Furthermore, by targeting pathways, such as AMPK/PGC-1α, PI3K/Akt/mTOR, and NF-κB, these botanicals disrupt the metabolism of cancer cells, prevent angiogenesis, limit the expression of genes linked to cancer cell growth, and induce apoptosis. They also boost the effectiveness of traditional cancer treatments by fortifying the immune responses. Adding phytochemicals to cancer treatment regimens offers a multimodal strategy that may improve medicine delivery to cancer cells and lessen adverse effects. To exploit the therapeutic potential of these medicinal plants and their constituents, future research should concentrate on clinical trials and the creation of innovative delivery systems. Thus, this chapter discusses how different botanicals and their components—such as terpenoids, alkaloids, flavonoids, and phenolic compounds—may alter important signaling pathways that contribute to cancer development.

Keywords

Cancer · Drug discovery · Medicinal plants · Phytoconstituents · Pathways · Gene expression

1 Introduction

Cancer remains one of the most significant health challenges worldwide, with an ever-increasing burden due to population aging and the prevalence of risk factors such as the use of tobacco, unhealthy dietary, and sedentary lifestyles, genetic mutations, and hormonal disorders among others (Iqbal et al. 2017). According to a 2020 GLOBOCAN estimate, there were about ten million cancer-related deaths, and 19.3 million new cancer cases detected. By 2040, it is anticipated that there will be 28.4 million cases of cancer (Sung et al. 2021). Lung cancer (11.4%), colorectal

A. A. Ahuchaogu
Department of Industrial Chemistry, Abia State University, Uturu, Nigeria
e-mail: ahuchaogu.aa@abiastateuniversity.edu.ng

A. Elbasyouni
Department of Molecular Biology and Biotechnology, Pan African University for Basic Sciences, Technology, and Innovation, Pausti, Kenya

A. A. Adegoke
Department of Pharmacognosy, Faculty of Pharmacy, University of Ibadan, Ibadan, Nigeria

O. A. Ambali
School of Medicine and Allied Health Sciences, University of the Gambia, Banjul, Gambia

T. L. Adetunji
Unit for Environmental Sciences and Management (UESM), Faculty of Natural and Agricultural Sciences, North-West University, Potchefstroom, South Africa

O. E. Awosola
Next Era Pharmacy, Ifako-Gbagada, Lagos, Nigeria

cancer (10.0%), prostate cancer (7.3%), stomach cancer (5.6%), and female breast cancer (11.7%) are now the most prevalent cancers diagnosed worldwide (Sung et al. 2021). Breast cancer is more common than lung cancer, but lung cancer still accounts for 1.8 million deaths (18%), followed by colorectal (9.4%), liver (8.3%), stomach (7.7%), and female breast cancer (6.9%) (Bray et al. 2018). Cancers that are most common in men are lung, prostate, and colorectal, while cancers that are most common in women are breast, colorectal, and lung. Over 60% of the world's cancer incidence and 70% of cancer-related deaths are caused by the top 10 cancers combined.

According to a study by the World Health Organization's International Agency for Research on Cancer (IARC), an estimated 20.0 million new cases of cancer were recorded globally in 2022 (19.96 million including non-melanoma skin cancer [NMSC] and 18.73 million excluding NMSC). Furthermore, 9.7 million fatalities were attributed to cancer (9.67 million excluding NMSC and 9.74 million including NMSC). The majority of cancer fatalities (56.1%) and nearly half of all cancer diagnoses (49.2%) were predicted to occur in Asia, which is home to 59.2% of the world's population. Because of the distribution of cancer types and higher case fatality rates, which are partially caused by late-stage diagnosis, the cancer mortality burden in Africa and Asia is disproportionately higher than the incidence burden in these regions. On the other hand, although making up only 9.6% of the world's population, Europe has a disproportionately high cancer incidence and mortality burden, contributing to around one-fifth of all cancer cases (22.4%) and deaths (20.4%). It is predicted that there will be 35 million cases annually by 2050, highlighting the urgent need for improved access to healthcare and preventive measures.

Initially, some cancer types were associated with certain regions or countries, for example, stomach, liver, and cervical cancers were profound in low- and middle-income countries while lung cancer was more prominent in high-income countries; however, in recent times, it has been discovered that these various cancers are now non-discriminatory (IRAC 2020), that is any of the cancers can occur in any country, whether low or middle-income countries. In 2022, the dominant cancer cases were reported to be lung, breast, and colorectal cancers. Lung cancer ranks as the most frequently diagnosed and deadliest form globally, contributing to 12.4% of new cases and 18.7% of deaths. This statistic is closely followed by breast cancer among women, while colorectal, prostate, and stomach cancers significantly impact public health across both genders. The main causes of the significantly higher cancer death rates in low-income nations are inadequate implementation of preventative measures and restricted access to early detection services. According to Gaobotse et al. (2023), cervical cancer continues to be the primary cause of cancer-related mortality among women in sub-Saharan Africa.

Cancer is the third most common cause of premature death in sub-Saharan Africa, taking the lives of about one in seven people. In 2020, there were an expected 520,158 cancer-related deaths in the area, along with 801,392 new cases. With 129,400 cases and 110,300 cases, respectively, breast cancer and cervical cancer account for 30% of all cancer diagnoses in women. In men, colon cancer (23,000 instances), liver cancer (24,700 cases), and prostate cancer (77,300 cases) are the most common malignancies diagnosed (Bray et al. 2021; Bray et al. 2024). Health disparities, higher healthcare expenditures, and decreased workforce productivity are some of the ways that cancer affects socioeconomic position. Over time, the cost of cancer medications has increased significantly, and changes to health insurance plans have more and more placed financial burdens on survivors—a concept referred to as "cost-sharing" (Wyman 2020). To tackle these issues, international approaches that prioritize prevention are needed, including lowering tobacco use, raising vaccination rates (for instance, against HPV), and encouraging lifestyle changes. More than half of cancer deaths worldwide may be avoided with these actions. These underline how urgently coordinated efforts are needed to advance public knowledge, establish fair healthcare systems, and give cancer research top priority. In addition to reducing the present cancer burden, such measures are necessary to make long-term progress in the fight against this deadly condition.

2 The Importance of Targeting Specific Cancer Pathways

Chemotherapy, radiation, and surgery are among the many techniques used to treat cancer. All of these methods, meanwhile, have their drawbacks (Karpuz et al. 2018). Radiation therapy, for example, has been shown to impact noncancerous cells in the vicinity during exposure (Delanian et al. 2012). Additionally, synthetic medications including lapatinib, carboplatin, raloxifene, and capecitabine are frequently used to treat breast cancer. Despite their efficacy, these drugs are linked to numerous side effects, including fatigue, headaches, dental complications, heart issues, secondary cancers, bone loss, osteoporosis, and cataracts (Wolff 2014). Current management strategies, including chemotherapy, hormone therapy, and targeted therapies, due to the accumulation of these synthetic drugs in the body, these methods are not the best to employ (Piedrola et al. 2001). As a result of these disadvantages posed by the use of these other forms of treatments, there is a need to explore other non-invasive with little or no side effects, yet effective methods. The precision targeting of cancer pathways represents a transformative approach to cancer treatment, driven by advances in molecular biology and genomics. Personalized medicine, also known as precision medicine, aims to treat patients using customized therapies based on traits unique to the individual or, ideally, traits specific to the population group the patient belongs to (Senft et al. 2017; Doroshow and Doroshow 2020; Pfohl et al. 2021).

Cancer is characterized by its ability to exploit cellular pathways to sustain uncontrolled growth, evade apoptosis, promote angiogenesis, and metastasize (Hanahan and Weinberg 2011; Kumar 2024). Modern oncology has transitioned from broad cytotoxic treatments to more targeted and effective therapies by identifying and focusing on specific pathways. Key cancer-related pathways, such as the PI3K/AKT/mTOR, RAS/RAF/MEK/ERK, and p53 signaling cascades, are often dysregulated in tumors. Therapeutics aimed at these pathways disrupt critical processes that fuel cancer progression. For instance, mTOR inhibitors, such as everolimus, have shown promise in the treatment of some types of cancer by inhibiting the production of proteins and growth of cells. Similarly, MEK inhibitors have shown promise in cancers with RAS mutations, including melanoma and lung cancer. The importance of this approach lies in its dual benefits: enhancing treatment efficacy and reducing systemic toxicity. By homing in on cancer-specific molecular alterations, targeted therapies spare normal tissues, significantly improving patients' quality of life compared to traditional chemotherapy. Furthermore, this approach makes combination therapy possible, which enables oncologists to address tumors from several angles and counteract resistance mechanisms. The trend towards precision oncology is essentially embodied in targeting cancer pathways, which offers a patient-centered, scientifically supported approach to transform cancer therapy. As investigations into the intricacies of cancer biology proceed, this methodology has the potential to convert molecular discoveries into interventions that can save lives. The development of state-of-the-art precision oncology treatments and significant advancements in molecular profiling of various cancers have not eliminated the need for histopathological or radiological analyses for prognostic and predictive evaluations of many tumor types (Smith and Sheltzer 2022; García-Figueiras et al. 2019; Mobadersany et al. 2018). However, many patients are still unable to afford precision medicine.

3 The Role of Medicinal Plants in Cancer Treatment

Only plant material derivatives may be used as active components in phytomedicines. This feature of the product is eliminated when an isolated substance from any source is added. The seasonality of the biochemistry, the soil and environment of the plant's growing area or occurrence, the processing method, and the common and scientific names of the plant must all be considered (Maciel 2002). To construct a phytomedicine, a plant species must be recognized, along with a clear botanical description (Barbosa et al. 2012). Cancer treatment consists of radiation, chemotherapy, and surgery. Chemotherapy aids patients by reducing relapses and metastases and increasing overall survival (Imtiaz et al. 2024). Many cytotoxic chemotherapeutic agents, including alkylating agents, anti-metabolites, anti-tumor antibiotics, topoisomerase inhibitors, and tubulin-binding medications, which target multiple cancer mechanisms to disrupt the cell cycle, are frequently the driving force behind the polypharmacy approach to cancer management (Dickens and Ahmed 2018). This strategy may result in additional adverse consequences, necessitating the development of fresh, more focused cancer treatment strategies. Traditional knowledge and herbal remedies present a novel and alluring way to look for possible new chemotherapies. Over 3000 plant species may have potential anticancer effects, while an estimated 70,000 plant species have historically been used medicinally (WHO 2020; Imtiaz et al. 2024). According to Alegbeleye et al. (2020), plants are the source of more than 60% of the chemotherapeutic medications currently used in clinical settings, including paclitaxel, vincristine, vinblastine, etoposide, irinotecan, and topotecan. Nigeria has a wealth of medicinal plants that are used to treat cancer and other illnesses. For many years, medicinal plants have been utilized to cure cancer, especially in Nigeria and most otherwise undeveloped nations (Akinwumi and Ambali 2023).

3.1 Historical Use of Medicinal Plants in Cancer Treatment

Over the years, people have learned about the therapeutic properties of plants and used this knowledge to create traditional treatments for a range of illnesses (Ouelbani et al. 2016). More recently, research on plants has focused on serious human diseases like HIV/AIDS and cancer (Nkwe et al. 2021). Medicinal plants are one of the most dependable sources of a wide variety of therapeutic medications, according to the World Health Organisation (WHO). Nowadays, between 65% and 80% of people in poor nations use phytochemicals from medicinal plants to treat a variety of ailments (AL Akeel et al. 2013; Cameron et al. 2011). With an estimated 300,000 plant species on Earth, only about 15% have been studied to identify their pharmacological properties (De Luca et al. 2012). This underlines the need for further exploration of biologically active phytochemicals, particularly in the fight against diseases such as cancer.

Ethnomedicinal traditions, which are frequently transmitted from one generation to the next, can contribute to the finding of plants with anticancer potential (Allen and Hatfield 2004; Akinwumi and Ambali 2023). Beginning with four World War II-related drug discovery programs, the impact of newly found medications catalyzed the establishment of the National Service Center for Chemotherapy against Cancer, a nationwide drug development initiative, in 1955. The ability of combination chemotherapy to cure advanced Hodgkin's disease and acute juvenile leukemia in the 1960s and early 1970s disproved the general pessimism about the effectiveness of natural remedies in treating advanced cancers. Additionally, this discovery aided the national cancer program and facilitated the study of adjuvant treatment (Devita and Chu 2008). Since the National Cancer Institute (NCI) identified the possibility of using natural products as anticancer agents in the 1950s, research has offered insightful information about the identification of novel anticancer agents that occur naturally. Research on novel anticancer drugs in plants and other species, with an emphasis on the world's tropical and subtropical regions, was driven by the advent of new screening technologies in the 1980s (Amaral et al. 2019).

3.2 Overview of Medicinal Plants in Cancer Treatment

These natural substances, including flavonoids, tannins, curcumin, resveratrol, gallocatechin alkaloids, flavonoids, terpenoids, phenolic compounds, anthraquinones, and saponins (Azmi et al. 2006), exhibit a wide range of mechanisms targeting cancer cells. The primary advantage of using phytochemicals as anticancer agents is their reduced risk of adverse effects while modulating critical pathways in carcinogenesis, such as apoptosis, cell cycle regulation, angiogenesis, and metastasis compared to synthetic drugs. Moreover, their affordability and high cost-effectiveness enhance their appeal as alternatives in cancer therapy (Majumder et al. 2019). Their antioxidant and anti-inflammatory properties also contribute to combating the oxidative stress and chronic inflammation that often accompany cancer progression. Given their diverse molecular targets and biological activities, phytochemicals are emerging as valuable candidates in the development of novel, more effective cancer therapies. These properties position phytochemicals as promising candidates in the ongoing quest for safer and more accessible cancer treatment options.

Key medicinal plants in cancer treatment include: *Taxus brevifolia*, *Catharanthus roseus*, *Podophyllum peltatum*, *Maclura pomifera*, *Camellia sinensis*, among others. Taxol (Paclitaxel) was initially extracted from the bark of the *Taxus brevifolia* Nutt (Moss 1998). The Pacific yew, or *T. brevifolia*, is found growing from the southeast coast of Alaska southward through western British Columbia and central California. The presence of taxol has led to a great deal of interest in the genus Taxus. According to Madalieva and Mamedov (2020) and Mishra and Verma (2017), taxol is one of the most often used medications for the treatment of ovarian, lung, and breast cancers. Its therapeutic qualities are highly prized, and its potential to treat a wide range of illnesses, including diabetes and cancer, has been thoroughly investigated. Raubasin, Reserpine, Catharanthine, Ajmalicine, Vincente, and Vineamine are frequently found in the roots of *C. roseus*, whereas vinblastine, vincristine, vindesine, actineoplastidemeric, and vindeline tabersonine are frequently found alkaloids in the aerial regions of the plant (Goswami et al. 2024). While medicinal plants offer significant promise, challenges such as variability in bioactive compound concentrations, extraction standardization, and limited clinical trials hinder their widespread application. Advances in phytochemistry, nanotechnology, and molecular biology are poised to overcome these limitations, fostering the development of more effective plant-based anticancer drugs. Medicinal plants serve as a cornerstone of cancer therapeutics, providing diverse bioactive compounds that complement and enhance modern oncology. Continued research into their mechanisms and therapeutic potential will expand their role in personalized and integrative cancer care.

4 Phytochemicals and Their Mechanisms of Action

Alkaloids have shown promise in cancer treatment, with recent studies focusing on processes, such as apoptosis induction, angiogenesis inhibition, and cell signaling pathway control, and overcoming multidrug resistance. According to Dhyani et al. (2022), alkaloids such as vinblastine, vincristine, vindesine, vinorelbine, and vincamine prevent the formation of mitotic spindles resulting in cell cycle arrest and death. The impact of plant-based alkaloids on critical signaling pathways associated with cancer cell death has been studied. According to Yakkala et al. (2023), topoisomerase I breaks DNA and kills cancer cells when it is blocked by camptothecin and other alkaloids. Alkaloids have also been found to aid cancer cells in overcoming their treatment resistance (Ferreira 2018). Through their ability to neutralize free radicals and activate antioxidant pathways, vincamine and other substances boost the cytotoxicity of chemotherapy medicines (Wang et al. 2024). These procedures reveal the potential of alkaloids as potent anticancer medications, either on their own or in combination with more conventional treatments.

Recent research suggests that they may treat cancer by altering important biological processes (Shi et al. 2024). The induction of programmed cell death, or apoptosis, is one of the main mechanisms. It has been demonstrated that two flavonoids, quercetin and kaempferol, cause cancer cells to undergo apoptosis, which results in cell death (Wendlocha et al. 2024). Another strategy is to prevent angiogenesis, which is the process by which new blood vessels form to supply tumors with nourishment (Yang et al. 2024). By inhibiting the migration and multiplication of endothelial cells, epigallocatechin gallate (EGCG) decreases tumor vascularization (Tanabe et al. 2023). Furthermore, flavonoids change cell signaling pathways that help cancer develop (Cai et al. 2024). Additionally, it has been demonstrated that flavonoids aid cancer cells in overcoming multidrug resistance (MDR) (Oršolić and Jembrek 2024). Naringenin enhances the cytotoxicity of chemotherapeutic medications by scavenging free radicals and initiating antioxidant pathways (Alhalmi et al. 2024). Additionally, recent research has examined the role of flavonoids in autophagy modulation. Flavonoids can either increase excessive autophagy or decrease autophagy flow, especially in cancer cells that are resistant to apoptosis (Zhang et al. 2021; Du et al. 2023). There are advantages and disadvantages to using flavonoids to treat cancer because of their dual function in regulating autophagy. Their anticancer qualities are being discovered by ongoing studies, opening the door for the creation of innovative treatment approaches (Chunarkar-Patil et al. 2024). A new hydroxyflavone [3,6-dihydroxy-2-(3,4-dimethylphenyl)-4H-chromen-4-one] recently showed an anti-proliferative effect on CHO 1 cells by deregulating p53 gene expression by binding to the mutant p53 structure and stabilizing the protein (Ogunlakin et al. 2023).

4.1 Medicinal Plants and Cancer Gene Expression

Changes in gene expression, which can result in either gene activation or silencing and so drastically affect the overall activity of the cell, are a hallmark of cancer (Qu et al. 2017). A variety of diseases, including cancer, may be treated by several bioactive compounds present in medicinal plants (Goswami et al. 2023; Situmorang et al. 2024). Furthermore, these metabolites alter immune system modulation and biological pathways, which inhibits cancer and other serious illnesses (Ali Reza et al. 2021; Ishara et al. 2021). Cytotoxicity, or the capacity to cause harm to cancer cells, apoptosis in cancer cells, halting tumor angiogenesis, and restricting the growth and multiplication of cancer cells are some of the anti-cancer qualities of medicinal plants (Khan et al. 2022; Jenca et al. 2024). Additionally, medicinal plant extracts have been demonstrated to stop the spread of several cancers by reducing MMP-2 production of cytochrome c (Cyt c), activating Cas-3 and -9, downregulating anti-apoptotic proteins like Bcl-XL, and then altering the ratio of Bcl2 family proteins

(Memarzia et al. 2023). HeLA cells have been shown to exhibit cytotoxicity, nuclear blebbing, and DNA breakage in response to medicinal extracts; these findings may suggest that the extracts trigger apoptosis in human cervical cell lines (Kuriakose et al. 2018; Nimlamool et al. 2022). Plant extracts decreased the expression of vascular endothelial growth factor (VEGF), a protein that promotes angiogenesis, carcinogenesis, and the spread of cancer (Hu et al. 2016), while upregulating the tumor suppressor gene p53 (Mbugua et al. 2019). A high level of serum VEGF indicates several other cancer-associated systemic diseases (Wang et al. 2021).

4.2 Case Studies of Specific Medicinal Plants Against Cancer Proliferation

Curcuma longa, also known as turmeric, is a perennial plant frequently grown in Asian countries. It plays a vital role in Ayurvedic treatment and serves as a spice or flavoring agent. Phytochemical constituents of *C. longa*, curcumin, and essential oils have shown pharmacological activities (Hosseini et al. 2018; Shashrajabian et al. 2024). Curcumin, being the most abundant compound, is a flavonoid that has anticancer effects, including apoptosis induction, inhibition of cancer cell proliferation, and modification of reactive oxygen species-scavenging enzyme activity (Mottaghi and Abbbaszadeh, 2021; Cozmin et al. 2024).

Curcumin treatment inhibited the migration and proliferation of cancer cells in a dose-dependent fashion (Nakayama et al. 2021; Mostofi et al. 2022). As a result of curcumin's reduction of SVCT2 expression in cancer cells (wild type p53), MMP2 activity was lowered, and apoptosis was inevitably raised (Jang et al. 2023). To achieve its antiproliferative and apoptotic effects, curcumin suppresses c-myc mRNA levels, inhibits protein tyrosine kinase activity, inhibits protein kinase C activity, and up-regulates B-cell lymphoma 2 (Bcl-2) mRNA expression (Lu et al. 1998). It also causes apoptosis, constitutive NF-kB transcriptional activity, sustained phosphorylation, activation of c-Jun N-terminal kinase (JNK), and p38 MAPK (Yao et al. 2015). The apoptotic and cell cycle pathways, along with the PI3K/Akt/mTOR, JAK/STAT, MAPK, NF-kB, p53, and Wnt/β-catenin pathways, are crucial signaling pathways for curcumin's anticancer action (Hasan et al. 2022). Phytocompounds such as catechins, and epicatechin-3-gallate isolated from *Camellia sinensis* (green tea) have several pharmacological effects such as cytotoxicity. The matrix metalloproteinases (MMPs), which are critical for tumor invasion and metastasis (Jiang et al. 2017), are blocked by catechins (Sarkar et al. 2016). It is well known that NF-κβ plays a crucial role in preventing cancer cells from dying (Verzella et al. 2020; Suhail et al. 2021). Treatment with epigallocatechin-3-gallate and EGCG reduced the activation of NF-κB in cancer cells (Luo et al. 2020; Li et al. 2024). Additionally, EGCG prevents cellular transformation and AP-1 activation in the epidermal cell line (Shirakami et al. 2018) as well as cell migration and proliferation in non-small cell lung cancer cell lines (Minnelli et al. 2021).

4.3 Epigenetic Modulation by Phytochemicals

The topic of epigenetic modification using phytochemicals is expanding quickly, combining knowledge of cancer biology with possible treatment approaches. The link between epigenetics and cancer emphasizes the significance of focused approaches that can correct aberrant epigenetic markers. Phytochemicals from medicinal plants offer a new way to control the epigenome, which in turn affects pathways crucial to the development of cancer (Fig. 11.1). Epigenetic modification, such as DNA methylation, histone modification, and microRNA (non-coding RNA) regulation, can alter gene expression without altering the underlying DNA sequence (Auvinen et al. 2024; Pathak et al. 2023; Baylin and Jones 2016). About the phytochemicals derived from medicinal plants, in particular, this reversible feature of epigenetic modification presents new prospects for therapeutic approaches.

Through a variety of processes, including DNA methylation, histone modifications, and microRNA dysregulation, epigenetic modulation plays a critical role in the development of cancer. The two most significant epigenetic changes that impact the progression of cancer are DNA methylation and histone modifications. These modifications are crucial for controlling gene expression, which ultimately impacts cellular function and aids in the development of tumors (Xu et al. 2023). Because these epigenetic changes are dynamic, they can either mute tumor suppressor genes or activate oncogenes, which promotes malignant transformation and the spread of cancer. DNA methylation, which is frequently linked to transcriptional silence, is

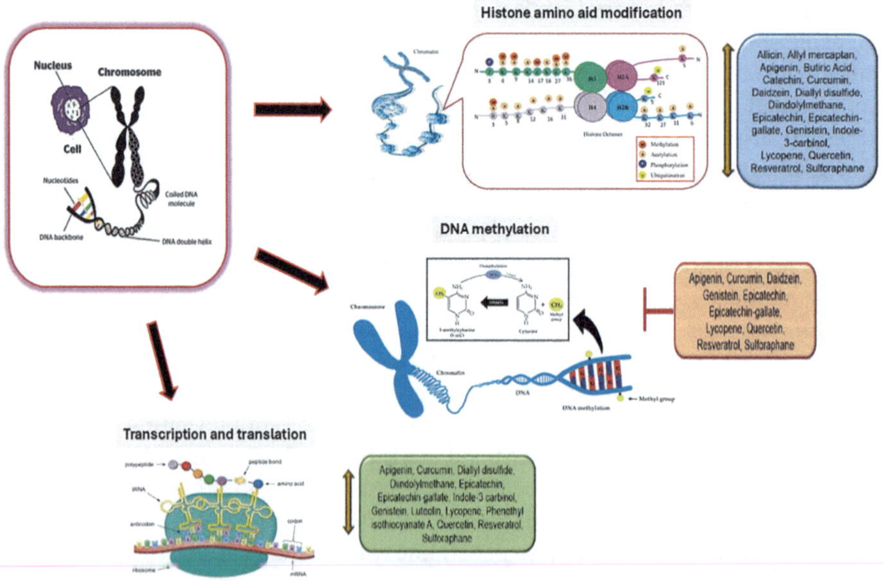

Fig. 11.1 Certain phytochemicals that affect histone modifications, DNA methylation patterns, and microRNA expression. (Modified from Shankar et al. 2013)

the addition of a methyl group (−CH3) to cytosine residues in DNA, especially in CpG dinucleotides (Moore et al. 2013). In DNA methylation, cancer cells exhibit site-specific CpG island promoter hypermethylation and genome-wide hypomethylation (Shen et al. 2007; Sincic and Herceg 2011). Hypermethylation at certain genes often affects promoter CpG-islands, deactivating transcription. Hypermethylation is observed at specific CpG islands. Promoter hypermethylation inactivates genes involved in various cellular pathways, including DNA repair [*hMLH1* (mismatch repair gene 1), *MGMT* (O6-methylguanine–DNA methyltransferase), *BRCA1* (breast cancer 1), *WRN* (Werner syndrome, RecQ helikase like)], Ras signaling {*RASSF1A* [Ras association (RalGDS/AF-6) domain family member 1], *NOREIA*}, apoptosis [*TMS1* (target of methylation-induced silencing 1), metastasis [cadherin 1 *(CDH1), CDH13, PCDH10, DAPK1* (death-associated protein kinase)], detoxification [*GSTP1* (glutathione S-transferase pi 1)], hormone response (*ESR1, ESR2*) among others. This gives tumor cells a growth opportunity while increasing their genetic instability and aggressiveness (Esteller 2007; Shen et al. 2007). The earliest identified modification in DNA methylation patterns in cancer cells was a regional reduction, which is now known as global DNA hypomethylation after genome-wide analyses (Feinberg and Vogelstein 1983; Ehrlich and Lacey 2012; Hansen et al. 2011; Berman et al. 2012; Bert et al. 2013). This methylation loss is thought to contribute to genomic instability and aneuploidy, both of which are characteristic of cancer. Low levels of the maintenance DNA methyltransferase, Dnmt1, results in higher mutation rates and tumor development, indicating that hypomethylation actively promotes chromosomal fragility (Chen et al. 1998; Narayan et al. 1998; Gaudet et al. 2003; Ehrlich and Lacey 2012). This loss may activate transcription repeats, transposable elements (TEs), and oncogenes, thereby resulting in increased chromosomal recombination (Jones and Baylin 2007; Ehrlich and Lacey 2012; Hur et al. 2014). Most CpGs are about 80% methylated in a healthy genome, but in cancer, this reduces to 40%–60%, with hypomethylation well localized in large genomic blocks (Hansen et al. 2011; Berman et al. 2012; Hon et al. 2012; Bert et al. 2013). The processes that cause DNA methylation loss are not entirely elucidated, however, they may be linked to variations in chromatin organization and chromatin mutation regulators impacting DNA methylation equilibrium, potentially involving the TET family and DNMT proteins (Baylin and Jones 2016). In addition to DNA methylation alterations, histone modification schemes are changed in human malignancies. Recent research has demonstrated that histone modification levels are indicative of gene expression (Kanwal and Gupta 2012). Histones, that wrap and organize DNA into structural units known as nucleosomes, undergo various post-translational modifications including methylation, acetylation, phosphorylation, and ubiquitination. Depending on the type of mutation, these modifications can either activate or suppress gene transcription (Liu et al. 2023). For example, tri-methylation of histone H3 at lysine 27 (H3K27me3) is commonly associated with transcriptional repression, which contributes to the silencing of crucial tumor suppressor genes in different cancers (Tom Charlesworth 2024).

4.4 Medicinal Plant-Derived Epigenetic Modifiers

Medicinal plants-derived phytochemicals such as curcumin, resveratrol, and epigallocatechin gallate (EGCG), have demonstrated substantial potential as modulators of epigenetic markers such as DNA methylation and histone acetylation. The investigation of dietary substances in the context of cancer therapy has garnered a lot of attention in recent years, especially in light of their epigenetic properties. Much research has been done on the epigenetic qualities of curcumin, the main component of turmeric (Choudhari et al. 2019). Additionally, studies reveal that curcumin can alter DNA methylation patterns, which helps reactivate tumor suppressor genes in many cancer types (Hassan et al. 2019). Furthermore, it has been demonstrated that resveratrol, a polyphenol present in red wine and grapes, can alter epigenetic processes, particularly in cancer, by influencing several pathways, such as histone acetylation and DNA methylation, which alter the epigenetic makeup of cancer cells (Jiang et al. 2015). The primary polyphenol in green tea, EGCG, has similar effects and can influence gene expression through epigenetic modifications. Research indicates that curcumin causes widespread genomic DNA hypo-methylation by inhibiting DNA methyltransferases (Hassan et al. 2019). By reinstating the expression of repressed tumor suppressor genes, this hypomethylation presents a promising technique for cancer chemoprevention. Among its many other actions, resveratrol promotes histone acetylation, which stimulates a more transcriptionally active chromatin structure. Furthermore, EGCG has been found to inhibit both DNMTs and HDACs, leading to increased expression of key tumor suppressor genes (Hassan et al. 2019). The compound's synergistic effects highlight their potential use in cancer therapy, highlighting the significance of addressing epigenetic modifications in cancer prevention and treatment. Understanding the basic processes by which these organic substances work can aid in the development of more effective cancer intervention techniques. Phytochemicals offer a promising approach to cancer prevention and treatment through epigenetic modifications. More research is required to completely comprehend their function and potential in clinical situations.

5 Clinical Studies and Trials on Botanicals

It has been found that the xerophyte *Aloe barbadensis* Miller (Aloe vera), which is a member of the genus Aloe, possesses anticancer properties. Its anti-proliferative and immunomodulatory properties are among its modes of action (Lissoni et al. 2009). Aloe vera mouthwash dramatically decreased the severity of stomatitis and inflammation in patients undergoing treatment for acute lymphocytic and acute myeloid leukemias in a clinical study (Mansouri et al. 2016). Aloe vera considerably reduced the impact of radiation-induced dermatitis in a different trial that included patients with head, pelvic, and breast cancer (Haddad et al. 2013). The presence of different phytochemicals in the roots and leaves of Withania somnifera has medicinal benefits such as antioxidant, cardioprotective, and anti-inflammatory potentials. In a 3-year study involving patients with advanced-stage breast cancer who experienced chemotherapy-induced fatigue, the root extract of Withania somnifera was found to alleviate cancer treatment-related fatigue and improve quality of life (Biswal et al. 2013). This suggested that *Aloe vera* may be effective in preventing chemotherapy-related side effects.

The bioactive components of *C. longa* L., known as curcuminoids, have been found to have antioxidant, anti-proliferative, anti-inflammatory, and anti-metastatic properties. Curcuminoid supplementation may enhance the quality of life for patients receiving chemotherapy for solid tumors, according to a study by Panahi et al. (2014). Throughout the 8-week trial, it was also found that curcuminoids significantly decreased the levels of inflammatory biomarkers in the patient's serum, including interleukin-6, C-reactive protein high-sensitivity, tumor necrosis factor-□, and tumor growth factor-β. Curcumin administration in pancreatic cancer patients improved the efficacy and safety of gemcitabine, a drug used in cancer treatment (Pastorelli et al. 2018). A phase II clinical trial also observed that complementary administration of curcumin with paclitaxel in patients with advanced, metastatic breast cancer had more efficacy compared with those who were administered only paclitaxel (Saghatelyan et al. 2020). Belcaro et al. (2014) investigated the effect of curcumin on radio- and chemotherapy-associated negative effects in lung, kidney, stomach, and ovarian cancer patients for 60 days. It was discovered that the negative effects, such as weight loss, vomiting, fatigue, throat and mouth ulcers, and diarrhea, were significantly reduced.

Resveratrol is found in grapes, blueberries, and blackberries, and it is shown to have anticancer and antiapoptotic activities. In recurrent prostate cancer patients, resveratrol significantly decreased the recurrence of prostate cancer for more than 5 months and even up to a year (Paller et al. 2015, 2018). Zingiber officinale is the source of gingerol, a phenolic molecule that has strong anticancer properties. Ginger extract improved antioxidant status and decreased oxidative stress markers in a pilot study involving recently diagnosed cancer, indicating that it may lessen side effects associated with chemotherapy (Danwilai et al. 2017). 6-gingerol also alleviated chemotherapy-related side effects such as nausea and vomiting in solid tumor patients, according to a phase II randomized clinical study (Konmun et al. 2017). Griola frondosa is a type of mushroom whose main bioactive ingredient is D-fraction, a β-glucan; it has been shown to suppress the growth of metastases and lower the expression of tumor markers (cancer antigen; CA15-3, CA19-9) in patients with lung, liver, or breast cancer (Hetland et al. 2020). Though some studies claim that the peel is the most active component of the fruit, pomegranate fruit—including the seed, peels, and arils—contains a variety of advantageous bioactive chemicals (Baradaran Rahimi et al. 2020).

According to a randomized double-blind phase II clinical study, pomegranate extract (1 or 3 g/day for 18 months) significantly improved PSA doubling time in men with prostate cancer. The study concluded that the extract could treat prostate cancer without metastases. In another clinical trial with colorectal cancer patients, pomegranate extract significantly modulated the activities of the microRNAs involved in colorectal cancer in the patient's colon tissues (Nuñez-Sánchez et al. 2015). Another study involving colorectal cancer patients reported low plasma levels of lipopolysaccharide-binding protein, a biomarker for endotoxemia, following administration of pomegranate extracts containing punicalagin, punicalin, and ellagic acid (EA) derivatives (González-Sarrías et al. 2018). Berberine was administered daily for 2 years to patients with colorectal adenoma in a clinical study, and it was discovered that the risk of colorectal recurrence adenoma was decreased (Chen et al. 2020). Vinblastine isolated from Madagascar periwinkle and *Cathatanthus roseus* G., along with vincristine, is effective in treating leukemias, lung, testicular, and breast cancers. Paclitaxel is derived from the bark of *Taxus brevifolia* Nutt. was among the first drugs to impact the treatment of cancer greatly and is used in treating non-small lung, ovarian, and breast cancers (Che et al. 2015; Greenwell and Rahman 2015). Elliptinium isolated from *Bleekeria vitensis A.* is used in breast cancer treatment, while Homoharringtonine isolated from *Cephalotaxus harrington* var. drupacea is used in treating acute myelogenous leukemia (Ashraf 2020).

Some challenges of anticancer clinical trials include: a low enrolment rate, successful preclinical trials might not be effective in clinical trials, clinical trials are expensive compared to preclinical trials leading to suspension of clinical studies due to insufficient funding, conflicting results in the studies, thereby preventing a solid conclusion on the use of some medicinal plants in clinical studies. The failure of some clinical studies has been attributed to the fact that some plant-derived compounds have low solubility rates and bioavailability. Due to the small sample sizes and short duration of the studies, conclusions on the clinical activity of the medicinal plants and their phytochemical constituents might be premature. Several studies have discovered the anticancer properties of medicinal plants, however, few have advanced to clinical studies, therefore, there is a need for further clinical trials to investigate the beneficial effects of the plants and discover the negative effects they may have on humans. A large sample size and longer duration of trials are recommended. A combination of medicinal plants and anti-cancer conventional drugs or two or more medicinal plants can also be considered.

6 Challenges and Future Perspectives

Many opportunities and challenges are associated with using medicinal plants and their phytochemicals to target specific cancer pathways. One of the main challenges is the complexity of phytochemicals, which frequently show variable bioavailability and stability (Atanasov et al. 2021). The complex interactions between phytochemicals and cancer pathways require detailed mechanistic studies, which are often hampered by the lack of standardized extraction and quantification methods. The shift from preclinical research to clinical trials is another challenge. While many phytochemicals exhibit potential in animal and in vitro models, they are difficult to apply to humans due to issues like toxicity, low solubility, and metabolic instability. The development of plant-derived compounds is also frequently constrained by intellectual property and regulatory frameworks. Future perspectives stress the integration of cutting-edge technologies, like metabolomics and artificial intelligence, to enhance the identification and validation of bioactive compounds. Ethnobotanists, pharmacologists, and oncologists must collaborate to develop effective, plant-based cancer therapeutics. In the end, addressing these issues will advance medicinal plants to the forefront of cancer research.

7 Limitations of Current Research on Medicinal Plants and Phytochemicals Targeting Cancer Pathways

Although medicinal plants and phytochemicals show promise in addressing particular cancer pathways, several restrictions prevent their widespread use. Research standardization, reproducibility, and quality control are complicated by the variability in the content of plant-derived extracts, which is influenced by harvesting methods, processing methods, and geographic location (Atanasov et al. 2021). Another disadvantage for many phytochemicals is the lack of comprehensive pharmacokinetic and pharmacodynamic studies. Lack of knowledge about their absorption, metabolism, and potential toxicity hinders their progress from preclinical research to clinical trials (Newman and Cragg 2020). Additionally, a large portion of research pertains to in vitro tests, which may not be a reliable indicator of interactions or in vivo efficacy due to the complex tumor environment.

Furthermore, high-quality randomized controlled trials (RCTs) often fail to validate the anticancer effects of these compounds in humans, and regulatory and financial barriers further impede the conversion of phytochemicals into therapeutic drugs due to the time-consuming and expensive nature of drug development (Newman and Cragg 2020). Collaboration between disciplines, cutting-edge research techniques, and ongoing financing are needed to close these gaps.

8 Prospects for Novel Therapeutic Developments

By examining medicinal plants and their phytochemicals to target specific cancer pathways, new therapeutic discoveries may be made. The fact that about 60% of contemporary chemotherapeutics are inspired by or derived from natural sources shows how long natural products have been used to make anticancer medications (Newman and Cragg 2020). Advances in phytochemical research, in conjunction with omics and bioinformatics technology, present an unprecedented opportunity to identify novel compounds and elucidate their mechanisms of action. Fares et al. (2020) state that phytochemicals such as green tea's epigallocatechin gallate (EGCG), curcumin from Curcuma longa, and resveratrol from grapes have shown promise in altering critical cancer pathways including angiogenesis, apoptosis, as well as immune evasion. Furthermore, secondary metabolites including alkaloids, terpenoids, and flavonoids are becoming recognized for their ability to inhibit carcinogenic signaling pathways like PI3K/AKT/mTOR and Wnt/β-catenin (Jayanthi et al. 2020). Despite enormous progress, much of the plant kingdom remains undiscovered, particularly species found in for biodiversity hotspots such as tropical rainforests. It may be possible to find bioactive compounds with improved selectivity and fewer side effects by fusing modern research with ancient knowledge. As a result, this area still has the potential for developing personalized cancer therapies.

Adding medicinal herbs to conventional cancer treatments is a good strategy to improve therapy efficacy and decrease adverse effects. Numerous medicinal plants include bioactive phytochemicals that operate in several ways, such as altering cancer signaling pathways, preventing angiogenesis, and inducing apoptosis (Chen et al. 2022). For instance, curcumin from Curcuma longa increases the therapeutic efficiency of chemotherapy by sensitizing cancer cells to drugs like doxorubicin and paclitaxel (Krishnan et al. 2022). Dzobo (2022) asserts that quercetin from onions and resveratrol from grapes complement conventional therapies to enhance outcomes and reduce toxicity. Pharmacokinetic studies and rigorous clinical trials are required to address issues such as bioavailability, standardization, and potential herb–drug interactions (Fares et al. 2020). Phytochemicals' potential as adjuvants is further underscored by their ability to target cancer stem cells, decrease drug resistance, and provide anti-inflammatory and antioxidant properties (Gholipour et al. 2022). A patient-centered, comprehensive strategy that improves overall survival and treatment tolerability is offered by combining standard therapies with medicinal plants. This cooperative strategy aligns with oncology's growing emphasis on customized drug regimens.

9 Conclusions

Medicinal plants and their constituents offer a potential cancer treatment option by concentrating on certain cancer pathways and processes. These natural compounds can increase the efficacy of conventional treatments while reducing their side effects by changing many signaling pathways. Recent research has shown that phytochemicals can inhibit cell division, induce apoptosis, and prevent metastasis. The application of these compounds originating from plants in cancer treatments may lead to more effective and customized therapy choices. Additionally, the use of plant-derived nanoparticles for targeted medication delivery is a unique approach to cancer treatment. Overall, there is a lot of potential for botanicals to enhance cancer treatment and improve patient outcomes.

References

Akinwumi IA, Ambali OA (2023) Traditional knowledge of cancer treatment in southwestern Nigeria: an ethnopharmacological review. J Chem Soc Nigeria 48(5):845–870

Alegbeleye BJ, Akpoveso OP, Mohammed RK (2020) The use of herbal medicines by cancer patients in contemporary African settings: a scoping review. Int J Sci Adv 1:49–73

Alhalmi A, Amin S, Ralli T, Ali KS, Kohli K (2024) Therapeutic role of naringin in cancer: molecular pathways, synergy with other agents, and nanocarrier innovations. Naunyn-Schmiedeb Arch Pharmacol 2024:1–21

Ali Reza ASM, et al (2021) Mechanistic insight into immunomodulatory effects of food-functioned plant secondary metabolites. 63(22): p. 5546–5576

Allen DE, Hatfield G (2004) Medicinal plants in folk tradition. Timber Press

Amaral RG, dos Santos SA, Andrade LN, Severino P, Carvalho AA (2019) Natural products as treatment against cancer: a historical and current vision. Clin Oncol 4:1562

Ashraf MA (2020) Phytochemicals as potential anticancer drugs: time to ponder nature's bounty. Biomed Res Int 2020(1):8602879

Atanasov AG, Zotchev SB, Dirsch VM, Supuran CT (2021) Natural products in drug discovery; advances and opportunities. Nat Rev Drug Discov 20(3):200–216

Auvinen P, Vehviläinen J, Rämö K, Laukkanen I, Marjonen-Lindblad H, Wallén E, Kaminen-Ahola N (2024) Genome-wide DNA methylation and gene expression in human placentas derived from assisted reproductive technology. Commun Med 4(1):267

Azmi A, Bhat S, Hanif S, Hadi S (2006) Plant polyphenols mobilize endogenous copper in human peripheral lymphocytes leading to oxidative DNA breakage: a putative mechanism for anticancer properties. FEBS Lett 580(2):533–538. https://doi.org/10.1016/j.febslet.2005.12.031

Baradaran Rahimi V, Ghadiri M, Ramezani M, Askari VR (2020) Antiinflammatory and anticancer activities of pomegranate and its constituent, ellagic acid: evidence from cellular, animal, and clinical studies. Phytother Res 34(4):685–720

Barbosa WLR, Nascimento MS, do Pinto LN, Maia FLC, Sousa AJA, Júnior JOCS, Monteiro MM, de Oliveira DR (2012) Selecting medicinal plants for development of phytomedicine and use in primary health care. In-Tech Open, Rijeka

Baylin SB, Jones PA (2016) Epigenetic determinants of cancer. Cold Spring Harb Perspect Biol 8(9):a019505. https://doi.org/10.1101/cshperspect.a019505

Belcaro G, Hosoi M, Pellegrini L, Appendino G, Ippolito E, Ricci A, Ledda A, Dugall M, Cesarone MR, Maione C, Ciammaichella G (2014) A controlled study of a lecithinized delivery system of curcumin (Meriva®) to alleviate the adverse effects of cancer treatment. Phytother Res 28(3):444–450

Berman BP, Weisenberger DJ, Aman JF, Hinoue T, Ramjan Z, Liu Y, Noushmehr H, Lange CP, Van Dijk CM, Tollenaar RA, Van Den Berg D (2012) Regions of focal DNA hypermethylation and long-range hypomethylation in colorectal cancer coincide with nuclear lamina–associated domains. Nat Genet 44(1):40–46

Bert SA, Robinson MD, Strbenac D, Statham AL, Song JZ, Hulf T, Sutherland RL, Coolen MW, Stirzaker C, Clark SJ (2013) Regional activation of the cancer genome by long-range epigenetic remodeling. Cancer Cell 23(1):9–22

Biswal BM, Sulaiman SA, Ismail HC, Zakaria H, Musa KI (2013) Effect of Withania somnifera (Ashwagandha) on the development of chemotherapy-induced fatigue and quality of life in breast cancer patients. Integr Cancer Ther 12(4):312–322

Bray F, Ferlay J, Soerjomataram I, Siegel RL, Torre LA, Jemal A (2018) Global cancer statistics 2018: GLOBOCAN estimates of incidence and mortality worldwide for 36 cancers in 185 countries. CA Cancer J Clin 68(6):394–424. https://doi.org/10.3322/caac.21492

Bray F, Laversanne M, Weiderpass E, Soerjomataram I (2021) The ever-increasing importance of cancer as a leading cause of premature death worldwide. Cancer 127(16):3029–3030. https://doi.org/10.1002/cncr.33587

Bray F, Laversanne M, Sung H, Ferlay J, Siegel RL, Soerjomataram I, Jemal A (2024) Global cancer statistics 2022: GLOBOCAN estimates of incidence and mortality worldwide for 36 cancers in 185 countries. CA Cancer J Clin 74(3):229–263. https://doi.org/10.3322/caac.21834

Cai J, Tan X, Hu Q, Pan H, Zhao M, Guo C, Zeng J, Ma X, Zhao Y (2024) Flavonoids and gastric cancer therapy: from signaling pathway to therapeutic significance. Drug Des Devel Ther 2024:3233–3253

Cameron A, Ewen M, Auton M, Abegunde D (2011) The world medicines situation: medicines prices, availability, and affordability. World Health Organization

Charlesworth T (2024) How histone modifications impact gene regulation | biomodal. biomodal. https://biomodal.com/blog/how-histone-modifications-impact-gene-regulation/

Che E, Gao Y, Wan L, Zhang Y, Han N, Bai J, Li J, Sha Z, Wang S (2015) Paclitaxel/gelatin coated magnetic mesoporous silica nanoparticles: preparation and antitumor efficacy in vivo. Microporous Mesoporous Mater 204:226–234

Chen RZ, Pettersson U, Beard C, Jackson-Grusby L, Jaenisch R (1998) DNA hypomethylation leads to elevated mutation rates. Nature 395(6697):89–93

Chen YX, Gao QY, Zou TH, Wang BM, Liu SD, Sheng JQ, Ren JL, Zou XP, Liu ZJ, Song YY, Xiao B (2020) Berberine versus placebo for the prevention of recurrence of colorectal adenoma: a multicentre, double-blinded, randomised controlled study. Lancet Gastroenterol Hepatol 5(3):267–275

Chen N, Qi Y, Ma X, Xiao X, Liu Q, Xia T et al (2022) Rediscovery of traditional plant medicine: an underestimated anticancer drug of chelerythrine. Front Pharmacol 13:906301

Choudhari AS, Mandave PC, Deshpande M, Ranjekar P, Prakash O (2019) Phytochemicals in cancer treatment: from preclinical studies to clinical practice. Front Pharmacol 10:1614

Chunarkar-Patil P, Kaleem M, Mishra R, Ray S, Ahmad A, Verma D, Bhayye S, Dubey R, Singh HN, Kumar S (2024) Anticancer drug discovery based on natural products: from computational approaches to clinical studies. Biomedicine 12(1):201

Cozmin M, et al (2024) Turmeric: from spice to cure. A review of the anti-cancer, radioprotective and anti-inflammatory effects of turmeric sourced compounds. Phytother Res, 11: p. 1399888

Danwilai K, Konmun J, Sripanidkulchai BO, Subongkot S (2017) Antioxidant activity of ginger extract as a daily supplement in cancer patients receiving adjuvant chemotherapy: a pilot study. Cancer Manag Res 2017:11–18

De Luca V, Salim V, Atsumi SM, Yu F (2012) Mining the biodiversity of plants: a revolution in the making. Science 336(6089):1658–1661. https://doi.org/10.1126/science.1217410

Devita VT, Chu E (2008) A history of cancer chemotherapy. Cancer Res 68(21):8643–8653

Dhyani P, Quispe C, Sharma E, Bahukhandi A, Sati P, Attri DC, Szopa A, Sharifi-Rad J, Docea AO, Mardare I, Calina D (2022) Anticancer potential of alkaloids: a key emphasis to colchicine, vinblastine, vincristine, vindesine, vinorelbine and vincamine. Cancer Cell Int 22(1):206

Dickens E, Ahmed S (2018) Principles of cancer treatment by chemotherapy. Surgery 36:134–138. https://doi.org/10.1016/j.mpsur.2017.12.002

Doroshow DB, Doroshow JH (2020) Genomics and the history of precision oncology. Surg Oncol Clin N Am 29(1):35–49. https://doi.org/10.1016/j.soc.2019.08.003

Du YX, Mamun AA, Lyu AP, Zhang HJ (2023) Natural compounds targeting the autophagy pathway in the treatment of colorectal cancer. Int J Mol Sci 24(8):7310

Dzobo K (2022) The role of natural products as sources of therapeutic agents for innovative drug discovery. Comprehen Pharmacol 2:408–422

Ehrlich M, Lacey M (2012) DNA hypomethylation and hemimethylation in cancer. In: Epigenetic alterations in oncogenesis, pp 31–56

Esteller M (2007) Epigenetic gene silencing in cancer: the DNA hypermethylome. Hum Mol Genet 16(R1):R50–R59

Fares J, Fares MY, Khachfe HH, Salhab HA, Fares Y (2020) Molecular principles of metastasis: a hallmark of cancer revisited. Signal Transduct Target Ther 5(1):28

Feinberg AP, Vogelstein B (1983) Hypomethylation distinguishes genes of some human cancers from their normal counterparts. Nature 301(5895):89–92

Ferreira MJ (2018) Natural product-derived compounds for targeting multidrug resistance in cancer and microorganisms. Int J Mol Sci 24(18):14321

García-Figueiras R, Baleato-González S, Padhani AR, Luna-Alcalá A, Vallejo-Casas JA, Sala E, Vargas HA (2019) How clinical imaging can assess cancer biology. Insights Imaging 10:28. https://doi.org/10.1186/s13244-019-0721-x

Gaudet F, Hodgson JG, Eden A, Jackson-Grusby L, Dausman J, Gray JW, Leonhardt H, Jaenisch R (2003) Induction of tumors in mice by genomic hypomethylation. Science 300(5618):489–492

Gholipour AR, Jafari L, Ranezanpour M, Evazalipour M, Chavoshi M, Yousefbeyk F, Moghaddam SJK, Kooshali MHY, Ramezanpour N, Daei P (2022) Apoptosis effects of *Oxalis corniculata* L. extract on human MCF-7 breast cancer cell line. Galen Med J 11:e2484

González-Sarrías A, Núñez-Sánchez MA, Ávila-Gálvez MA, Monedero-Saiz T, Rodríguez-Gil FJ, Martínez-Díaz F, Selma MV, Espín JC (2018) Consumption of pomegranate decreases plasma lipopolysaccharide-binding protein levels, a marker of metabolic endotoxemia, in patients with newly diagnosed colorectal cancer: a randomized controlled clinical trial. Food Funct 9(5):2617–2622

Goswami M, et al (2023) A comprehensive update on phytochemistry, analytical aspects, medicinal attributes, specifications and stability of stigmasterol. Steroids, 196: p. 109244

Goswami S, Ali A, Prasad ME, Singh P (2024) Pharmacological significance of *Catharanthus roseus* in cancer management: a review. Pharmacol Res Mod Chin Med 11:100444. https://doi.org/10.1016/j.prmcm.2024.100444

Greenwell M, Rahman PKSM (2015) Medicinal plants: their use in anticancer treatment. Int J Pharm Sci Res 6(10):4103

Haddad P, Amouzgar–Hashemi F, Samsami S, Chinichian S, Oghabian MA (2013) *Aloe vera* for prevention of radiation-induced dermatitis: a self-controlled clinical trial. Curr Oncol 20(4):e345

Hanahan D, Weinberg RA (2011) Hallmarks of cancer: the next generation. Cell 144(5):646–674. https://doi.org/10.1016/j.cell.2011.02.013

Hansen KD, Timp W, Bravo HC, Sabunciyan S, Langmead B, McDonald OG, Wen B, Wu H, Liu Y, Diep D, Briem E (2011) Increased methylation variation in epigenetic domains across cancer types. Nat Genet 43(8):768–775

Hasan AA, V Tatarskiy and E Kalinina (2022) Synthetic Pathways and the Therapeutic Potential of Quercetin and Curcumin. 23(22)

Hassan F, Rehman MS, Khan MS, Ali MA, Javed A, Nawaz A, Yang C (2019) Curcumin as an alternative epigenetic modulator: mechanism of action and potential effects. Front Genet. https://doi.org/10.3389/fgene.2019.00514

Hetland G, Tangen JM, Mahmood F, Mirlashari MR, Nissen-Meyer LSH, Nentwich I, Therkelsen SP, Tjønnfjord GE, Johnson E (2020) Antitumor, anti-inflammatory and antiallergic effects of *Agaricus blazei* mushroom extract and the related medicinal basidiomycetes mushrooms, *Hericium erinaceus* and *Grifola frondosa*: a review of preclinical and clinical studies. Nutrients 12(5):1339

Hu C and X Jiang (2016) Role of NRP-1 in VEGF-VEGFR2-Independent Tumorigenesis. Target Oncol, 11(4): p. 501–5

Hon GC, Hawkins RD, Caballero OL, Lo C, Lister R, Pelizzola M, Valsesia A, Ye Z, Kuan S, Edsall LE, Camargo AA (2012) Global DNA hypomethylation coupled to repressive chromatin domain formation and gene silencing in breast cancer. Genome Res 22(2):246–258

Hur K, Cejas P, Feliu J, Moreno-Rubio J, Burgos E, Boland CR, Goel A (2014) Hypomethylation of long interspersed nuclear element-1 (LINE-1) leads to activation of proto-oncogenes in human colorectal cancer metastasis. Gut 63(4):635–646

Hosseini A and H Hosseinzadeh (2018) Antidotal or protective effects of Curcuma longa (turmeric) and its active ingredient, curcumin, against natural and chemical toxicities: A review. Biomed Pharmacother99: p. 411–421

Imtiaz I, Schloss J, Bugarcic A (2024) Traditional and contemporary herbal medicines in management of cancer: a scoping review. J Ayurveda Integr Med 15:100904

Ishara J, et al (2021) Nutraceutical potential of mushroom bioactive metabolites and their food functionality. Crit Rev Food Sci Nutr, 46(1): p. e14025

Jang SY, et al (2023) Curcumin inhibits human cancer cell growth and migration through downregulation of SVCT2. 41(6): p. 696–703

Jayanthi M, Gpkulanathan A, Haribalan P, Ashakiran K, Kumar CD, Kamal D, Girija S, Renukadevi BS (2020) Plumbagin from two *Plumbago* species inhibits the growth of stomach and breast cancer cell lines. Crop Prot 146:112147

Jiang A, Wang X, Shan X, Li Y, Wang P, Jiang P, Feng Q (2015) Curcumin reactivates silenced tumor suppressor gene *RARβ* by reducing DNA methylation. Phytother Res. https://doi.org/10.1002/ptr.5373

Jiang C, et al (2017) Membranous type matrix metalloproteinase 16 induces human prostate cancer metastasis. Oncol Lett, 14(3): p. 3096–3102

Jenca, A, et al (2024) Herbal Therapies for Cancer Treatment: A Review of Phytotherapeutic Efficacy. Biologics, 18: p. 229–255

Jones PA, Baylin SB (2007) The epigenomics of cancer. Cell 128(4):683–692

Kanwal R, Gupta S (2012) Epigenetic modifications in cancer. Clin Genet 81(4):303–311

Karpuz M, Silindir-Gunay M, Ozer AY (2018) Current and future approaches for effective cancer imaging and treatment. Cancer Biother Radiopharm 33(1):39–51. https://doi.org/10.1089/cbr.2017.2416

Khan MI, et al (2022) Therapeutic Effects of Saponins for the Prevention and Treatment of Cancer by Ameliorating Inflammation and Angiogenesis and Inducing Antioxidant and Apoptotic Effects in Human Cells. International Journal of Molecular Sciences, 23(18)

Konmun J, Danwilai K, Ngamphaiboon N, Sripanidkulchai B, Sookprasert A, Subongkot S (2017) A phase II randomized double-blind placebo-controlled study of 6-gingerol as an anti-emetic in solid tumor patients receiving moderately to highly emetogenic chemotherapy. Med Oncol 34:1–10

Krishnan V, Subramaniam S, Chia-Chuan C, Venkatachalam B, Thomas Cheeran A, Chi-Ying FH (2022) Anticancer activity of *Leonurus sibiricus* L.: possible involvement of intrinsic apoptotic pathway. Nutr Cancer 74(1):225–236

Kuriakose GC, et al (2018) Extract of Penicillium sclerotiorum an endophytic fungus isolated from Cassia fistula L. induces cell cycle arrest leading to apoptosis through mitochondrial membrane depolarization in human cervical cancer cells. Biomed Pharmacother, 105: p. 1062–1071

Lissoni P, Rovelli F, Brivio F, Zago R, Colciago M, Messina G, Mora A, Porro G (2009) A randomized study of chemotherapy versus biochemotherapy with chemotherapy plus *Aloe arborescens* in patients with metastatic cancer. In Vivo 23(1):171–175

Liu R, Wu J, Guo H, Yao W, Li S, Lu Y, Jia Y, Liang X, Tang J, Zhang H (2023) Post-translational modifications of histones: mechanisms, biological functions, and therapeutic targets. MedComm 4(3):e292

Li X, et al (2024) S100A4/NF-kappaB axis mediates the anticancer effect of epigallocatechin-3-gallate in platinum-resistant ovarian cancer. iScience, 27(2): p. 108885

Lu L-H et al (1998) The possible mechanisms of the antiproliferative effect of fullerenol, polyhydroxylated C60, on vascular smooth muscle cells. British journal of pharmacology, 123(6): p. 1097

Luo KW et al (2020) EGCG Enhanced the Anti-tumor Effect of Doxorubicine in Bladder Cancer via NF-kappaB/MDM2/p53 Pathway. Front Cell Dev Biol, 8: p. 606123

Maciel MAM (2002) Plantas medicinais: a necessidade de estudos multidisciplinares. Quim Nova 25(3):429–438

Mansouri P, Haghighi M, Beheshtipour N, Ramzi M (2016) The effect of aloe vera solution on chemotherapy-induced stomatitis in clients with lymphoma and leukemia: a randomized controlled clinical trial. Int J Community Based Nurs Midwifery 4(2):119

Mbugua RW, et al (2019) Gene expression mediated antiproliferative potential and safety of selected medicinal plants against cancerous and normal cells. BioRxiv, p. 578948

Memarzia A, et al (2023) Therapeutic effects of medicinal plants and their constituents on lung cancer, in vitro, in vivo and clinical evidence. Journal of Cellular and Molecular Medicine,. 27(19): p. 2841–2863

Minnelli C, et al (2021) Effect of Epigallocatechin-3-Gallate on EGFR Signaling and Migration in Non-Small Cell Lung Cancer. 22(21)

Mishra J, Verma N (2017) A brief study on *Catharanthus roseus*: a review. Int J Res Pharm Sci 2(2):20–23

Mobadersany P, Yousefi S, Amgad M, Gutman DA, Barnholtz-Sloan JS, Velázquez Vega JE et al (2018) Predicting cancer outcomes from histology and genomics using convolutional networks. Proc Natl Acad Sci USA 115(13):E2970–E2979. https://doi.org/10.1073/pnas.1717139115

Moore LD, Le T, Fan G (2013) DNA methylation and its basic function. Neuropsychopharmacology 38(1):23–38

Moss RW (1998) Herbs against cancer: history and controversy. Equinox Press Inc, 300 p

Mostofi S, et al (2022) Anti-migratory effect of curcumin on A-549 lung cancer cells via epigenetic reprogramming of RECK/ matrix metalloproteinase axis. Horm Mol Biol Clin Investig, 43(4): p. 455–461

Mottaghi S and H Abbaszadeh (2021) The anticarcinogenic and anticancer effects of the dietary flavonoid, morin: Current status, challenges, and future perspectives. 35(12): p. 6843–6861

Nakayama A, et al (2021) Effects of Curcumin Combined With the 5-alpha Reductase Inhibitor Dutasteride on LNCaP Prostate Cancer Cells. In Vivo, 35(3): p. 1443–1450

Narayan A, Ji W, Zhang XY, Marrogi A, Graff JR, Baylin SB, Ehrlich M (1998) Hypomethylation of pericentromeric DNA in breast adenocarcinomas. Int J Cancer 77(6):833–838

Newman DJ, Cragg GM (2020) Natural products as sources of new drugs over the nearly four decades from 01/1981 to 09/2019. J Nat Prod 83(3):770–803

Nimlamool W, et al (2022) The Leaf Extract of Mitrephora chulabhorniana Suppresses Migration and Invasion and Induces Human Cervical Cancer Cell Apoptosis through Caspase-Dependent Pathway. 2022: p. 2028082

Nkwe DO, Lotshwao B, Rantong G, Matshwele J, Kwape TE, Masisi K et al (2021) Anticancer mechanisms of bioactive compounds from Solanaceae: an update. Cancers (Basel) 13(19):4989. https://doi.org/10.3390/cancers13194989

Nuñez-Sánchez MA, Dávalos A, González-Sarrías A, Casas-Agustench P, Visioli F, Monedero-Saiz T et al (2015) MicroRNAs expression in normal and malignant colon tissues as biomarkers of colorectal cancer and in response to pomegranate extracts consumption: critical issues to discern between modulatory effects and potential artefacts. Mol Nutr Food Res 59(10):1973–1986

Ogunlakin AD, Sonibare MA, Yeye OE, Gyebi GA, Ayokunle DI, Arigbede OE et al (2023) Isolation and characterization of novel hydroxyflavone from *Kigelia africana* (Lam.) Benth. fruit ethyl acetate fraction against CHO 1 and HeLa cancer cell lines: In vitro and in silico studies. J Mol Struct 1282:135180

Oršolić N, Jembrek MJ (2024) Potential strategies for overcoming drug resistance pathways using propolis and its polyphenolic/flavonoid compounds in combination with chemotherapy and radiotherapy. Nutrients 16(21):3741

Ouelbani R, Bensari S, Mouas TN, Khelifi D (2016) Ethnobotanical investigations on plants used in folk medicine in the regions of Constantine and Mila (North-East of Algeria). J Ethnopharmacol 194:196–218. https://doi.org/10.1016/j.jep.2016.08.016

Paller CJ, Ye X, Wozniak PJ, Gillespie BK, Sieber PR, Greengold RH et al (2013) A randomized phase II study of pomegranate extract for men with rising PSA following initial therapy for localized prostate cancer. Prostate Cancer Prostatic Dis 16(1):50–55

Paller CJ, Rudek MA, Zhou XC, Wagner WD, Hudson TS, Anders N et al (2015) A phase I study of muscadine grape skin extract in men with biochemically recurrent prostate cancer: safety, tolerability, and dose determination. Prostate 75(14):1518–1525

Paller CJ, Zhou XC, Heath EI, Taplin ME, Mayer T, Stein MN et al (2018) Muscadine grape skin extract (MPX) in men with biochemically recurrent prostate cancer: a randomized, multicenter, placebo-controlled clinical trial. Clin Cancer Res 24(2):306–315

Panahi Y, Saadat A, Beiraghdar F, Sahebkar A (2014) Adjuvant therapy with bioavailability-boosted curcuminoids suppresses systemic inflammation and improves quality of life in patients with solid tumors: a randomized double-blind placebo-controlled trial. Phytother Res 28(10):1461–1467

Pastorelli D, Fabricio AS, Giovanis P, D'Ippolito S, Fiduccia P, Soldà C et al (2018) Phytosome complex of curcumin as complementary therapy of advanced pancreatic cancer improves safety and efficacy of gemcitabine: results of a prospective phase II trial. Pharmacol Res 132:72–79

Pathak A, Tomar S, Pathak S (2023) Epigenetics and cancer: a comprehensive review. Asian Pac J Cancer Biol 8(1):75–89

Pfohl U, Pflaume A, Regenbrecht M, Finkler S, Graf Adelmann Q, Reinhard C et al (2021) Precision oncology beyond genomics: the future is here—it is just not evenly distributed. Cells 10(4):928. https://doi.org/10.3390/cells10040928

Piedrola G, Novo E, Escobar F, Garcia-Robles R (2001) White blood cell count and insulin resistance in patients with coronary artery disease. Diabetes Res Clin Pract 53(3):205–211. https://doi.org/10.1016/S0168-8227(01)00266-3

Qu Y, et al 2017 Cancer-specific changes in DNA methylation reveal aberrant silencing and activation of enhancers in leukemia. Blood, 129(7): p. e13–e25

Sarkar J, et al (2016) Inhibition of MMP-9 by green tea catechins and prediction of their interaction by molecular docking analysis. Biomed Pharmacother, 84: p. 340–347

Saghatelyan T, Tananyan A, Janoyan N, Tadevosyan A, Petrosyan H, Hovhannisyan A et al (2020) Efficacy and safety of curcumin in combination with paclitaxel in patients with advanced, metastatic breast cancer: a comparative, randomized, double-blind, placebo-controlled clinical trial. Phytomedicine 70:153218

Senft D, Leiserson MDM, Ruppin E, Ronai ZA (2017) Precision oncology: the road ahead. Trends Mol Med 23(10):874–898. https://doi.org/10.1016/j.molmed.2017.08.003

Shahrajabian MH and W Sun (2024) The Golden Spice for Life: Turmeric with the Pharmacological Benefits of Curcuminoids Components, Including Curcumin, Bisdemethoxycurcumin, and Demethoxycurcumins. Curr Org Synth, 21(5): p. 665–683

Shankar S, Kumar D, Srivastava RK (2013) Epigenetic modifications by dietary phytochemicals: implications for personalized nutrition. Pharmacol Ther 138(1):1–7

Shen L, Kondo Y, Guo Y, Zhang J, Zhang L, Ahmed S et al (2007) Genome-wide profiling of DNA methylation reveals a class of normally methylated CpG Island promoters. PLoS Genet 3(10):e181

Shi X, Wang X, Yao W, Shi D, Shao X, Lu Z et al (2024) Mechanism insights and therapeutic intervention of tumor metastasis: latest developments and perspectives. Signal Transduct Target Ther 9(1):192

Shirakami Y and M (2018) Shimizu Possible mechanisms of green tea and its constituents against cancer. Molecules, 23(9): p. 2284

Sincic N, Herceg Z (2011) DNA methylation and cancer: ghosts and angels above the genes. Curr Opin Oncol 23(1):69–76

Situmorang PC, et al (2024)Prospects of compounds of herbal plants as anticancer agents: a comprehensive review from molecular pathways. Front Pharmacol, 15: p. 1387866

Smith JC, Sheltzer JM (2022) Genome-wide identification and analysis of prognostic features in human cancers. Cell Rep 38(2):110569. https://doi.org/10.1016/j.celrep.2022.110569

Suhail M, et al (2021) A Critical Transcription Factor NF-kappaB as a Cancer Therapeutic Target and its Inhibitors as Cancer Treatment Options. Curr Med Chem, 28(21): p. 4117–4132

Sung H, Ferlay J, Siegel RL, Laversanne M, Soerjomataram I, Jemal A et al (2021) Global cancer statistics 2020: GLOBOCAN estimates of incidence and mortality worldwide for 36 cancers in 185 countries. CA Cancer J Clin 71(3):209–249. https://doi.org/10.3322/caac.21660

Tanabe H, Suzuki T, Ohishi T, Isemura M, Nakamura Y, Unno K (2023) Effects of epigallocatechin-3-gallate on matrix metalloproteinases in terms of its anticancer activity. Molecules 28(2):525

Verzella D, et al (2020) Life, death, and autophagy in cancer: NF-kappaB turns up everywhere. Cell Death Dis, 11(3): p. 210

Wang G, et al (2021) The expression and diagnostic value of serum levels of EphA2 and VEGF-A in patients with colorectal cancer. Cancer Biomark, 31(4): p. 399–408

Wang P, Chen C, Shan M (2024) Vincamine alleviates brain injury by attenuating neuroinflammation and oxidative damage in a mouse model of Parkinson's disease through the NF-κB and Nrf2/HO-1 signaling pathways. J Biochem Mol Toxicol 38(5):e23714

Wendlocha D, Kubina R, Krzykawski K, Mielczarek-Palacz A (2024) Selected flavonols targeting cell death pathways in cancer therapy: the latest achievements in research on apoptosis, autophagy, necroptosis, pyroptosis, ferroptosis, and cuproptosis. Nutrients 16(8):1201

Wolff A (2014) The impact of breast cancer treatment on your long-term health. Johns Hopkins Medicine. https://www.hopkinsmedicine.org/kimmel_cancer_center/centers/breast_cancer_program/survivorship/side_effects.html

World Health Organization (2020) All cancers fact sheet. International Agency for Research on Cancer, Lyon

Wyman O (2020) El impacto económico y social del cáncer en España: Estudio elaborado para la Asociación Española Contra el Cáncer (AECC). https://www.contraelcancer.es/sites/default/files/content-file/Informe-Los-costes-cancer.pdf

Xu X, Peng Q, Jiang X, Tan S, Yang Y, Yang W et al (2023) Metabolic reprogramming and epigenetic modifications in cancer: from the impacts and mechanisms to the treatment potential. Exp Mol Med 55(7):1357–1370

Yakkala PA, Penumallu NR, Shafi S, Kamal A (2023) Prospects of topoisomerase inhibitors as promising anti-cancer agents. Pharmaceuticals 16(10):1456

Yang Y, Guo J, Li M, Chu G, Jin H, Ma J et al (2024) Cancer stem cells and angiogenesis. Pathol Res Pract 253:155064

Yao Q et al (2015) Curcumin induces the apoptosis of A549 cells via oxidative stress and MAPK signaling pathways. Int J Mol Med, 36(4): p. 1118–26

Zhang Z, Shi J, Nice EC, Huang C, Shi Z (2021) The multifaceted role of flavonoids in cancer therapy: leveraging autophagy with a double-edged sword. Antioxidants 10(7):1138

Medicinal Plants in Combination Therapies for Cancer

12

Urja Desai, Lipi Poojara, Dweipayan Goswami, Medha Pandya, Shreya Modi, Rachana Shukla, and D. K. Acharya

Abstract

Cancer continues to be a leading cause of death globally, prompting the need for effective and less toxic therapies. Medicinal plants, with their diverse chemical compositions and biological activities, have emerged as promising candidates for cancer treatment. Recent research indicates that combining natural compounds with standard chemotherapeutic drugs can enhance their therapeutic effectiveness while reducing side effects. Preclinical and clinical trials have shown that the synergistic effects of medicinal plants and chemotherapy drugs can enhance anticancer treatment efficacy, while overcoming drug resistance mechanisms and reducing side effects like nausea and fatigue. Advances in drug delivery systems, such as nano-drug delivery systems, have improved the stability, bioavailability, and target-specific activity of natural bioactive compounds. Developments such as liposomes, nanoparticles, and micelles enable targeted delivery to tumor sites,

U. Desai (✉)
Department of Zoology, Biomedical Technology, Human Genetics, Wildlife Biology & Conservation, University School of Sciences, Gujarat University, Ahmedabad, Gujarat, India

L. Poojara
Department of Biochemistry and Forensic Science, University School of Sciences, Gujarat University, Ahmedabad, Gujarat, India

D. Goswami
Department of Microbiology and Biotechnology, University School of Sciences, Gujarat University, Ahmedabad, Gujarat, India

M. Pandya
Raan Pharmacy, Islip Terrace, NY, USA

S. Modi · R. Shukla · D. K. Acharya
Department of Microbiology, Gandhinagar Institute of Science, Gandhinagar University, Kalol, Gujarat, India
e-mail: shreya.modi@gandhinagaruni.ac.in; rachana.shukla@gandhinagaruni.ac.in

K. Kuca et al. (eds.), *Medicinal Plants and Their Bioactives in Human Diseases*,
https://doi.org/10.1007/978-3-032-01356-9_12

reducing side effects and enhancing therapeutic efficacy. However, challenges remain in large-scale production, stability, physiological barriers, and regulatory issues. The development of natural product-based hybrids, such as antibody-drug conjugates (ADCs), proteolysis-targeting chimeras (PROTACs), and autophagy-targeting chimeras (AUTOTACs), has revolutionized cancer therapy by combining the therapeutic benefits of natural compounds with innovative drug delivery systems. These advancements hold great promise for personalized, less toxic cancer therapies, providing a more patient-centered approach. Ongoing research is crucial to overcoming challenges and moving these innovations from the lab to clinical application, offering hope for improved patient outcomes and quality of life in cancer care.

Keywords

Medicinal plant · Cancer treatment · Combination therapy · Bioactive compound · Drug delivery system · Synergistic effect

1 Introduction

Cancer encompasses a group of diseases characterized by the uncontrolled growth and spread of abnormal cells. When cancer cells spread to other parts of the body—a process known as metastasis—it can become life-threatening if left uncontrolled. The development of cancer is influenced by a combination of external factors, such as tobacco use, exposure to harmful chemicals, radiation, and infectious agents, and internal factors, such as inherited genetic mutations, hormonal imbalances, immune system dysfunctions, and random genetic changes (Patel et al. 2005; Asma et al. 2022). The causes of cancer are complex, multifaceted, and not yet fully understood. However, several risk factors are known to increase its likelihood. These include unhealthy dietary habits, certain infections, physical inactivity, obesity, and exposure to environmental pollutants. These factors can work together to initiate or accelerate the process of carcinogenesis, leading to the development of cancer (Patel et al. 2004; Friedenreich et al. 2021).

In India, cancer has emerged as a significant cause of mortality. At any given time, an estimated 2–2.5 million individuals are affected by cancer. Each year, the country sees over 700,000 new cases and approximately 300,000 deaths due to the disease. Furthermore, nearly 1.5 million patients require diagnostic, treatment, and follow-up care at any given time, highlighting the pressing need for robust healthcare infrastructure and resources to address this growing burden (Sathishkumar et al. 2022). Carcinogens are substances that directly contribute to DNA damage, thereby promoting or triggering the development of cancer. Examples of carcinogens include tobacco, asbestos, arsenic, gamma and X-ray radiation, sunlight, and chemicals found in vehicle exhaust. When the body is exposed to these harmful agents, free radicals are generated. These unstable molecules seek to steal electrons from other molecules, leading to cellular damage. This damage can impair normal cell function and increase the risk of cancer (Kirave et al. 2020).

Population-based measures of cancer incidence and mortality were analyzed across different regions of the world, with estimates covering the period from 2005 to 2010. In 2008, approximately 12.66 million people worldwide were diagnosed with cancer, which corresponds to about 188 cases per 100,000 individuals (Sung et al. 2021). The number of new cases varied significantly by region, ranging from 67,000 in Middle Africa to 3.72 million in Eastern Asia (Sung et al. 2021). Given Asia's large population, it accounted for the highest proportion of cases globally, with nearly half (48%) of all diagnoses occurring in the region. Four types of cancer—lung, breast (in females), colorectal, and stomach—made up 41% of the global cancer burden. In the UK, the most prevalent cancer types were breast, lung, colorectal, and prostate cancer, which together represented over half (54%) of all cancer cases in 2008 (Sung et al. 2021).

1.1 Role of Medicinal Plants in Cancer Treatment

Medicinal plants have been recognized for their significant role in the development of cancer treatments due to the diverse bioactive compounds they possess, which exhibit anti-cancer properties. Many modern anti-cancer drugs have been derived or inspired by phytochemicals found in plants. These natural compounds can be used as direct therapeutic agents or as templates for the development of synthetic drugs. The use of medicinal plants in cancer therapy has a long history, dating back thousands of years and spanning across various cultures and civilizations (Asma et al. 2022). Traditional medicine systems have relied on the healing properties of plants to treat a wide variety of ailments, including tumors, long before modern cancer treatments were developed. In ancient Egyptian medicine, texts such as the Ebers Papyrus (1550 BCE) mention the use of herbal remedies for treating tumors, with plants like *Aloe vera* and *Helleborus* being used (Ji et al. 2009). Traditional Chinese medicine has used medicinal plants such as *Camptotheca acuminata* and *Artemisia annua* for centuries. Similarly, Indian Ayurveda has used plants like *Withania somnifera*, *Curcuma longa*, and *Tinospora cordifolia* to manage growths and improve overall health (Patwardhan et al. 2005). In Greek and Roman medicine, plants like *Chelidonium majus* and *Vinca* species were documented for their use in tumor-related conditions (Rowinsky 2003).

Furthermore, indigenous knowledge has played a significant role in identifying plants with anticancer properties. Native American medicine used plants such as *Taxus brevifolia* and *Podophyllum peltatum* to treat various ailments, including cancer-like conditions. African traditional medicine has used plants such as *Catharanthus roseus* and *Annona muricata* for their anticancer and healing properties. The traditional use of medicinal plants has laid the foundation for the discovery of many modern anticancer drugs (Adico et al. 2024). For example, Taxol (paclitaxel) was derived from the Pacific yew tree (*Taxus brevifolia*), vincristine and vinblastine were derived from *Catharanthus roseus*, and camptothecin was derived from *Camptotheca acuminata* (Gurib-Fakim 2006). Medicinal plants remain an essential part of cancer research, bridging ancient wisdom and modern science to develop more effective, less toxic therapies. Traditional medicine has been instrumental in identifying plants with anticancer properties, and many traditional systems emphasize balancing the body, which aligns with modern integrative cancer therapies. Ethnobotanical studies continue to uncover plants with potential anticancer properties, preserving traditional knowledge while advancing modern medicine (Gezici and Şekeroğlu 2019). Table 12.1 provides a comprehensive summary of medicinal plants, their active compounds, mechanisms of action, and therapeutic benefits in combination therapies, highlighting their potential to enhance cancer treatment outcomes.

1.2 Significance of Combination Therapies

Combination therapy is significant in the treatment of cancer, infectious diseases, and chronic disorders because single-agent therapies often have limitations. The advantages of combination therapies include:

1.2.1 Enhanced Efficacy

Combination therapies improve treatment effectiveness by targeting different pathways or mechanisms. For example, combining chemotherapy with immunotherapy can both attack rapidly dividing cells and boost the immune system. Agents may also work together to enhance each other's effects, achieving better outcomes than single treatments alone (Vanneman and Dranoff 2012).

1.2.2 Reduced Resistance

Combination therapy reduces the likelihood of resistance by attacking the disease through multiple mechanisms. For example, in HIV treatment, a cocktail of antiretroviral drugs helps prevent the virus from mutating and evading treatment (Domingo-Fernández et al. 2023).

1.2.3 Lower Toxicity

Combination therapy often allows for lower doses of each drug, minimizing side effects and toxicity while maintaining efficacy. In cancer, combining targeted therapies with lower-dose chemotherapy can reduce the harsh side effects of high-dose chemotherapy alone (Lopez and Banerji 2017).

1.2.4 Broader Spectrum of Action

Combination therapies, which use agents with complementary actions, can tackle the heterogeneity of diseases like cancer, which often involves diverse cell populations with different vulnerabilities. For instance, antibiotic combinations are used to treat infections involving multiple bacterial strains (Alaoui Mdarhri et al. 2022).

1.2.5 Improved Patient Outcomes

Combination therapies attack diseases from multiple angles, leading to faster recovery, prolonged remission, or better disease control. In diseases such as cancer,

Table 12.1 Medicinal plants, active compounds, and their roles in enhancing cancer combination therapies

Plant name	Active compound(s)	Mechanism of action	Combination benefits
Curcuma longa	Curcumin	Anti-inflammatory, antioxidant, inhibits NF-κB, MAPK, PI3K/Akt pathways	Enhances chemotherapy efficacy, reduces drug resistance, mitigates toxicity
Vitis vinifera	Resveratrol	Induces apoptosis, inhibits NF-κB, PI3K/Akt, modulates oxidative stress	Reduces chemotherapy toxicity, enhances DNA repair, improves radiotherapy outcomes
Allium cepa	Quercetin	Inhibits angiogenesis, arrests cell cycle, downregulates anti-apoptotic proteins	Synergizes with chemotherapeutic drugs, reduces nephrotoxicity and cardiotoxicity
Berberis species	Berberine	Modulates apoptotic pathways, inhibits NF-κB, enhances mitochondrial dysfunction	Enhances efficacy of cisplatin and doxorubicin, protects normal cells from oxidative damage
Camellia sinensis	Epigallocatechin gallate (EGCG)	Targets VEGF, PI3K/Akt, inhibits angiogenesis, reduces metastasis	Potentiates effects of paclitaxel and cisplatin, protects normal tissues during radiotherapy
Withania somnifera	Withaferin A	Proteasome inhibition, modulation of heat shock proteins	Enhances tumor shrinkage, reduces systemic toxicity, synergizes with gemcitabine and paclitaxel
Brassica oleracea	Sulforaphane	Induces phase II detox enzymes, inhibits histone deacetylases, modulates immune response	Sensitizes cancer cells to chemotherapy, protects against oxidative stress and inflammation
Taxus brevifolia	Paclitaxel	Promotes microtubule stabilization, arrests cell division	Combines well with targeted therapies to reduce drug resistance and improve survival
Artemisia annua	Artemisinin	Generates ROS, induces apoptosis	Synergizes with radiotherapy, enhances effects of chemotherapeutics
Tinospora cordifolia	Tinosporide, Berberine derivatives	Immunomodulatory, antioxidant	Improves tolerance to chemotherapy, enhances immune response
Catharanthus roseus	Vinblastine, Vincristine	Inhibits tubulin polymerization, induces mitotic arrest	Used in multi-drug regimens for leukemia and lymphoma
Annona muricata	Annonaceous acetogenins	Inhibits mitochondrial complex I, induces oxidative stress	Targets drug-resistant cancer cells

(continued)

Table 12.1 (continued)

Plant name	Active compound(s)	Mechanism of action	Combination benefits
Podophyllum peltatum	Podophyllotoxin	DNA topoisomerase II inhibition	Basis for semi-synthetic anticancer drugs like etoposide and teniposide
Azadirachta indica	Nimbolide, Azadirachtin	Induces apoptosis, ROS generation, immune modulation	Enhances chemotherapeutic sensitivity in breast and colon cancers
Zingiber officinale	Gingerol, Shogaol	Anti-inflammatory, antioxidant, induces apoptosis	Reduces nausea during chemotherapy, synergizes with paclitaxel
Camptotheca acuminata	Camptothecin	Topoisomerase I inhibition	Backbone of drugs like irinotecan, used in colon and lung cancer therapy
Glycyrrhiza glabra	Glycyrrhizin, Liquiritigenin	Anti-inflammatory, induces apoptosis, inhibits NF-κB	Enhances drug sensitivity in lung and liver cancers
Nigella sativa	Thymoquinone	Antioxidant, inhibits PI3K/Akt pathway	Reduces chemotherapy resistance in breast cancer
Curcuma aromatica	Curcuminoids	Modulates Wnt/β-catenin signaling	Enhances chemotherapeutic efficacy in gastrointestinal cancers
Piper nigrum	Piperine	Enhances bioavailability of other compounds, induces apoptosis	Synergizes with curcumin to increase anticancer activity
Eclipta alba	Wedelolactone	Antioxidant, inhibits NF-κB, induces apoptosis	Potential in hepatocellular carcinoma therapies
Phyllanthus emblica	Gallic acid, Ellagic acid	Antioxidant, anti-inflammatory	Enhances effects of radiation therapy
Ocimum sanctum	Eugenol	Anti-inflammatory, modulates oxidative stress	Protective against chemotherapy-induced toxicity
Cymbopogon citratus	Citral	Anti-proliferative, induces apoptosis	Enhances sensitivity to chemotherapy
Triphala	Gallic acid, Chebulinic acid	Antioxidant, induces apoptosis, inhibits metastasis	Improves overall outcomes in combination with radiotherapy

combining therapies can lower the risk of recurrence by eliminating residual disease (Liu et al. 2024).

1.2.6 Versatility Across Diseases

Combinations of chemotherapy, radiation, immunotherapy, and targeted drugs improve survival and quality of life in cancer patients. Drug cocktails effectively manage diseases like tuberculosis, malaria, and HIV/AIDS. In conditions such as diabetes or hypertension, combining medications addresses multiple aspects of the disease, leading to better control (Babar et al. 2022). While combination therapies offer significant advantages, they also present challenges such as increased cost and complexity of treatment, the potential for drug–drug interactions, and difficulty in optimizing dosage and timing. Advances in personalized medicine, computational modeling, and clinical trials continue to refine combination therapies (Goetz and Schork 2018).

2 Historical Background

The use of medicinal plants in cancer therapy has deep roots, spanning thousands of years and diverse cultures. Traditional medicine systems across the globe have relied on the healing properties of plants to treat a wide array of ailments, including tumors and related conditions, long before the advent of modern cancer treatments (Desai et al. 2008).

2.1 Ancient Civilizations and Early Records

The use of medicinal plants in cancer therapy dates back to ancient civilizations, with evidence found in various historical records. In ancient Egyptian medicine, texts like the Ebers Papyrus (1550 BCE) document the use of herbal remedies for treating tumors. Plants such as *Aloe vera* and *Helleborus* were commonly employed for their healing and purgative properties (Metwaly et al. 2021). Traditional Chinese Medicine (TCM) has also utilized medicinal plants for centuries, including *Camptotheca acuminata* (source of camptothecin) and *Artemisia annua* (source of artemisinin). These plants were traditionally used to treat fever, inflammation, and conditions resembling cancer (Graziose et al. 2010). Indian Ayurveda, one of the oldest medical systems, has used plants like *Withania somnifera* (Ashwagandha), *Curcuma longa* (Turmeric), and *Tinospora cordifolia* to manage growths and improve overall health. Ayurvedic texts described concepts like "Arbuda" (tumors) and "Granthi" (nodules), with treatments involving plant-based formulations. Similarly, in Greek and Roman medicine, Hippocrates and Dioscorides documented the use of plants like *Chelidonium majus* (greater celandine) and *Vinca* species (source of vincristine and vinblastine) for tumor-related conditions (Desai et al. 2008). These early records highlight the long-standing reliance on medicinal plants for addressing various health issues, including those related to cancer.

2.2 Indigenous Knowledge

Indigenous peoples around the world have long utilized medicinal plants for various health conditions, including those that resemble cancer. In Native American medicine, plants such as *Taxus brevifolia* (Pacific yew) and *Podophyllum peltatum* (mayapple) were used to treat a variety of ailments, including cancer-like conditions. Similarly, in African Traditional Medicine, plants like *Catharanthus roseus* (Madagascar periwinkle) and *Annona muricata* (soursop) have been historically used for their anticancer and healing properties. This traditional knowledge represents a valuable resource for identifying potential new treatments and underscores the importance of preserving these practices. The continued exploration of these traditional uses can contribute significantly to the development of modern therapies (Yeshi et al. 2022).

2.3 Transition to Modern Medicine

The traditional use of medicinal plants has been pivotal in the discovery of many modern anticancer drugs. The knowledge and practices of ancient and indigenous cultures have laid the groundwork for the development of several key pharmaceuticals used in cancer treatment today. For instance, Taxol (paclitaxel), a widely used chemotherapy drug, was derived from the Pacific yew tree (*Taxus brevifolia*) (Chunarkar-Patil et al. 2024). Similarly, vincristine and vinblastine, alkaloids used to treat leukemia and lymphomas, were isolated from the Madagascar periwinkle (*Catharanthus roseus*). Additionally, camptothecin, a compound crucial in the development of topoisomerase inhibitors, was derived from *Camptotheca acuminata* (Mazumder et al. 2022). These examples highlight the direct influence of traditional medicine on contemporary cancer therapy, demonstrating how the study of medicinal plants continues to bridge ancient practices with modern scientific advancements. The identification of these key compounds from plants has not only led to effective treatments but also underscores the importance of continued exploration of natural resources in the search for novel therapeutic agents.

2.4 Significance of Traditional Knowledge in Contemporary Cancer Therapy

The significance of traditional knowledge in contemporary cancer therapy is multifaceted, stemming from its role as a natural reservoir of bioactive compounds and its holistic approach to health and healing. Traditional medicine systems have long identified and utilized plants with anticancer properties, providing a rich source of potential therapeutic agents for modern medicine. This knowledge is invaluable because it guides researchers to specific plants and compounds that have demonstrated efficacy over centuries of use, often through empirical observation. This has led to the discovery of many modern anticancer drugs, including Taxol (paclitaxel), vincristine, vinblastine, and camptothecin, all derived from plants that were part of traditional medical practices (Mazumder et al. 2022). The continued study of these plants and their compounds not only provides direct therapeutic agents but also serves as a template for the development of synthetic drugs.

Furthermore, traditional medical systems often emphasize a holistic approach to health, which focuses on balancing the body and addressing the root causes of disease rather than solely treating symptoms. This aligns with modern integrative cancer therapies, which seek to complement conventional treatments with supportive approaches that enhance overall well-being. In addition, traditional practices recognize the complex interplay of various factors in the development and progression of cancer, and this perspective encourages a more comprehensive approach to treatment. This holistic view can be particularly beneficial in managing the side effects of conventional treatments and improving patients' quality of life (Gureje et al. 2015).

Moreover, the exploration of ethnobotanical knowledge is essential for the continued discovery of new anticancer agents. Ethnobotanical studies involve the documentation and analysis of traditional uses of plants by different cultures, which helps identify plants with potential anticancer properties. By preserving traditional knowledge, researchers gain access to a wealth of information that can significantly enhance the development of modern medicine. These studies provide leads for laboratory research and clinical trials, ultimately leading to more effective and less toxic cancer therapies. This also honors and respects traditional practices, recognizing the valuable contributions of indigenous and ancient cultures to scientific advancements (Domingo-Fernández et al. 2023). Therefore, traditional knowledge serves not only as a historical foundation but also as a vital guide for the ongoing development of cancer therapies in contemporary medicine.

3 Mechanisms of Action

3.1 Synergistic Effects with Conventional Therapies

After the cardiac diseases, cancer has the second highest rate. In 2022, there were about 20 million new cases of cancer, including nonmelanoma skin cancers (NMSCs), and 9.7 million cancer-related deaths. According to estimates, one in five men and women will get cancer at some point in their lives, while one in nine men and one in 12 women will pass away from the disease. With about 2.5 million new cases, or one in eight cancers worldwide (12.4% of all cancers), lung cancer was the most commonly diagnosed cancer in 2022. It was followed by stomach (4.9%), colorectum (9.6%), prostate (7.3%), and female breast (11.6%) cancers. With an anticipated 1.8 million deaths (18.7%), lung cancer was also the most common cause of cancer-related mortality. Colorectal (9.3%), liver (7.8%), female breast (6.9%), and stomach (6.8%) cancers were next in line (Bray et al. 2024).

Research in cancer mainly relies on medicinal herbs (Mehta et al. 2018). The number of plants derived secondary metabolites has been reported for cancer treatment, but their limited solubility and unfavourable stability limit their applications in therapeutic research. For instance, phenolic phytoconstituents exhibit strong antioxidant properties but are unstable under experimental conditions. Apart from this, bioactive substances like curcumin and paclitaxel have poor solubility and bioavailability despite their high potential activity (Chavda et al. 2023). The application of multi-drug therapy is a useful tactic that focuses on both the activation of the body's defences or repair systems and the direct stopping or killing of harmful agents responsible for causing cancer or other infections. The fundamental goal of

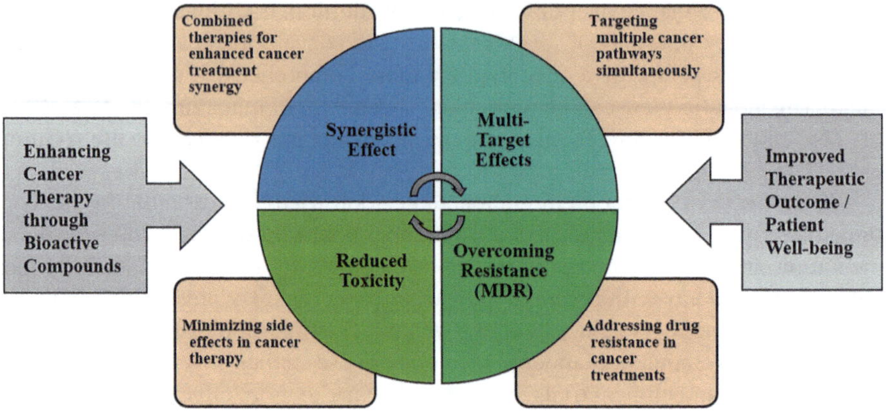

Fig. 12.1 Strategy for enhancing cancer therapy using bioactive compounds with conventional treatments

combining different therapeutic agents follows established principles, often aiming to achieve synergistic effects, overcome drug resistance, or reduce toxicity associated with single high-dose treatments (Desai et al. 2015, 2023). Synergism or supra-additivity is the combination of two or more products to enhance the efficiency. They may be utilized more frequently in combination with traditional cancer treatments as scientific knowledge advances in order to improve patient results and provide new hope to patients. More qualitatively, the combination of two drugs/phytochemicals results in improved effect that is better compared to anticipated. Such kind of synergism typically lowers the dosage of the majority of medications, which lowers the risk of adverse effects (Sharma et al. 2019). The overall strategy for utilizing medicinal plants in combination with conventional treatments, outlining the potential synergistic mechanisms and desired outcomes, is depicted in Fig. 12.1. Herbal products may interact with conventional drugs or biological molecules to produce synergistic effects (Pezzani et al. 2019). The combination treatments can produce and their potential positive outcomes, such as boosting therapeutic effect efficacy, lowering dosage while preserving or increasing efficacy to prevent toxicity, and reducing the emergence of drug resistance, which are some advantages of using them (Castañeda et al. 2022). Illustrating this potential, research has focused on enhancing the cytotoxic effects of conventional chemotherapy agents, such as Cytarabine, through various combination strategies (Desai et al. 2015).

The combination of *citrus reticulate* peel black tea (CRPBT) and black tea significantly inhibited phosphorylation of PI3K and AKT, and up-regulated the ratio of Bax/Bcl-2, and suppressed the expression of MMP2/9, N-cadherin, and Vimetin proteins in liver cancer cells (Wen et al. 2020). Breast cancer has first overtaken lung cancer as the top cause of cancer-related death among women globally, with an expected 2,261,419 new cases and 684,996 deaths in 2020. The combination of aspirin with resveratrol (polyphenol- 3,4,5-Trihydroxy-trans-stilbene) reduces cancer cell viability and tumor growth via inhibiting nuclear factor kappa B (NFkB) activation, cytokine production, the growth rate of cancer cells, and also enable to alleviate intestinal inflammation (Salla et al. 2020).

3.2 Combination of Plant Extract with Nanomaterials

The utilization of nano formulations that contain natural agents has become a promising approach to address the various drawbacks of both the existing cancer treatment techniques and the low absorption of natural substances. Nanomaterials are classified as nanocarriers as they target and transport anticancer drugs/chemicals to cancer cells. Apart from this, nanomaterials themselves exhibit anticancer properties (Saavedra-Leos et al. 2021). Lipid vesicles, which were initially reported in the 1960s, were the first nanoscale drug delivery systems. Since then, research has advanced the creation of different nanomaterials with multifunctional characteristics (Huang et al. 2024). Berberine, an isoquinoline alkaloid, represents advanced therapeutical and pharmacological properties against cancer, but due to lower water solubility and poor oral bioavailability, has been combined with liquid crystalline nanoparticles for the treatment of human lung epithelial lung cancer (Paudel et al. 2022). In another study, liquid crystalline nanoparticles loaded with berberine and phytotriol (BP-LCNs) were synthesized to evaluate their anti-cancer synergistic properties in vitro using the human lung adenocarcinoma A549 cell line. The in vitro study results reveal the utility of the BP-LCN formulation in downregulating proteins involved in cancer cell migration and proliferation as well as in controlling the expression of P53, PTEN, and KRT18 genes linked to cancer (Alnuqaydan et al. 2022). Without causing any obvious off-target side effects, delivery of DOX by nanodiscs caused immunogenic cell death in cancer cells and demonstrated antitumor effectiveness. In addition to expanding their epitope recognition to tumor-associated antigens, neoantigens, and intact entire tumor cells, "priming" tumors with DOX-carrying nanodiscs produced strong anticancer CD8+ T cell responses. It had been revealed that 80–88% of mice had their established CT26 and MC38 colon carcinoma tumors completely regressed after receiving combination chemoimmunotherapy with nanodiscs and anti-programmed death therapy. Survivors were also shielded from tumor recurrence (Kuai et al. 2018). Table 12.2 summarizes the application of nanomaterials as carriers for co-delivery drug combinations, highlighting their enhanced cytotoxicity, tumor inhibition, and synergistic effects in various cancer types. Figure 12.2 illustrates the evolution and advancements in combining plant derivatives with nanomaterials for cancer treatment, highlighting the integration of innovative drug delivery systems to enhance therapeutic outcomes.

4 Prominent Medicinal Plants in Combination Therapies

Combination therapies utilizing prominent medicinal plants have emerged as a promising approach in modern medicine. Bioactive compounds such as curcumin, resveratrol, and quercetin, derived from natural sources, are at the forefront of this

Table 12.2 Nano-formulations with combination of different drugs

Nanomaterial as carrier	Co-delivery drug combination	Cancer type	Important findings	References
Pegylated nanoliposomes	Resveratrol and 5-fluorouracil	Head and neck squamous cell carcinoma	Co-encapsulation of drugs represented different effects on different genes and enhanced the cytotoxicity than free drugs	Annaji et al. (2021)
Shell crosslinked zein nanocapsules	Exemestane and resveratrol	Breast cancer	Nanocapsules enhanced the cytotoxicity against MCF-7 and 4T1 breast cancer cells and reduced the tumor volume by 2.4-fold compared to the free drug combination	Elzoghby et al. (2017)
Folic acid conjugated nanoparticles	Resveratrol and docetaxel	Prostate cancer	Nanoparticles downregulated the expression of NF-kB p65, cox-2, and antiapoptotic genes and exhibited additional cytotoxic effects with the downregulation of survivin and upregulation of	Singh et al. (2018)
Alginate nanoparticles	Curcumin and resveratrol	Prostate cancer	Curcumin was found to have good cellular uptake from both the solutions, as well as nanoparticles. Whereas resveratrol showed poor cellular uptake	Saralkar and Dash (2017)
Novel peptide-cationic liposomal nanocarrier	Resveratrol and P53 gene	Cervical cancer and breast cancer	Co-delivery system revealed enhanced tumor inhibition and apoptosis-inducing activity compared to resveratrol liposomes or p53 gene liposomes	Xu et al. (2019)

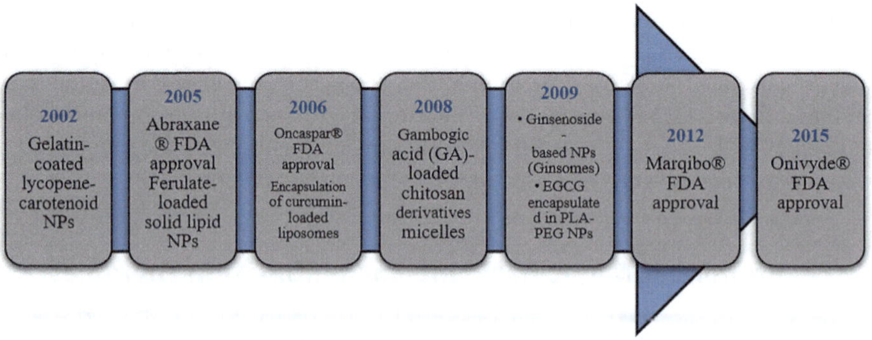

Fig. 12.2 The evolution of combined therapies of plant derivatives and nanomaterials for cancer treatment. (Kim et al. 2021; Kawish et al. 2024)

innovation. Curcumin, extracted from *Curcuma longa* (turmeric), is renowned for its potent anti-inflammatory and antioxidant properties. Resveratrol, found in grapes, berries, and peanuts, exhibits cardioprotective, anticancer, and neuroprotective effects. Quercetin, a flavonoid abundant in onions, apples, and green tea, is celebrated for its immune-modulating and antihistamine activities (Prasathkumar et al. 2021). These compounds can also be synthesized through chemical or biotechnological methods to meet industrial demand and ensure purity. For instance, resveratrol can be synthesized via chemical processes or biotransformation using engineered microorganisms. Such advancements enable scalable production while maintaining bioactivity. Collectively, these compounds, whether naturally sourced or synthetically produced, offer versatile applications in combating oxidative stress, chronic inflammation, and other pathophysiological conditions, making them invaluable in combination therapies for complex diseases (Thapa et al. 2019).

4.1 Curcumin

Curcumin, the active polyphenolic compound extracted from the rhizome of *Curcuma longa* (turmeric), has garnered significant attention for its broad-spectrum anticancer properties. Traditionally used in Ayurveda and Traditional Chinese Medicine, its therapeutic potential is now supported by extensive scientific evidence (Sharifi-Rad et al. 2020). Structurally, curcumin is a diarylheptanoid with a conjugated heptadiene chain and keto-enol tautomerism, which contributes to its chemical stability and biological activity. Its non-toxic nature, even at relatively high doses, makes curcumin a suitable candidate for standalone treatments and as an adjunct in combination therapies. Curcumin exhibits diverse anticancer mechanisms, including antiproliferative, anti-angiogenic, anti-metastatic, and pro-apoptotic activities (Rege et al. 2019). These properties have been demonstrated in various malignancies, including breast, colorectal, prostate, and lung cancers. Furthermore, curcumin modulates several molecular pathways, such as NF-κB, MAPK, PI3K/Akt, and STAT3, which are crucial for cancer progression. In combination therapies, curcumin enhances the efficacy of conventional chemotherapeutic agents while mitigating their toxicity, underscoring its role as a multifaceted therapeutic agent (Islam et al. 2024). Illustrating a specific molecular mechanism of synergy, the combination of Curcumin with Cytarabine has been shown to effectively downregulate multi-drug resistance genes in acute myeloid leukemia cell lines (Shah et al. 2015).

One of the key advantages of curcumin is its ability to reduce chemotherapy-induced adverse effects. By protecting normal tissues from oxidative damage and reducing systemic inflammation, curcumin alleviates symptoms like nausea, fatigue, and myelosuppression. These properties significantly improve the quality of life for patients undergoing cancer treatment (Liu et al. 2018). For instance, when used alongside cisplatin, curcumin enhances cytotoxicity by promoting apoptosis and suppressing tumor growth. It also modulates oxidative stress and restores the tumor-suppressor function of p53, thereby increasing cisplatin sensitivity. Similarly, curcumin's co-administration with paclitaxel in breast cancer has been shown to enhance mitochondrial-mediated apoptosis and alleviate systemic toxicities such as peripheral neuropathy (Zou et al. 2018).

In colorectal cancer, curcumin synergizes with 5-fluorouracil (5-FU) to inhibit cancer cell proliferation by targeting NF-κB and Wnt/β-catenin pathways, amplifying DNA damage responses (Gao et al. 2021). Beyond enhancing drug efficacy, curcumin addresses multidrug resistance (MDR), a major challenge in chemotherapy. By inhibiting the expression of P-glycoprotein and other ABC transporters, it increases the intracellular retention of chemotherapeutic agents. Additionally, curcumin suppresses MDR-associated signalling pathways, such as PI3K/Akt and STAT3, to enhance drug-induced apoptosis (Xue and Liang 2012; Ashrafizadeh et al. 2020).

Extensive preclinical studies validate curcumin's ability to potentiate the effects of chemotherapeutic agents. In vivo and in vitro studies consistently demonstrate its role in inhibiting tumor growth, inducing apoptosis, and reducing metastasis (Islam et al. 2024). Clinical trials further corroborate these findings. For instance, a Phase II trial in pancreatic cancer revealed that oral curcumin enhanced the efficacy of gemcitabine, leading to reduced tumor burden and improved survival. In advanced breast cancer, curcumin combined with docetaxel showed superior therapeutic efficacy (Dhillon et al. 2008). Curcumin's versatility in cancer therapy lies in its simultaneous modulation of multiple signalling pathways. By inhibiting NF-κB activation, curcumin disrupts cancer cell proliferation and survival (Tong et al. 2016). It also suppresses the MAPK/ERK pathway (Shi et al. 2024), which regulates cell growth and apoptosis, and downregulates the PI3K/Akt pathway, a critical contributor to drug resistance. Additionally, curcumin targets the STAT3 signalling pathway, thereby reducing tumor growth and metastasis. These multifaceted actions make curcumin a potent enhancer of existing cancer therapies and a cornerstone of next-generation oncological strategies (Farghadani and Naidu 2021). In summary, curcumin represents a cost-effective, biologically potent, and non-toxic adjuvant to traditional cancer therapies. Its ability to enhance drug efficacy, reduce side effects, and overcome resistance underscores its potential as a transformative agent in combination therapies. Continued research and clinical evaluation are essential to fully realize curcumin's therapeutic promise in improving cancer treatment outcomes.

4.2 Resveratrol

Resveratrol, a polyphenolic compound found in grapes, berries, and peanuts, has garnered significant attention for its chemopreventive properties. Its role in cancer management has been particularly highlighted in combination therapies, where it acts synergistically with other agents to target multiple molecular pathways, improve drug bioavailability, and reduce toxicity. This combinatorial approach not only enhances therapeutic efficacy but also addresses multidrug resistance (MDR), a critical challenge in cancer treatment that often results in chemotherapy failure and poor patient outcomes. By reversing MDR and sensitizing tumor cells to cytotoxic agents, resveratrol presents a promising avenue for advancing oncological treatments (Ko et al. 2017).

Recent advancements have demonstrated the effectiveness of resveratrol in combination with chemotherapeutic drugs like paclitaxel (PTX). Nanocarrier systems, such as PEGylated liposomes, have been employed to co-deliver resveratrol and PTX, yielding remarkable results in preclinical models of drug-resistant tumors like MCF-7/Adr. These systems enhance drug solubility, ensure sustained release, and leverage the enhanced permeability and retention (EPR) effect for preferential tumor accumulation (Najafiyan et al. 2024). PEGylation further extends the circulation time of the drugs, leading to improved therapeutic outcomes. In vivo studies corroborate the potential of such combinations, showing significant tumor growth inhibition with minimal systemic toxicity (Alrashidi et al. 2024). Resveratrol's integration into nanocarrier-based systems addresses key limitations of traditional therapies, such as poor bioavailability and rapid clearance. The optimal size (~50 nm) of these liposomes facilitates effective tumor penetration while evading macrophage clearance. Additionally, the co-delivery of resveratrol and PTX ensures a precise drug ratio, critical for achieving synergistic effects. This innovative approach aligns with the broader objectives of targeted cancer therapies and reinforces resveratrol's role in combating drug-resistant cancers (Annaji et al. 2021).

Resveratrol exhibits a multifaceted mechanism of action that contributes to its anticancer properties. It induces apoptosis by activating caspases and downregulating anti-apoptotic proteins like Bcl-2, thereby facilitating programmed cell death in tumor cells (Jang et al. 2022). Furthermore, it arrests the cell cycle at G1 and G2/M phases by modulating cyclins and CDK inhibitors such as p21 and p27. Resveratrol's antioxidant properties also play a pivotal role by reducing reactive oxygen species (ROS) levels and mitigating oxidative DNA damage, a hallmark of cancer progression (Ahmad et al. 2001). Beyond these effects, resveratrol modulates key signal transduction pathways, including NF-κB, PI3K/Akt, and MAPK, which are implicated in cell survival, angiogenesis, and metastasis. It also impacts the tumor microenvironment by suppressing pro-inflammatory cytokines, angiogenic factors like VEGF, and tumor-promoting signals. This comprehensive targeting of cancer-related pathways underscores its potential as a therapeutic agent in combination regimens (Ko et al. 2017). Preclinical studies provide robust evidence for resveratrol's antitumor activity. In murine models, resveratrol co-delivered with paclitaxel in PEGylated liposomes demonstrated potent cytotoxicity against drug-resistant tumors (Meng et al. 2016). These studies revealed enhanced bioavailability and improved tumor retention of the drugs, translating into superior treatment outcomes. In other preclinical experiments, resveratrol reduced lung metastasis and delayed tumor progression in HER-2/neu transgenic mice, further validating its potential in cancer management (Provinciali et al. 2005). Clinical investigations have shown that resveratrol is well-tolerated and enhances the efficacy of standard chemotherapy agents like cisplatin and doxorubicin (Cocetta et al. 2021). Its natural origin and favourable safety profile make it an attractive alternative to traditional chemosensitizers, which are often associated with severe side effects. Early-phase trials have demonstrated promising results, highlighting its translational potential in overcoming MDR and improving therapeutic outcomes in resistant cancers (Rezk et al. 2006). In addition to enhancing therapeutic efficacy, resveratrol mitigates chemotherapy-associated toxicities, contributing to its appeal as a supportive agent. Its antioxidant properties protect normal tissues from oxidative stress induced by chemotherapeutic drugs, thereby reducing collateral damage (Xiao et al. 2019). Resveratrol also exhibits gastroprotective effects, preventing mucositis and preserving gut barrier integrity often compromised during cancer treatment. Furthermore, it has cardioprotective and hepatoprotective properties, shielding the heart and liver from drug-induced toxicity through the modulation of detoxifying enzymes and inhibition of pro-inflammatory pathways (Chow et al. 2010).

Despite its promising therapeutic potential, resveratrol's clinical application is hindered by poor solubility and bioavailability. Advanced nanotechnological approaches have been developed to address these challenges. Nano formulations, including liposomes, solid lipid nanoparticles, and metallic nanoparticles, stabilize resveratrol's active form, enhance solubility, and facilitate targeted delivery to tumor sites. For instance, mitochondria-targeting liposomes loaded with resveratrol have shown the ability to induce apoptosis in drug-resistant cancer cells, while transferrin-functionalized liposomes selectively target glioblastoma cells (Jhaveri et al. 2018). Emerging strategies like the use of folic acid-conjugated nanoparticles combine targeting capabilities with improved bioavailability, further advancing the therapeutic efficacy of resveratrol (Sharifi-Rad et al. 2021). However, challenges such as scalability and toxicity assessment of these nanoformulations need to be addressed for their successful clinical translation. Resveratrol's integration into combination therapies represents a significant advancement in cancer treatment, particularly for drug-resistant tumors. Its dual role as a therapeutic agent and chemosensitizer, coupled with innovative delivery systems, underscores its potential to revolutionize oncology. Future research should focus on elucidating the exact mechanisms of its synergistic effects, optimizing drug ratios, and validating its efficacy in large-scale clinical trials. With the advent of personalized medicine and immunotherapy, resveratrol could play a pivotal role in shaping the future of cancer management, offering hope for more effective and less toxic treatment strategies.

4.3 Quercetin

Quercetin, a plant-derived flavonoid widely found in fruits, vegetables, and tea, has gained significant attention in oncology due to its extensive pharmacological profile. Known for its antioxidant, anti-inflammatory, and anti-cancer properties, quercetin targets various molecular pathways that drive cancer progression and metastasis (Lotfi et al. 2023). Recent advances highlight its potential to enhance therapeutic outcomes when used in combination with chemotherapeutic agents, radiotherapy, or other natural compounds. This synergistic approach not only improves efficacy but also reduces the adverse effects associated with conventional therapies. Quercetin exerts its anti-cancer effects by influencing several key cellular processes. It induces apoptosis by upregulating pro-apoptotic proteins like Bax and caspases while suppressing anti-apoptotic proteins such as Bcl-2 (Tang et al. 2020). Additionally, quercetin causes cell cycle arrest at the G2/M phase, inhibiting tumor cell proliferation. A crucial mechanism involves its ability to modulate the PI3K/Akt/mTOR pathway, which plays a central role in cell survival, growth, and drug resistance (Asgharian et al. 2022). In ovarian cancer, quercetin downregulates mTOR signalling, promoting apoptosis and reducing tumor growth. For example, a study demonstrated that quercetin-treated ovarian cancer cells exhibited increased cleavage of caspase-3 and reduced expression of cyclin D1, leading to enhanced apoptosis and reduced proliferation rates (Ji et al. 2024). Quercetin's role in angiogenesis and metastasis is equally important. By inhibiting vascular endothelial growth factor (VEGF) and matrix metalloproteinases (MMPs), quercetin disrupts neovascularization and prevents the invasion of cancer cells into surrounding tissues (Pratheeshkumar et al. 2012). This mechanism has been validated in in vivo models of breast and prostate cancers, where quercetin significantly reduced metastatic burden and angiogenesis.

Combination therapies involving quercetin and conventional chemotherapeutics have shown remarkable promise in overcoming drug resistance (Xu et al. 2021). Quercetin enhances the efficacy of agents like cisplatin, paclitaxel, and doxorubicin by sensitizing cancer cells to their cytotoxic effects (Sharma et al. 2023). This is achieved through the inhibition of drug efflux pumps such as P-glycoprotein (P-gp) and modulation of survival pathways like NF-κB and STAT3. A study by Li et al. (2019) investigated the combination of quercetin with cisplatin in ovarian cancer and revealed a significant increase in cytotoxicity compared to cisplatin alone. Quercetin suppressed the Akt/mTOR signalling pathway, overcoming cisplatin resistance while reducing its nephrotoxic effects through antioxidant activity. Moreover, in triple-negative breast cancer (TNBC), quercetin has been shown to enhance the efficacy of doxorubicin by downregulating NF-κB and inhibiting epithelial-mesenchymal transition (EMT), which is critical for metastasis and drug resistance (Wang et al. 2024). Preclinical studies demonstrated that the quercetin-doxorubicin combination reduced tumor volume by 70% compared to monotherapy, highlighting its potential in aggressive and drug-resistant cancers (Lv et al. 2016).

Radiotherapy remains a cornerstone in cancer treatment, but its clinical utility is often limited by toxicity to normal tissues and radioresistance in tumors. Quercetin addresses these challenges by sensitizing cancer cells to radiation while protecting healthy cells from radiation-induced damage. Studies have shown that quercetin enhances radiosensitivity in lung and cervical cancer cells by inhibiting DNA repair mechanisms and increasing ROS levels, leading to greater DNA damage and apoptosis in tumor cells (Chen and Kuo 2017). Simultaneously, its antioxidant properties protect normal tissues, reducing the risk of radiation-induced fibrosis, mucositis, and cardiotoxicity. For instance, a study on cervical cancer demonstrated that quercetin combined with radiotherapy significantly enhanced tumor cell apoptosis and reduced radiation-associated oxidative stress markers in normal tissues. This dual role makes quercetin an excellent adjunct to radiotherapy.

One of the primary limitations of quercetin is its low bioavailability due to poor water solubility and rapid metabolism. To overcome, nano formulations have emerged as a solution, enhancing its pharmacokinetic profile and targeted delivery to tumors. Polymeric nanoparticles, liposomes, and solid lipid nanoparticles have been developed to encapsulate quercetin, improving its stability and cellular uptake (Tomou et al. 2023). A notable example is the use of PEGylated liposomes co-encapsulating quercetin and doxorubicin, which demonstrated enhanced cytotoxicity against multi-drug-resistant breast cancer cells while reducing systemic toxicity in vivo. In ovarian cancer, quercetin-loaded chitosan nanoparticles have shown significant tumor growth inhibition and increased apoptosis compared to free quercetin. These nano formulations also enable sustained release, maintaining therapeutic levels of quercetin over an extended period (Lv et al. 2016).

Quercetin's combination with other natural compounds like curcumin and resveratrol offers a synergistic approach to cancer therapy. Studies have shown that quercetin and curcumin co-administration suppresses the PI3K/Akt/mTOR and NF-κB pathways more effectively than either compound alone, resulting in enhanced apoptosis and reduced tumor growth in colorectal cancer models (Park et al. 2013). Similarly, the combination of quercetin and resveratrol has been shown to inhibit EMT and reduce metastatic potential in breast cancer. These combinations leverage the complementary mechanisms of action of each compound, targeting multiple hallmarks of cancer simultaneously (Guo et al. 2021). Preclinical studies provide compelling evidence for quercetin's efficacy in combination therapies. In a study involving hepatocellular carcinoma, quercetin combined with sorafenib, a first-line treatment, enhanced apoptosis and reduced angiogenesis by downregulating VEGF and HIF-1α. This combination also mitigated sorafenib-induced hepatotoxicity, improving overall treatment outcomes. Clinical trials are gradually validating these findings (Abdu et al. 2022). In a Phase II trial, quercetin supplementation in patients undergoing chemotherapy for non-small cell lung cancer (NSCLC) improved quality of life scores and reduced chemotherapy-induced fatigue and nausea. While larger studies are needed, these early results highlight quercetin's potential as an adjuvant in clinical oncology (Jeong et al. 2009).

Despite its promising potential, several challenges must be addressed for the clinical translation of quercetin. These include standardizing its dosage, optimizing its delivery systems, and ensuring scalability of its nano formulations. Furthermore, rigorous clinical trials are needed to establish its efficacy and safety across different cancer types and patient populations. The integration of quercetin into immunotherapy and precision oncology represents a frontier in cancer research. By modulating immune checkpoints and enhancing T-cell-mediated responses, quercetin could complement immune checkpoint inhibitors and adoptive cell therapies. Quercetin's multifaceted role in cancer therapy, encompassing chemo sensitization, radioprotection, and direct anti-tumor effects, positions it as a cornerstone of future combination therapies. With ongoing research and innovation, quercetin holds immense potential to transform the landscape of cancer treatment.

4.4 Other Key Medicinal Plants

Apart from curcumin, quercetin, and resveratrol, numerous other medicinal plants have shown promise in enhancing cancer therapies when used in combination with conventional treatments. Berberine, a bioactive compound derived from *Berberis* species, has garnered attention for its potent anticancer properties. It enhances the efficacy of chemotherapeutic agents such as cisplatin and doxorubicin by modulating apoptotic pathways and reducing the expression of drug resistance proteins like P-glycoprotein. Studies have shown that berberine induces mitochondrial dysfunction and oxidative stress, leading to cancer cell apoptosis. Additionally, its ability to inhibit the mTOR and NF-κB pathways makes it a valuable adjunct to standard therapies in cancers such as colorectal, breast, and hepatocellular carcinoma. Importantly, berberine's role in protecting normal cells from chemotherapy-induced oxidative damage adds a layer of safety to its therapeutic profile (Xiong et al. 2022). Epigallocatechin gallate (EGCG), a polyphenol found in green tea (*Camellia sinensis*), has demonstrated synergistic effects with chemotherapy and radiotherapy in cancers such as breast, ovarian, and prostate. By targeting multiple molecular pathways, including PI3K/Akt and VEGF, EGCG enhances the cytotoxic effects of paclitaxel, cisplatin, and doxorubicin while suppressing angiogenesis and metastasis (Negri et al. 2018). Its antioxidant properties also protect normal cells from radiation-induced damage. Clinical studies have highlighted EGCG's ability to potentiate treatment efficacy while reducing systemic toxicity, making it a strong candidate for integrative oncology. Other compounds such as Withaferin A, derived from *Withania somnifera* (Ashwagandha), exhibit significant potential in combination therapies. Withaferin A disrupts key cancer cell survival mechanisms, including proteasome inhibition and the modulation of heat shock proteins, which are crucial for tumor growth and metastasis. Synergistic effects have been observed when combined with drugs like gemcitabine in pancreatic cancer and paclitaxel in breast cancer. These combinations have shown improved tumor shrinkage and reduced systemic toxicity in preclinical models (Sari et al. 2020).

Additional notable compounds include Sulforaphane, a sulfur-containing compound found in cruciferous vegetables like broccoli, has also shown remarkable potential in cancer combination therapies. By inducing phase II detoxifying enzymes and inhibiting histone deacetylases, sulforaphane sensitizes cancer cells to chemotherapeutics such as cisplatin and doxorubicin. In addition to its cytotoxic effects, sulforaphane enhances the immune response and protects normal cells from oxidative stress and inflammation caused by cancer treatments (Jabbarzadeh Kaboli et al. 2020). In Fig. 12.3 illustrates the integration of various plant-derived compounds with standard chemotherapeutic agents, highlighting their significant impact on patient treatment outcomes in the context of combination therapy. These natural compounds not only enhance the therapeutic efficacy of conventional treatments but also mitigate their adverse effects, improving the quality of life for patients. While preclinical and early-phase clinical studies are promising, further research and large-scale clinical trials are crucial to fully understand their mechanisms of action, optimal dosages, and long-term safety. Their ability to target diverse pathways and

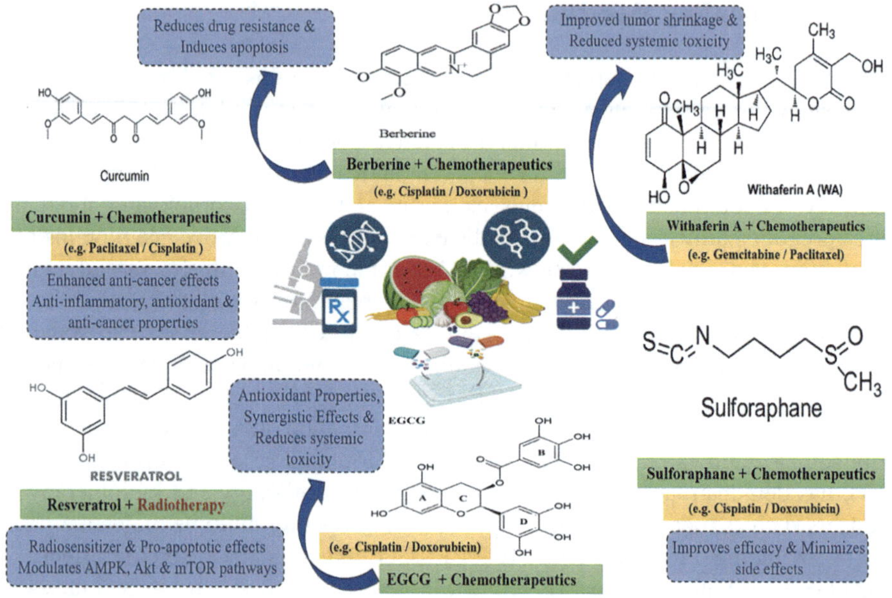

Fig. 12.3 Integration of plant-derived compounds with standard chemotherapeutic agents, and their impact on patient treatment outcomes

offer protective effects highlights their potential as integral components of modern cancer treatment regimens.

5 Clinical Evidence and Research

5.1 Preclinical Studies

Preclinical studies serve as a critical foundation for the development of cancer therapies, including the integration of medicinal plants in combination therapies. These studies, conducted using in vitro models (cell lines) and in vivo models (animal experiments), offer invaluable insights into the mechanisms, safety, and efficacy of bioactive compounds derived from medicinal plants. By assessing the pharmacological potential of these compounds, researchers can identify their synergistic effects when combined with conventional cancer therapies, paving the way for clinical applications (Guo et al. 2024).

5.1.1 In Vitro Studies for Exploring Cellular Mechanisms

In vitro studies have extensively demonstrated the potential of medicinal plant-derived compounds to combat cancer. For instance, curcumin, a polyphenolic compound extracted from *Curcuma longa*, has shown remarkable cytotoxic effects against various cancer cell lines, such as breast, lung, and colon cancer (Islam et al.

2024). These effects are mediated through the inhibition of critical pathways, including NF-κB, PI3K/AKT, and MAPK, which are often dysregulated in cancer (Zhang et al. 2023). Similarly, resveratrol, a natural stilbene from grapes (*Vitis vinifera*), has demonstrated the ability to induce apoptosis, inhibit angiogenesis, and suppress metastasis in melanoma and prostate cancer cell lines (Kursvietiene et al. 2023). Quercetin, a flavonoid found in many fruits and vegetables, has been shown to enhance the cytotoxic effects of standard chemotherapeutic agents like doxorubicin and cisplatin. It achieves this by modulating oxidative stress, downregulating anti-apoptotic proteins, and sensitizing cancer cells to chemotherapy-induced damage (Shahbaz et al. 2023). Other bioactive compounds, such as berberine, apigenin, and epigallocatechin gallate (EGCG), have also been extensively studied, highlighting their potential in targeting cancer stem cells, reversing drug resistance, and promoting chemosensitivity (Nguyen et al. 2024).

5.1.2 In Vivo Studies for Validation in Animal Models

In vivo studies further validate the findings from in vitro experiments by providing a more complex biological system to study the interactions of medicinal plant compounds with cancer cells and the host environment. Animal models have revealed the profound impact of these compounds in combination with chemotherapeutic agents on tumor growth, metastasis, and survival outcomes (Li et al. 2024). For example, studies involving curcumin combined with paclitaxel in murine models of breast cancer have demonstrated enhanced tumor regression compared to monotherapy. This synergy is attributed to curcumin's ability to downregulate multidrug resistance proteins and enhance the intracellular accumulation of paclitaxel (Farghadani and Naidu 2022). Similarly, the co-administration of resveratrol and 5-fluorouracil in colorectal cancer models has been shown to inhibit tumor growth more effectively than either agent alone, largely due to the modulation of apoptotic and autophagic pathways (Moutabian et al. 2022). Quercetin, when combined with cisplatin, has been shown to reduce nephrotoxicity in animal models while maintaining or even improving the anticancer efficacy of cisplatin (Sanchez-Gonzalez et al. 2011). This highlights the dual role of medicinal plant compounds in enhancing therapeutic outcomes and mitigating adverse effects associated with conventional cancer treatments.

5.1.3 Synergistic Effects and Drug Delivery Innovations

Recent advancements in drug delivery systems have further amplified the preclinical success of medicinal plant compounds. Nano-formulations, such as curcumin-loaded nanoparticles and liposomal resveratrol, have shown improved bioavailability, stability, and tumor-specific accumulation. For instance, curcumin nanoparticles combined with gemcitabine in pancreatic cancer models have demonstrated significantly higher tumor suppression rates compared to free drug combinations (Hafez Ghoran et al. 2022). In addition, natural product-based hybrids, such as antibody–drug conjugates (ADCs) and proteolysis-targeting chimeras (PROTACs), have shown promise in preclinical settings. These innovations leverage the therapeutic potential of medicinal plant compounds while overcoming pharmacokinetic limitations and enhancing specificity to cancer cells (Liu et al. 2022).

5.1.4 Challenges in Preclinical Studies

Despite their promise, preclinical studies also face limitations, including the translation of results from animal models to humans. Physiological differences between species and the complexity of human cancer biology often hinder direct extrapolation of findings. Moreover, variability in the extraction, standardization, and purity of medicinal plant compounds can affect the reproducibility and reliability of preclinical results (Guo et al. 2024).

5.2 Clinical Trials and Their Outcomes

Clinical trials play a pivotal role in bridging the gap between promising preclinical findings and the development of effective cancer therapies for patients. By evaluating the safety, efficacy, and therapeutic potential of medicinal plant-derived compounds in human populations, these trials provide critical evidence to support their integration into standard cancer treatment regimens. In the context of combination therapies, clinical trials highlight the synergistic effects of medicinal plant compounds when paired with conventional chemotherapeutics or radiotherapy, offering insights into their potential to improve treatment outcomes (Wang et al. 2023b).

5.2.1 Evaluating Synergy in Clinical Settings

One of the most well-documented examples of medicinal plants in clinical cancer therapy is the use of curcumin. Several clinical trials have investigated its efficacy in combination with standard chemotherapeutic agents. For instance, a Phase II trial involving patients with advanced colorectal cancer treated with curcumin and FOLFOX (5-fluorouracil, leucovorin, and oxaliplatin) demonstrated improved progression-free survival compared to chemotherapy alone (Lubner et al. 2010; Howells et al. 2019). Curcumin's anti-inflammatory properties and its ability to downregulate NF-κB signaling are thought to enhance the cytotoxic effects of FOLFOX while mitigating side effects like gastrointestinal toxicity (Howells et al. 2019). Similarly, resveratrol has been evaluated in clinical settings for its potential to augment the effects of chemotherapy and radiotherapy. In a pilot study involving breast cancer patients undergoing radiotherapy, oral supplementation with resveratrol was associated with reduced oxidative stress markers and enhanced DNA repair capacity (Alavi et al. 2021). These findings suggest that resveratrol not only improves therapeutic efficacy but also minimizes collateral damage to healthy tissues.

5.2.2 Enhancing Chemosensitivity and Reducing Toxicity

Another area where clinical trials have shown promise is in reducing drug resistance and toxicity. Quercetin, for example, has been studied for its ability to improve the efficacy of chemotherapeutic drugs like cisplatin and doxorubicin (Sharma et al. 2023). In a clinical trial involving patients with non-small cell lung cancer, the addition of quercetin to cisplatin therapy resulted in higher overall response rates and reduced nephrotoxicity. This dual benefit of enhancing therapeutic effects while protecting against adverse side effects underscores the potential of medicinal plants in cancer combination therapies (Asgharian et al. 2022). Berberine, an alkaloid from *Berberis* species, has also been the subject of clinical research for its anticancer properties. In a randomized controlled trial involving patients with colorectal adenomas, berberine supplementation significantly reduced polyp recurrence rates, suggesting a role in chemoprevention. When combined with chemotherapeutic agents, berberine has shown the ability to overcome multidrug resistance in cancers like leukemia and breast cancer (Jiang et al. 2022).

5.2.3 Clinical Trials of Nano-formulations

Advancements in drug delivery systems have been translated into clinical trials, further enhancing the therapeutic potential of medicinal plant compounds. Liposomal curcumin formulations, for instance, have been evaluated in Phase I and II trials for their safety and efficacy in pancreatic and ovarian cancers. These formulations improve bioavailability and target specificity, leading to enhanced tumor regression with fewer systemic side effects (Chen et al. 2024). Nanoparticle-based delivery systems for resveratrol and quercetin are also under clinical investigation. These approaches aim to address the pharmacokinetic limitations of these compounds, such as poor solubility and rapid metabolism, thereby improving their therapeutic index in cancer patients (Ali et al. 2024).

5.2.4 Reviews of Clinical Outcomes

Systematic reviews and meta-analyses of clinical trials further substantiate the potential of medicinal plants in cancer combination therapies. A meta-analysis of clinical studies on curcumin as an adjuvant to chemotherapy highlighted significant improvements in patient-reported outcomes, including pain reduction, fatigue alleviation, and enhanced quality of life (De Waure et al. 2023). Similarly, reviews of resveratrol and quercetin clinical trials have reported consistent evidence of their ability to reduce inflammation, enhance immune responses, and improve overall treatment tolerability (Li et al. 2023).

5.2.5 Challenges in Clinical Trials

Despite encouraging results, clinical trials involving medicinal plants face several challenges. Variability in the preparation, dosage, and standardization of plant-based compounds can lead to inconsistent outcomes. Furthermore, the lack of large-scale, multicenter trials limits the generalizability of findings. Regulatory hurdles, such as obtaining approval for plant-derived compounds, also pose significant barriers to the widespread adoption of these therapies (Wang et al. 2023a).

5.3 Systematic Reviews and Meta-analyses

Systematic reviews and meta-analyses play a crucial role in synthesizing evidence from clinical trials and observational studies, offering a comprehensive understanding of the effectiveness and safety of medicinal plant-based combination therapies in cancer treatment. By critically analyzing a broad spectrum of studies, these reviews provide valuable insights into the potential therapeutic benefits, underlying mechanisms, and limitations of incorporating medicinal plant-derived compounds into oncology practices (Al-Khabori and Rasool 2022).

5.3.1 Curcumin, Evidence of Synergistic Effects

Numerous systematic reviews and meta-analyses have assessed the role of curcumin as an adjunct in cancer therapy. A meta-analysis of randomized controlled trials (RCTs) involving curcumin as a complementary therapy demonstrated improved treatment outcomes, including reduced tumor progression and enhanced quality of life (Ferguson et al. 2021). The study highlighted curcumin's ability to enhance the efficacy of chemotherapeutic agents, such as 5-fluorouracil and gemcitabine, by modulating pathways like NF-κB and STAT3 (Cacciola et al. 2023). Additionally, curcumin was shown to alleviate chemotherapy-induced side effects, including gastrointestinal toxicity and fatigue, emphasizing its dual role as a therapeutic enhancer and a symptom mitigator (Ruiz De Porras et al. 2023).

5.3.2 Broad-Spectrum Benefits of Resveratrol

Systematic reviews of resveratrol in cancer therapy have highlighted its broad-spectrum anticancer activities, particularly in combination with standard treatments. A meta-analysis of clinical trials involving resveratrol supplementation in cancer patients reported significant reductions in oxidative stress markers, improved DNA repair capacity, and enhanced treatment tolerability (Alam et al. 2024). The synergistic effects of resveratrol with radiotherapy and chemotherapy were attributed to its ability to sensitize cancer cells to treatment while protecting normal cells from damage. However, the analysis also noted variability in study designs and dosages, indicating the need for standardized protocols in future research (Alam et al. 2024).

5.3.3 Emerging Evidence of Quercetin and Other Flavonoids

Quercetin, another extensively studied compound, has been the subject of systematic reviews evaluating its role in cancer prevention and therapy. A comprehensive review of clinical and preclinical studies found that quercetin enhances the efficacy of chemotherapeutic drugs, such as cisplatin and doxorubicin, by modulating apoptotic pathways and reducing drug resistance (Xu et al. 2021). Meta-analyses have also reported reduced chemotherapy-induced nephrotoxicity and cardiotoxicity in patients receiving quercetin as an adjunct. The review emphasized the importance of further clinical trials to validate these findings and explore optimal dosages for therapeutic application (Sanchez-Gonzalez et al. 2011). In addition to quercetin, systematic reviews have identified other flavonoids, such as apigenin and luteolin, as potential candidates for cancer combination therapies. These compounds have shown promise in preclinical models and early-phase clinical trials, particularly in targeting angiogenesis, metastasis, and inflammation (Sahoo et al. 2023).

5.3.4 Combined Evidence Across Medicinal Plants

Several systematic reviews have aggregated evidence across multiple medicinal plants, providing a holistic perspective on their role in cancer therapy (Jenča et al. 2024). For instance, studies on medicinal plants, including curcumin, resveratrol, and berberine, revealed consistent evidence of their ability to enhance chemotherapy efficacy, reduce toxicity, and improve patient outcomes. The analysis highlighted common mechanisms of action, such as the inhibition of pro-inflammatory cytokines, the induction of apoptosis, and the modulation of oxidative stress (Meybodi et al. 2023).

5.3.5 Challenges in Systematic Reviews and Meta-analyses

Despite the valuable insights they offer, systematic reviews and meta-analyses face several challenges, including heterogeneity in study designs, variations in the quality of included trials, and differences in dosages and formulations of medicinal plant compounds. These factors can lead to inconsistencies in findings and limit the generalizability of conclusions. Additionally, the lack of large-scale, high-quality clinical trials poses a barrier to drawing robust conclusions about the efficacy and safety of these therapies (Lorenc et al. 2016).

6 Challenges and Considerations

The lack of established dosages is one of the main obstacles to using herbal medicine. In contrast to pharmaceutical drugs, which undergo extensive clinical trials to determine exact dosages, herbal products frequently lack this consistency. Due to variations in plant species, growing environments, harvesting practices, and processing methods, the concentration of active ingredients can vary greatly. This variability can result in either underdosing, which reduces efficacy, or overdosing, which increases the risk of adverse effects.

Standardization of herbal products involves ensuring that a consistent amount of the active compounds is present in each dose. However, this process can be complicated because many herbs contain multiple active constituents, and their synergistic effects contribute to their therapeutic benefits. Establishing clear guidelines and regulatory standards for dosage and standardization is essential to improve the reliability and safety of herbal medicines. In the context of cancer combination therapies, the dosage and standardization of medicinal plants present new challenges. Cancer treatment frequently combines traditional therapies, such as chemotherapy or radiation, with herbal remedies to improve efficacy or reduce side effects. However, the absence of standardized dosages for medicinal plants can complicate treatment protocols and result in inconsistent results. The bioavailability and pharmacokinetics of plant-derived compounds, such as curcumin from turmeric or epigallocatechin gallate (EGCG) from green tea, can differ depending on preparation methods and patient-specific factors (Gupta et al. 2013; Cheng et al. 2020).

Furthermore, using medicinal plants combined with conventional cancer treatments requires a thorough understanding of potential interactions and cumulative toxicities. Standardization efforts should include the development of guidelines for the co-administration of herbal and traditional therapies that are supported by rigorous clinical studies. Analytical advances, such as high-performance liquid chromatography (HPLC) and mass spectrometry, can help quantify active compounds and ensure consistency in herbal formulations (Wang et al. 2023b). Establishing dosage recommendations for medicinal plants in combination therapies demands collaboration between researchers, clinicians, and regulatory bodies. This approach has the potential to help bridge the gap between traditional medicine and modern oncology, opening the door to safer and more effective integrative cancer treatments.

7 Conclusions

The integration of medicinal plants into combination cancer therapies represents a significant evolution from traditional remedies towards scientifically validated integrative oncology. This manuscript has highlighted the substantial preclinical and growing clinical evidence supporting the synergistic potential of plant-derived bioactive compounds, such as curcumin, resveratrol, and quercetin, when used alongside conventional treatments like chemotherapy and radiotherapy. These natural agents demonstrate multifaceted mechanisms, including modulating key oncogenic pathways, inducing apoptosis, reducing inflammation, mitigating oxidative stress, and potentially overcoming multidrug resistance—a major obstacle in cancer care. Furthermore, their ability to protect normal tissues from treatment-induced toxicity addresses a critical need for improving patients' quality of life.

The advent of advanced drug delivery systems, particularly nanotechnology-based approaches like liposomes and nanoparticles, has begun to overcome the inherent limitations of poor bioavailability and stability associated with many phytochemicals, enhancing their therapeutic index and targeted delivery. While traditional knowledge provided the crucial foundation for identifying promising candidates, modern research is essential for validating efficacy, understanding complex interactions, and ensuring safety. Despite the compelling potential, significant challenges remain. Rigorous standardization of plant extracts, precise dosage determination, and a deeper understanding of pharmacokinetic and pharmacodynamic interactions are paramount. There is a critical need for more large-scale, well-designed randomized controlled clinical trials to definitively establish the efficacy and safety of these combination therapies across diverse cancer types and patient populations. Navigating regulatory pathways for complex botanical products also requires careful consideration. Ultimately, harnessing the therapeutic wealth of medicinal plants through evidence-based combination strategies offers a promising pathway towards developing more effective, personalized, and less toxic cancer treatments. Continued interdisciplinary research, bridging ethnobotany, pharmacology, nanotechnology, and clinical oncology, is crucial to fully realize this potential and translate these ancient remedies into tangible benefits for modern cancer care, offering renewed hope for improved patient outcomes.

References

Abdu S, Juaid N, Amin A et al (2022) Effects of Sorafenib and quercetin alone or in combination in treating hepatocellular carcinoma: in vitro and in vivo approaches. Molecules 27:8082. https://doi.org/10.3390/molecules27228082

Adico MDW, Bayala B, Bunay J et al (2024) Contribution of Sub-Saharan African medicinal plants to cancer research: scientific basis 2013–2023. Pharmacol Res 202:107138. https://doi.org/10.1016/j.phrs.2024.107138

Ahmad N, Adhami VM, Afaq F et al (2001) Resveratrol causes WAF-1/p21-mediated G(1)-phase arrest of cell cycle and induction of apoptosis in human epidermoid carcinoma A431 cells. Clin Cancer Res 7:1466–1473

Alam MK, Alqhtani NR, Alnufaiy B et al (2024) A systematic review and meta-analysis of the impact of resveratrol on oral cancer: potential therapeutic implications. BMC Oral Health 24:412. https://doi.org/10.1186/s12903-024-04045-8

Alaoui Mdarhri H, Benmessaoud R, Yacoubi H et al (2022) Alternatives therapeutic approaches to conventional antibiotics: advantages, limitations and potential application in medicine. Antibiotics 11:1826. https://doi.org/10.3390/antibiotics11121826

Alavi M, Farkhondeh T, Aschner M, Samarghandian S (2021) Resveratrol mediates its anti-cancer effects by Nrf2 signaling pathway activation. Cancer Cell Int 21:579. https://doi.org/10.1186/s12935-021-02280-5

Ali M, Benfante V, Di Raimondo D et al (2024) Recent developments in nanoparticle formulations for resveratrol encapsulation as an anticancer agent. Pharmaceuticals 17:126. https://doi.org/10.3390/ph17010126

Al-Khabori M, Rasool W (2022) Introduction to systematic reviews and meta-analyses of therapeutic studies. Oman Med J 37:e428–e428. https://doi.org/10.5001/omj.2022.42

Alnuqaydan A, Almutary A, Azam M et al (2022) Evaluation of the cytotoxic activity and anti-migratory effect of berberine–phytantriol liquid crystalline nanoparticle formulation on non-small-cell lung cancer in vitro. Pharmaceutics 14:1119. https://doi.org/10.3390/pharmaceutics14061119

Alrashidi AA, Alavi SZ, Koohi M et al (2024) Synergistic strategies for enhanced liver cancer therapy with sorafenib/resveratrol PEGylated liposomes *in vitro* and *in vivo*. J Drug Deliv Sci Technol 96:105703. https://doi.org/10.1016/j.jddst.2024.105703

Annaji M, Poudel I, Boddu SHS et al (2021) Resveratrol-loaded nanomedicines for cancer applications. Cancer Rep 4:e1353. https://doi.org/10.1002/cnr2.1353

Asgharian P, Tazekand AP, Hosseini K et al (2022) Potential mechanisms of quercetin in cancer prevention: focus on cellular and molecular targets. Cancer Cell Int 22:257. https://doi.org/10.1186/s12935-022-02677-w

Ashrafizadeh M, Zarrabi A, Hashemi F et al (2020) Polychemotherapy with curcumin and doxorubicin via biological Nanoplatforms: enhancing antitumor activity. Pharmaceutics 12:1084. https://doi.org/10.3390/pharmaceutics12111084

Asma ST, Acaroz U, Imre K et al (2022) Natural products/bioactive compounds as a source of anticancer drugs. Cancer 14:6203. https://doi.org/10.3390/cancers14246203

Babar Q, Saeed A, Tabish TA et al (2022) Novel epigenetic therapeutic strategies and targets in cancer. Biochim Biophys Acta (BBA) – Mol Basis Dis 1868:166552. https://doi.org/10.1016/j.bbadis.2022.166552

Bray F, Laversanne M, Sung H et al (2024) Global cancer statistics 2022: GLOBOCAN estimates of incidence and mortality worldwide for 36 cancers in 185 countries. CA Cancer J Clin 74:229–263. https://doi.org/10.3322/caac.21834

Cacciola NA, Cuciniello R, Petillo GD et al (2023) An overview of the enhanced effects of curcumin and chemotherapeutic agents in combined cancer treatments. Int J Mol Sci 24:12587. https://doi.org/10.3390/ijms241612587

Castañeda AM, Meléndez CM, Uribe D, Pedroza-Díaz J (2022) Synergistic effects of natural compounds and conventional chemotherapeutic agents: recent insights for the development of cancer treatment strategies. Heliyon 8:e09519. https://doi.org/10.1016/j.heliyon.2022.e09519

Chavda VP, Nalla LV, Balar P et al (2023) Advanced phytochemical-based nanocarrier systems for the treatment of breast cancer. Cancer 15:1023. https://doi.org/10.3390/cancers15041023

Chen HHW, Kuo MT (2017) Improving radiotherapy in cancer treatment: promises and challenges. Oncotarget 8:62742–62758. https://doi.org/10.18632/oncotarget.18409

Chen J, Hu S, Sun M et al (2024) Recent advances and clinical translation of liposomal delivery systems in cancer therapy. Eur J Pharm Sci 193:106688. https://doi.org/10.1016/j.ejps.2023.106688

Cheng Z, Zhang Z, Han Y et al (2020) A review on anti-cancer effect of green tea catechins. J Funct Foods 74:104172. https://doi.org/10.1016/j.jff.2020.104172

Chow H-HS, Garland L, Hsu C-H et al (2010) Resveratrol modulates drug and carcinogen metabolizing enzymes in a healthy volunteer study. Cancer Prev Res (Phila Pa) 3:1168–1175. https://doi.org/10.1158/1940-6207.CAPR-09-0155

Chunarkar-Patil P, Kaleem M, Mishra R et al (2024) Anticancer drug discovery based on natural products: from computational approaches to clinical studies. Biomedicine 12:201. https://doi.org/10.3390/biomedicines12010201

Cocetta V, Quagliariello V, Fiorica F et al (2021) Resveratrol as Chemosensitizer agent: state of art and future perspectives. Int J Mol Sci 22:2049. https://doi.org/10.3390/ijms22042049

De Waure C, Bertola C, Baccarini G et al (2023) Exploring the contribution of curcumin to cancer therapy: a systematic review of randomized controlled trials. Pharmaceutics 15:1275. https://doi.org/10.3390/pharmaceutics15041275

Desai A, Qazi G, Ganju R et al (2008) Medicinal plants and cancer chemoprevention. Curr Drug Metab 9:581–591. https://doi.org/10.2174/138920008785821657

Desai U, Shah K, Mirza S et al (2015) Enhancement of the cytotoxic effects of Cytarabine in synergism with Hesperidine and Silibinin in acute myeloid leukemia: An in-vitro approach. J Cancer Res Ther 11:352. https://doi.org/10.4103/0973-1482.157330

Desai U, Pandya M, Saiyed H, Rawal R (2023) Modulation of drug resistance in leukemia using phytochemicals: an in-silico, in-vitro, and in-vivo approach. In: Recent frontiers of phytochemicals. Elsevier, pp 583–599

Dhillon N, Aggarwal BB, Newman RA et al (2008) Phase II trial of curcumin in patients with advanced pancreatic cancer. Clin Cancer Res 14:4491–4499. https://doi.org/10.1158/1078-0432.CCR-08-0024

Domingo-Fernández D, Gadiya Y, Mubeen S et al (2023) Modern drug discovery using ethnobotany: a large-scale cross-cultural analysis of traditional medicine reveals common therapeutic uses. iScience 26:107729. https://doi.org/10.1016/j.isci.2023.107729

Elzoghby AO, El-Lakany SA, Helmy MW et al (2017) Shell-crosslinked zein nanocapsules for oral codelivery of exemestane and resveratrol in breast cancer therapy. Nanomedicine 12:2785–2805. https://doi.org/10.2217/nnm-2017-0247

Farghadani R, Naidu R (2021) Curcumin: modulator of key molecular signaling pathways in hormone-independent breast cancer. Cancer 13:3427. https://doi.org/10.3390/cancers13143427

Farghadani R, Naidu R (2022) Curcumin as an enhancer of therapeutic efficiency of chemotherapy drugs in breast cancer. Int J Mol Sci 23:2144. https://doi.org/10.3390/ijms23042144

Ferguson JJA, Abbott KA, Garg ML (2021) Anti-inflammatory effects of oral supplementation with curcumin: a systematic review and meta-analysis of randomized controlled trials. Nutr Rev 79:1043–1066. https://doi.org/10.1093/nutrit/nuaa114

Friedenreich CM, Ryder-Burbidge C, McNeil J (2021) Physical activity, obesity and sedentary behavior in cancer etiology: epidemiologic evidence and biologic mechanisms. Mol Oncol 15:790–800. https://doi.org/10.1002/1878-0261.12772

Gao J, Hou D, Hu P, Mao G (2021) Curcumol increases the sensitivity of colon cancer to 5-FU by regulating Wnt/β-catenin signaling. Transl. Cancer Res 10:2437–2450. https://doi.org/10.21037/tcr-21-689

Gezici S, Şekeroğlu N (2019) Current perspectives in the application of medicinal plants against cancer: novel therapeutic agents. Anti Cancer Agents Med Chem 19:101–111. https://doi.org/10.2174/1871520619666181224121004

Goetz LH, Schork NJ (2018) Personalized medicine: motivation, challenges, and progress. Fertil Steril 109:952–963. https://doi.org/10.1016/j.fertnstert.2018.05.006

Graziose R, Ann Lila M, Raskin I (2010) Merging traditional Chinese medicine with modern drug discovery technologies to find novel drugs and functional foods. Curr Drug Discov Technol 7:2–12. https://doi.org/10.2174/157016310791162767

Guo K, Feng Y, Zheng X et al (2021) Resveratrol and its analogs: potent agents to reverse epithelial-to-mesenchymal transition in tumors. Front Oncol 11:644134. https://doi.org/10.3389/fonc.2021.644134

Guo H, Xu X, Zhang J et al (2024) The pivotal role of preclinical animal models in anti-cancer drug discovery and personalized cancer therapy strategies. Pharmaceuticals 17:1048. https://doi.org/10.3390/ph17081048

Gupta SC, Patchva S, Aggarwal BB (2013) Therapeutic roles of curcumin: lessons learned from clinical trials. AAPS J 15:195–218. https://doi.org/10.1208/s12248-012-9432-8

Gureje O, Nortje G, Makanjuola V et al (2015) The role of global traditional and complementary systems of medicine in the treatment of mental health disorders. Lancet Psychiatry 2:168–177. https://doi.org/10.1016/S2215-0366(15)00013-9

Gurib-Fakim A (2006) Medicinal plants: traditions of yesterday and drugs of tomorrow. Mol Asp Med 27:1–93. https://doi.org/10.1016/j.mam.2005.07.008

Hafez Ghoran S, Calcaterra A, Abbasi M et al (2022) Curcumin-based Nanoformulations: a promising adjuvant towards cancer treatment. Molecules 27:5236. https://doi.org/10.3390/molecules27165236

Howells LM, Iwuji COO, Irving GRB et al (2019) Curcumin combined with FOLFOX chemotherapy is safe and tolerable in patients with metastatic colorectal cancer in a randomized phase IIa trial. J Nutr 149:1133–1139. https://doi.org/10.1093/jn/nxz029

Huang X, He T, Liang X et al (2024) Advances and applications of nanoparticles in cancer therapy. MedComm Oncol 3:e67. https://doi.org/10.1002/mog2.67

Islam MR, Rauf A, Akash S et al (2024) Targeted therapies of curcumin focus on its therapeutic benefits in cancers and human health: molecular signaling pathway-based approaches and future perspectives. Biomed Pharmacother 170:116034. https://doi.org/10.1016/j.biopha.2023.116034

Jabbarzadeh Kaboli P, Afzalipour Khoshkbejari M, Mohammadi M et al (2020) Targets and mechanisms of sulforaphane derivatives obtained from cruciferous plants with special focus on breast cancer – contradictory effects and future perspectives. Biomed Pharmacother 121:109635. https://doi.org/10.1016/j.biopha.2019.109635

Jang JY, Im E, Kim ND (2022) Mechanism of resveratrol-induced programmed cell death and new drug discovery against cancer: a review. Int J Mol Sci 23:13689. https://doi.org/10.3390/ijms232213689

Jenča A, Mills D, Ghasemi H et al (2024) Herbal therapies for cancer treatment: a review of phytotherapeutic efficacy. Biol Targets Ther 18:229–255. https://doi.org/10.2147/BTT.S484068

Jeong J-H, An JY, Kwon YT et al (2009) Effects of low dose quercetin: cancer cell-specific inhibition of cell cycle progression. J Cell Biochem 106:73–82. https://doi.org/10.1002/jcb.21977

Jhaveri A, Deshpande P, Pattni B, Torchilin V (2018) Transferrin-targeted, resveratrol-loaded liposomes for the treatment of glioblastoma. J Control Release 277:89–101. https://doi.org/10.1016/j.jconrel.2018.03.006

Ji H, Li X, Zhang H (2009) Natural products and drug discovery: can thousands of years of ancient medical knowledge lead us to new and powerful drug combinations in the fight against cancer and dementia? EMBO Rep 10:194–200. https://doi.org/10.1038/embor.2009.12

Ji H, Zhang Z, Chen C et al (2024) The impact of quercetin and paclitaxel combination on ovarian cancer cells. iScience 27:110434. https://doi.org/10.1016/j.isci.2024.110434

Jiang X, Jiang Z, Jiang M, Sun Y (2022) Berberine as a potential agent for the treatment of colorectal cancer. Front Med 9:886996. https://doi.org/10.3389/fmed.2022.886996

Kawish SM, Sharma S, Gupta P et al (2024) Nanoparticle-based drug delivery platform for simultaneous administration of phytochemicals and chemotherapeutics: emerging trends in cancer management. Part Part Syst Charact 41:2400049. https://doi.org/10.1002/ppsc.202400049

Kim B, Park J-E, Im E et al (2021) Recent advances in nanotechnology with Nano-phytochemicals: molecular mechanisms and clinical implications in cancer progression. Int J Mol Sci 22:3571. https://doi.org/10.3390/ijms22073571

Kirave P, Gondaliya P, Kulkarni B et al (2020) Exosome mediated miR-155 delivery confers cisplatin chemoresistance in oral cancer cells via epithelial-mesenchymal transition. Oncotarget 11:1157

Ko J-H, Sethi G, Um J-Y et al (2017) The role of resveratrol in cancer therapy. Int J Mol Sci 18:2589. https://doi.org/10.3390/ijms18122589

Kuai R, Yuan W, Son S et al (2018) Elimination of established tumors with nanodisc-based combination chemoimmunotherapy. Sci Adv 4:eaao1736. https://doi.org/10.1126/sciadv.aao1736

Kursvietiene L, Kopustinskiene DM, Staneviciene I et al (2023) Anti-cancer properties of resveratrol: a focus on its impact on mitochondrial functions. Antioxidants 12:2056. https://doi.org/10.3390/antiox12122056

Li X, Guo S, Xiong X-K et al (2019) Combination of quercetin and cisplatin enhances apoptosis in OSCC cells by downregulating xIAP through the NF-κB pathway. J Cancer 10:4509–4521. https://doi.org/10.7150/jca.31045

Li C, Xu Y, Zhang J et al (2023) The effect of resveratrol, curcumin and quercetin combination on immuno-suppression of tumor microenvironment for breast tumor-bearing mice. Sci Rep 13:13278. https://doi.org/10.1038/s41598-023-39279-z

Li L, Wazir J, Huang Z et al (2024) A comprehensive review of animal models for cancer cachexia: implications for translational research. Genes Dis 11:101080. https://doi.org/10.1016/j.gendis.2023.101080

Liu Z, Huang P, Law S et al (2018) Preventive effect of curcumin against chemotherapy-induced side-effects. Front Pharmacol 9. https://doi.org/10.3389/fphar.2018.01374

Liu M, Martyn AP, Quinn RJ (2022) Natural product-based PROteolysis TArgeting chimeras (PROTACs). Nat Prod Rep 39:2292–2307. https://doi.org/10.1039/D2NP00038E

Liu B, Zhou H, Tan L et al (2024) Exploring treatment options in cancer: tumor treatment strategies. Signal Transduct Target Ther 9:175. https://doi.org/10.1038/s41392-024-01856-7

Lopez JS, Banerji U (2017) Combine and conquer: challenges for targeted therapy combinations in early phase trials. Nat Rev Clin Oncol 14:57–66. https://doi.org/10.1038/nrclinonc.2016.96

Lorenc T, Felix L, Petticrew M et al (2016) Meta-analysis, complexity, and heterogeneity: a qualitative interview study of researchers' methodological values and practices. Syst Rev 5:192. https://doi.org/10.1186/s13643-016-0366-6

Lotfi N, Yousefi Z, Golabi M et al (2023) The potential anti-cancer effects of quercetin on blood, prostate and lung cancers: an update. Front Immunol 14:1077531. https://doi.org/10.3389/fimmu.2023.1077531

Lubner SJ, LoConte NK, Holen KD et al (2010) A phase II study of Oxaliplatin, 5-fluorouracil, Leucovorin, and high-dose Capecitabine in patients with metastatic colorectal cancer. Clin Colorectal Cancer 9:157–161. https://doi.org/10.3816/CCC.2010.n.021

Lv L, Liu C, Chen C et al (2016) Quercetin and doxorubicin co-encapsulated biotin receptor-targeting nanoparticles for minimizing drug resistance in breast cancer. Oncotarget 7:32184–32199. https://doi.org/10.18632/oncotarget.8607

Mazumder K, Aktar A, Roy P et al (2022) A review on mechanistic insight of plant derived anticancer bioactive phytocompounds and their structure activity relationship. Molecules 27:3036. https://doi.org/10.3390/molecules27093036

Mehta LS, Watson KE, Barac A et al (2018) Cardiovascular disease and breast cancer: where these entities intersect: a scientific statement from the American Heart Association. Circulation 137. https://doi.org/10.1161/CIR.0000000000000556

Meng J, Guo F, Xu H et al (2016) Combination therapy using co-encapsulated resveratrol and paclitaxel in liposomes for drug resistance reversal in breast cancer cells in vivo. Sci Rep 6:22390. https://doi.org/10.1038/srep22390

Metwaly AM, Ghoneim MM, IbrahimH E et al (2021) Traditional ancient Egyptian medicine: a review. Saudi J Biol Sci 28:5823–5832. https://doi.org/10.1016/j.sjbs.2021.06.044

Meybodi SM, Rezaei P, Faraji N et al (2023) Curcumin and its novel formulations for the treatment of hepatocellular carcinoma: new trends and future perspectives in cancer therapy. J Funct Foods 108:105705. https://doi.org/10.1016/j.jff.2023.105705

Moutabian H, Majdaeen M, Ghahramani-Asl R et al (2022) A systematic review of the therapeutic effects of resveratrol in combination with 5-fluorouracil during colorectal cancer treatment: with a special focus on the oxidant, apoptotic, and anti-inflammatory activities. Cancer Cell Int 22:142. https://doi.org/10.1186/s12935-022-02561-7

Najafiyan B, Bokaii Hosseini Z, Esmaelian S et al (2024) Unveiling the potential effects of resveratrol in lung cancer treatment: mechanisms and nanoparticle-based drug delivery strategies. Biomed Pharmacother 172:116207. https://doi.org/10.1016/j.biopha.2024.116207

Negri A, Naponelli V, Rizzi F, Bettuzzi S (2018) Molecular targets of epigallocatechin—Gallate (EGCG): a special focus on signal transduction and cancer. Nutrients 10:1936. https://doi.org/10.3390/nu10121936

Nguyen MH, Nguyen TYN, Le THN et al (2024) Medicinal plants as a potential resource for the discovery of novel structures towards cancer drug resistance treatment. Heliyon 10:e39229. https://doi.org/10.1016/j.heliyon.2024.e39229

Park W, Amin ARMR, Chen ZG, Shin DM (2013) New perspectives of curcumin in cancer prevention. Cancer Prev Res (Phila) 6:387–400. https://doi.org/10.1158/1940-6207.CAPR-12-0410

Patel PS, Shah MH, Jha FP et al (2004) Alterations in plasma lipid profile patterns in head and neck cancer and oral precancerous conditions. Indian J Cancer 41:25–31

Patel BP, Shah PM, Rawal UM et al (2005) Activation of MMP-2 and MMP-9 in patients with oral squamous cell carcinoma. J Surg Oncol 90:81–88. https://doi.org/10.1002/jso.20240

Patwardhan B, Warude D, Pushpangadan P, Bhatt N (2005) Ayurveda and traditional Chinese medicine: a comparative overview. Evid Based Complement Alternat Med 2:465–473. https://doi.org/10.1093/ecam/neh140

Paudel KR, Mehta M, Yin GHS et al (2022) Berberine-loaded liquid crystalline nanoparticles inhibit non-small cell lung cancer proliferation and migration in vitro. Environ Sci Pollut Res 29:46830–46847. https://doi.org/10.1007/s11356-022-19158-2

Pezzani R, Salehi B, Vitalini S et al (2019) Synergistic effects of plant derivatives and conventional chemotherapeutic agents: an update on the cancer perspective. Medicina (Mex) 55:110. https://doi.org/10.3390/medicina55040110

Prasathkumar M, Anisha S, Dhrisya C et al (2021) Therapeutic and pharmacological efficacy of selective Indian medicinal plants—a review. Phytomed Plus 1:100029. https://doi.org/10.1016/j.phyplu.2021.100029

Pratheeshkumar P, Budhraja A, Son Y-O et al (2012) Quercetin inhibits angiogenesis mediated human prostate tumor growth by targeting VEGFR- 2 regulated AKT/mTOR/P70S6K signaling pathways. PLoS One 7:e47516. https://doi.org/10.1371/journal.pone.0047516

Provinciali M, Re F, Donnini A et al (2005) Effect of resveratrol on the development of spontaneous mammary tumors in HER-2/neu transgenic mice. Int J Cancer 115:36–45. https://doi.org/10.1002/ijc.20874

Rege SA, Arya M, Momin SA (2019) Structure activity relationship of tautomers of curcumin: a review. Ukr Food J 8:45–60. https://doi.org/10.24263/2304-974X-2019-8-1-6

Rezk YA, Balulad SS, Keller RS, Bennett JA (2006) Use of Resveratrol to improve the effectiveness of cisplatin and doxorubicin: study in human gynecologic cancer cell lines and in rodent heart. Am J Obstet Gynecol 194:e23–e26. https://doi.org/10.1016/j.ajog.2005.11.030

Rowinsky E (2003) The Vinca alkaloids. In: Holland-Frei cancer medicine, 6th edn. BC Decker

Ruiz De Porras V, Figols M, Font A, Pardina E (2023) Curcumin as a hepatoprotective agent against chemotherapy-induced liver injury. Life Sci 332:122119. https://doi.org/10.1016/j.lfs.2023.122119

Saavedra-Leos MZ, Jordan-Alejandre E, López-Camarillo C et al (2021) Nanomaterial complexes enriched with natural compounds used in cancer therapies: a perspective for clinical application. Front Oncol 11:664380. https://doi.org/10.3389/fonc.2021.664380

Sahoo S, Mohapatra P, Sahoo SK (2023) Flavonoids for the treatment of breast cancer, present status and future prospective. Anti Cancer Agents Med Chem 23:658–675. https://doi.org/10.2174/1871520623666221024114521

Salla M, Pandya V, Bhullar KS et al (2020) Resveratrol and resveratrol-aspirin hybrid compounds as potent intestinal anti-inflammatory and anti-tumor drugs. Molecules 25:3849. https://doi.org/10.3390/molecules25173849

Sanchez-Gonzalez PD, Lopez-Hernandez FJ, Perez-Barriocanal F et al (2011) Quercetin reduces cisplatin nephrotoxicity in rats without compromising its anti-tumour activity. Nephrol Dial Transplant 26:3484–3495. https://doi.org/10.1093/ndt/gfr195

Saralkar P, Dash AK (2017) Alginate nanoparticles containing curcumin and resveratrol: preparation, characterization, and in vitro evaluation against DU145 prostate cancer cell line. AAPS PharmSciTech 18:2814–2823. https://doi.org/10.1208/s12249-017-0772-7

Sari AN, Bhargava P, Dhanjal JK et al (2020) Combination of Withaferin-a and CAPE provides superior anticancer potency: bioinformatics and experimental evidence to their molecular targets and mechanism of action. Cancer 12:1160. https://doi.org/10.3390/cancers12051160

Sathishkumar K, Chaturvedi M, Das P et al (2022) Cancer incidence estimates for 2022 & projection for 2025: Result from National Cancer Registry Programme, India. Indian J Med Res 156:598–607. https://doi.org/10.4103/ijmr.ijmr_1821_22

Shah K, Mirza S, Desai U et al (2015) Synergism of curcumin and Cytarabine in the down regulation of multi-drug resistance genes in acute myeloid leukemia. Anti Cancer Agents Med Chem 16:128–135. https://doi.org/10.2174/1871520615666150817115718

Shahbaz M, Naeem H, Momal U et al (2023) Anticancer and apoptosis inducing potential of quercetin against a wide range of human malignancies. Int J Food Prop 26:2590–2626. https://doi.org/10.1080/10942912.2023.2252619

Sharifi-Rad J, Rayess YE, Rizk AA et al (2020) Turmeric and its major compound curcumin on health: bioactive effects and safety profiles for food, pharmaceutical, biotechnological and medicinal applications. Front Pharmacol 11:01021. https://doi.org/10.3389/fphar.2020.01021

Sharifi-Rad J, Quispe C, Mukazhanova Z et al (2021) Resveratrol-based Nanoformulations as an emerging therapeutic strategy for cancer. Front Mol Biosci 8. https://doi.org/10.3389/fmolb.2021.649395

Sharma M, Dwivedi J, Kumar B et al (2019) Plant-based secondary metabolites for health benefits. In: Suleria HAR, Goyal MR, Butt MS (eds) Phytochemicals from medicinal plants, Series statement: innovations in plant science for better health: from soil to fork, 1st edn. Apple Academic Press, pp 33–64

Sharma S, Cwiklinski K, Mahajan SD et al (2023) Combination modality using quercetin to enhance the efficacy of docetaxel in prostate cancer cells. Cancer 15:902. https://doi.org/10.3390/cancers15030902

Shi A, Liu L, Li S, Qi B (2024) Natural products targeting the MAPK-signaling pathway in cancer: overview. J Cancer Res Clin Oncol 150:6. https://doi.org/10.1007/s00432-023-05572-7

Singh SK, Lillard JW, Singh R (2018) Reversal of drug resistance by planetary ball milled (PBM) nanoparticle loaded with resveratrol and docetaxel in prostate cancer. Cancer Lett 427:49–62. https://doi.org/10.1016/j.canlet.2018.04.017

Sung H, Ferlay J, Siegel RL et al (2021) Global cancer statistics 2020: GLOBOCAN estimates of incidence and mortality worldwide for 36 cancers in 185 countries. CA Cancer J Clin 71:209–249. https://doi.org/10.3322/caac.21660

Tang S-M, Deng X-T, Zhou J et al (2020) Pharmacological basis and new insights of quercetin action in respect to its anti-cancer effects. Biomed Pharmacother 121:109604. https://doi.org/10.1016/j.biopha.2019.109604

Thapa SB, Pandey RP, Park YI, Sohng JK (2019) Biotechnological advances in resveratrol production and its chemical diversity. Molecules 24:2571. https://doi.org/10.3390/molecules24142571

Tomou E-M, Papakyriakopoulou P, Saitani E-M et al (2023) Recent advances in Nanoformulations for quercetin delivery. Pharmaceutics 15:1656. https://doi.org/10.3390/pharmaceutics15061656

Tong W, Wang Q, Sun D, Suo J (2016) Curcumin suppresses colon cancer cell invasion via AMPK-induced inhibition of NF–κB, uPA activator and MMP9. Oncol Lett 12:4139–4146. https://doi.org/10.3892/ol.2016.5148

Vanneman M, Dranoff G (2012) Combining immunotherapy and targeted therapies in cancer treatment. Nat Rev Cancer 12:237–251. https://doi.org/10.1038/nrc3237

Wang H, Chen Y, Wang L et al (2023a) Advancing herbal medicine: enhancing product quality and safety through robust quality control practices. Front Pharmacol 14:1265178. https://doi.org/10.3389/fphar.2023.1265178

Wang X, Chan YS, Wong K et al (2023b) Mechanism-driven and clinically focused development of botanical foods as multitarget anticancer medicine: collective perspectives and insights from preclinical studies, IND applications and early-phase clinical trials. Cancer 15:701. https://doi.org/10.3390/cancers15030701

Wang S, Wang K, Li C et al (2024) Role of flavonoids in inhibiting triple-negative breast cancer. Front Pharmacol 15. https://doi.org/10.3389/fphar.2024.1411059

Wen S, Sun L, An R et al (2020) A combination of Citrus reticulata peel and black tea inhibits migration and invasion of liver cancer via PI3K/AKT and MMPs signaling pathway. Mol Biol Rep 47:507–519. https://doi.org/10.1007/s11033-019-05157-z

Xiao Q, Zhu W, Feng W et al (2019) A review of resveratrol as a potent Chemoprotective and synergistic agent in cancer chemotherapy. Front Pharmacol 9:1534. https://doi.org/10.3389/fphar.2018.01534

Xiong R-G, Huang S-Y, Wu S-X et al (2022) Anticancer effects and mechanisms of Berberine from medicinal herbs: an update review. Molecules 27:4523. https://doi.org/10.3390/molecules27144523

Xu X, Liu A, Bai Y et al (2019) Co-delivery of resveratrol and p53 gene via peptide cationic liposomal nanocarrier for the synergistic treatment of cervical cancer and breast cancer cells. J Drug Deliv Sci Technol 51:746–753. https://doi.org/10.1016/j.jddst.2018.05.008

Xu W, Xie S, Chen X et al (2021) Effects of quercetin on the efficacy of various chemotherapeutic drugs in cervical cancer cells. Drug Des Devel Ther 15:577–588. https://doi.org/10.2147/DDDT.S291865

Xue X, Liang X-J (2012) Overcoming drug efflux-based multidrug resistance in cancer with nanotechnology. Chin J Cancer 31:100–109. https://doi.org/10.5732/cjc.011.10326

Yeshi K, Turpin G, Jamtsho T, Wangchuk P (2022) Indigenous uses, phytochemical analysis, and anti-inflammatory properties of Australian tropical medicinal plants. Molecules 27:3849. https://doi.org/10.3390/molecules27123849

Zhang X, Zhu L, Wang X et al (2023) Basic research on curcumin in cervical cancer: Progress and perspectives. Biomed Pharmacother 162:114590. https://doi.org/10.1016/j.biopha.2023.114590

Zou J, Zhu L, Jiang X et al (2018) Curcumin increases breast cancer cell sensitivity to cisplatin by decreasing FEN1 expression. Oncotarget 9:11268–11278. https://doi.org/10.18632/oncotarget.24109

13 Emerging Research on Bioactive Compounds and Cancer Stem Cells

Mouvanal Sajana, Mutthuraj Dasegowda, M. Brunda, K. R. Arjun, and Kanthesh M. Basalingappa

Abstract

Cancer stem cells (CSCs) belong to a small subset of tumor cells with self-renewal and differentiation capacity, implicating the initiation and progression of cancer and metastasis. CSCs are relatively resistant to traditional cancer treatments like chemotherapy and radiation. These are the very cells that initiate tumor development, progression, metastasis, and recurrence, making cancer therapy a daunting task. Bioactive compounds (naturally derived molecules from plants mainly belonging to the medicinal plant family) come into the picture as potential candidates for CSCs. Such phytochemicals (such as terpenoids, alkaloids, and flavonoids) have been reported to disrupt specific signaling pathways indispensable for the proper stemness, self-renewal, and survival of CSCs. Key pathways are pivotal in CSCs maintenance, i.e., Wnt/β-catenin, Notch, and Hedgehog. These pathways can be targeted by bioactive compounds such as resveratrol, curcumin, and hesperetin. The interaction between immune evasion, the role of tumor-associated macrophages (TAMs), and other constituents of CSC-stromal cells remains a challenge for targeting the CSC population. Three innovative approaches to counteract prerequisite cancer stem cell resistance are bioactive-based therapeutics associated with conventional drugs, nanotechnology-based delivery techniques, and immunotherapy.

M. Sajana · M. Dasegowda · M. Brunda · K. R. Arjun · K. M. Basalingappa (✉)
Division of Molecular Biology, School of Life Sciences, JSS Academy of Higher Education & Research, Mysuru, Karnataka, India
e-mail: sajanam@jssuni.edu.in; mutthurajd@jssuni.edu.in; 24plm028@jssuni.edu.in; 24plm055@jssuni.edu.in; kantheshmb@jssuni.edu.in

K. Kuca et al. (eds.), *Medicinal Plants and Their Bioactives in Human Diseases*,
https://doi.org/10.1007/978-3-032-01356-9_13

Keywords

Cancer stem cells · Microenvironment · Phytochemicals · Self-renewal · Signalling · Therapeutics

1 Introduction

A constantly shifting condition that produces a vast array of cells with varying morphologies, genotypes, and phenotypes is what defines the course of cancer. There are two main models. This heterogeneity is explained by the concept of clonal evolution and the cancer stem cell (CSC) hypothesis. The CSC model states that malignancies are a diverse group of genetically distinct subclones grouped in a well-organized hierarchy, with CSCs at the top (Burrell et al. 2013; Meacham and Morrison 2013). Only a portion of cancer cells are responsible for the origin and spread of tumors, according to the stem cell hypothesis of cancer (Reya et al. 2001). The ability to self-renew and specialize into various cell types are two of the main functional traits of CSCs that are comparable to those of conventional stem cells. CSCs have a higher capacity to cause tumors and are more resistant to both traditional and targeted treatments. The ability to self-renew and specialize into various cell types are two of the main functional traits of CSCs that are comparable to those of conventional stem cells. CSCs have a higher capacity to cause tumors and are more resistant to both traditional and targeted treatments (Aiello et al. 2019). According to recent research, some substances obtained from plant parts such as bark, roots, leaves, seeds, and fruits have anti-carcinogenic qualities. Through the promotion of apoptosis and cell cycle arrest, the use of natural sources like cruciferous vegetables, soybeans, tomatoes, garlic, turmeric, ginger, olives, whole grains, mushrooms, pomegranates, berries, grapes, plums, and green tea can prevent the development and carcinogenesis of colon cancer (Tao et al. 2018). The approximately 35,000 herbal bioactive components derived from plants, marine life, and other sources reduce the adverse side effects of contemporary cancer treatment methods like chemotherapy and radiation therapy (Gupta et al. 2022a, b, c). Medicinal plants are the most dependable source of bioactive chemicals for natural remedies that supplement medications in alternative systems as a sustainable method of treating colorectal cancer. With fewer dangers and adverse effects, alkaloids, flavonoids, phenolics, quinones, volatile oils, terpenoids, and saponins have a strong cytotoxic impact on cancer cells (Edgar et al. 2007). Cancer therapies that enhance a patient's quality of life include surgery, chemotherapy, and radiotherapy.

Furthermore, side effects from traditional treatment methods are one of the main issues with the cancer treatment procedure (Barabadi et al. 2017). Since most conventional cancer therapies target neoplastic cells that are developing excessively quickly, current cancer treatments have a poor survival rate. Thus, CSCs may survive because of their strong drug resistance and slower rate of proliferation, which results in treatment failure. Several molecular mechanisms, including signaling pathways, regulate the generation of stem cells; aberrant activation of these pathways regulates the development, differentiation, and proliferation of these CSCs, all of which are critical for the carcinogenicity of these cells. A typical tissue stem cell may give rise to CSCs when it undergoes oncogenic somatic mutations and is impacted by external microenvironmental stimuli. The process by which tumor cells progress to metastasis is intricate and multi-stage. Traditionally, they undergo epithelial-mesenchymal transition (EMT), which enables them to spread from the primary tumor and enter the bloodstream as a local invasion. Later, they must survive in the bloodstream before extravasating and colonizing foreign tissues. Following their adaptation to the microenvironment, these cells create macro metastases, meaning that CSC may develop from transformed epithelial cells that undergo an EMT process to acquire migratory and metastatic properties. CSCs, also known as tumor stem cells and cancer starting cells, are the primary cause of tumor growth, invasion, metastasis, and recurrence. Distinct signaling pathways essential to the growth process control how CSCs differentiate and produce different types of tumor cells (Holohan et al. 2013; Lage 2008; Venkatesh et al. 2018) (Fig. 13.1).

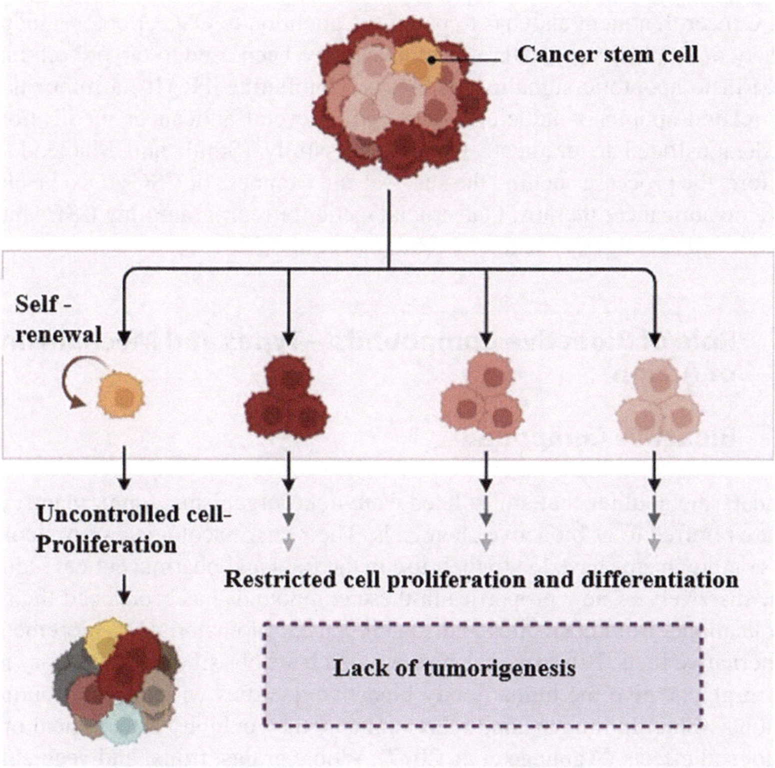

Fig. 13.1 Stem cell proliferation

1.1 Cancer Cell Survival and Maintenance

Autophagy has a crucial role in CSC maintenance and aggression. The ability of CSCs to differentiate and self-renew compared to conventional stem cells is what distinguishes them. Conversely, pluripotency is a key feature of CSCs that enables undifferentiated cell maintenance and unrestricted growth and division. The control of cancer cells and CSCs depends on the management of autophagy. Using fluorescence-activated cell sorting, CSCs were initially discovered in AML based on the expression of CD34 and CD38 (CD34+ and CD38-), two cell surface markers (Yang et al. 2015). According to other studies, the Forkhead box protein O (FOXO) may be involved in controlling the fate of CSCs. Both adult and embryonic stem cell homeostasis depend on FOXO-mediated transcriptional control. Cellular plasticity, very effective DNA damage repair, production of genes linked to multi-drug resistance, and apoptosis prevention are some of the several biochemical pathways that contribute to CSC resistance to drug therapy (Rahman et al. 2020a, b). Many human malignancies, including cancers of the brain, breast, pancreas, melanoma, leukemia, and colorectal cancers, have also been linked to CSCs and medication resistance. Cancer treatment also has to pay more attention to CSC apoptosis induction. A variety of synthetic and natural chemicals have been used to target both intrinsic and extrinsic apoptotic signaling pathways. Combining TRAIL, a tumor necrosis factor-related apoptosis-inducing ligand, with several anticancer medications has been demonstrated to eradicate CSCs successfully (Smith and Macleod 2019). Therefore, the processes behind the survival and stemness of CSCs must be clarified to develop anticancer therapy. One crucial tactic for reprogramming CSCs into regular stem cells is targeted treatment (Rahman et al. 2020a, b).

2 Role of Bioactive Compounds—Types and Mechanism of Action

2.1 Bioactive Compounds

Secondary metabolites that are isolated from microorganisms, fungi, plants, or animals are referred to as bioactive chemicals. Their pharmacological or toxicological effects on organisms have led to their use in the food and pharmaceutical industries, but the discovery of new properties in these compounds has broadened their range of applications, from cosmetics and functionalized biomaterials to bioremediation and alternative fuels. Bioactive substances, which are classified as secondary metabolites in plants, give the human body bioactive qualities without contributing any nutritional value. In humans and other animals, they exhibit toxicological or pharmacological effects (Agnhage et al. 2017). Whole grains, fruits, and vegetables are some of the foods that contain these chemicals. Polyphenols, triterpenes and phytosterols, terpenoids, polysaccharides, capsaicinoids, carotenoids and tocopherols, alkaloids, saponins, glucosinolates, and other substances are examples of natural bioactive chemicals (Makkar et al. 2021). They are vital to the human body and exhibit anti-inflammatory, anti-allergic, anti-carcinogenic, antimicrobial, and anti-mutagenic qualities.

Additionally, they are essential to the food, chemical, and pharmaceutical industries (Câmara et al. 2020). According to multiple pharmacological research, bioactive substances have many health benefits and aid in the prevention of various illnesses and metabolic disorders (Chhikara et al. 2019). Vegetable oils, meat and fish products, bakery goods, dairy products, and other products are processed using bioactive substances such as essential oils, flavonoids, tannins, phenolic acids, carotenoids, organosulfur compounds, phytosterols, tocopherols (Fig. 13.2) (Chhikara et al. 2019). Over time, a variety of extraction techniques have been used, considering characteristics like the source matrix's nature, the bioactive compounds, and the solvents' relative solubility, structure, and chemical characteristics, as well as the impact of temperature, pressure, pH, and other factors on the extraction process's duration, yield, and selectivity (Shirahigue and Antonini 2020). Traditional extraction techniques, including maceration and decoction, have been extensively employed in both the model laboratory-scale and domestic extraction of essential oils and other bioactive components. To help extract substances from samples placed in a thimble, the Soxhlet extraction process uses slightly higher temperatures to circulate solvent inside an apparatus (Azmir et al. 2013; Mutthuraj et al. 2020; Muala et al. 2021). However, when relative yields and specificity of extracted chemicals are considered, these methods' shortcomings include lengthy extraction times, the use of bulk solvents, and lower efficiency. For instance, maceration takes 2–7 days to achieve a decent extraction, and in the case of decoction (hot water extraction), the solvent-to-crude extract ratios might range from 4:1 to even about 50:1. Pure, costly, and hazardous solvents are frequently needed. Certain techniques use high temperatures that are inappropriate for extracting thermally unstable

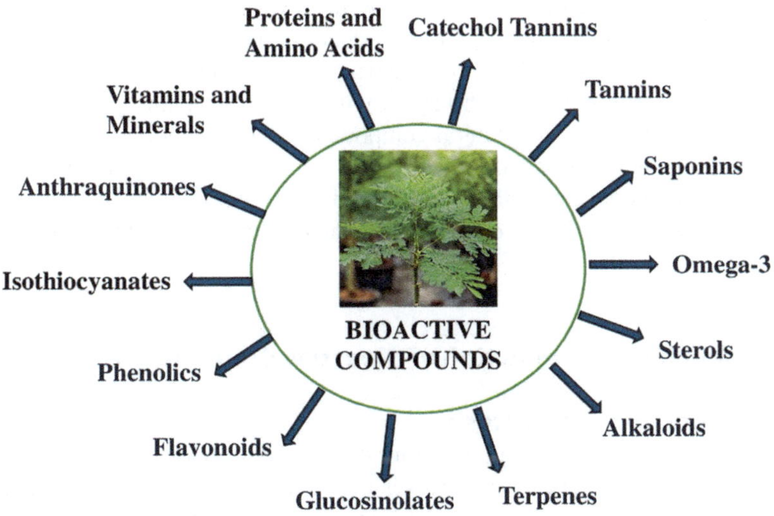

Fig. 13.2 Major types of bioactive compounds

substances (Raynie 2019). The restrictions have been addressed in several ways, as well as by creating a number of novel solvents that are more selective, economical, less hazardous, and "green." For example, deep eutectic solvents, ionic liquids, and aqueous two-phase systems have been used to boost extraction speed and yield and help extract molecules that were too challenging to extract using traditional solvents. Several characteristics, including the Hansen parameters and solvation parameter models, could be used to compare and optimize the interactions between the solvent and bioactive substances (Li et al. 2020a, b; Pal and Jadeja 2020).

2.2 Role of Bioactive Compounds in Cancer Research

Humans have investigated a wide range of plant species to heal diseases and enhance health from the beginning of time. Consequently, they have discovered a vast array of bioactive chemicals in plants with a wide range of therapeutic applications. Among the most extensively studied plant compounds are flavonoids, carotenoids, alkaloids, and phenolics; they have a variety of therapeutic uses, including anticancer action (García-Vaquero and Rajauria 2018; Pai et al. 2022). Research shows that plant anticancer compounds demonstrate low toxicity and safety properties. Multiple studies reveal that some phytochemicals trigger programmed cell death in cancer cells utilizing specific molecular mechanisms. Though not every plant-derived compound shows natural non-toxicity, researchers should evaluate them meticulously to investigate their healing capacity alongside their safety risks.

Many substances found in plants, such as ricin, can be hazardous. As a result, mining the possible compounds needs careful consideration (Pai et al. 2022). Cancer is an intricate collection of diseases involving the growth of cells uncontrollably, the invasion of tissues, and the ability to spread to different body parts. Even with the adoption of modern methods such as chemotherapy, surgery, and radiotherapy, cancer is estimated to be the second cause of death in the world. Therefore, there is a need to explore new methods of management, and synthetic or natural compounds called bioactive compounds with targets within cells are gaining prominence for cancer management. Several natural compounds, including polyphenols, saponins, alkaloids, fatty acids, terpenoids, and some synthetic ones, share the ability to restrain the carcinogenesis pathways. These bioactive compounds' activity is multifaceted in that they act during cancer initiation, development, and spread (Sajana M et al. 2024).

2.3 Mechanisms of Action of Bioactive Compounds in Cancer

2.3.1 Induction of Apoptosis

Apoptosis is a natural cell death program that helps our bodies eliminate damaged or unneeded cells. In cancer, apoptosis is one mechanism that is often suppressed, thus allowing the malignant cells to survive and, consequently, proliferate. The apoptotic process in cancer cells can be reestablished by bioactive compounds that are concerned with the regulation of key apoptotic proteins, including Bcl-2 and Bax. Bcl-2 is an anti-apoptotic protein that plays a role in preventing the cell from dying, while the pro-apoptotic protein is Bax, which promotes apoptosis. Bioactive compounds, such as curcumin and resveratrol, have shown that they can upregulate Bax expression and downregulate Bcl-2, thereby shifting the balance toward cellular death. Caspases are a group of proteases that serve as the key executors of apoptosis. The bioactive compounds can activate caspases, especially caspase-9 and caspase-3, to induce apoptosis. Curcumin triggers apoptosis in breast cancer cells by altering Bax/Bcl-2 ratio, mitochondria dysfunction, and caspase activation Pai et al. (2022).

2.3.2 Inhibition of Cell Proliferation

Cancer cells acquire the capability of these rapidly dividing cell cycles, which is a primary reason for the emergence of tumors due to mutations in genes controlling the cell cycle. Bioactive compounds can also deal with cell cycle progression by ceasing cell division in the G1 phase, thus preventing tumor growth. Cyclin-Dependent Kinases (CDKs) regulate the cell cycle's transition from one phase to another. By inhibiting CDKs, we can stop the cell cycle during certain checks (G1, S, G2/M). For instance, bioactive compounds like quercetin are highly effective and block CDK activity completely. Quercetin, one of the flavonoid compounds, has been shown to cause cell cycle arrest at the G1 phase in cancer cells, thus far inhibiting the proliferation of breast cancer cells Mahato et al. (2019), García-Vaquero et al. (2018), and Pai et al. (2022).

2.3.3 Inhibition of Angiogenesis

Angiogenesis is the building of new blood vessels from already existing ones. This process is critical for tumors to develop and disseminate. Bioactive compounds can inhibit it by blocking the key pro-angiogenic factors like vascular endothelial growth factor (VEGF). Compared with the angiogenic factor, VEGF, HIF-1α which regulates the expression of VEGF under low oxygen conditions, is the hypoxia-inducible factor. Some bioactive compounds, for example, resveratrol, can inhibit VEGF expression by the HIF-1α pathway. Resveratrol, a compound that impairs blood vessel formation, inhibits breast cancer cells by downregulating the expression of VEGF and HIF-1α, thus limiting the blood supply to the tumor (Sharma and Gescher 2005; Baur and Sinclair 2006; Gupta and Sung 2013).

2.3.4 Inhibition of Metastasis

The primary cause of cancer-related mortality is metastasis, or the spread of cancer cells to other areas of the body. Cancer metastasis can be avoided by using bioactive substances, such as those that stop the migration, invasion, and epithelial-to-mesenchymal transition (EMT) of cancer cells. Matrix metalloproteinases (MMPs) are proteolytic enzymes that cleave the extracellular matrix, thereby enabling cancer cell invasion. Green tea polyphenol compounds such as EGCG can inhibit MMP activity and arrest metastasis. EGCG inhibits the mobility and infiltration of gastric cancer cells, which is done through the downregulation of MMPs and by preventing EMT (Vassallo and Zamboni 2005; Wang and Cao 2011).

2.3.5 Modulation of Inflammation

Persistent inflammation is an essential element in cancer development and progression. Anti-inflammatory bioactive compounds may interfere with the inflammatory environment by blocking the activity of the inflammatory mediators' cytokines and transcription factors. NF-κB is a transcription factor that induces the production of pro-inflammatory cytokines such as TNF-α. Curcumin is a compound that dwindles N-κB activity; thus, inflammation is minimized, and tumor growth is inhibited. For example, curcumin prevents NF-κB activation in cancer cells, reducing the synthesis of pro-inflammatory cytokines and making detecting cancer mortality easier (Aggarwal and Harikumar 2009).

2.3.6 Modulation of the Immune System

Various bioactive compounds can augment the immune system's resistance to tumor cells. They can drive immune cell activity, including dendritic cells, T lymphocytes, and natural killer (NK) cells. Beta-glucans are polysaccharides that have mushrooms in them. They become a target for cancer cells to be killed by activated immune cells through the production of cytokines and reactive oxygen species (ROS) (Terao 2011). Beta-glucans reinforce the body's anti-tumor and immune response and escalate the defense system against cancer by turning on natural killer cells and macrophages. In addition, beta-glucans have been reported as the catalysts of radiation and chemotherapy regimens for cancer treatment. Successful cancer treatment will be a scenario that cancer patients can see. With the help of these beta-glucans Strategies, the immune system will work harder, thus complementing current cancer treatments. This proves how beta-glucans, for instance, can be used with conventional treatments to increase their natural efficacy (Linnewiel-Hermoni et al. 2015; Moshiri et al. 2016; Cosme et al. 2020).

2.4 Therapeutic Implications of Bioactive Compounds in Cancer Treatment

The pharmaceutical development of bioactive compounds, i.e., the natural substances with medicinal, pharmacological, or fluxional properties in cancer therapy, means health care will move from cancer prevention to adjunctive therapy advised to be used with other traditional treatment methods. Here are some implications of bioactive compounds in cancer care:

2.4.1 Curcumin (Turmeric)

Source: Curcumin is an extract from the rhizome of *Curcuma longa*, a plant known worldwide as turmeric.

Role in Cancer: Curcumin has been proven to have strong anti-inflammatory and antioxidant effects and is known to have the potential to inhibit cancer. They have been researching on their ability to act as chemoprevention and chemotherapy for numerous types of cancer, such as breast, colon, and prostate cancers.

Mechanism of Action

Induces Apoptosis: The current researches reveal that curcumin induces apoptosis by altering apoptotic-related proteins, including Bax/Bcl-2, disrupting the mitochondria and caspase activation (Zhu et al. 2015).

Inhibits Metastasis: Curcumin inhibits cancer's metastatic potential by downregulating the genes of matrix metalloproteinase (MMPs) and tissue Inhibitors of Metalloproteinases (TIMPs) (Hassan and Daghestani 2012).

Modulates NF-κB: NFκB is a transcription factor bowed by curcumin whose products include cytokines and survival genes in tumor cells that heighten inflammation and cancer progression (Aggarwal and Harikumar 2009; Sharma and Gescher 2005; Gupta and Sung 2013).

2.4.2 Resveratrol

Source: Available in grapes, red wine, berries, and peanuts.

Role in Cancer: Resveratrol is a polyphenolic compound that plays chemopreventive and anticancer roles. Consequently, it is effective in cancer cell proliferation, metastasis, and angiogenesis.

Mechanism of Action

Inhibits VEGF: It inhibits the expression of vascular endothelial growth factor (VEGF), which plays a crucial role in angiogenesis, thereby cutting off cancer's relations with blood vessels (Li et al. 2023).

Activates Sirtuins: Sirtuins enhance the stability of the genome, repair DNA damage, and also prevent the formation of tumors (Matsuno et al. 2020).

Suppresses NF-κB: It has been established that resveratrol suppresses the action of NF-κB, a pro-inflammatory factor, which is essential in the survival of tumor cells (Vassallo and Zamboni 2005; Baur and Sinclair 2006; Wang et al. 2011).

2.4.3 Epigallocatechin Gallate (EGCG)

Source: EGCG is the predominant polyphenol present in green tea.

Role in Cancer: EGCG has been found to possess excellent antioxidant activity and inflammation-preventing capability, but it can hinder cancer cell migration. It is instrumental in the prevention of breast, prostate, and lung cancer.

Mechanism of Action

Inhibits MMPs: Having anti-angiogenic effects, EGCG also suppresses MMPs, which are responsible for cancer cell invasion and metastasis (Wei and Zhang 2024).

Suppression of PI3K/Akt Pathway: EGCG also suppressed the cell survival and proliferation marker PI3K/Akt signaling pathway (Gu et al. 2018).

Induces Apoptosis: EGCG induces apoptosis through caspase-3 and enhances the expression ratio of the Bax/Bcl-2 protein (Zaveri 2006; Pan and Lai 2015; Shafiee and Salehi 2019).

2.4.4 Quercetin

Source: Dense in apples, onions, citrus fruits, and tea.

Role in Cancer: Quercetin is a flavin that has been found to have the virtues of an anticancer agent, antioxidant, and anti-inflammatory agent. Research has proven it to slow the growth of cancer cells, especially breast, lung, and colon cancer cells.

Mechanism of Action

Blocks Cell Cycle Progression: Quercetin halts cancerous cell division by preventing the kinases (CDKs) that control the process from the G1 phase of the cell cycle (Tang et al. 2020).

Promotes Apoptosis: They stimulated pro-apoptotic proteins Bax and caspases and suppressed anti-apoptotic proteins, including Bcl-2 (Catalano and Barone 2009; Mendez and de Guillebon 2015).

2.4.5 Sulforaphane

Source: They are found in foods such as broccoli, cabbage, and cauliflower, which are part of the cruciferous group.

Role in Cancer: Sulforaphane acts as a promising chemoprevention because it protects the body against different types of tumors, such as breast, prostate, and liver tumors.

Mechanism of Action

Activates Nrf2: Sulforaphane is known to modulate the Nrf2 pathway and increase the levels of phase II detoxifying enzymes, neutralizing many carcinogens (Houghton et al. 2016).

Induced Epigenetic Modifications: Sulforaphane reduces HDAC and, hence, the degree of histone deacetylation with the consequent reactivation of tumor suppressor genes (Zhang and Talalay 1994; Saha and Sharma 2016; Johnson and Encarnación 2016).

2.4.6 Lycopene

Source: It is available in tomatoes, watermelon, and red pepper.

Role in Cancer: Lycopene is a carotenoid that is an antioxidant and has been found to lower the chances of prostate, lung, and stomach cancers.

Mechanism of Action

Inhibits NF-κB: Lycopene inhibits NF-kappa-B, which plays a role in inflammatory response and tumor growth (Hadad and Levy 2012).

Regulates Growth Factor Receptors: Lycopene suppresses the effects of growth factors like epidermal growth factor receptor (EGFR) to minimize tumor cell growth (Giovannucci 2002; Rao and Rao 2007; Murillo and Stacewicz-Sapuntzakis 2015).

2.4.7 Beta-Glucans

Source: Present in shiitake, reishi, and oats, among other foods and medicines.

Role in Cancer: Beta-glucans are the polysaccharides that have received great interest due to their immunomodulating activity and ability to prevent metastasis.

Mechanism of Action

Immune Modulation: Beta glucans enhance the functions of macrophages and NK cells and improve the phagocytosis of cancer cells (Mishra et al. 2023).

Inhibition of MMPs: This study also found that beta-glucans suppress the activity of MMPs, which leads to cancer cell migration and invasion (Yang and Liu 2013; Chen and Li 2013; Vetvicka and Vannucci 2015).

2.4.8 Taxol (Paclitaxel)

Source: Extracted from the outer surface of the tree called Pacific Yew.

Role in Cancer: Paclitaxel is an antimicrotubular agent that belongs to the chemotherapeutic drugs used to treat various cancers such as breast, ovarian, and lung.

Mechanism of Action

Microtubule Stabilization: Paclitaxel has actions that include binding to microtubules, inhibiting their depolymerization, preventing the formation of the mitotic spindle, inhibiting cell division, and promoting programmed cell death (Rowinsky 1995; Lee and Kim 2019; Comis and Lippman 2006).

2.4.9 Vincristine

Source: Extracted from the Madagascar periwinkle plant, also called Catharanthus roseus.

Role in Cancer: It treats hematologic cancers, including leukemia and lymphoma.

Mechanism of Action

Inhibition of Microtubule Formation: Vincristine inhibits the formation of cell microtubules, stopping cell division and leading to cellular death in cancerous cells (Vaidya and Walker 1993; Kantarjian and Thomas 2012; Pendergast and Godwin 2012).

2.4.10 Berberine

Source: It is available in the Goldenseal, Barberry, and Oregon grape plants.

Role in Cancer: With regards to anticancer properties, berberine has shown efficacy in arresting cancer cell proliferation in addition to inducing apoptosis in colorectal, lung, and breast cancer cells.

Mechanism of Action

Regulates AMPK Pathway: Berberine has been found to act through the activation of the metabolic suppressor, AMPK, which is able to inhibit cancer cell proliferation and promote apoptosis (Li et al. 2015).

Inhibition of Inflammatory Pathways: Berberine suppresses the production of inflammatory cytokines such as TNF-α to decrease the tumor-promoting inflammation (Wu and Zhang 2011; Zhang and Luo 2017; Li and Wu 2016).

2.4.11 Bromelain

Source: Part obtained from the stems of the pineapple plant.

Role in Cancer: Bromelain possesses anti-inflammatory, anticancer, and immunomodulatory properties.

Mechanism of Action

Enhances Immune Activity: Finally, bromelain increases the ability of immune cells as macrophages and boosts cytokine production. Some of the cancer remedies provided by bromelain include interfering with the capability of cancer cells to migrate by breaking down fibrin, a protein used by the cancer cells as support (Ricke and Kelley 2008; Srinivasan and Muthukumar 2012; Kaushik and Pathak 2018).

2.4.12 Gingerol

Source: Extracted from ginger (*Zingiber officinale*).

Role in Cancer: Gingerol has anticancer ability because of the impact on decreasing inflammation response, apoptosis regulation, and new blood vessel formation.

Mechanism of Action

Inhibits NF-κB: Gingerol inhibits the NF-κB signaling cascade, decreases inflammatory markers' production, and induces cancer cells' death (Liang et al. 2018).

Angiogenesis Inhibition: Gingerol is capable of inhibiting the growth of fresh blood vessels that are essential to the development of tumors (Simopoulos 2002; Zick and Sloane 2009; Wang and Zhang 2011; Choi and Lee 2012).

2.4.13 Omega-3 Fatty Acids

Source: Available in fish meat, especially fatty fish such as salmon, mackerel, and flax seeds.

Role in Cancer: Omega 3 is known to lower inflammation, slow down cancer cell growth, and prevent cancer metastasis, especially breast and prostate cancer.

Mechanism of Action

Regulation of Inflammation: Omega-3 PUFA decreases the release of pro-inflammatory cytokines and down-regulates COX-2 and NF-κB (Calder 2013).

Induction of Apoptosis: They increase the effects of apoptosis sensitivity in cancer cells, which is frequently done together with chemotherapy (Simopoulos 2002; Ahmed and Pandey 2012; Brenner and Hauner 2013).

3 Targeting Cancer Stem Cell (CSC) Pathways through Bioactive Compounds

Cancer stem cells (CSCs) are a few cells in a tumor characterized by certain properties like self-renewal, the ability to differentiate into other cell types, and some extent of therapy resistance, including chemotherapy and radiation. Cancer stem cells are believed to be active in the development of the primary tumor, metastasis, cancer relapse, and resistance to therapies (Chu et al. 2024). Due to their responsibility for tumor heterogeneity and tumor progression, CSCs have emerged as a novel therapeutic strategy. Other small molecules, especially plant-sourced phytochemicals, have been identified as potential candidates for modulating CSCs and their networks.

3.1 Wnt/β-Catenin Pathway and Phytochemicals

Wnt/β-catenin signaling is essential in cell self-renewal and differentiation and thus has a significant role in stem cell function. This pathway is commonly dysregulated in CSCs, contributing to the cytoplasmic retention and build-up of β-catenin. This β-catenin migrates to the nucleus, as a transcriptional factor and proposes target genes that enhance cell proliferation and survival, including MYC, Cyclin D1, and

Survivin. Wnt/β-catenin signaling plays a potent role in promoting CSC properties, including drug-resistant traits, metastatic features, and tumor relapse (Valkenburg et al. 2011).

3.1.1 Curcumin

Literature reveals that curcumin, the polyphenolic compound derived from turmeric, can act on the Wnt/β-catenin pathway by regulating β-catenin levels along with the downregulation of genes it promotes, such as Myc, Cyclin D1, and others. These changes result in the attenuation of CSC features and diminished tumorigenic potential based on various cancer models such as colorectal and breast tumors. It also inhibits numerous proteins associated with cell cycle regulation and modulates cell survival from cyclin to anti-apoptotic proteins (Wang et al. 2018).

3.1.2 Epigallocatechin Gallate (EGCG)

EGCG, green tea polyphenol, reduces the level of β-catenin and restores it from cytoplasm to nucleus contact during interaction with Wnt, therefore interrupting the Wnt/β-catenin signaling pathway. EGCG inhibits stem cell proliferation and self-renewal by down-regulating some transcription factors, such as MYC and LEF1 (Mokra et al. 2023).

3.2 Notch Signaling Pathway and Phytochemicals

Notch signaling continues to be important in cell destiny choices, differentiation, and stem cell self-renewal. Some of the original characteristics of CSCs include activating Notch signaling to promote self-renewal and drug resistance. Spirituality can be defined as the integration of the subtle physical elements of the body, such as the ligand (e.g., Jagged or Delta) binding to the Notch receptor, which leads to a proteolytic cleavage that releases the Notch intracellular domain (NICD). The NICD translocates to the nucleus, where it binds with transcription factors to activate the genes required for cell survival and stemness. Notch signaling is overactivated in many types of cancers and is associated with cancer advance and poor survival outcomes (Xiao et al. 2017).

3.2.1 Quercetin

Quercetin, a flavonoid obtained from fruits and vegetables, restrains the Notch1 signaling pathway and suppresses the Notch target genes, including HES1, which controls the maintenance of stem cells. Quercetin can decrease CSC-like characteristics, including self-renewal and chemoresistance in breast and lung cancers (Krishna et al. 2019).

3.2.2 Curcumin

Additionally, curcumin downregulates the Notch signaling by restraining Notch receptors' activation and lowering the levels of target genes. In CSCs of breast cancer, the authors have reported that curcumin inhibited self-renewal capacity as well as invasion and metastasis properties by down-regulating the Notch signaling pathway (Pathak et al. 2019).

3.3 Hedgehog (Hh) Pathway and Phytochemicals

The Hedgehog signaling pathway is critical for stem cells' proper functioning and differentiation and is an essential factor for teratogenesis. Under physiological conditions, the Hedgehog signaling pathway is kept essentially checkpointed; however, in CSCs, its function is dysregulated, promoting self-renewal, proliferation, metastasis, and chemotherapeutic resistance. Sonic Hedgehog, Indian Hedgehog, and Patched interact, encoding to Smoothened where the inhibition is lifted, hence encouraging the activation of the downstream transcription factors Gli1 and Gli2 in the irregularity of genes that characterize the stem cell's multiplication and presence. The Hedgehog signaling pathway is currently considered the oncogenic signaling pathway responsible for neither basal cell carcinoma nor medulloblastoma nor glioblastoma (Cochrane et al. 2015).

3.3.1 Sulforaphane

Sulforaphane, extracted from cruciferous vegetables such as broccoli, was known to prevent the working of Hedgehog signaling through the down-regulation of Gli1. This transcription factor comes downstream of Smoothened. Hedgehog pathway disruption has inhibited sulforaphane levels in glioblastoma and particulate apoptosis of CSC and prostate cancer (Rizzo et al. 2019).

3.3.2 Curcumin

Curcumin inhibits the Hedgehog pathway by reducing the levels of Gli1 and, therefore, decreasing the CSC proliferation. In models of breast cancer, curcumin has adverse effects on CSC migration and metastasis as a result of modulating the Hedgehog signaling pathway (Wang et al. 2021).

3.4 PI3K/Akt/mTOR Pathway and Phytochemicals

The PI3K/Akt/mTOR molecular signal transduction cascade involves essential cellular functions, including cell survival, metabolism, growth, and differentiation. This pathway is often initiated in CSCs due to an increased chance of survival, and the SIN3A protein promotes proliferation and inhibits apoptosis. The mTOR pathway is significant for stress anabolic processes controlling CSC metabolism and proliferation under hypoxia or nutrient limitation. Therefore, dysregulation of the PI3K/Akt/mTOR signal pathway is well-known as the major cause of high-tumor behaviors, therapy resistance, and poor prognosis in many cancers (Karami Fath et al. 2022).

3.4.1 Berberine

Berberine, an alkaloid derived from medicinal plants, inhibited PI3K/Akt/mTOR signaling, reduced CSC proliferation, and increased susceptibility to chemotherapy treatments. Berberine also significantly reduced the activation of some of the components of this pathway in both colorectal and liver cancer models while enhancing cancer cell death and reducing CSC survival (Zhang et al. 2020).

3.4.2 Resveratrol

An antioxidant derived from grapes, resveratrol has been demonstrated to suppress the PI3K/Akt/mTOR signaling pathway, resulting in CSCs diminished survival and mobility. The information regarding resveratrol has been tested on different models of cancers, such as breast and colon, and can pinpoint CSCs to the conventional treatment regimen (Wang et al. 2020a, b).

3.5 JAK/STAT Pathway and Phytochemicals

The JAK/STAT pathway is involved in the signaling by cytokines and growth factors and greatly influences immune response, cell survival, and differentiation processes. In CSCs, it is necessary for the JAK/STAT signaling to be active for stemness and tumorigenicity. Among the downstream targets of IL-6, STAT3 is often upregulated in CSCs and promotes tumorigenesis, immune checkpoint, and therapy resistance (Liu et al. 2021).

3.5.1 Gingerol

Gingerol, derived from ginger, is known to modulate the JAK2/STAT3 signaling pathway, and it analyzed that gingerol attenuates JAK2/STAT3 signaling through the reduction of the phosphorylation of JAK2/STAT3 proteins. These effects inhibit CSC proliferation and promote apoptosis in various cancers, including gastric and breast carcinoma.

3.5.2 Curcumin

Curcumin is a potent inhibitor of the JAK2/STAT3 signaling pathway, and its modulation has been reported in cancers such as breast, colon, and liver cancers. In this short review, we showed that the effect of curcumin on STAT3 suppression decreases CSC stemness and migration and invasive abilities and increases the efficiency of chemotherapy (Yang et al. 2019; Futakuchi et al. 2019; Liu et al. 2021).

3.6 TGF-β/SMAD Pathway and Phytochemicals

The Transforming Growth Factor-β (TGF-β) pathway is vital in determining cells' differentiation, proliferation, and survival. In CSCs, TGF-β signaling is responsible for epithelial-mesenchymal transition (EMT), which is necessary for migration, invasion, and ultimately metastasis. TGF-β-induced EMT, however, is also attributed to chemotherapy and radiation resistance. Thus, the inhibition of TGF-β signaling may reduce metastasis and recurrence of cancer stem cells (CSCs) (Futakuchi et al. 2019).

3.6.1 EGCG

EGCG has been proven to inhibit TGF-β-induced EMT by breaking down the expression of mesenchymal markers (e.g., Snail, Twist) and reviving epithelial markers. This inhibition leads to a decrease in mobility, aggressiveness, and metastasis of CSCs in breast and ovarian tumors (Li et al. 2021a, b).

3.6.2 Curcumin

Curcumin has been shown to inhibit the TGF-β/SMAD pathway and lessen colon cancer-associated cancer stem cell cisternal mobility and invasion. Curcumin contains active principles that prevent the EMT process, reducing the aggressive nature of CSCs and inhibiting the ability of those cells to spread (Pandey et al. 2020).

3.7 NF-kB Signaling Pathway and Phytochemicals

The NF-κB pathway (Nuclear Factor kappa B) involves many physiological processes, such as inflammation, cell survival, and proliferation. The persistent activation of NF-κB characterizes multiple cancers and helps maintain the cancer stem cells to overcome treatment and metastasize. In cancer stem cells, the pathway of NF-kB is activated in relation to the transcription of a group of genes associated with inflammation, proliferative cell homeostasis, and inhibition of programmed cell death (Kaltschmidt et al. 2019).

3.7.1 Quercetin

Quercetin is a potent inhibitor of NF-kB signaling, resulting in lower CSC survival and self-renewal. Quercetin showed success in preclinical chemotherapy enhancement by reducing NF-kB-mediated inflammation in lung, cervical, and pancreatic cancer model systems and enhancing chemosensitivity.

3.7.2 Sulforaphane

Sulforaphane has also been shown to inhibit NF-kB transcription, which in turn inhibits CSC growth. Suppression of CSC growth by this compound has been attributed to the downregulation of inflammatory cytokine synthesis and inhibition of in vivo tumor formation in acanthosis cancer models (Satapathy et al. 2021; Li et al. 2022; Gupta et al. 2022a, b, c).

4 Interaction with the Tumor Microenvironment

Cancer is more than the uncontrolled growth of abnormal cells. Cancer is the central issue, and the environment has played a key role in this complexity through intricate interactions within its cells. Cancer stem cells (CSCs) are central to this complexity and pivotal for tumor initiation, growth, metastasis, and drug resistance. Cancer stem cells (CSCs) are central to this complexity and play a crucial role in tumor initiation, growth, metastasis, and treatment resistance (Galassi et al. 2024a, b; Katoh 2017). Importantly, CSCs do not function in isolation; they interact closely with the tumor microenvironment (TME), which includes various immune cells, fibroblasts, endothelial cells, extracellular matrix (ECM), and various signaling molecules (Hou et al. 2019). Such an interaction significantly determines CSC behavior, shaping essential cancer characteristics such as immune evasion, continued proliferative signaling, and metastasis (Song et al. 2019). The development of efficient treatments for cancer depends mainly on the dynamics between cancer stem cells (CSCs) and their tumor microenvironments (TMEs). This chapter addresses the strategies that allow cancer stem cells (CSCs) to survive and resist conventional treatments (Fortunato et al. 2020). The chapter explores these survival mechanisms, highlighting how CSCs evade the immune system and interact with tumor-associated macrophages (TAMs) in increasingly complex ways.

4.1 Immune Evasion Mechanisms

One of the leading setbacks in targeting CSCs is their ability to escape immune surveillance. This immune evasion allows CSCs to persist, leading to relapse and metastasis even after aggressive treatments (Wilson et al. 2011). The strategies for immune escape are sophisticated and include antigen downregulation, the expression of immune checkpoint molecules, and the secretion of immunosuppressive factors. Tumor immune evasion is an essential hallmark of cancer, where the tumor actively exploits the surrounding microenvironment to evade the immune system. Immune evacuation mechanisms involve the inhibition of certain checkpoints, cytokine signaling, and alteration of antigen presentation (Toh et al. 2017). Their appeal to immunosuppressive cell populations is a noticeable feature. The chapter also expands on how these mechanisms contribute to abetting resistance to therapies, such as immune checkpoint inhibitors (Knudson et al. 2018).

4.1.1 Downregulation of Tumor Antigens

For the immune system to effectively recognize cancer cells, these cells must present antigens that immune cells can identify. CSCs employ reversible epigenetic changes, such as DNA methylation and histone modification, to inhibit the expression of tumor antigens, effectively hiding from immune detection (Katoh et al. 2017). Transcriptional Regulation: Aberrant activation of pathways like Wnt/β-catenin further suppresses tumor antigen expression, aiding immune evasion (Katoh

2017; Song et al. 2019). CSCs may ubiquitinate and degrade antigens through proteasomal pathways, ensuring antigens do not appear on the cell surface.

4.1.2 Expression of Immune Checkpoint Molecules

CSCs further evade immune responses by expressing immune checkpoint molecules such as PD-L1, which inhibit immune cell activity by engaging with receptors like PD-1 on T cells. The tumor microenvironment (TME) is a complex and dynamic milieu that plays a critical role in immune evasion, particularly by promoting the expression of immune checkpoint molecules like PD-L1 on cancer stem cells (CSCs). Hypoxia, a common feature of the tumor microenvironment (TME) caused by poor blood flow, prompts cancer cells to adapt in several ways, including increasing the levels of immune checkpoint proteins like PD-L1. Under low oxygen conditions, hypoxia-inducible factors (HIFs), particularly HIF-1α, become active. These factors bind directly to the promoters of immune-suppressive genes, including PD-L1, enhancing the tumor's ability to evade immune detection (Hou et al. 2019). Elevated PD-L1 expression shields CSCs from immune surveillance, allowing them to evade destruction by cytotoxic T lymphocytes (CTLs). This adaptive mechanism is critical for tumor progression and resistance to immunotherapy (Hou et al. 2019). Under hypoxic conditions, the heightened expression of PD-L1 strengthens the tumor's ability to evade the immune system. It supports the survival and maintenance of cancer stem cells (CSCs), ultimately aiding in tumor spread and metastasis (Knudson et al. 2018).

4.1.3 Immune Checkpoints and Chronic Inflammation

Chronic inflammation within the tumor microenvironment (TME) is crucial in driving the increased expression of immune checkpoints. TME, in turn, helps tumors evade the body's immune system, creating an environment where they can continue to grow and survive (Ghalehbandi et al. 2023). Tumor-associated macrophages (TAMs), neutrophils, and myeloid-derived suppressor cells (MDSCs) are among the immune cells that infiltrate tumors and cause persistent inflammatory responses (Wilson et al. 2011). These cells release pro-inflammatory cytokines, such as TNF-α, IL-1β, and interleukin-6 (IL-6), which trigger signaling pathways such as the STAT3 and NF-κB pathways. These pathways directly regulate the expression of immune checkpoint molecules such as PD-L1. Chronic inflammation in the TME fosters an immunosuppressive environment by promoting the polarization of immune cells into an anti-inflammatory, pro-tumor phenotype, effectively creating a barrier to effective immune response and allowing tumor cells to evade immune destruction. Inflammatory cytokines also facilitate CSC self-renewal, further promoting tumorigenesis (Yi et al. 2024).

4.1.4 Genetic Alterations and Immune Evasion

Genetic mutations in key signaling pathways, such as PI3K/Akt, have been implicated in the upregulation of immune checkpoint molecules (Wilson et al. 2011). The PI3K/Akt pathway is frequently activated in cancer due to mutations in tumor suppressors like PTEN or activating mutations in oncogenes like PIK3CA. This pathway supports cell survival, growth, and metabolism while playing a key role in immune evasion. Notably, by activating several transcription factors and inducing epigenetic modifications, the PI3K/Akt pathway raises PD-L1 expression (Yang et al. 2020a, b). The expression of PD-L1 leads to an enhanced immune checkpoint response, enabling tumor cells, including cancer stem cells (CSCs), to evade immune destruction. Moreover, genetic alterations in the TME can result in a feedback loop where immune suppression facilitates further genetic mutations, leading to an increasingly aggressive and immune-resistant tumor phenotype. These genetic alterations make creating efficient, targeted therapies more difficult by enhancing the tumor's resistance to therapy and immune system evasion (Ghalehbandi et al. 2023).

4.1.5 Secretion of Immunosuppressive Factors

CSCs evade immune surveillance through immune checkpoint upregulation and actively contribute to remodeling the immune landscape within the tumor microenvironment (TME). One of the key ways they achieve this is through the secretion of immunosuppressive factors, which help create an environment that supports tumor growth while inhibiting effective immune responses (Li et al. 2020a, b). Here is an in-depth exploration of the secretion of two major immunosuppressive factors—TGF-β and IL-10—and their role in immune evasion:

(a) *TGF-β Secretion*

Transforming growth factor-beta (TGF-β) is a multifaceted cytokine with crucial roles in regulating immune responses and tumor progression (Yi et al. 2024). In the context of CSCs, TGF-β acts as a key immunosuppressive factor that influences both the innate and adaptive immune systems.

(b) *Inhibition of Cytotoxic T Cells*

One of the primary effects of TGF-β secretion by CSCs is the suppression of cytotoxic T lymphocyte (CTL) function. TGF-β inhibits the activation and proliferation of CTLs, which are essential for recognizing and destroying tumor cells (Wilson et al. 2011; Yang et al. 2020a, b). By dampening CTL activity, TGF-β helps to protect CSCs from immune-mediated killing. TGF-β achieves this by reducing the expression of key effector molecules in CTLs, such as granzyme B and perforin, which are critical for inducing target cell apoptosis. Moreover, TGF-β signaling can induce the expression of immune checkpoint molecules like PD-L1 on CTLs, further weakening their ability to mount a strong anti-tumor response (Ghalehbandi et al. 2023).

(c) *Promotion of Regulatory T Cells (Tregs)*

TGF-β also plays a pivotal role in the expansion and activation of regulatory T cells (Tregs), a subset of T cells that suppress immune responses and help maintain immune tolerance. Within the TME, Tregs are often recruited and activated by TGF-β secretion, creating an immunosuppressive niche. Tregs inhibit the function of effector T cells and promote immune tolerance, allowing CSCs to persist and evade immune detection. This TGF-β-driven Treg accumulation contributes to establishing a suppressive environment that favors tumor growth and metastasis while limiting the effectiveness of anti-tumor immunity.

(d) *IL-10 Production*

Interleukin-10 (IL-10) is another key immunosuppressive cytokine produced by CSCs, contributing to the immune evasion strategy by promoting immune tolerance and enhancing the survival of CSCs (Fortunato et al. 2020). IL-10 has well-established immunosuppressive effects that inhibit the activation and function of several immune cell types, including dendritic cells (DCs), macrophages, and effector T cells. IL-10 can directly suppress the maturation and antigen-presenting functions of DCs, thereby preventing the initiation of an effective adaptive immune response. It also dampens the production of pro-inflammatory cytokines, such as TNF-α and IL-12, crucial for activating immune responses against tumor cells. Additionally, IL-10 inhibits the activity of natural killer (NK) cells, further promoting immune evasion (Wang et al. 2020a, b)

(e) *Activation of Survival Pathways*

In addition to its immune-modulatory role, IL-10 can activate intracellular survival pathways in CSCs, such as the PI3K/Akt pathway. This pathway promotes cell survival, proliferation, and resistance to apoptosis. IL-10 enhances CSC persistence by regulating survival pathways, enabling them to resist conventional therapies and contribute to tumor recurrence. The IL-10-mediated activation of the PI3K/Akt pathway supports the self-renewal and maintenance of the CSC population, which is central to tumor initiation, progression, and metastasis (Masoumi et al. 2021; Shiravand et al. 2022).

4.2 Role of Tumor-Associated Macrophages

Tumor-associated macrophages (TAMs) belong to the tumor microenvironment (TME) components and are involved in managing tumor growth, metastasis, and therapy response. Macrophages belong to the innate immune system and, depending on various signals, can polarize and acquire functions characteristic of at least two groups, namely, M1 and M2 macrophages. M1 macrophages are involved in pro-inflammatory response and possess anti-tumor characteristics; on the other hand, M2 macrophages repair and are immunosuppressive and promote tumor characteristics. TAMs are often far from unbiased towards the M2-like phenotype within the TME, enhancing immunosuppression and creating a specialized atmosphere for tumor advancement. This article discusses the interaction of TAMs to control cancer stem cell function. In this article, the author describes how TAMs affect CSC behaviors such as self-renewal, therapeutics resistance, and metastasis (Gupta et al. 2022a, b, c).

4.2.1 TAMs and Their Polarization in the Tumor Microenvironment

Macrophages are from monocyte lineage and possess plasticity to adapt to environmental signals of the TME. According to the stimuli from cytokines from M2 and TGF-β, IL-4, and IL-13 of the tumor microenvironment, macrophages in the context of cancer are often polarized towards M2. M2 macrophages, which are defined by the production of anti-inflammatory cytokines, including IL-10 and TGF β, possess modifications of the tumor microenvironment promoting angiogenesis, immune suppression, and enhancing cancer cell migration. Such characteristics point to M2 macrophages, and consequently TAMs, as key regulators of cancer progression, including control of CSCs (Wang et al. 2023).

4.2.2 Influence TAMs in Maintaining and Renewing CSCs

CSCs are discrete groups of cancer cells possessing the capacity of self-renewal and differentiative potential, forming various lineages of cancer cells and with broad resistance to cancer treatment, including chemotherapy and radiotherapy. These cells are believed to be responsible for the initiation of tumor formation, metastasizing, and relapse. In this study, the evidence for the engagement of TAMs with CSCs in fashioning their maintenance and their self-renewal capabilities favorable for tumor advancement is presented (Li et al. 2020a, b).

4.2.3 TAM's Impact on CSC Maintenance and Self Renewal

Cancer stem cells or cancer-initiating cells (CICs) are a subset of cells in a tumor capable of self-renewal and differentiation into other cell types within a tumor, as well as resistance to conventional treatments such as chemotherapy and radiotherapy. These cells are believed to be responsible for primary tumorigenesis, metastases, and relapse. Earlier studies have indicated that they have crosstalk with CSCs in sectors that aid CSCs in maintenance and their quality of self-renewal, thus helping the growth and progression of the tumor (Wang et al. 2023). The following factors facilitate the initiation, progression, and maintenance of CSC, which TAMs secrete to support CSC survival and proliferation. For instance, TAMs release vascular endothelial growth factor (VEGF), which promotes angiogenesis and CSC survival since tumor mass requires prompt blood supply, for example, to support the growth and development of CSCs. Furthermore, chemokines like TGF-β and IL-6 released by TAM promote the functions of CSC stem cell-like properties via a network of signaling pathways that includes STAT3 and Smad 2/3. These pathways modulate the levels of stem cell markers and transcription factors, including OCT4, SOX2, and Nanog, which are crucial for CSCs' functions of self-renewal and pluripotency. TAMs also regulate CSCs through exosomes, small particles outside cell membranes that contain proteins, lipids, and RNA. TAM exosomes can induce CSC pro-tumoral/survival phenotypes by transferring potent 'signatures' that amplify the

CSC stem-like features and resistance to immune/TAM meta-stability and therapies. For example, exosomes from TAM can transfer miRNAs and proteins that control the molecular regulators of CSCs, including transcription factors that boost the CSC's properties of self-renewal and survival (Rascio et al. 2021).

4.2.4 TAMs and CSC Therapy Resistance

Another essential aspect of CSCs is their drug resistance to standard treatment protocols such as chemotherapy, radiotherapy, and immunotherapy. TAMs contribute to this resistance by secreting factors that protect CSCs from damage caused by the treatment. In addition, TAMs can also improve the capabilities of CSCs in adapting to high-stress environments resulting from chemotherapeutic or radiation treatments. Some evidence suggests that TAMs produce large amounts of survival factors, including IL-6 and HGF, which are powerful, and MET signaling pathways in CSCs and possibly contribute to quantitative apoptosis resistance. TAMs also increase the epithelial-to-mesenchymal transition (EMT), in which CSCs gain migratory and invasive properties and resist cell death. EMT is linked with increased levels of the anti-apoptotic proteins of Bcl-2 and survivin, which help CSCs to evade apoptosis induced by therapy. TAMs also work actively to maintain CSCs in dormant conditions where CSCs are less cycling and less sensitive to treatments that target actively growing cells. The lower oxygen levels that prevail in tumor tissues, which can be further accentuated by TAM infiltration, are widely known to promote CSC dormancy by inducing HIF that stabilizes the expression of genes that protect CSCs from oxidative stress or metabolic shifts for the purpose of chemotherapy or radiation (Gupta et al. 2022a, b, c; Li et al. 2020a, b).

4.2.5 TAMs and CSC Metastasis

The spread of tumor cells to different organs of the body without the formation of a new tumor is called metastasis, and it is one of the main reasons for cancer fatalities. It was believed that CSCs are responsible for metastasis because of their inherent ability to move, invade, and proliferate in distant tissues. The interactions of TAMs are especially significant in supporting the metastatic process of CSCs by releasing cytokines, chemokines, and proteases that stimulate the migration of CSCs. Some MMPs, including MMP-9 and MMP-2, increase the tumor cell motility and facilitate the invasive properties of CSCs. TAMs also release chemokines, including CCL2 and CXCL12, that home in on CSCs by binding to CSC chemokine receptors, including CCR2 and CXCR4. The process of recruitment described here enhances the invasion and implantation in distant tissues and organs, which is associated with the metastatic process. Also, TAMs release both VEGF and other factors that contribute to neo-angiogenesis, which enhances the capability of CSCs to spread from the primary tumor and form metastatic sites. Moreover, by promoting the formation of new blood vessels, TAMs also increase the supply of oxygen and nutrients crucial for the survival of CSCs during the metabolic phase of cancer cells (Ghalehbandi et al. 2023).

4.3 Role of Tumor-Associated Macrophages (TAMs)

Tumor-associated macrophages (TAMs) are a crucial component of the tumor microenvironment (TME) and play a significant role in promoting and suppressing tumor progression. TAMs demonstrate impressive plasticity, polarizing into two main subtypes: M1 macrophages, which generally attack tumors, and M2 macrophages, which support tumor growth (Ghalehbandi et al. 2023). In most tumors, TAMs predominantly polarize toward the M2 phenotype, actively fostering an environment that promotes tumor growth, metastasis, and immune evasion. Cancer stem cells (CSCs) actively recruit and reprogram macrophages toward this M2 phenotype, thus enhancing their ability to promote tumor progression (Ghalehbandi et al. 2023).

4.4 TAM Polarization and Its Role in Tumor Progression

4.4.1 Angiogenesis Support

One of the most important contributions of M2-polarized TAMs to tumor progression is their role in promoting angiogenesis—the formation of new blood vessels—which is crucial for tumor growth and metastasis (Huang et al. 2022a, b). TAMs secrete various pro-angiogenic factors, most notably vascular endothelial growth factor (VEGF) (Zhang et al. 2018). VEGF stimulates endothelial cells to form new blood vessels that supply the growing tumor with essential nutrients and oxygen. This vascularization is critical for the expansion of the tumor and its ability to metastasize to distant organs (Zhang et al. 2018). TAM-derived VEGF supports the tumor's growth and enhances the survival of CSCs by ensuring a continuous supply of oxygen and nutrients, making the tumor microenvironment more conducive to CSC self-renewal and resistance to therapy (Zhang et al. 2018).

4.4.2 Survival Pathway Activation

TAMs also contribute to CSC survival by secreting pro-inflammatory cytokines such as IL-6 and TNF-α. These cytokines activate key survival and proliferative pathways in CSCs, including the PI3K/Akt and NF-κB pathways (Shibata and Hoque 2019). IL-6 is a potent cytokine that activates the JAK/STAT3 signaling pathway, which promotes cell survival, immune evasion, and tumor progression. In CSCs, IL-6 signaling through STAT3 is linked to enhanced self-renewal capacity, chemoresistance, and evading apoptosis (Hashem et al. 2022). TNF-α can activate the NF-κB pathway, which regulates inflammation, immune suppression, and cell survival genes. NF-κB activation helps CSCs resist apoptosis and adapt to the hostile tumor microenvironment, often including oxidative stress, low oxygen, and nutrient deprivation (Valkenburg et al. 2018). These cytokines and other factors secreted by TAMs contribute to a robust survival signaling network in CSCs, increasing their resistance to conventional therapies such as chemotherapy and radiation therapy.

4.5 Immune Suppression

M2-polarized TAMs are known for creating an immunosuppressive microenvironment that shields CSCs from immune surveillance. One of the ways they achieve this is by secreting anti-inflammatory cytokines such as IL-10 and TGF-β. IL-10 is a potent immunosuppressive cytokine that dampens the activity of various immune cells, including dendritic cells, T cells, and macrophages. By reducing the ability of these immune cells to mount an effective anti-tumor response, IL-10 helps to protect CSCs from immune-mediated elimination. TGF-β not only supports the conversion of T cells into regulatory T cells (Tregs) and suppresses the cytotoxicity of natural killer (NK) cells and CTLs. This cytokine aids in the formation of a tolerogenic immune environment that allows CSCs to persist and thrive despite the presence of immune cells. In addition to cytokine secretion, TAMs also express molecules such as PD-L1, which directly interact with T cells and inhibit their anti-tumor activity. By promoting immune evasion through cytokine release and direct cell interactions, TAMs significantly contribute to the persistence of CSCs and tumor progression (Liu et al. 2022).

4.6 CSC-TAM Interaction and Tumor Progression

CSCs are key players in the reprogramming of TAMs toward the M2 phenotype. These stem-like cells release signals, such as C-C motif chemokine ligand 2 (CCL2), that recruit monocytes to the tumor site, where they can differentiate into macrophages. Once at the tumor site, CSCs influence the polarization of these macrophages toward the M2 phenotype through the secretion of cytokines, growth factors, and extracellular matrix remodeling proteins. It creates a positive feedback loop where TAMs support the survival and expansion of CSCs. In turn, CSCs continue to recruit and reprogram more macrophages to maintain the tumor-promoting M2 state.

5 Therapeutic Strategies and Challenges

5.1 Current Treatment Approaches

Standard cancer treatments like chemotherapy and radiation effectively reduce tumor bulk but often fail to target CSCs. This limitation leads to treatment resistance, relapse, and metastasis. Immune checkpoint inhibitors, though promising, have limited efficacy against CSCs due to their immune evasion tactics (Vasudev and Reynolds 2014).

5.2 Emerging Approaches

5.2.1 Epigenetic Therapies

Cancer stem cells often use reversible epigenetic changes to downregulate tumor antigens, thus avoiding immune surveillance. The therapeutic drugs targeting these epigenetic modifications are trying to reactivate the expression of tumor antigens and make CSCs more recognizable and sensitive to immune attacks (Nayak et al. 2022). Examples include drugs inhibiting DNA methylation or histone modification agents, which may act as reprogramming agents for CSCs into a more differentiated and less malignant phenotype, hence making them more susceptible to conventional therapies. Such treatments increase the effectiveness of treatment, reduce the chance of recurrence and metastasis, and act against the survival mechanisms of CSCs (Wang et al. 2022a, b).

5.2.2 Combination Therapies

Combining immune checkpoint inhibitors with agents that target CSC-specific pathways provides a synergistic approach to overcoming CSC-mediated treatment resistance (Shiravand et al. 2022). For instance, drugs that inhibit Wnt/β-catenin, Notch, or Hedgehog signaling pathways are significant for CSC survival and self-renewal. It would thus break the survival mechanism of CSCs if these were combined with immune checkpoint inhibitors, thus boosting the immune system's capability to destroy CSCs by neutralizing their evasive mechanisms. This two-pronged approach shrinks the bulk of the tumor and tackles CSCs to avoid recurrence and metastasis (Masoumi et al. 2021).

5.2.3 Bioactive Compounds

Natural compounds like resveratrol (found in grapes) and curcumin (from turmeric) have garnered attention for their ability to target CSC pathways with minimal toxicity. These compounds modulate CSCs through signaling pathways, such as NF-κB, PI3K/AKT, and STAT3. They also modulate the tumor microenvironment through reduced inflammation and enhanced immune activity against CSCs (Masoumi et al. 2021; Shiravand et al. 2022). Their antioxidant and anti-inflammatory properties make them promising compounds for other therapeutic agents, thus providing a better and less invasive treatment modality.

5.2.4 Additional Emerging Approaches

(a) *Immunotherapies Beyond Checkpoints*: While immune checkpoint inhibitors have revolutionized cancer therapy, their effectiveness against CSCs remains limited due to the immune evasion strategies employed by these cell approaches (Asghari et al. 2019). Emerging immunotherapies are focusing on approaches that go beyond checkpoints. Chimeric antigen receptor T-cell therapy is adapted to target antigens expressed on CSCs, such as CD44, CD133, or EpCAM. These engineered T cells can recognize and kill CSCs, potentially overcoming the challenges of relapse and metastasis (Zhao et al. 2021). Therapeutic vaccines are being developed to train the immune system to identify and attack CSC-specific proteins or markers, such as ALDH or specific integrins (Luna et al. 2017). NK cells are being explored in their natural form, and CSCs can be recognized for lysis without any prior antigen exposure. Genetic modification or combination therapies to enhance their function are promising approaches (Asghari et al. 2019).

(b) *Targeted Nanotherapies*: Nanotechnology enables the targeted delivery of drugs and therapeutic agents to CSCs, reducing systemic side effects and enhancing the treatment outcome (Yasuda et al. 2021). Nanotechnology enables the precise delivery of drugs and therapeutic agents to CSCs, minimizing systemic side effects and improving treatment outcomes. Nanoparticles are designed to carry chemotherapy, RNA-based therapeutics, or small-molecule inhibitors directly to CSCs. These nanoparticles can be functionalized with ligands or antibodies that bind specifically to CSC markers, ensuring targeted delivery (Vasudev and Reynolds 2014). Some nanoparticles, such as gold or carbon-based nanostructures, are engineered to generate heat or reactive oxygen species when exposed to specific wavelengths of light. This can selectively kill CSCs without harming surrounding healthy tissues (Valkenburg et al. 2018).

(c) *Metabolic Reprogramming*: CSCs have distinct metabolic preferences that differ from bulk tumor cells. Unlike many cancer cells that rely on glycolysis (Warburg effect), CSCs often prefer oxidative phosphorylation (OXPHOS) (Jones et al. 2021; Nayak et al. 2022). OXPHOS inhibitors targeting the mitochondrial function of CSCs disrupt their energy production and self-renewal capacity. Drugs like metformin or phenformin, which inhibit mitochondrial complex I, are being repurposed to target CSC metabolism (Wang et al. 2022a, b). CSCs often rely on lipid metabolism for energy and survival. Targeting fatty acid oxidation pathways or blocking lipid uptake can starve CSCs and impair their ability to thrive. CSCs also exhibit unique dependencies on glutamine for biosynthesis and energy. Glutaminase inhibitors are being investigated as a means to disrupt CSC metabolism selectively.

(d) *Microenvironment Modulation*: The tumor microenvironment (TME) plays a critical role in supporting CSCs. Strategies targeting the TME aim to disrupt this supportive niche (Wang et al. 2022a, b). Drugs inhibiting fibroblasts or extracellular matrix components can make the TME less conducive to survival for CSC (Wang et al. 2022a, b). CSCs often reside in hypoxic niches supported by abnormal blood vessels. Anti-angiogenic agents, such as bevacizumab, disrupt the blood supply to CSC niches, impairing their function and survival (Wu et al. 2023). Modulating the immune environment within the tumor through cytokines, chemokines, or T regulatory cell depletion can enhance immune responses against CSCs (Wang et al. 2022a, b).

(e) *Gene Editing Technologies*: CRISPR-Cas9 and other gene editing tools are employed to target genes critical for CSC survival and self-renewal selectively (Huang et al. 2022a, b). Gene editing can silence oncogenes or activate tumor suppressor genes specifically within CSCs. CRISPR systems can be precisely delivered to CSCs using nanoparticles or viral vectors, ensuring minimal off-target effects (Xiao et al. 2022). Gene editing can potentially reprogram CSCs into less malignant states, reducing their ability to drive tumor progression and metastasis (Charbe et al. 2020).

(f) *Therapeutic RNA Molecules*: RNA-based therapies, such as small interfering RNA (siRNA) or microRNA (miRNA), are being developed to target CSC-specific pathways (Fu et al. 2021). siRNA Therapies can silence key genes involved in CSC maintenance, such as Wnt or Notch pathway components. miRNA acting as tumor suppressors can be reintroduced to counteract CSC survival mechanisms. For example, miR-34a has been shown to inhibit CSC self-renewal and promote differentiation (Li et al. 2021a, b).

Despite the tremendous gains of modern medicine in terms of better diagnostic methods, the efficacy of chemotherapy protocols, and precise adjuvant and multimodal treatments, it is surprising that there exists constant tumor growth and recurrence in patients. Cancer stem cells are a notable factor behind tumor persistence and recurrence because they are CSCs endowed with self-renewal, proliferation, and differentiation into any lineage. These traits make CSCs particularly resistant to standard treatments, implying an urgent need for new strategies to target such a resilient subpopulation. To date, methods to isolate and enrich CSCs are primarily based on identifying specific surface markers that may be separable from FACS and MACS. However, searching for universal markers has been a significant challenge in CSC research. This chapter will be a general overview of the obstacles and limitations encountered in CSC studies, from the absence of universal markers to resistance to therapy to the problems in finding effective drugs targeting CSCs. Promising tools have emerged from cutting-edge technologies to address these challenges. Single-cell-level, comprehensive analyses of the genome, epigenome, and transcriptome provide an unprecedented understanding of the heterogeneity and plasticity of CSCs. Genome-wide CRISPR-Cas9 screens also promise to identify novel pathway-based targets to create precision therapies. These methods underscore the need for further refinement of clinical trial designs to effectively validate therapies and ensure their successful translation into clinical practice.

6 Future Research Directions

It is now important to resolve the complex molecular mechanisms governing CSC biology to advance CSC research. High-throughput technologies such as NGS will allow researchers to identify effective bioactive compounds targeting CSCs and specifics. Improving our knowledge of CSC diversity and adaptability requires a multiomics approach based on genomics, transcriptomics, and proteomics to develop personalized therapeutic strategies. The future also holds promise with drug delivery systems developed based on nanotechnology. Nanoparticle carriers like exosomes and liposomes have better solubility, stability, and cellular uptake, reducing secondary effects and toxicity. These systems could overcome the challenges of the tumor microenvironment in order to improve the bioavailability and targeted delivery of therapeutic agents.

Immunotherapy has also emerged as a potential approach to targeting CSCs. The next step for further research is using immune checkpoint inhibitors, CAR-T cell therapies, and cancer vaccines to fight CSC mechanisms that lead to immune evasion. Understanding how CSCs interact with the immune system is necessary to design combination therapies that tap the body's immune response to eliminate CSC populations effectively. Despite these achievements, the hurdle of translating findings from preclinical studies into clinically successful applications continues to exist. Future efforts must include well-designed clinical trials that assess the safety and efficacy of CSC-targeted therapies. Effective collaboration among scientists, clinicians, pharmaceutical companies, and regulatory agencies will be key to making these processes more streamlined.

7 Conclusions

Targeting cancer stem cells represents one of the promising frontiers in the oncology area, which would involve the exploration of bioactive compounds. CSCs have been pointed out as having an important role in the initiation, progression, metastasis, and therapy resistance of tumors. They present challenging issues for traditional treatment modalities. Bioactive compounds derived from natural sources act on different pathways of cell death and interfere with the self-renewal pathways or modification of the tumor microenvironment. Targeting CSC pathways, such as Wnt/β-catenin, Hedgehog, and Notch signaling, holds promise for disrupting the stemness and survival of CSCs. Interactions with the tumor microenvironment, including immune evasion strategies and the influence of tumor-associated macrophages, highlight the complexity of CSC biology and the need for combinatory approaches. Emerging therapeutic strategies, such as nanoparticle delivery systems and immunotherapy, offer improved precision and efficacy, but challenges associated with drug resistance, specificity, and systemic toxicity remain. In conclusion, bioactive compounds have tremendous potential to overcome the limitations of CSCs for more effective cancer treatments. The sustained efforts in filling this gap between basic science and clinical applications can truly propel innovative therapies toward improving patient outcomes and survival rates.

References

Aggarwal BB, Harikumar KB (2009) Potential therapeutic effects of curcumin, the anti-inflammatory agent, on human diseases. Indian J Exp Biol 47(1):10–18

Agnhage T, Zhou Y, Guan J, Chen G, Perwuelz A, Behary N, Nierstrasz V (2017) Bioactive and multifunctional textile using plant-based madder dye: characterization of UV protection ability and antibacterial activity. Fibers Polym 18(11):2170–2175. https://doi.org/10.1007/s12221-017-7294-z

Ahmed MM, Pandey SK (2012) The role of omega-3 fatty acids in cancer: a review. Crit Rev Food Sci Nutr 52(9):741–746. https://doi.org/10.1080/10408398.2011.621575

Aiello P, Sharghi M, Mansourkhani SM, Ardekan AP, Jouybari L, Daraei N, Peiro K, Mohamadian S, Rezaei M, Heidari M, Peluso I (2019) Medicinal plants in the prevention and treatment of colon cancer. Oxidative Med Cell Longev 2019:2075614. https://doi.org/10.1155/2019/2075614

Asghari F, Khademi R, Esmaeili Ranjbar F, Veisi Malekshahi Z, Faridi Majidi R (2019) Application of nanotechnology in targeting of cancer stem cells: a review. Int J Stem Cells 12:227–239. https://doi.org/10.15283/ijsc18102

Azmir J, Sarker MZI, Rahman MM, Khan MS, Awang MB, Ferdosh S, Jahurul MHA, Ghafoor K, Norulaini NAN, Omar AKM (2013) Techniques for extraction of bioactive compounds from plant materials: a review. J Food Eng 117(4):426–436. https://doi.org/10.1016/j.jfoodeng.2013.01.014

Barabadi H, Ovais M, Shinwari ZK, Saravanan M (2017) Anti-cancer green bionanomaterials: present status and future prospects. Green Chem Lett Rev 10(4):285–314. https://doi.org/10.1080/17518253.2017.1398135

Baur JA, Sinclair DA (2006) Therapeutic potential of resveratrol: the in vivo evidence. Nat Rev Drug Discov 5(6):493–506. https://doi.org/10.1038/nrd2060

Brenner D, Hauner H (2013) Omega-3 fatty acids and cancer: an overview of their role in cancer prevention and treatment. J Cancer Res Clin Oncol 139(3):353–359. https://doi.org/10.1007/s00432-012-1329-3

Burrell RA, McGranahan N, Bartek J, Swanton C (2013) The causes and consequences of genetic heterogeneity in cancer evolution. Nature 501(7467):338–345. https://doi.org/10.1038/nature12625

Calder PC (2013) Omega-3 polyunsaturated fatty acids and inflammatory processes: nutrition or pharmacology? Br J Clin Pharmacol 75(3):645–662. https://doi.org/10.1111/j.1365-2125.2012.04374.x

Câmara JS, Albuquerque BR, Aguiar J, Corrêa R, Gonçalves JL, Granato D, Pereira J, Barros L, Ferreira I (2020) Food bioactive compounds and emerging techniques for their extraction: polyphenols as a case study. Foods 10(1):37. https://doi.org/10.3390/foods10010037

Carlson KN, Pavan-Guimaraes J, Verhagen JC, Chlebeck P, Verhoven B, Jennings H, Al-Adra DP (2021) Interleukin-10 and transforming growth factor-β cytokines decrease immune activation during normothermic ex vivo machine perfusion of the rat liver. Liver Transplant 27(11):1577–1591. https://doi.org/10.1002/lt.26199

Catalano A, Barone R (2009) Curcumin and cancer. Biochim Biophys Acta 1792(5):352–357. https://doi.org/10.1016/j.bbadis.2009.01.007

Cha JH, Chan LC, Li CW, Hsu JL, Hung MC (2019) Mechanisms controlling PD-L1 expression in cancer. Mol Cell 76(3):359–370. https://doi.org/10.1016/j.molcel.2019.09.030

Charbe NB, Amnerkar ND, Ramesh B, Tambuwala MM, Bakshi HA, Aljabali AA et al (2020) Small interfering RNA for cancer treatment: overcoming hurdles in delivery. Acta Pharm Sin B 10:2075–2109. https://doi.org/10.1016/j.apsb.2020.07.006

Chen YH, Li L (2013) Biological activities of polysaccharides from medicinal mushrooms. Biol Act Polysaccharides 11(3):219–227

Chhikara N, Kushwaha K, Sharma P, Gat Y, Panghal A (2019) Bioactive compounds of beetroot and utilization in food processing industry: a critical review. Food Chem 272:192–200. https://doi.org/10.1016/j.foodchem.2018.08.478

Choi JS, Lee D (2012) Pharmacological activities of dietary polyphenols in inflammation and cancer. J Clin Pharmacol Ther 15(2):60–70

Chu X, Tian W, Ning J, Xiao G, Zhou Y, Wang Z, Zhai Z, Tanzhu G, Yang J, Zhou R (2024) Cancer stem cells: advances in knowledge and implications for cancer therapy. Signal Transduct Target Ther 9(1):170. https://doi.org/10.1038/s41392-024-01851-y

Cochrane CR, Szczepny A, Watkins DN, Cain JE (2015) Hedgehog signaling in the maintenance of cancer stem cells. Cancer 7(3):1554–1585. https://doi.org/10.3390/cancers7030860

Comis RL, Lippman SM (2006) Chemoprevention: progress and promise. J Clin Oncol 24(26):4173–4182. https://doi.org/10.1200/JCO.2006.06.7808

Cosme P, Rodríguez AB, Espino J, Garrido M (2020) Plant phenolics: bioavailability as a key determinant of their potential health-promoting applications. Antioxidants 9(12):1263. https://doi.org/10.3390/antiox9121263

Edgar AD, Levin R, Constantinou CE, Denis L (2007) A critical review of the pharmacology of the plant extract of *Pygeum africanum* in the treatment of LUTS. Neurourol Urodyn 26(4):458–463. https://doi.org/10.1002/nau.20376

Fortunato O, Belisario DC, Compagno M et al (2020) CXCR4 inhibition counteracts immunosuppressive properties of metastatic NSCLC stem cells. Front Immunol 11:2168. https://doi.org/10.3389/fimmu.2020.02168

Fu Z, Wang L, Li S, Chen F, Au-Yeung KKW, Shi C (2021) MicroRNA as an important target for anticancer drug development. Front Pharmacol 12:736323. https://doi.org/10.3389/fphar.2021.736323

Futakuchi M, Lami K, Tachibana Y, Yamamoto Y, Furukawa M, Fukuoka J (2019) The effects of TGF-β signaling on cancer cells and cancer stem cells in the bone microenvironment. Int J Mol Sci 20(20):5117. https://doi.org/10.3390/ijms20205117

Galassi C, Chan TA, Vitale I, Galluzzi L (2024a) The hallmarks of cancer immune evasion. Cancer Cell 42(11):1825–1863. https://doi.org/10.1016/j.ccell.2024.09.010

Galassi C, Esteller M, Vitale I, Galluzzi L (2024b) Epigenetic control of immunoevasion in cancer stem cells. Trends Cancer. https://doi.org/10.1016/j.trecan.2024.02.004

García-Vaquero M, Rajauria G (2018) Analytical techniques for phytochemical estimation in fruit juices. In: Fruit juices: extraction, composition, quality and analysis. Elsevier, pp 669–692. https://doi.org/10.1016/B978-0-12-802230-6.00037-6

Ghalehbandi S, Yuzugulen J, Pranjol MZI, Pourgholami MH (2023) The role of VEGF in cancer-induced angiogenesis and research progress of drugs targeting VEGF. Eur J Pharmacol 949:175586. https://doi.org/10.1016/j.ejphar.2023.175586

Giovannucci E (2002) Epidemiologic studies of tomatoes and lycopene in cancer prevention. J Natl Cancer Inst 94(7):391–398. https://doi.org/10.1093/jnci/94.7.391

Gu JJ, Qiao KS, Sun P, Chen P, Li Q (2018) Study of EGCG induced apoptosis in lung cancer cells by inhibiting PI3K/Akt signaling pathway. Eur Rev Med Pharmacol Sci 22(14):4557–4563. https://doi.org/10.26355/eurrev_201807_15511

Guo Q, Jin Y, Chen X, Ye X, Shen X, Lin M, Zhang J (2024) NF-κB in biology and targeted therapy: new insights and translational implications. Signal Transduct Target Ther 9(1):53. https://doi.org/10.1038/s41392-024-01793-7

Gupta SC, Sung B (2013) Cancer chemoprevention with curcumin: the emerging molecular targets. Antioxid Redox Signal 19(3):413–442. https://doi.org/10.1089/ars.2012.4774

Gupta J, Ahuja A, Gupta R (2022a) Green approaches for cancers management: an effective tool for health care. Anti-Cancer Agents Med Chem 22(1):101–114. https://doi.org/10.2174/1871520621666210705104650

Gupta S, Soni A, Roy S et al (2022b) Sulforaphane targets NF-κB signaling to inhibit stemness and metastasis in breast cancer. Cancer Lett 526:1–10. https://doi.org/10.1016/j.canlet.2021.11.031

Gupta G, Merhej G, Saravanan S, Chen H (2022c) Cancer resistance to immunotherapy: what is the role of cancer stem cells? Cancer Drug Resist 5(4):981. https://doi.org/10.20517/cdr.2022.61

Hadad N, Levy R (2012) The cytotoxic effect of quercetin on leukemia cell lines is mediated through inhibition of glucose uptake. Free Radic Biol Med 53(7):1381–1391. https://doi.org/10.1016/j.freeradbiomed.2012.07.078

Hashem S, Ali TA, Akhtar S, Nisar S, Sageena G, Ali S et al (2022) Targeting cancer signaling pathways by natural products: exploring promising anti-cancer agents. Biomed Pharmacother 150:113054. https://doi.org/10.1016/j.biopha.2022.113054

Hassan ZK, Daghestani MH (2012) Curcumin effect on MMPs and TIMPs genes in a breast cancer cell line. Asian Pac J Cancer Prev 13(7):3259–3264. https://doi.org/10.7314/apjcp.2012.13.7.3259

Holohan C, Van Schaeybroeck S, Longley DB, Johnston PG (2013) Cancer drug resistance: an evolving paradigm. Nat Rev Cancer 13(10):714–726. https://doi.org/10.1038/nrc3599

Hou YC, Chao YJ, Hsieh MH et al (2019) Low CD8+ T cell infiltration and high PD-L1 expression are associated with level of CD44+/CD133+ cancer stem cells and predict an unfavorable prognosis in pancreatic cancer. Cancer 11(4):541. https://doi.org/10.3390/cancers11040541

Houghton CA, Fassett RG, Coombes JS (2016) Sulforaphane and other nutrigenomic Nrf2 activators: can the clinician's expectation be matched by the reality? Oxidative Med Cell Longev 2016:7857186. https://doi.org/10.1155/2016/7857186

Hu X, Li J, Fu M, Zhao X, Wang W (2021) The JAK/STAT signaling pathway: from bench to clinic. Signal Transduct Target Ther 6(1):402. https://doi.org/10.1038/s41392-021-00731-3

Huang B, Lang X, Li X (2022a) The role of IL-6/JAK2/STAT3 signaling pathway in cancers. Front Oncol 12:1023177. https://doi.org/10.3389/fonc.2022.1023177

Huang J, Zhou Y, Li J, Lu A, Liang C (2022b) CRISPR/Cas systems: delivery and application in gene therapy. Front Bioeng Biotechnol 10:942325. https://doi.org/10.3389/fbioe.2022.942325

Johnson IT, Encarnación S (2016) Targeting dietary polyphenols in cancer prevention: current status and future perspectives. Food Chem Toxicol 94:46–53. https://doi.org/10.1016/j.fct.2016.05.004

Jones CL, Inguva A, Jordan CT (2021) Targeting energy metabolism in cancer stem cells: progress and challenges in leukemia and solid tumors. Cell Stem Cell 28:378–393. https://doi.org/10.1016/j.stem.2021.01.013

Kaltschmidt C, Banz-Jansen C, Benhidjeb T et al (2019) A role for NF-κB in organ-specific cancer and cancer stem cells. Cancer 11(5):655. https://doi.org/10.3390/cancers11050655

Kantarjian HM, Thomas DA (2012) Targeted therapy in leukemia: new insights and perspectives. J Clin Oncol 30(24):3251–3258. https://doi.org/10.1200/JCO.2011.40.6076

Karami Fath M, Ebrahimi M, Nourbakhsh E, Zia Hazara A, Mirzaei A, Shafieyari S, Salehi A, Hoseinzadeh M, Payandeh Z, Barati G (2022) PI3K/Akt/mTOR signaling pathway in cancer stem cells. Pathol Res Pract 237:154010. https://doi.org/10.1016/j.prp.2022.154010

Katoh M (2017) Canonical and non-canonical WNT signaling in cancer stem cells and their niches: cellular heterogeneity, omics reprogramming, targeted therapy and tumor plasticity. Int J Oncol 51(5):1357–1369

Kaushik N, Pathak R (2018) Role of phytochemicals in apoptosis and cancer therapy. J Clin Exp Pathol 8(1):1–6. https://doi.org/10.4172/2161-0681.1000358

Knudson KM, Hicks KC, Luo X et al (2018) M7824, a novel bifunctional anti-PD-L1/TGFβ trap fusion protein, promotes anti-tumor efficacy as monotherapy and in combination with vaccine. Onco Targets Ther 7(5):e1426519. https://doi.org/10.1080/2162402X.2017.1426519

Krishna BM, Jana S, Singhal J, Horne D, Awasthi S, Salgia R, Singhal SS (2019) Notch signaling in breast cancer: from pathway analysis to therapy. Cancer Lett 461:123–131. https://doi.org/10.1016/j.canlet.2019.07.016

Lage H (2008) An overview of cancer multidrug resistance: A still unsolved problem. Cell Mol Life Sci 65:3145–3167. https://doi.org/10.1007/s00018-008-8111-5

Lee CH, Kim DS (2019) Inhibitory effects of ginsenosides on cancer metastasis. Anticancer Res 39(1):19–26. https://doi.org/10.21873/anticanres.13007

Lei MML, Lee TKW (2021) Cancer stem cells: emerging key players in immune evasion of cancers. Front Cell Dev Biol 9:692940. https://doi.org/10.3389/fcell.2021.692940

Li X, Wu M (2016) Natural product-based drug design and therapeutic strategies for cancer. Front Pharmacol 7:232. https://doi.org/10.3389/fphar.2016.00232

Li W, Hua B, Saud SM, Lin H, Hou W, Matter MS, Jia L, Colburn NH, Young MR (2015) Mechanisms by which dietary phytochemicals suppress inflammation and cancer. Mol Carcinog 54(10):1096–1109. https://doi.org/10.1002/mc.22179

Li J, Niu D, Zhang Y, Zeng XA (2020a) Physicochemical properties, antioxidant and antiproliferative activities of polysaccharides from *Morinda citrifolia* L. (Noni) based on different extraction methods. Int J Biol Macromol 150:114–121. https://doi.org/10.1016/j.ijbiomac.2020.01.261

Li L, Yu R, Cai T et al (2020b) Effects of immune cells and cytokines on inflammation and immunosuppression in the tumor microenvironment. Int Immunopharmacol 88:106939. https://doi.org/10.1016/j.intimp.2020.106939

Li X, Liao Z, Yang L et al (2021a) EGCG inhibits TGF-β-induced epithelial-mesenchymal transition and enhances chemosensitivity in breast cancer stem cells. Front Pharmacol 12:617407. https://doi.org/10.3389/fphar.2021.617407

Li WJ, Wang Y, Liu R, Kasinski AL, Shen H, Slack FJ, Tang DG (2021b) MicroRNA-34a: potent tumor suppressor, cancer stem cell inhibitor, and potential anticancer therapeutic. Front Cell Dev Biol 9:640587. https://doi.org/10.3389/fcell.2021.640587

Li Z, Wang J, Li Y et al (2022) Quercetin suppresses NF-κB-mediated inflammation and enhances chemosensitivity in pancreatic cancer stem cells. Front Pharmacol 13:896939. https://doi.org/10.3389/fphar.2022.896939

Li X, Zhou J, Wang X, Li C, Ma Z, Wan Q, Peng F (2023) New advances in the research of clinical treatment and novel anticancer agents in tumor angiogenesis. Biomed Pharmacother 163:114806. https://doi.org/10.1016/j.biopha.2023.114806

Liang N, Sang Y, Liu W, Yu W, Wang X (2018) Anti-inflammatory activities of natural compounds. Inflammation 41(3):835–845. https://doi.org/10.1007/s10753-018-0737-3

Linnewiel-Hermoni K, Khanin M, Danilenko M, Zango G, Amosi Y, Levy J, Sharoni Y (2015) The anti-cancer effects of carotenoids and other phytonutrients resides in their combined activity. Arch Biochem Biophys 572:28–35. https://doi.org/10.1016/j.abb.2015.02.018

Liu Y, Zhang W, Zhang W et al (2021) Gingerol inhibits the JAK2/STAT3 signaling pathway to suppress stemness and drug resistance in glioblastoma stem cells. Phytomedicine 83:153482. https://doi.org/10.1016/j.phymed.2020.153482

Liu J, Xiao Q, Xiao J, Niu C, Li Y, Zhang X et al (2022) Wnt/β-catenin signalling: function, biological mechanisms, and therapeutic opportunities. Signal Transduct Target Ther 7:3. https://doi.org/10.1038/s41392-021-00831-5

Luna JI, Grossenbacher SK, Murphy WJ, Canter RJ (2017) Targeting cancer stem cells with natural killer cell immunotherapy. Expert Opin Biol Ther 17:313–324. https://doi.org/10.1080/14712598.2017.1280461

Luo X, Zhan W, Zhang Q et al (2021) Quercetin inhibits notch signaling and enhances chemosensitivity in non-small cell lung cancer cells. Oncol Lett 21(6):403. https://doi.org/10.3892/ol.2021.12673

Mahato N, Sinha M, Sharma K, Koteswararao R, Cho MH (2019) Modern extraction and purification techniques for obtaining high purity food-grade bioactive compounds and value-added co-products from citrus wastes. Foods 8(11):523. https://doi.org/10.3390/foods8110523

Mai Y, Su J, Yang C, Xia C, Fu L (2023) The strategies to cure cancer patients by eradicating cancer stem-like cells. Mol Cancer 22:171. https://doi.org/10.1186/s12943-023-01885-5

Makkar R, Behl T, Kumar A, Nijhawan P, Arora S (2021) Emerging therapeutic effects of herbal plants in rheumatoid arthritis. Endocr Metab Immune Disord Drug Targets 21(4):617–625. https://doi.org/10.2174/1871530320666200724120136

Masoumi J, Jafarzadeh A, Abdolalizadeh J, Khan H, Philippe J, Mirzaei H, Mirzaei HR (2021) Cancer stem cell-targeted chimeric antigen receptor (CAR)-T cell therapy: challenges and prospects. Acta Pharm Sin B 11:1721–1739. https://doi.org/10.1016/j.apsb.2021.03.015

Matsuno Y, Atsumi Y, Alauddin M, Rana MM, Fujimori H, Hyodo M, Shimizu A, Ikuta T, Tani H, Torigoe H, Nakatsu Y, Tsuzuki T, Komai M, Shirakawa H, Yoshioka KI (2020) Resveratrol and its related polyphenols contribute to the maintenance of genome stability. Sci Rep 10(1):5388. https://doi.org/10.1038/s41598-020-62292-5

Meacham CE, Morrison SJ (2013) Tumour heterogeneity and cancer cell plasticity. Nature 501(7467):328–337. https://doi.org/10.1038/nature12624

Mendez D, de Guillebon A (2015) Anti-cancer potential of quercetin: mechanisms of action and its role in the modulation of molecular pathways. Curr Cancer Drug Targets 15(2):102–110. https://doi.org/10.2174/1568009615666150126153603

Mishra V, Tripathi V, Yadav P, Singh MP (2023) Therapeutic applications of resveratrol-loaded nanoparticles in cancer. Int J Biol Macromol 234:123617. https://doi.org/10.1016/j.ijbiomac.2023.123617

Mokra D, Joskova M, Mokry J (2023) Therapeutic effects of green tea polyphenol (–)-epigallocatechin-3-gallate (EGCG) in relation to molecular pathways controlling inflammation, oxidative stress, and apoptosis. Int J Mol Sci 24(1):340. https://doi.org/10.3390/ijms24010340

Moshiri M, Hamid F, Etemad L (2016) Ricin toxicity: clinical and molecular aspects. Rep Biochem Mol Biol 4:60

Muala W, Desobgo Z, Jong NE (2021) Optimization of extraction conditions of phenolic compounds from *Cymbopogon citratus* and evaluation of phenolics and aroma profiles of extract. Heliyon 7(4):e06788. https://doi.org/10.1016/j.heliyon.2021.e06788

Murillo G, Stacewicz-Sapuntzakis M (2015) Role of lycopene and tomato products in prostate cancer prevention. Anti Cancer Agents Med Chem 15(5):428–438. https://doi.org/10.2174/1871520615666150302201117

Mutthuraj D, Vinutha T, Gopenath T, Kaginelli B, Karthikeyan M, Ashok G et al (2020) Inhibition of pro-inflammatory molecules by ginger (*Zingiber officinale* Roscoe) and its anti-inflammatory effects on arthritis patients. J Drug Deliv Ther 10:125–139. https://doi.org/10.22270/jddt.v10i2-s.3963

Nayak A, Warrier NM, Kumar P (2022) Cancer stem cells and the tumor microenvironment: targeting the critical crosstalk through nanocarrier systems. Stem Cell Rev Rep 18:2209–2233. https://doi.org/10.1007/s12015-022-10357-9

Pai S, Hebbar A, Selvaraj S (2022) A critical look at challenges and future scopes of bioactive compounds and their incorporations in the food, energy, and pharmaceutical sector. Environ Sci Pollut Res 29:35518–35541. https://doi.org/10.1007/s11356-021-17938-8

Pal C, Jadeja GC (2020) Microwave-assisted extraction for recovery of polyphenolic antioxidants from ripe mango (Mangifera indica L.) peel using lactic acid/sodium acetate deep eutectic mixtures. Food Sci Technol Int 26(1):78–92. https://doi.org/10.1177/1082013219864545

Pan MH, Lai CS (2015) Epigallocatechin-3-gallate and cancer chemoprevention: molecular mechanisms and therapeutic potential. Nat Prod Commun 10(7):1303–1312. https://doi.org/10.1177/1934578X1501000804

Pandey MK, Kumar R, Sethi G (2020) Curcumin inhibits TGF-β signaling and reduces stemness properties of cancer cells. Mol Cancer 19(1):115. https://doi.org/10.1186/s12943-020-01226-w

Pathak A, Kumar A, Kumar R et al (2019) Curcumin suppresses Notch signaling and induces differentiation in human colorectal cancer stem cells. Cell Oncol 42(4):447–457. https://doi.org/10.1007/s13402-019-00433-y

Pendergast J, Godwin J (2012) Chemotherapy resistance in breast cancer: mechanisms and novel strategies. Cancer Chemother Pharmacol 69(3):421–432. https://doi.org/10.1007/s00280-011-1744-9

Priyadarshi S, Balaraman M, Naidu MM (2020) Ionic liquid-based microwave-assisted extraction of Heneicos-1-ene from coriander foliage and optimizing yield parameters by response surface methodology. Prep Biochem Biotechnol 50(3):246–251. https://doi.org/10.1080/10826068.2019.1641531

Rahman MA, Saha SK, Rahman MS, Uddin MJ, Uddin MS, Pang MG, Rhim H, Cho SG (2020a) Molecular insights into therapeutic potential of autophagy modulation by natural products for cancer stem cells. Front Cell Dev Biol 8:283. https://doi.org/10.3389/fcell.2020.00283

Rahman MA, Rahman MR, Zaman T, Uddin MS, Islam R, Abdel-Daim MM, Rhim H (2020b) Emerging potential of naturally occurring autophagy modulators against neurodegeneration. Curr Pharm Des 26(7):772–779. https://doi.org/10.2174/1381612825666191112143755

Rao AV, Rao LG (2007) Carotenoids and human health. Pharmacol Res 55(3):207–216. https://doi.org/10.1016/j.phrs.2007.01.012

Rascio F, Spadaccino F, Rocchetti MT, Castellano G, Stallone G, Netti GS, Ranieri E (2021) The pathogenic role of PI3K/AKT pathway in cancer onset and drug resistance: an updated review. Cancer 13(16):3949. https://doi.org/10.3390/cancers13163949

Raynie DE (2019) Looking at the past to understand the future: Soxhlet extraction. LC GC N Am 37(8):510–513. https://www.chromatographyonline.com/view/looking-past-understand-future-soxhlet-extraction

Reya T, Morrison SJ, Clarke MF, Weissman IL (2001) Stem cells, cancer, and cancer stem cells. Nature 414(6859):105–111. https://doi.org/10.1038/35102167
Ricke SC, Kelley R (2008) Natural plant extracts in antimicrobial resistance: advances and prospects. Nat Prod Rep 25(3):402–412. https://doi.org/10.1039/B710114K
Rizzo S, Rovati GE, Chieco P et al (2019) Sulforaphane regulates the Hedgehog signaling pathway in lung cancer stem-like cells. Cancers (Basel) 11(4):548. https://doi.org/10.3390/cancers11040548
Rowinsky EK (1995) The development and clinical utility of paclitaxel (Taxol). N Engl J Med 332(15):1004–1014. https://doi.org/10.1056/NEJM199504133321507
Saha S, Sharma A (2016) Flavonoids as cancer chemopreventive agents: a review. Anticancer Res 36(12):6085–6092
Sajana M, Gopenath TS, Kanthesh BM (2024) A comprehensive review of phytochemical approaches in treatment of acute myeloid leukemia: associated pathways and molecular mechanisms, vol 17. Chin Herbal Med, p 41
Satapathy SR, Jena M, Prusty D et al (2021) Curcumin downregulates NF-κB pathway and sensitizes cancer stem cells to chemotherapy in colorectal cancer. Mol Carcinog 60(7):413–425. https://doi.org/10.1002/mc.23287
Shafiee M, Salehi B (2019) Green tea catechins as anticancer agents: a mechanistic review. Molecules 24(1):48. https://doi.org/10.3390/molecules24010048
Sharma RA, Gescher AJ (2005) Curcumin: the story so far. Eur J Cancer 41(13):1955–1965. https://doi.org/10.1016/j.ejca.2005.05.009
Shibata M, Hoque MO (2019) Targeting cancer stem cells: a strategy for effective eradication of cancer. Cancer 11(5):732. https://doi.org/10.3390/cancers11050732
Shirahigue L, Antonini S (2020) Agro-industrial wastes as sources of bioactive compounds for food and fermentation industries. Cienc Rural 50(4):e20190721. https://doi.org/10.1590/0103-8478cr20190721
Shiravand Y, Khodadadi F, Kashani SMA, Hosseini-Fard SR, Hosseini S, Sadeghirad H et al (2022) Immune checkpoint inhibitors in cancer therapy. Curr Oncol 29:3044–3060. https://doi.org/10.3390/curroncol29050245
Simopoulos AP (2002) Omega-3 fatty acids in health and disease and in growth and development. Am J Clin Nutr 71(1):1–9. https://doi.org/10.1093/ajcn/71.1.1
Smith AG, Macleod KF (2019) Autophagy, cancer stem cells and drug resistance. J Pathol 247(5):708–718. https://doi.org/10.1002/path.5222
Song MK, Park BB, Uhm J (2019) Understanding immune evasion and therapeutic targeting associated with PD-1/PD-L1 pathway in diffuse large B-cell lymphoma. Int J Mol Sci 20(6):1326
Srinivasan M, Muthukumar K (2012) Immunomodulatory role of dietary compounds in inflammation and cancer. Immunopharmacology 16(3):211–219
Tang SM, Deng XT, Zhou J, Li QP, Ge XX, Miao L (2020) Pharmacological basis and new insights of quercetin action in respect to its anti-cancer effects. Biomed Pharmacother 121:109604. https://doi.org/10.1016/j.biopha.2019.109604
Tao J, Li Y, Li S, Li HB (2018) Plant foods for the prevention and management of colon cancer. J Funct Foods 42:95–110. https://doi.org/10.1016/j.jff.2018.01.018
Terao J (2011) Dietary flavonoids as antioxidants. Mol Asp Med 32(1):39–45. https://doi.org/10.1016/j.mam.2010.09.008
Tie Y, Tang F, Wei YQ, Wei XW (2022) Immunosuppressive cells in cancer: mechanisms and potential therapeutic targets. J Hematol Oncol 15(1):61. https://doi.org/10.1186/s13045-022-01236-5
Toh TB, Lim JJ, Chow EKH (2017) Epigenetics in cancer stem cells. Mol Cancer 16:61. https://doi.org/10.1186/s12943-017-0631-2
Vaidya P, Walker D (1993) Anthracyclines and their role in cancer therapy. Pharmacol Rev 45(1):39–62
Valkenburg KC, Graveel CR, Zylstra-Diegel CR, Zhong Z, Williams BO (2011) Wnt/β-catenin signaling in normal and cancer stem cells. Cancer 3(2):2050–2079. https://doi.org/10.3390/cancers3022050

Valkenburg KC, de Groot AE, Pienta KJ (2018) Targeting the tumour stroma to improve cancer therapy. Nat Rev Clin Oncol 15:366–381. https://doi.org/10.1038/s41571-018-0007-1

Vassallo M, Zamboni P (2005) Resveratrol in cancer prevention and therapy: an update. Int J Cancer 116(1):1–10. https://doi.org/10.1002/ijc.20981

Vasudev NS, Reynolds AR (2014) Anti-angiogenic therapy for cancer: current progress, unresolved questions and future directions. Angiogenesis 17:471–494. https://doi.org/10.1007/s10456-014-9420-y

Venkatesh V, Nataraj R, Thangaraj GS, Karthikeyan M, Gnanasekaran A, Kaginelli SB, Kuppanna G, Kallappa CG, Basalingappa KM (2018) Targeting Notch signalling pathway of cancer stem cells. Stem Cell Investig 5:5. https://doi.org/10.21037/sci.2018.04.01

Vetvicka V, Vannucci L (2015) β-glucan in cancer immunotherapy. Cancer Immunol Immunother 64(5):467–478. https://doi.org/10.1007/s00262-015-1664-2

Wang Z, Cao Y (2011) Resveratrol and its potential to inhibit cancer: a systematic review. Anti Cancer Agents Med Chem 11(9):1230–1237. https://doi.org/10.2174/187152011796150351

Wang JY, Zhang H (2011) Ginger and its anticancer effects. J Cancer Res Clin Oncol 137(3):445–453. https://doi.org/10.1007/s00432-010-0907-5

Wang JY, Wang X, Wang XJ, Zheng BZ, Wang Y, Wang X, Liang B (2018) Curcumin inhibits the growth via Wnt/β-catenin pathway in non-small-cell lung cancer cells. Eur Rev Med Pharmacol Sci 22(21):7492–7499

Wang C, Zhozu Y, Zhang Z et al (2020a) Resveratrol modulates the PI3K/Akt/mTOR pathway in cancer stem cells to inhibit tumorigenesis. Int J Mol Sci 21(18):6692. https://doi.org/10.3390/ijms21186692

Wang Y, Lu J, Jiang B, Guo J (2020b) The roles of curcumin in regulating the tumor immunosuppressive microenvironment (review). Oncol Lett 19:3059–3070. https://doi.org/10.3892/ol.2020.11422

Wang X, Yao J, Zhang Y et al (2021) Curcumin inhibits Hedgehog signaling and suppresses cancer stem cell-like properties in glioblastoma multiforme cells. Biochem Biophys Res Commun 570:101–108. https://doi.org/10.1016/j.bbrc.2021.01.087

Wang Z, Wang X, Xu Y et al (2022a) Mutations of PI3K-AKT-mTOR pathway as predictors for immune cell infiltration and immunotherapy efficacy in dMMR/MSI-H gastric adenocarcinoma. BMC Med 20(1):133. https://doi.org/10.1186/s12916-022-02316-w

Wang SW, Gao C, Zheng YM, Yi L, Lu JC, Huang XY et al (2022b) Current applications and future perspective of CRISPR/Cas9 gene editing in cancer. Mol Cancer 21:57. https://doi.org/10.1186/s12943-022-01533-3

Wang Q, Shao X, Zhang Y et al (2023) Role of tumor microenvironment in cancer progression and therapeutic strategy. Cancer Med 12(10):11149–11165. https://doi.org/10.1002/cam4.6052

Wei Q, Zhang YH (2024) Flavonoids with anti-angiogenesis function in cancer. Molecules 29(7):1570. https://doi.org/10.3390/molecules29071570

Wilson EB, El-Jawhari JJ, Neilson AL et al (2011) Human tumour immune evasion via TGF-β blocks NK cell activation but not survival allowing therapeutic restoration of anti-tumour activity. PLoS One 6(9):e22842. https://doi.org/10.1371/journal.pone.0022842

Wu H, Zhang J (2011) Targeting signal transducer and activator of transcription 3 (STAT3) pathway in cancer. Mol Cancer Ther 10(2):289–296. https://doi.org/10.1158/1535-7163.MCT-10-0820

Wu B, Shi X, Jiang M, Liu H (2023) Cross-talk between cancer stem cells and immune cells: potential therapeutic targets in the tumor immune microenvironment. Mol Cancer 22:38. https://doi.org/10.1186/s12943-023-01733-6

Xiao W, Gao Z, Duan Y, Yuan W, Ke Y (2017) Notch signaling plays a crucial role in cancer stem-like cells maintaining stemness and mediating chemotaxis in renal cell carcinoma. J Exp Clin Cancer Res 36:1. https://doi.org/10.1186/s13046-016-0477-3

Xiao X, Chen H, Yang L, Xie G, Shimuzu R, Murai A (2022) Concise review: cancer cell reprogramming and therapeutic implications. Transl Oncol 24:101503. https://doi.org/10.1016/j.tranon.2022.101503

Yang H, Liu Y (2013) Polysaccharides as immune modulators in cancer therapy. Immunopharmacol Immunotoxicol 35(3):377–384. https://doi.org/10.3109/08923973.2013.812045

Yang MC, Wang HC, Hou YC, Tung HL, Chiu TJ, Shan YS (2015) Blockade of autophagy reduces pancreatic cancer stem cell activity and potentiates the tumoricidal effect of gemcitabine. Mol Cancer 14:179. https://doi.org/10.1186/s12943-015-0455-0

Yang Z, Yang J, Li X et al (2019) Curcumin inhibits the JAK2/STAT3 pathway to attenuate chemoresistance in colon cancer stem cells. Biol Pharm Bull 42(2):302–310. https://doi.org/10.1248/bpb.b18-00999

Yang L, Shi P, Zhao G, Xu J, Peng W, Zhang J, Cui H (2020a) Targeting cancer stem cell pathways for cancer therapy. Signal Transduct Target Ther 5(1):8. https://doi.org/10.1038/s41392-019-0095-2

Yang Q, Guo N, Zhou Y, Chen J, Wei Q, Han M (2020b) The role of tumor-associated macrophages (TAMs) in tumor progression and relevant advance in targeted therapy. Acta Pharm Sin B 10(11):2156–2170. https://doi.org/10.1016/j.apsb.2020.04.008

Yang Y, Zhang M, Wang Y (2022) The roles of histone modifications in tumorigenesis and associated inhibitors in cancer therapy. J Natl Cancer Center 2:277–290. https://doi.org/10.1016/j.jncc.2022.09.001

Yasuda T, Ishimoto T, Baba H (2021) Conflicting metabolic alterations in cancer stem cells and regulation by the stromal niche. Regen Ther 17:8–12. https://doi.org/10.1016/j.reth.2020.12.001

Yi M, Li T, Niu M et al (2024) Blockade of CCR5+ T cell accumulation in the tumor microenvironment optimizes anti-TGF-β/PD-L1 bispecific antibody. Adv Sci 11(43):2408598. https://doi.org/10.1002/advs.202408598

Zaveri NT (2006) Green tea and its polyphenolic catechins: medicinal uses in cancer and other diseases. Life Sci 78(18):2073–2080. https://doi.org/10.1016/j.lfs.2005.12.006

Zhang Z, Luo Q (2017) Therapeutic potential of flavonoids in cancer: focus on molecular mechanisms. Curr Med Chem 24(26):2981–2991. https://doi.org/10.2174/0929867324666170412154700

Zhang Y, Talalay P (1994) Anticarcinogenic activities of organic isothiocyanates: chemistry and mechanisms. Cancer Res 54(1):216–220

Zhang S, Yang X, Wang L, Zhang C (2018) Interplay between inflammatory tumor microenvironment and cancer stem cells. Oncol Lett 16(1):679–686. https://doi.org/10.3892/ol.2018.8715

Zhang H, Zhang X, Sun Y et al (2020) Berberine inhibits the PI3K/Akt/mTOR pathway to suppress colorectal cancer stem cell self-renewal. Sci Rep 10:21393. https://doi.org/10.1038/s41598-020-78233-1

Zhao Y, Baldin AV, Isayev O, Werner J, Zamyatnin AA Jr, Bazhin AV (2021) Cancer vaccines: Antigen selection strategy. Vaccines 9:85. https://doi.org/10.3390/vaccines9020085

Zhu L, Han MB, Gao Y, Wang H, Dai L, Wen Y, Na LX (2015) Curcumin triggers apoptosis via upregulation of Bax/Bcl-2 ratio and caspase activation in SW872 human adipocytes. Mol Med Rep 12(1):1151–1156. https://doi.org/10.3892/mmr.2015.3450

Zick SM, Sloane R (2009) Clinical trials of plant-based interventions in cancer. Cancer Chemother Pharmacol 63(3):411–421. https://doi.org/10.1007/s00280-008-0767-5

Safety, Dosage, and Regulatory Aspects in the Use of Phytochemicals Against Cancer

14

Ipsita Priyadarsini Samall, Sameer Jena, B. Jyotirmayee, Gyanranjan Mahalik, Itishree Jogamaya Das, and Himansu Bhusan Samal

Abstract

At the nexus of conventional medicine and contemporary pharmacological science, the therapeutic potential of phytochemicals in oncology offers a complicated paradigm. Current research on safety limits, pathological factors, and legal frameworks controlling phytochemical interventions in cancer treatment is summarized in this thorough review. Recent developments in bioavailability-enhancing technology and analytical chemistry have clarified essential elements influencing these substances' therapeutic index, such as drug–herb interactions, matrix effects, and metabolic transformation. The review includes meta-analyses of clinical trials, emphasizing the influence of genetic variants on metabolism and interindividual variability in response patterns. A comparative examination of worldwide norms explores regulatory problems, highlighting the necessity for standardized quality assessment methodologies and jurisdictional differences. The safety concerns, dosage optimization, and regulatory frameworks controlling the therapeutic use of phytochemicals in cancer treatment are all rigorously discussed here. Despite the promising anticancer characteristics of plant-derived drugs, standardization, bioavailability, and potential toxicity pose significant obstacles to their clinical application. The impact of extraction techniques on drug purity, potency and the pharmacokinetic diversity across various patient

I. P. Samall · S. Jena · B. Jyotirmayee · G. Mahalik
Department of Botany, School of Applied Sciences, Centurion University of Technology and Management, Gyanranjan Mahalik, Odisha, India
e-mail: gyanranjan.mahalik@cutm.ac.in

I. J. Das
Faculty of Medical Science and Research, Sai Nath University, Ranchi, Jharkhand, India

H. B. Samal (✉)
Department of Pharmaceutics, School of Pharmacy and Life Science, Centurion University of Technology and Management, Gyanranjan Mahalik, Odisha, India

K. Kuca et al. (eds.), *Medicinal Plants and Their Bioactives in Human Diseases*,
https://doi.org/10.1007/978-3-032-01356-9_14

populations are given special consideration. We also discuss the difficulties in determining bioequivalence and performing safety evaluations for botanical preparations that contain many components. In addition to offering evidence-based suggestions for maximizing the therapeutic use of plant-derived chemicals in oncology, our findings highlight the necessity of standardized procedures and harmonized regulatory frameworks in phytochemical research.

Keywords

Phytochemical · Oncological applications · Pharmacokinetic · Pathology · Biotransformation · Pharmacodynamics

1 Introduction

The use of phytochemicals, or molecules derived from plants, in treating cancer has attracted much interest in recent years. It is crucial to carefully address the safety profile, dosage optimization, and regulatory issues related to the clinical application of phytochemicals as this field of study develops. A broad class of bioactive substances originating from plants and phytochemicals has attracted much interest in cancer research because of their possible therapeutic uses. Fruits, vegetables, herbs, and other plant-based foods include these naturally occurring compounds, which have shown impressive potential in preventing, suppressing, and possibly treating various cancers (Vikram et al. 2022). Because of their intricate molecular structures, which might interact with biological processes to potentially inhibit the growth and advancement of cancer cells, phytochemicals are gaining attention.

Traditional medical methods have given way to advanced molecular research in the scientific study of phytochemicals' capacity to fight cancer, which has shown complex mechanisms by which these substances may be able to do so. Numerous phytochemicals, such as polyphenols, flavonoids, terpenoids, and alkaloids, have been found by researchers to possess anticancer characteristics (Mtewa et al. 2021). These compounds can each target distinct components of the genesis of cancer. These substances have demonstrated encouraging capacities to strengthen the body's defenses against the growth of cancer cells by inducing apoptosis, inhibiting angiogenesis, and modifying cellular signaling (Fatma and Siddique 2023).

As contemporary medicine continues to look for less intrusive and more tailored cancer treatments, phytochemicals present an intriguing research avenue that unites cutting-edge scientific expertise with traditional botanical knowledge. In addition to their direct anticancer properties, these plant-derived chemicals have the potential to complement traditional cancer treatments, which could lessen side effects and enhance patient outcomes overall (Dehelean et al. 2021). The therapeutic potential of natural chemicals in the fight against one of the most challenging diseases of our time is highlighted by this technique, which reflects a comprehensive approach to cancer treatment.

1.1 Historical Perspective of Phytochemicals in Cancer Treatment

From traditional herbal medicine to contemporary scientific studies, the use of phytochemicals in cancer treatment has a rich history. Ancient societies such as the Greeks, Chinese, and Egyptians used herbal remedies for various illnesses after realizing the therapeutic benefits of plants. However, the twentieth century saw scientific advances that revolutionized our knowledge of the potential of plant-derived compounds to treat cancer. Notable findings included the isolation of paclitaxel from the Pacific yew tree and the methodical study of plant-based anticancer agents (Choudhari et al. 2020; Kumar et al. 2022; Kumari et al. 2022; Mohamed et al. 2022).

By the late twentieth and early twenty-first centuries, scientists were able to understand better the mechanisms of phytochemicals and how certain compounds interact with cancer cells at the proteome and genomic levels, thanks to sophisticated molecular research methodologies. Numerous plant-derived compounds, including curcumin, camptothecin, vinca alkaloids, and epigallocatechin gallate (EGCG), were discovered by researchers to exhibit promising anticancer properties. As a result, targeted therapies and combination treatments were developed to reduce side effects and improve patient outcomes (Claro et al. 2024; Wang et al. 2018; Wanjari et al. 2010; Yallapu et al. 2010). The enormous potential of phytochemicals in cancer treatment is still being investigated by modern research, which focuses on customized medicine strategies and comprehends the intricate relationships between substances produced by plants and cellular processes. The development of this field offers hope for less intrusive and more successful cancer treatments that utilize the therapeutic potential of natural chemicals, as it represents a unique convergence of traditional botanical knowledge and state-of-the-art scientific study.

1.2 Significance in Modern Oncology

1.2.1 Safety Considerations

The possibility of adverse effects is one of the main worries regarding the application of phytochemicals in cancer treatment. Although substances originating from plants are frequently thought of as "natural" and benign, they may be poisonous or mixed with other drugs to cause unfavorable side effects. The safety profile of phytochemicals can be significantly impacted by dosage, administration method, and patient characteristics (Van Dyk et al. 2013).

Thorough safety evaluations, including in vitro, in vivo, and clinical research, are crucial for comprehending the hazards of applying phytochemicals. Concerns like mutagenicity, genotoxicity, acute and chronic toxicity, and possible combinations with traditional cancer treatments must all be covered in these investigations. For instance, research has shown that excessive amounts of curcumin, a polyphenol present in turmeric, can result in diarrhea, nausea, and gastrointestinal distress. Furthermore, many phytochemicals have the potential to interact with chemotherapeutic medications, changing their pharmacokinetics and possibly decreasing or raising their toxicity (Das et al. 2022; Jyotirmayee and Mahalik 2022).

To guarantee the safe administration of phytochemicals to cancer patients, biomarker identification and the establishment of suitable safety monitoring measures are essential. Healthcare professionals can identify and control any adverse effects that might occur during phytochemical-based therapies using biomarkers, such as inflammatory markers or liver enzymes (Ulbricht and Chao 2010; Prasad et al. 2011). Patients, medical professionals, and researchers must work closely together and conduct routine monitoring to guarantee the safe and efficient application of phytochemicals in cancer treatment (Hembram et al. 2020; Ibrahim et al. 2021; Jain et al. 2015).

1.2.2 Dosage Considerations

The correct dosage and formulation are crucial for successfully applying phytochemicals in cancer treatment. Phytochemicals can display numerous biological activities, and their effectiveness as a treatment is frequently dose-dependent. Choosing the best dosage schedule is difficult since it requires considering several variables, including the phytochemical's pharmacokinetic profile, bioavailability, and the particular type of cancer and stage. The correct dosage ranges for phytochemicals in cancer treatment are primarily determined by preclinical research and clinical trials (Segal et al. 2017). Issues like determining the least effective dose, the maximum tolerable dose, and the possibility of dose-dependent toxicity must all be included in these studies. For example, because of the intricate and occasionally biphasic dose–response interactions of phytochemicals, research has indicated that modest doses of green tea polyphenols may be more efficient than large doses for preventing the proliferation of cancer cells (Dhupal and Chowdhury 2020; Kaur et al. 2022; Kesharwani et al. 2022; Khan et al. 2021).

Furthermore, the consistency and reproducibility of phytochemical-based therapies depend on creating standardized and thoroughly described formulations. Several factors, including the extraction technique, solvent selection, and the presence of synergistic compounds, can influence the bioavailability and effectiveness of phytochemicals. To maximize phytochemicals' therapeutic potential and reduce the possibility of side effects, researchers and manufacturers must collaborate closely to optimize their formulation and administration (Rudzińska et al. 2023).

1.2.3 Regulatory Considerations

Different regulatory systems that differ between nations and regions also apply to incorporating phytochemicals into cancer treatment. Rather than being categorized as pharmaceutical pharmaceuticals, phytochemicals are frequently categorized as dietary supplements or traditional/herbal remedies in several countries. Clinical trial methodologies, quality control standards, and marketing and labeling laws are only a few of the regulatory requirements that this classification may significantly impact (Thakur et al. 2020).

For phytochemical-based cancer treatments to be developed and implemented successfully, navigating the regulatory environment is crucial. Researchers and healthcare professionals must understand their countries' relevant laws, rules, and approval procedures. For instance, the US Food and Drug Administration (FDA) has a distinct regulatory structure for dietary supplements, including items based on phytochemicals, than it does for pharmaceutical medications. This mandates that producers follow Good Manufacturing Practices (GMPs) and present proof of the product's quality and safety (Hossain et al. 2022). Cooperation with regulatory bodies and adherence to set standards are essential to guarantee phytochemical-based cancer treatments' safety, effectiveness, and quality. To help researchers and producers navigate the complicated regulatory environment, regulatory agencies, including the FDA, the European Medicines Agency (EMA), and the World Health Organization (WHO), have created frameworks and guidelines. Early interaction with these regulatory bodies can assist in guaranteeing that cancer treatments based on phytochemicals fulfill the requirements and can be effectively released onto the market (Frankos 2009). In conclusion, there is much potential for using phytochemicals in cancer treatment; however, addressing the critical issues of safety, dose, and regulations is crucial. Realizing the full potential of phytochemicals in the battle against cancer will require further research, thorough safety evaluations, carefully planned clinical trials, and adherence to regulatory frameworks.

1.3 Scope of the Chapter

The best practices and available data for the safe and efficient application of phytochemicals in oncology are thoroughly reviewed in this chapter. We describe the main categories of phytochemicals with anticancer properties and how they work. We next examine common side effects, organ-specific toxicities, and particular population concerns as we dig into the safety factors. The intricacies of dosage optimization, including formulation techniques, pharmacokinetic considerations, and therapeutic window determination, are then covered. We also look at the current international norms and quality control requirements because of the inherent heterogeneity in phytochemical composition and the particular regulatory issues of botanical medicines. Lastly, we discuss the practical aspects of incorporating phytochemicals into traditional cancer treatment, offering evidence-based suggestions for clinical application, surveillance procedures, and risk mitigation techniques.

2 Documented Clinical Applications of Phytochemicals in Cancer Treatment

Numerous substances have shown encouraging outcomes in documented clinical trials and research studies, marking a substantial evolution in the clinical applications of phytochemicals in cancer treatment. One of the most effective phytochemical-based cancer treatments is Taxol (paclitaxel), which is extracted from the Pacific yew tree and is frequently used to treat pancreatic, breast, lung, and ovarian cancers. The practical integration of plant-derived substances into conventional medical therapies is demonstrated by the widespread use of vinca alkaloids, which are isolated from periwinkle plants, in chemotherapy regimens for a variety of lymphomas and leukemias (Hosseini and Ghorbani 2015; Saeed et al. 2017). The analysis of phytochemicals across various cancer types has been broadened by contemporary clinical research, with substances such as curcumin, resveratrol, and epigallocatechin gallate (EGCG) becoming the subject of in-depth study. In clinical trials, several substances have demonstrated promise in slowing the spread of cancer, lowering adverse drug reactions, and maybe improving the efficacy of traditional cancer treatments. Their anti-inflammatory and antioxidant qualities have drawn particular attention from researchers who have demonstrated their promise in controlling cancer at different phases, from prevention to supportive care during treatment (Altay & Gürer 2023; Amin et al. 2021; Anand et al. 2010; Choudhury et al. 2017).

Clinical uses of phytochemicals go beyond the direct treatment of cancer; they are also becoming more widely acknowledged for their potential to improve cancer patients' quality of life and provide supportive care. Research has examined their potential to lower the risk of cancer recurrence, maintain immunological function, and manage side effects associated with treatment. The reported clinical uses show the great potential of phytochemicals as complementary treatments in comprehensive cancer care, connecting traditional botanical knowledge with current medical techniques (Arora and Suvartan 2020), even though more thorough research is required (Kotecha et al. 2016; Yarla et al. 2016; Yashin et al. 2017; Yin et al. 2011).

3 Safety Considerations and Risk Assessment

Although phytochemicals have demonstrated encouraging outcomes in cancer treatment, there are several obstacles to their therapeutic application. To address the possible dangers associated with these natural chemicals, thorough safety investigations are essential (Miller and Snyder 2012).

3.1 Common Safety Concerns

Intrinsic toxicity, contamination problems, and standardization difficulties are possible safety concerns with phytochemicals. The existence of intrinsically harmful substances or the limited therapeutic range of specific phytochemicals can result in intrinsic toxicity. Certain alkaloids, like aristolochic acid, have been linked to nephrotoxicity and carcinogenicity, necessitating careful monitoring and dosage modifications. Significant safety problems can also arise from pesticide contamination, heavy metals, or microorganisms, especially in plant preparations. Furthermore, because the amounts of active chemicals can change significantly between product batches, a lack of consistency in phytochemical composition and quality control might result in inconsistent and dangerous effects (Arora and Suvartan 2020).

3.2 Population-Specific Safety Considerations

Phytochemicals' safety profiles can vary depending on the patient population. Age, genetic background, and comorbidities can affect these drugs' pharmacokinetics and pharmacodynamics. Vulnerable populations need special attention, including children, the elderly, and people with weakened immune systems or impaired organ function. Pharmacogenomic variables, such as genetic variations in transporters and enzymes that break down drugs might also influence how each person reacts to phytochemicals, requiring customized dosage plans (Basu et al. 2007).

3.3 Known Adverse Effects

Using phytochemicals can have several negative effects, such as gastrointestinal issues, allergic reactions, liver and kidney damage, and possible conflicts with traditional cancer treatments (Fig. 14.1). For example, hepatotoxicity has been associated with some polyphenols, such as green tea catechins, mainly when used in large quantities or over an extended period. Careful observation and reporting of adverse events are necessary to create a thorough safety profile and find any idiosyncratic reactions (Johnson 2007).

3.4 Risk–Benefit Assessment Approaches

Determining the proper clinical use of phytochemicals requires assessing the risk–benefit ratio. The extent of cancer, the patient's general health, the possibility of adverse or synergistic interactions with conventional treatments, and the existing

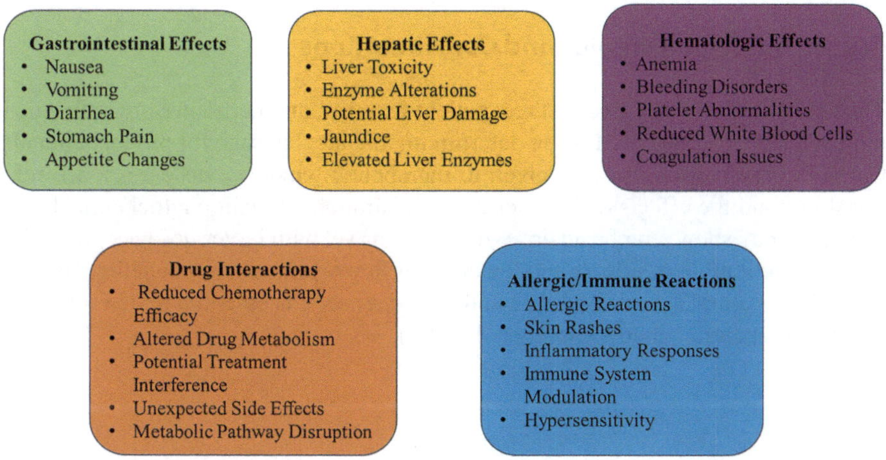

Fig. 14.1 Common adverse effects associated with phytochemical use in cancer treatment

data regarding the safety and effectiveness of the particular phytochemical intervention should all be considered during this evaluation. The potential advantages of phytochemicals may occasionally exceed the hazards, especially in situations with advanced or refractory cancer. To make an informed choice, however, a careful assessment of the scientific evidence and the clinical characteristics of each patient is required (Bao et al. 2014).

4 Dosage Optimization and Bioavailability

Phytochemical dosing regimens must effectively achieve the intended therapeutic effects while lowering the risk of side effects. An in-depth understanding of these substances' pharmacokinetic and pharmacodynamic characteristics is necessary to optimize the dosage.

4.1 Factors Affecting Bioavailability

Numerous factors, such as the method of administration, interactions between food and drugs, metabolic transformation, and efflux transporters, might affect the bioavailability of phytochemicals. For instance, many polyphenolic chemicals have poor water solubility, which can limit their absorption. New delivery vehicles or adjuvants may need to be developed to improve bioavailability. Likewise, the systemic exposure and distribution of phytochemicals to target tissues can significantly influence their metabolism by Phase I and II enzymes and active efflux by membrane transporters (Table 14.1). Improving phytochemicals' distribution, absorption, and therapeutic effectiveness requires addressing these variables (Aqil et al. 2013).

4.2 Dosing Strategies and Considerations

The target cancer type, illness stage, patient features, and particular phytochemical molecule are all considered when determining the ideal dosage for phytochemicals. Careful consideration must be given to therapeutic windows, dose–response relationships, and the effects of long-term administration. Certain phytochemicals, for example, may show a biphasic dose–response curve, with larger dosages producing cytotoxic or inhibitory effects and lower doses having stimulatory benefits. To guarantee these natural substances' safe and efficient use, it is essential to determine the proper therapeutic window (Scott et al. 2009).

Table 14.1 Factors affecting the bioavailability of phytochemicals

Factor category	Specific factors	Impact on bioavailability	Examples	Mitigation strategies
Physicochemical properties	Molecular weight	Determines absorption capacity	Large molecules: difficult absorption, small molecules: easier absorption	Micronization, nanoformulation
	Chemical structure	Influences solubility and membrane permeability	Polarity, hydrogen bonding	Structural modifications, analogs
	Lipophilicity	It affects membrane crossing ability	Highly lipophilic compounds cross membranes easier	Lipid-based delivery systems
Physiological factors	Gastrointestinal pH	Impacts chemical stability and absorption	Acidic/alkaline environments alter compound structure	pH-responsive encapsulation
	Enzymatic activity	Metabolizes or transforms compounds	Liver enzymes, gut microbiota	Targeted delivery, enzyme inhibitors
	Gut microbiome composition	Metabolizes and transforms phytochemicals	Different bacterial populations affect metabolism	Probiotic/prebiotic interventions
Dietary interactions	Food matrix	Traps or releases phytochemicals	Protein binding, fiber interactions	Processing techniques, extraction methods
	Co-consumed nutrients	Enhances or inhibits absorption	Absorption: Calcium can reduce iron absorption	Strategic meal composition
	Cooking methods	Alters chemical structure	Heat can degrade or transform compounds	Minimal processing, raw consumption
Genetic factors	Metabolic enzyme variants	Differences in compound metabolism	CYP450 enzyme polymorphisms	Personalized dosing
	Transporter protein variations	It affects absorption and excretion	ABC transporters, organic anion transporters	Genetic screening
Formulation techniques	Delivery systems	Improves stability and absorption	Liposomes, nanoparticles, micelles	Advanced drug delivery technologies
	Complexation	Enhances solubility and stability	Cyclodextrin complexes	Molecular encapsulation

(continued)

Table 14.1 (continued)

Factor category	Specific factors	Impact on bioavailability	Examples	Mitigation strategies
Physiological conditions	Age	Metabolic efficiency changes	Reduced absorption in elderly	Age-specific formulations
	Health status	Impacts absorption mechanisms	Inflammatory conditions, digestive disorders	Targeted therapeutic approaches
Environmental factors	Storage conditions	Affects chemical stability	Light, temperature, oxygen exposure	Proper storage, protective packaging
	Processing methods	Transforms chemical structure	Drying, extraction techniques	Optimized processing protocols

4.3 Novel Delivery Systems

Novel delivery methods have demonstrated promise in improving phytochemicals' bioavailability and targeted distribution, including liposomes, nanoparticles, and self-emulsifying drug delivery systems. Adjusting the compound's solubility, stability, and tissue distribution can enhance the pharmacokinetic profile and reduce possible toxicity. For instance, it has been shown that encapsulating curcumin in nanoparticles increases its bioavailability and anticancer activity compared to the free substance (Lagoa et al. 2020).

4.4 Therapeutic Window Determination

The safe and efficient use of phytochemicals depend on determining the therapeutic window or the range between the minimal effective dose and the maximum tolerated dose. Combining pre-clinical research, clinical trials, and post marketing surveillance information determines the best dosage schedule. To make sure that the therapeutic benefits of using phytochemicals outweigh any possible dangers, considerable consideration must be given to factors like the length of treatment, drug interactions, and patient-specific variables (Xie et al. 2016).

5 Drug Interactions and Clinical Implications

Using phytochemicals and traditional cancer treatments at the same time may lead to complicated medication interactions that could affect the overall safety and effectiveness of the treatment plan.

5.1 Interactions with Chemotherapy

Through various processes, including activating or inhibiting enzymes, phytochemicals may modify the pharmacokinetics and pharmacodynamics of chemotherapeutic agents. This can result in changes to the conventional medications' absorption, distribution, metabolism, and excretion. For example, it has been demonstrated that some polyphenols, such as resveratrol, limit the activity of cytochrome P450 enzymes, which regulate the metabolism of numerous chemotherapy drugs. As a result, traditional medications may become more hazardous and have a higher systemic exposure. On the other hand, some phytochemicals, like curcumin, can cause the development of enzymes that break down drugs, which would decrease the co-administered chemotherapy's bioavailability and effectiveness (Sundaram et al. 2019; Pan et al. 2015; Rachmawati et al. 2022).

5.2 Interactions with Other Medications

Other drugs that are frequently prescribed to cancer patients, such as immunosuppressants, anticoagulants, and antidepressants, may also interact with phytochemicals. These interactions may raise the chance of adverse events or change the effectiveness of treatment. Supplementing garlic, for instance, has been linked to an increased risk of bleeding when used with warfarin, a frequently prescribed anticoagulant. These concurrent drugs' pharmacokinetics may be affected by phytochemicals' modulation of drug-metabolizing enzymes and transporters, which may call for careful monitoring and possible dose modifications (Ayaz et al. 2019; Shao et al. 2020; Thomas & Subash-Babu 2022; Vafadar et al. 2020).

5.3 Clinical Management Strategies

Careful patient monitoring, therapeutic medication monitoring, dose modifications, and the application of evidence-based recommendations for the safe co-administration of phytochemicals and traditional cancer treatments are all methods for handling drug interactions (Fig. 14.2). To make wise choices and reduce the possibility of unfavorable results, healthcare professionals should keep up-to-date on the possible interactions between phytochemicals and traditional cancer treatments. The safe and efficient incorporation of phytochemicals into cancer treatment can be facilitated by creating and distributing clinical practice guidelines and decision-support systems (Rizeq et al. 2020).

6 Regulatory Framework and Quality Control

Different regulatory criteria and quality control standards apply to incorporating phytochemicals into cancer treatment, and these range depending on the country.

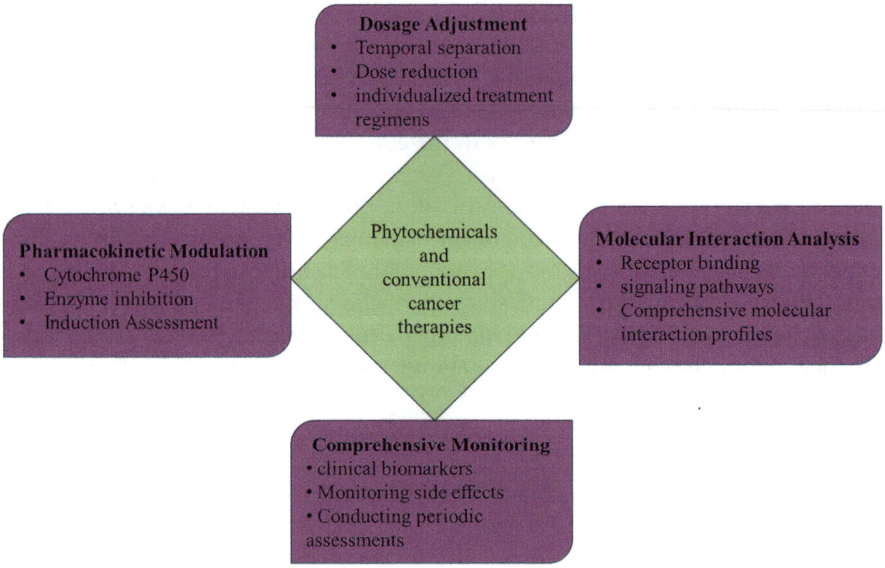

Fig. 14.2 Strategies for managing drug interactions between phytochemicals and conventional cancer therapies

6.1 International Regulatory Standards

There are differences across nations and regions in the complicated regulatory environment around phytochemicals in oncology. Important regulatory agencies have set criteria and rules that need to be followed, including the World Health Organization (WHO), the European Medicines Agency (EMA), and the US Food and Drug Administration (FDA). These contain specifications for product characterization, safety and efficacy information, clinical trial design, and production procedures. For developing and commercializing phytochemical-based cancer therapies, navigating the disparate regulatory regimes across various markets can present serious obstacles (Raza et al. 2023).

6.2 Quality Assurance Requirements

One of the biggest challenges is guaranteeing phytochemical products' consistency, quality, and purity. Robust analytical procedures, standardization strategies, and adherence to good manufacturing practices are necessary to satisfy the quality control standards for botanical medicines. This covers the detection of possible impurities, the identification and measurement of active chemicals, and the development of reliable and repeatable production procedures. The regulatory clearance and clinical application of phytochemical-based cancer treatments may be made more difficult by the absence of uniform quality requirements throughout foreign jurisdictions (Yau et al. 2015).

6.3 Documentation and Compliance

Extensive documentation, such as thorough product characterization, preclinical and clinical data, and post-marketing monitoring reports, is frequently needed in the regulatory procedure for the clinical application of phytochemicals. Maintaining adherence to these standards is essential for securing market authorization and the safe and efficient application of phytochemicals. In order to successfully integrate phytochemicals into cancer therapy, researchers and healthcare professionals need to be aware of the particular documentation and reporting requirements in their various locations (Maru et al. 2016).

6.4 Market Authorization Procedures

The process by which phytochemical-based cancer treatments can be approved for sale varies by jurisdiction. Depending on the situation, phytochemicals can be regulated as conventional herbal remedies or dietary supplements, or they might be regarded as experimental novel pharmaceuticals or biologics. Successful commercialization and clinical uptake of these natural chemicals depend on knowing the particular requirements and negotiating the regulatory environment. Working together, researchers, healthcare providers, and regulatory bodies can overcome particular difficulties and expedite the market authorization procedure (Calixto, 2000).

7 Clinical Implementation

A comprehensive strategy that considers these natural substances' particular difficulties and factors is needed to incorporate phytochemicals into traditional cancer treatment (Fig. 14.3).

7.1 Integration with Conventional Therapy

In addition to conventional cancer treatments, including radiation therapy, chemotherapy, and targeted therapies, phytochemicals may be employed. To maximize the overall treatment plan, thoroughly assess any possible antagonistic or synergistic interactions. For example, certain phytochemicals, such as curcumin, have shown the capacity to sensitize cancer cells to these agents to increase the effectiveness of chemotherapeutic treatments (Fig. 14.3). According to Efferth et al. (2017), some phytochemicals, on the other hand, can disrupt how traditional treatments work. Therefore, their incorporation should be done with greater caution.

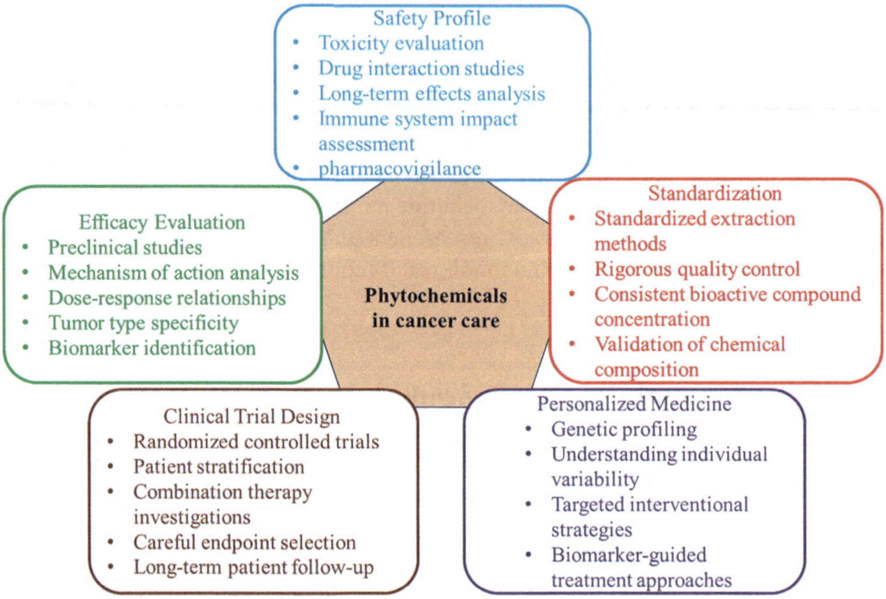

Fig. 14.3 Key considerations for the clinical implementation of phytochemicals in cancer care

7.2 Monitoring Protocols

Thorough monitoring procedures are necessary to guarantee the safe and efficient administration of phytochemicals to cancer patients. This involves routinely evaluating clinical results, side effects, and compliance with the recommended action. In order to identify any organ-specific toxicities linked to phytochemicals, biomarker monitoring, such as the assessment of liver and kidney function, may be required. In order to add to the increasing amount of empirical data on the therapeutic application of these natural substances, healthcare providers must also be careful to report any adverse events or unexpected results (Ranjan et al. 2019).

7.3 Patient Education and Compliance

Promoting treatment adherence and controlling patient expectations require educating patients on phytochemical therapies' proper application, possible advantages, and potential hazards. Patients should be informed about the necessity of vigilant monitoring, timely reporting of any side effects, and the significance of telling their healthcare professionals about all concomitant drugs and dietary supplements. The effective incorporation of phytochemicals into the overall cancer care plan can be improved by encouraging candid dialogue and collaborative decision-making between patients and their medical team (Estrela et al. 2017).

7.4 Documentation Requirements

To satisfy regulatory standards and add to the expanding body of knowledge in this area, careful documentation of the clinical application of phytochemicals is required, including patient records, adverse event reporting, and outcome data. To guarantee thorough data collection and make translating research findings into clinical practice easier, healthcare providers must be aware of their jurisdictions' particular documentation and reporting requirements (Lee et al. 2011).

8 Future Perspectives

Several new developments and potential avenues for future research in phytochemicals and cancer treatment show promise for expanding the clinical use of these organic substances.

8.1 Emerging Research Directions

To maximize the therapeutic application of these natural products, ongoing research is looking into new phytochemical substances, combination medicines, customized dosing plans, and the incorporation of cutting-edge technology like artificial intelligence and machine learning. Researchers are investigating the synergistic effects of combining phytochemicals with immunotherapies or targeted medicines to improve overall therapeutic efficacy and combat drug resistance. Furthermore, based on a person's genetic and clinical profile, computer models and prediction algorithms may help determine the best dosage schedules and customize phytochemical therapies (Issa et al. 2006).

8.2 Technological Advances

Improved safety, effectiveness, and clinical translation of phytochemical-based cancer treatments are anticipated due to advancements in analytical chemistry, drug delivery methods, and bioavailability enhancement strategies (Saranya et al. 2022). Thanks to advanced analytical techniques like mass spectrometry and high-performance liquid chromatography, complex phytochemical mixtures will be easier to identify, quantify, and check for quality. The solubility, stability, and targeted transport of phytochemicals to the tumor site can be improved by creating innovative drug delivery platforms, such as nanoparticles, liposomes, and intelligent materials, which may raise their therapeutic index. Additionally, the large-scale manufacture of standardized phytochemical extracts with consistent composition and potency may be made more accessible by biotechnological methods such as plant cell and tissue culture systems Johnson (2007).

8.3 Regulatory Evolution

Further improvement of the regulatory environment for phytochemicals in oncology is anticipated, with the possibility of harmonizing global standards and creating more efficient market authorization processes. The development of international quality standards and the acceptance of phytochemicals as valid medicinal substances may significantly ease the clinical application of these substances. Furthermore, adding post-marketing monitoring data and real-world evidence to regulatory decision-making procedures should aid in resolving the particular difficulties brought on by the intrinsic unpredictability of botanical medicines (Samec et al. 2019).

8.4 Clinical Practice Implications

Incorporating phytochemicals into standard clinical practice will become more crucial as data supporting their efficacy in cancer treatment grows. Creating thorough treatment plans, interdisciplinary cooperation, and explicit rules for medical personnel are all necessary. Educating oncologists, pharmacists, and other healthcare professionals on the proper usage, possible advantages, and safety concerns of these natural compounds will be essential to integrating phytochemicals into complete cancer care programs in addition to traditional medications. Additionally, they might be made more accessible and easily incorporated into the cancer care continuum by opening specialized clinics or integrative oncology centers that focus on the evidence-based use of phytochemicals (Harrison et al. 2015).

9 Conclusion

Polyphenols, alkaloids, terpenoids, and organosulfur compounds are phytochemicals encouraging anticancer effects through various methods, including preventing tumor development, triggering apoptosis, and reducing angiogenesis. Their therapeutic application, however, is intricate and multidimensional, encompassing several crucial factors related to safety, dose, drug interactions, regulatory requirements, and incorporation into cancer treatment regimens. When it comes to using phytochemicals to treat cancer, safety is the most crucial consideration. Standardization difficulties, possible contamination problems, and inherent toxicity must all be carefully considered by researchers. Population-specific variables and recognized side effects, such as gastrointestinal issues, allergic reactions, and organ-specific toxicities, must be carefully evaluated to guarantee patient safety further. A thorough understanding of the variables influencing bioavailability, such as metabolism, efflux transporters, and the route of administration, is necessary to optimize phytochemical dose. Maximizing therapeutic potential while lowering possible hazards requires developing suitable dosage techniques and establishing the therapeutic window. The possibility of medication interactions between phytochemicals and traditional cancer treatments must be carefully considered. The effectiveness and safety of the entire treatment plan can be significantly impacted by altered pharmacokinetics and pharmacodynamics, necessitating careful observation and calculated action. In cancer, regulatory frameworks and quality control standards for phytochemicals pose particular difficulties in many jurisdictions. Successful clinical implementation requires much paperwork, adherence to good manufacturing techniques, and negotiating various market authorization processes, including phytochemicals in traditional cancer treatment calls for a thorough, multidimensional strategy. This entails creating robust monitoring systems, educating patients, and keeping thorough records to satisfy regulatory standards and advance our understanding of these drugs scientifically.

Phytochemicals appear to have a bright future in cancer treatment as research into new substances, combination treatments, customized dosage plans, and possible harmonization of global standards continues. Regulatory changes and technological advancements are constantly increasing the potential of these natural substances in oncological treatment. The importance of comprehensive safety assessments, individualized dosing plans, careful drug interaction monitoring, regulatory compliance, patient education, thorough documentation, and cooperative approaches between regulatory authorities, researchers, and healthcare professionals is all emphasized in clinical practice recommendations. By tackling these diverse obstacles and prospects, scientists and medical professionals can strive for a more seamless incorporation of phytochemicals into standard cancer treatment, which could provide patients with fresh hope and therapeutic alternatives for this intricate illness.

References

Altay A, Gürer US (2023) Phytochemicals in cancer treatment: current strategies and recent trends. In: Medicinal plants and natural product research. Springer, pp 287–308. https://doi.org/10.1007/978-981-19-7235-6_15

Amin F, Rehman MU, Mir SA, Ahmad A, Rashid SM, Rizwan A et al (2021) Role of phytochemicals in regulation of oxidative stress and apoptosis in cancer therapy: current status and future perspectives. Curr Pharm Des 27(42):4404–4415. https://doi.org/10.2174/1381612827666210818115454

Anand P, Nair HB, Sung B, Kunnumakkara AB, Yadav VR, Tekmal RR, Aggarwal BB (2010) Design of curcumin-loaded PLGA nanoparticles formulation with enhanced cellular uptake, and increased bioactivity in vitro and superior bioavailability in vivo. Biochem Pharmacol 79(3):330–338. https://doi.org/10.1016/j.bcp.2009.08.020

Arora S, Suvartan R (2020) Phytochemicals: benefits, concerns and challenges. Advan Func Food Ingred 225

Aqil F, Munagala R, Jeyabalan J, Vadhanam MV (2013) Bioavailability of phytochemicals and its enhancement by drug delivery systems. Cancer lett 334(1):133–141. https://doi.org/10.1016/j.canlet.2013.02.032

Ayaz M, Ullah F, Sadiq A, Ullah F, Ovais M, Ahmed J, Devkota HP (2019) Synergistic interactions of phytochemicals with antimicrobial agents: Potential strategy to counteract drug resistance. Chem-Biol Interact 308:294–303. https://doi.org/10.1016/j.cbi.2019.05.050

Basu SK, Thomas JE, Acharya SN (2007) Prospects for growth in global nutraceutical and functional food markets: a Canadian perspective. Aust J Basic Appl Sci 1(4):637–649

Bao Y, Wang W, Zhou Z, Sun C (2014) Benefits and risks of the hormetic effects of dietary isothiocyanates on cancer prevention. PLoS One 9(12):e114764. https://doi.org/10.1371/journal.pone.0114764

Calixto JB (2000) Efficacy, safety, quality control, marketing and regulatory guidelines for herbal medicines (phytotherapeutic agents). Braz J Med Biol Res, 33:179–189. https://doi.org/10.1590/S0100-879X2000000200004

Choudhari AS, Mandave PC, Deshpande M, Ranjekar P, Prakash O (2020) Phytochemicals in cancer treatment: from preclinical studies to clinical practice. Front Pharmacol 10:1614. https://doi.org/10.3389/fphar.2019.01614

Choudhury H, Gorain B, Pandey M, Chatterjee LA, Sengupta P, Das A et al (2017) Recent update on nanoemulgel as topical drug delivery system. J Pharm Sci 106(7):1736–1751. https://doi.org/10.1016/j.xphs.2017.03.015

Claro AE, Palanza C, Mazza M, Schuenemann GEUM, Rigoni M, Pontecorvi A, Muti P (2024) Historical use of medicinal plants and future potential from phytotherapy to phitochemicals. Ann Bot 14(1). https://doi.org/10.13133/2239-3129/18564

Das IJ, Dawes SL, Dominello MM, Kavanagh B, Miyamoto CT, Pawlicki T et al (2022) Quality and safety considerations in stereotactic radiosurgery and stereotactic body radiation therapy: an ASTRO safety white paper update. Pract Radiat Oncol 12(4):e253–e268. https://doi.org/10.1016/j.prro.2022.03.001

Dehelean CA, Marcovici I, Soica C, Mioc M, Coricovac D, Iurciuc S et al (2021) Plant-derived anticancer compounds as new perspectives in drug discovery and alternative therapy. Molecules 26(4):1109. https://doi.org/10.3390/molecules26041109

Dhupal M, Chowdhury D (2020) Phytochemical-based nanomedicine for advanced cancer theranostics: perspectives on clinical trials to clinical use. Int J Nanomedicine 15:9125–9157. https://doi.org/10.2147/IJN.S259628

Efferth T, Saeed ME, Mirghani E, Alim A, Yassin Z, Saeed E, Daak S (2017) Integration of phytochemicals and phytotherapy into cancer precision medicine. Oncotarget 8(30):50284. https://doi.org/10.18632/oncotarget.17466

Estrela JM, Mena S, Obrador E, Benlloch M, Castellano G, Salvador R, Dellinger RW (2017) Polyphenolic phytochemicals in cancer prevention and therapy: bioavailability versus bioefficacy. J Med Chem 60(23):9413–9436. https://doi.org/10.1021/acs.jmedchem.6b01026

Fatma H, Siddique HR (2023) Research and patents status of selected phytochemicals against cancer: how close and how far? Recent Pat Anticancer Drug Discov 18(4):428–447. https://doi.org/10.2174/1574892818666221107113648

Frankos VH (2009) FDA's Dietary Supplement Good Manufacturing Practice Regulatory Requirements for Globally Marketed Botanicals. Planta Medica 75(04):S–38. https://doi.org/10.1055/s-2009-1216429

Harrison AM, Heritier F, Childs BG, Bostwick JM, Dziadzko MA (2015) Systematic review of the use of phytochemicals for management of pain in cancer therapy. Bio Med Res Int2015(1):506327. https://doi.org/10.1155/2015/506327

Hembram KC, Prabha S, Paramanantham P, Padma VVR, Sivakumar M (2020) Nano-curcumin therapeutics for cancer. In: Curcumin for neurological and psychiatric disorders. Academic, pp 279–292. https://doi.org/10.1016/B978-0-12-817114-1.00019-5

Hossain CM, Gera ME, Ali KA (2022) Current status and challenges of herbal drug development and regulatory aspect: a global perspective. Asian J Pharm Clin Res 15:31–41

Hosseini A, Ghorbani A (2015) Cancer therapy with phytochemicals: evidence from clinical studies. Avicenna J Phytomed 5(2):84

Ibrahim A, Rukayya M, Sandai D, Abdulkadir M (2021) Phytochemicals and antimicrobial agents: an overview of health benefits and their modes of action. In: Discovery and development of antidiabetic agents from natural products. Elsevier, pp 99–113. https://doi.org/10.1016/B978-0-12-819573-4.00006-2

Issa AY, Volate SR, Wargovich MJ (2006) The role of phytochemicals in inhibition of cancer and inflammation: New directions and perspectives. J Food Compos Anal 19(5):405–419. https://doi.org/10.1016/j.jfca.2006.02.009

Jain A, Jain A, Gulbake A, Shilpi S, Hurkat P, Jain SK (2015) Recent advances in lipid-based vesicles and particulate carriers for topical and transdermal application. J Pharm Sci 106(2):423–445. https://doi.org/10.1016/j.xphs.2016.11.014

Johnson IT (2007) Phytochemicals and cancer. Proc Nutr Soc 66(2):207–215. https://doi.org/10.1017/S0029665107005459

Jyotirmayee B, Mahalik G (2022) Traditional uses and variation in curcumin content in varieties of Curcuma – the saffron of India. Ambient Sci 9(1):6–12. https://doi.org/10.21276/ambi.2022.09.1.rv01

Kaur J, Tikoo K, Sandhu M, Kukreja S, Manuja A, Kumar B et al (2022) Biomedical applications of functionalized nanocarriers in cancer. In: Biomedical applications of nanoparticles. Elsevier, pp 203–222. https://doi.org/10.1016/B978-0-323-85444-6.00011-3

Kesharwani P, Iyer AK, Gajbhiye V (2022) Polymeric micelles of curcumin: current status and opportunities for cancer therapy. In: Advanced drug delivery systems in the management of cancer. Academic, pp 347–368. https://doi.org/10.1016/B978-0-12-821389-6.00016-5

Khan H, Belwal T, Efferth T, Farooqi AA, Sethi G, Devkota HP et al (2021) Targeting inflammation by flavonoids: novel therapeutic strategy for metabolic disorders. Int J Mol Sci 22(16):9124. https://doi.org/10.3390/ijms22169124

Kotecha R, Takami A, Espinoza JL (2016) Dietary phytochemicals and cancer chemoprevention: a review of the clinical evidence. Oncotarget 7(32):52517. https://doi.org/10.18632/oncotarget.9593

Kumar R, Sarma H, Sreedharan S, Kumar V, Gopinath P, Biswas R, Maiti P (2022) Targeted cancer therapy using phytochemicals and nanoparticles: current trends and future perspectives. In: Biogenic nano-particles and their use in agro-ecosystems. Elsevier, pp 369–387. https://doi.org/10.1016/B978-0-323-85262-6.00010-1

Kumari N, Chauhan DS, Awasthi A, Gupta MK, Roy P, Rawat M (2022) Nanoparticle-based drug delivery system for cancer therapy. In: Nanotechnology-based targeted drug delivery systems for lung cancer. Academic, pp 275–302. https://doi.org/10.1016/B978-0-323-85170-4.00009-2

Lee KW, Bode AM, Dong Z (2011) Molecular targets of phytochemicals for cancer prevention. Nat Rev Cancer 11(3):211–218. https://doi.org/10.1038/nrc3017v

Maru GB, Hudlikar RR, Kumar G, Gandhi K, Mahimkar MB (2016) Understanding the molecular mechanisms of cancer prevention by dietary phytochemicals: From experimental models to clinical trials. World J Biol Chem7(1):88. https://doi.org/10.4331/wjbc.v7.i1.88

Miller PE, Snyder DC (2012) Phytochemicals and cancer risk: a review of the epidemiological evidence. Nutr Clin Pract 27(5):599–612. https://doi.org/10.1177/0884533612456043

Lagoa R, Silva J, Rodrigues JR, Bishayee A (2020) Advances in phytochemical delivery systems for improved anticancer activity. Biotechnol Adv 38:107382. https://doi.org/10.1016/j.biotechadv.2019.04.004

Mohamed IN, Jayapaul J, Aravind SR, Mani K (2022) Nanoformulations for phytochemical-based anticancer therapy. In: Phytochemicals as lead compounds for new drug discovery. Elsevier, pp 389–416. https://doi.org/10.1016/B978-0-323-85541-2.00013-1

Mtewa AG, Egbuna C, Beressa TB, Ngwira KJ, Lampiao F (2021) Phytopharmaceuticals: efficacy, safety, and regulation. In: Preparation of phytopharmaceuticals for the management of disorders. Academic, pp 25–38. https://doi.org/10.35516/jjps.v17i2.1768

Pan MH, Lai CS, Tsai ML, Ho CT (2015) Chemoprevention of nonalcoholic fatty liver disease by dietary natural compounds. Mol Nutr Food Res 59(1):49–66. https://doi.org/10.1002/mnfr.201400426

Prasad S, Yadav VR, Sung B, Reuter S, Kannappan R, Deorukhkar A et al (2011) Targeting inflammatory pathways by flavonoids for prevention and treatment of cancer. Planta Med 77(11):1084–1093. https://doi.org/10.1055/s-0030-1270930

Rachmawati H, Yanda YL, Rahma A, Mase N (2022) Recent advances on delivery system of curcumin for cancer therapy: part I. J Drug Deliv Sci Technol 67:102970. https://doi.org/10.1016/j.jddst.2021.102970

Ranjan A, Ramachandran S, Gupta N, Kaushik I, Wright S, Srivastava S, Srivastava SK (2019) Role of phytochemicals in cancer prevention. Int J Mol Sci 20(20):4981. https://www.mdpi.com/1422-0067/20/20/4981#

Raza W, Singh J, Hussain Y, Luqman S, Meena A (2023) FDA Approval for Phytochemicals in the Treatment of Cancer. In Phytonutrients in the Treatment of Gastrointestinal Cancer. Bentham Science Publishers, pp. 58–82. https://doi.org/10.2174/9789815049633123010l

Rizeq B, Gupta I, Ilesanmi J, AlSafran M, Rahman MM, Ouhtit A (2020) The power of phytochemicals combination in cancer chemoprevention. J Cancer 11(15):4521. https://doi.org/10.7150/jca.34374

Rudzińska A, Juchaniuk P, Oberda J, Wiśniewska J, Wojdan W, Szklener K, Mańdziuk S (2023) Phytochemicals in cancer treatment and cancer prevention—review on epidemiological data and clinical trials. Nutrients 15(8):1896. https://doi.org/10.3390/nu15081896

Saeed M, Naveed M, BiBi J, Kamboh AA, Mohamed K, Hussain T et al (2017) Phytochemicals and oxidation status in humans: a review of literature. Curr Res Nutr Food Sci 5(3):316–326. https://doi.org/10.12944/CRNFSJ.5.3.08

Samec M, Liskova A, Koklesova L, Mestanova V, Franekova M, Kassayova M, Kubatka P (2019) Fluctuations of histone chemical modifications in breast, prostate, and colorectal cancer: An implication of phytochemicals as defenders of chromatin equilibrium. Biomolecules 9(12):829. https://doi.org/10.3390/biom9120829v

Saranya V, Radhika Rajasree S, Priya Dharshini R, Vijayakumar G, Senthil Kumar N (2022) Recent development in nanotechnology-based drug delivery systems for the treatment of various cancers. J Drug Deliv Sci Technol 67:103038. https://doi.org/10.1016/j.jddst.2021.103038

Scott EN, Gescher AJ, Steward WP, Brown K (2009) Development of dietary phytochemical chemopreventive agents: biomarkers and choice of dose for early clinical trials. Cancer Preven Res 2(6):525–530. https://doi.org/10.1158/1940-6207.CAPR-08-0223

Segal R, Zwaal C, Green E, Tomasone JR, Loblaw A, Petrella T (2017) Exercise for people with cancer: a clinical practice guideline. Curr Oncol 24(1):40. https://doi.org/10.3747/co.24.3376

Shao J, Xue J, Dai Y, Liu H, Chen N, Jia L (2020) Applications of nanotechnology in the diagnosis and treatment of cancer. In: Biomedical engineering tools for management for patients with COVID-19. Elsevier, pp 175–204. https://doi.org/10.1016/B978-0-12-824473-9.00015-1

Sundaram MK, Raina R, Afroze N, Dhupkar A, Kaur NP, Arte A, Hussain A (2019) Combinational use of phytochemicals and chemotherapeutic drugs enhance their therapeutic potential on human cervical cancer cells. Int J Cancer Manag 12(6). https://doi.org/10.5812/ijcm.91783

Thakur M, Singh K, Khedkar R (2020) Phytochemicals: extraction process, safety assessment, toxicological evaluations, and regulatory issues. In: Functional and preservative properties of phytochemicals. Academic, pp 341–361. https://doi.org/10.1016/B978-0-12-818593-3.00011-7

Thomas C, Subash-Babu P (2022) Advances in nanotechnology for cancer treatment. In: Biopolymeric nanomaterials. Elsevier, pp 437–456. https://doi.org/10.1016/B978-0-323-85837-6.00011-6

Ulbricht CE, Chao W (2010) Phytochemicals in the oncology setting. Curr Treat Options in Oncol 11:95–106. https://doi.org/10.1007/s11864-010-0130-4

Vafadar A, Shabaninejad Z, Movahedpour A, Mohammadi S, Ghasemi Y, Savardashtaki A, Mirzaei H (2020) Quercetin and cancer: new insights into its therapeutic effects on ovarian cancer cells. Cell Biosci 10:32. https://doi.org/10.1186/s13578-020-00400-3

Van Dyk J, Battista JJ, Bauman GS (2013) Accuracy and uncertainty considerations in modern radiation oncology. Mod Technol Radiat Oncol 3:361–412

Vikram ENT, Ilavarasan R, Kamaraj R (2022) Anti-cancer activities of schedule E1 drugs used in ayurvedic formulations. J Ayurveda Integr Med 13(2):100545. https://doi.org/10.1016/j.jaim.2022.100545

Wang D, Liang G, Wang B, Zhou X, Huang Z, Zhao Y et al (2018) Recent advances in bioactive natural products for targeting the hallmarks of cancer. Anti Cancer Agents Med Chem 18(16):2111–2133. https://doi.org/10.2174/1871520618666181022150107

Wanjari MM, Harle UN, Gaikwad NJ (2010) Validated RP-HPLC method for quantification of curcumin in Curcuma longa and pharmacokinetic study in rats. J AOAC Int 93(6):1392–1397. https://doi.org/10.1093/jaoac/93.6.1392

Xie J, Yang Z, Zhou C, Zhu J, Lee RJ, Teng L (2016) Nanotechnology for the delivery of phytochemicals in cancer therapy. Biotechnol Adv 34(4):343–353. https://doi.org/10.1016/j.biotechadv.2016.04.002

Yallapu MM, Chauhan N, Othman SF, Khalilzad-Sharghi V, Kumar D, Jaggi M, Chauhan SC (2010) Implications of nanotechnology for cancer therapy: a brief review. Rev Recent Clin Trials 5(2):84–91. https://doi.org/10.2174/157488710791304735

Yarla NS, Bishayee A, Sethi G, Reddanna P, Kalle AM, Dhananjaya BL et al (2016) Targeting arachidonic acid pathway by natural products for cancer prevention and therapy. Semin Cancer Biol 40–41:48–81. https://doi.org/10.1016/j.semcancer.2016.05.004

Yashin A, Yashin Y, Xia X, Nemzer B (2017) Antioxidant activity of spices and their impact on human health: a review. Antioxidants 6(3):70. https://doi.org/10.3390/antiox6030070

Yau WP, Goh CH, Koh HL (2015) Quality control and quality assurance of phytomedicines: Key considerations, methods, and analytical challenges. Phytotherapies: efficacy, safety, and regulation. pp18–48. https://doi.org/10.1002/9781119006039.ch2

Yin H, Xu L, Porter NA (2011) Free radical lipid peroxidation: mechanisms and analysis. Chem Rev 111(10):5944–5972. https://doi.org/10.1021/cr200084z

Phytotherapy for Upper Respiratory Infections (URIs) 15

Aisha Shitu Sa'id

Abstract

Upper respiratory infections (URIs) are among the world's most common infectious diseases, impacting millions of people every year leading to significant discomfort and economic burden. Conventional treatments often focused on relieving symptoms, but interest in phytotherapy has increased due to the development of antibiotic resistance and the need for alternative solutions. The therapeutic potential of medicinal plants in treating common URIs, such as the common cold, sinusitis, laryngitis, and pharyngitis, is explored in this chapter. We evaluated the key phytochemicals with antiviral, antibacterial, and anti-inflammatory activities, including phenolic acid, flavonoids, terpenoids, and alkaloids, in terms of their pharmacological properties. Utilizing a combination of clinical research and traditional usage, we critically assessed the effectiveness of several plant-based remedies, including ginger, garlic, peppermint, thyme, licorice root, elderberry, and Echinacea. Furthermore, the mechanisms of action, suitable doses, and possible adverse effects of these phytotherapeutic compounds have been addressed. This chapter aims to provide a comprehensive overview of phytotherapy as a viable alternative or complementary approach to conventional URI treatments by integrating scientific evidence and traditional knowledge, emphasizing its role in improving patient outcomes and promoting sustainable healthcare practices.

A. S. Sa'id (✉)
Department of Microbiology, Faculty of Life Sciences, Modibbo Adama University, Yola, Adamawa State, Nigeria
e-mail: aishashitu@mau.edu.ng

K. Kuca et al. (eds.), *Medicinal Plants and Their Bioactives in Human Diseases*,
https://doi.org/10.1007/978-3-032-01356-9_15

Keywords

Clinical application · Echinacea · Elderberry · Phytotherapy · Plant-based remedies · Upper respiratory infections

1 Introduction

Upper respiratory infections (URIs) are among the world's most common infectious diseases that influence millions of lives each year (James et al. 2018; Banoon et al. 2021). These infections affect the upper respiratory tract, involving the nose, sinuses, pharynx, and larynx. URIs usually occur by viruses; however, they may also be triggered by bacteria, fungi, or helminthic infections (Chen et al. 2019). Rhinoviruses, coronaviruses, adenoviruses, and respiratory syncytial virus, or RSV, constitute the most common viral agents. Bacterial causes include *Streptococcus pyogenes*, *Haemophilus influenza*, *Moraxella catarrhalis*, *Klebsiella pneumoniae*, and *Streptococcus pneumoniae* (Thapa et al. 2017; Hu et al. 2020; Dutu et al. 2022). URIs may come with a variety of symptoms, including a runny nose, sore throat, cough, nasal congestion, and sneezing. Common symptoms include fever, headache, and malaise. The diagnosis is mostly clinical, relying on symptoms and a physical examination (Bartlow et al. 2022). Certain pathogens may occasionally be identified by laboratory testing like polymerase chain reaction (PCR) assays, fast antigen detection tests, or throat cultures (Garcia and Patel 2024). Because of productivity losses and medical expenses, these infections represent a substantial financial burden. A global report of 17.2 billion URI cases was reported in 2015 (Smith and Brown 2024). Even while the majority of URIs resolve on their own, complications like sinusitis, otitis media, and lower respiratory tract infections can occur, especially in susceptible groups including young children, elderly, and immunodeficient people (Geffen 2006; Metz et al. 2023). The main goal of URI management is to alleviate symptoms. Commonly used over-the-counter drugs include decongestants, antipyretics, analgesics, and antihistamines. To prevent the development of drug resistance, antibiotics are only used for bacterial infections. Good hand cleanliness, preventing close contact with diseased people, and, when necessary, vaccination—such as the influenza vaccine—are all preventive methods.

Traditional therapies for URIs largely rely on symptomatic alleviation and, in certain situations, antibiotic medication. Nevertheless, these techniques have many drawbacks that impair their overall effectiveness and safety (Little et al. 2023). To relieve the symptoms of URIs, over-the-counter drugs such cough suppressants, decongestants, and antihistamines are frequently used. Such drugs may not treat the root source of the infection, even though they may offer short-term respite (Smith et al. 2022). Additionally, these drugs might have negative impacts like drowsiness, dry mouth, and elevated blood pressure, and their effectiveness is frequently restricted. For example, research conducted by Smith et al. (2022) showed that the length and intensity of symptoms are not significantly affected by various over-the-counter cold remedies.

Although the majority of URIs are viral in nature and do not respond to antibiotic treatment, antibiotics are nonetheless commonly recommended for these infections. Misuse of antibiotics often leads to the rising problem of resistant microbes, which constitutes a substantial public health threat (WHO 2023). A study by Fleming-Dutra et al. (2022) found that over 30% of antibiotic prescriptions for URIs are unwarranted, which can have negative side effects and raise medical expenses. Adverse drug reactions, varying from minor gastrointestinal problems to serious allergic response, can result from the overuse and misuse of antibiotics (Chouhan 2024). Furthermore, the extensive use of antibiotics has sped up the emergence of bacteria resistant to them, making it more difficult to properly treat bacterial illnesses. A review by Ventola (2022) stressed the urgent need to control antibiotic misuse to prevent the escalation of resistance. While antiviral drugs can be beneficial against certain viral URIs, such as Influenza, their usage is limited to certain infections, and they must be provided early in the course of the illness to be effective. Moreover, the emergence of resistance to antiviral medications is a growing worry. Research carried out by Hayden et al. (2023) emphasized the limits of current antiviral medicines and the need for novel therapeutic options. The primary aim of this chapter is to explore the potential of phytotherapy as a complementary and alternative approach to managing URIs. By emphasizing the therapeutic roles of medicinal plants, this chapter seeks to bridge the gap between traditional remedies and modern medical practices. This chapter provides an overview of the prevalence and impact of URIs, highlighting the urgent need for innovative and effective treatment approaches. This chapter then identifies key medicinal plants, their phytocompounds, and mechanism of action in the management of URIs. This chapter also discusses the clinical efficacy and safety profiles of specific phytotherapeutic agents through existing research and case studies. Finally, challenges and opportunities associated with integrating phytotherapy into modern healthcare systems for URI management were also discussed.

2 Phytotherapy and Respiratory Health

Phytotherapy, the application of plant-derived treatments for therapeutic objectives, has garnered considerable interest in recent years for its possible advantages in regulating respiratory health (Popova et al. 2021). This technique uses the bioactive chemicals present in medicinal plants to ease symptoms, reduce inflammation, and boost overall respiratory functions. Several plant-based products used in phytotherapy offer significant anti-inflammatory and antioxidant effects that are helpful in controlling respiratory disorders such as asthma, chronic obstructive pulmonary disease (COPD), and upper respiratory infections (Popova et al. 2021). A study by Wadhwa et al. (2019) underscores the significance of phytomedicine in regulating inflammatory pathways and diminishing oxidative stress, thereby enhancing respiratory health. These qualities help in decreasing chronic inflammation and the tissue damage linked with respiratory distress. Plant-based medicines generally contain bioactive compounds that display antibacterial and anti-inflammatory activities (Kha and Le 2021).

2.1 Key Medicinal Plants Used for the Management of URIs

Throughout history, medicinal plants have been integral to the treatment of URIs, offering natural remedies rooted in traditional practices. Ancient civilizations such as Egyptians, Greeks, and Chinese utilized plants such as garlic (*Allium sativum*), thyme (*Thymus vulgaris*), licorice (*Glycyrrhiza glabra*), ginger (*Zingiber officinale*), elderberry (*Sambucus nigra*), echinacea (*Echinacea purpurea*), elderberry (*Sambucus nigra*), oregano (*Origanum vulgare*), peppermint (*Mentha piperita*), among others, for their antimicrobial, anti-inflammatory, and soothing properties (Fig. 15.1).

Native American tribes and Europeans have been using echinacea and elderberry to prevent and cure upper respiratory infections since the eighteenth century. These plants acquired popularity in the nineteenth and early twentieth centuries for treating common cold, flu, laryngitis, and other numerous diseases including lower respiratory infections (Marshall 2010). They possess antiviral, anti-inflammatory, and immune-boosting characteristics, making them beneficial for respiratory health (Zakay-Rones et al. 2004).

Licorice root, ginger, and garlic have been used in traditional Chinese medicine since ages for over 4000 years to treat infections including respiratory disorders (Rivlin, 2001; Fiore et al. 2008). These plants have garnered significant global attention for their therapeutic properties. These botanicals have long been recognized as integral components in the treatment of URIs across diverse cultural and geographical context, owing to their anti-inflammatory, antimicrobial, and antioxidant effect (Viljoen et al. 2014) (Table 15.1). Their widespread use highlights their enduring relevance in both traditional and complementary medicine frameworks.

Thyme, peppermint, and oregano have long been integral to traditional practices for medicinal and culinary uses. Thyme was historically employed for its antiseptic and antimicrobial properties, particularly in purification rituals in ancient Greece and Romes, and is now appreciated for supporting respiratory health (Table 15.1). Peppermint, known for its soothing effects, has been a staple in traditional remedies for digestion and respiratory issue (McKay and Blumberg 2020). Similarly, oregano has been celebrated since ancient times for its antibacterial and anti-inflammatory properties, often used to treat respiratory infections and digestive disorders (Nurzynska-Wierdak and Wlasek-Janusz 2025) (Table 15.1). Today, these plants not only remain key ingredients in Mediterranean and global cuisines but also gain popularity but also for their health benefits. Table 15.1 shows some instances of some key medicinal plants and their therapeutic benefits.

2.2 Mechanisms of Action of Phytocompounds

These bioactive chemicals can operate through various methods as stated below:

Fig. 15.1 Images of some medicinal plants used as traditional remedies for upper respiratory infections

2.2.1 Cell Wall Disruption

Many plant extracts contain chemicals that can disrupt the bacterial cell wall resulting to cell lysis and death. For example, essential oils from plants like oregano and thyme contain thymol and carvacrol, which have been found to damage the cell walls of bacteria by integrating into the lipid bilayer of bacterial cell membranes, causing structural integration. In gram negative bacteria, carvacrol can cause the release of lipopolysaccharides (LPS) from the outer membrane. This increases membrane permeability and compromises the integrity of the bacterial cell wall (Kha and Le 2021).

Table 15.1 Phytochemical constituents, effects, and uses of some medicinal plants for Upper respiratory Infections (URIs)

Medicinal plants	Phytochemical constituents	Uses	Benefits	URI Treated	References
Ginger (*Zingiber officinale*)	Gingerol, Shogaol, Zingiberene	Antimicrobial, anti-inflammatory, anti-emetic	Relieve sore throat and cough	Common cold, bronchitis	Harun and Mohamad (2023)
Garlic (*Allium sativum*)	Allicin, saponin, flavonoids	Antimicrobial, antiviral	Enhances immune responses	Common cold, flu	Banerjee et al. (2022)
Licorice (*Glycyrrhiza glabra*)	Anthocyanins, flavonoids	Antiviral, anti-inflammatory	Soothes throat. Reduce inflammation	Bronchitis, sore throat	Wahab et al. (2021)
Turmeric (*Curcuma longa*)	Curcumin, demethoxycurcumin	Antiviral, anti-inflammatory	Boost immunity, reduces inflammation	Sinusitis, common cold	Pal and Bareth (2023)
Elderberry (*Sambucus nigra*)	Anthocyanins, flavonoids	Antiviral, immune boosting	Reduces severity of symptoms	Flu, common cold	Strugala et al. (2018)
Eucalyptus (*Eucalyptus globulus*)	Eucalyptol, tannin, flavonoids	Expectorants, antimicrobial	Clears nasal congestion	Sinusitis, bronchitis, laryngitis	Pal and Bareth (2023)
Peppermint (*Mentha piperita*)	Menthol, flavonoids, tannins	Decongestant, antimicrobial, antiseptic	Sooth throats, relieve cough	Common cold, sinusitis	McKay and Blumberg (2020)
Oregano (*Origanum vulgare*)	Carvacrol, thymol, rosmarinic acid	Antimicrobial, antiviral	Relieves congestion, fight infection	Sinusitis, bronchitis	Nurzynska-Wierdak and Wlasek-Janusz (2025)
Thyme (*Thymus vulgaris*)	Thymol, carvacrol, flavonoids,	Antimicrobial, expectorants, antiseptic	Clears mucus, soothes throat	Bronchitis, common cold	Marchese et al. (2016)
Echinacea (*Echinacea purpurea*)	Alkamides, caffeic acid derivatives	Immune boosting, antiviral	Reduces duration of symptoms	Flu, common cold	Shah et al. (2007)
Pelargonium sidoides	Coumarins, polyphenols	Antibacterial, expectorants	Clear mucus, relieves cough	Acute bronchitis	
Holy basil (*Ocimum sanctum*)	Eugenol, tannin, flavonoids	Antimicrobial, anti-inflammatory	Boost immunity, relieves cough	Common cold, flu	Njoku and Obi (2009)

Aloevera (*Aloe barbadensis miller*)	Anthraquinone, flavonoids	Antimicrobial, anti-inflammatory	Sooth throat irritation, boost immunity	Pharyngitis, sore throat	Mendicino et al. (2022)
Mustard (*Brassica juncea*)	Glucosinolates, flavonoids, alkaloids	Antimicrobial, expectorants	Clear mucus, reduce inflammation	Sinusitis, bronchitis	Lucero et al. (2025)
Chamomile (*Matricaria recutita*)	Apigenin, bisabolol, flavonoids	Antioxidant, anti-inflammatory	Relieves cough, soothes throat	Common cold, flu	Massingill et al. (2021)
Calendula (*Calendula officinalis*)	Flavonoids, carotenoids, saponin	Antimicrobial, anti-inflammatory	Reduces inflammation, enhances immunity	Sore throat, tonsilitis	Mendicino et al. (2022)
Lemon balm (*Melissa officinalis*)	Terpenoids, polyphenols flavonoids	Antiviral expectorant	Ease congestion, fights viral infections	Influenza, rhinovirus infection	Lucero et al. (2025)
Celandine (*Chelidonium majus*)	Flavonoids, alkaloids, saponins	Antibacterial, bronchodilator	Improves breathing, kills pathogens	Common cold, flu	Massingill et al. (2021)
Lavender (*Lavandula angustifola*)	Linalool, terpenoids, flavonoids	Antimicrobial, relaxant	Reduces inflammation, relieves cough	Asthma, sore throat	Mendicino et al. (2022)
Dandelion (*Taraxacum officinale*)	Sesquiterpenes, polyphenols flavonoids	Antioxidant, immune booster	Enhances respiratory health, reduces inflammation,	Common cold, flu	Lucero et al. (2025)
Rosemary (*Rosmarinus officinalis*)	Rosmarinic acid flavonoids, terpenoids	Antimicrobial, expectorants	Clears airways, fight bacterial infections	Sinusitis, bronchitis	Massingill et al. (2021)
Yarrow (*Achillea millefolium*)	Flavonoids, alkaloids, tannins	Antimicrobial, anti-inflammatory	Reduce mucus production, ease congestion	Pharyngitis, sinusitis	Mendicino et al. (2022)
Cowberry (*Vaccinium vitis-idaea*)	Anthocyanins, polyphenols, fFlavonoids	Antiviral, immune booster	Strengthen immunity, fights viral infections	Flu, rhinovirus infection	Lucero et al. (2025)

2.2.2 Cytoplasmic Membrane Damage

Some plant chemicals can integrate into the cytoplasmic membrane, increasing its permeability and triggering leakage of cellular contents (Nazzaro et al. 2013). This is a frequent mechanism for phenolic chemicals present in plants including echinacea, elderberry, tea tree, and eucalyptus (Strugala et al. 2018; Yaps et al. 2021).

2.2.3 Inhibition of Nucleic Acid Production

Certain plant-derived chemicals can hinder the production of DNA and RNA in bacteria. For instance, alkaloids from plants such as garlic and licorice have been proven to interfere with microbial DNA and RNA synthesis. Alkaloids intercalate between DNA base pairs, altering the natural structure of DNA and slowing the replication and transcription process necessary for nucleic acid production (Banoon et al. 2021).

2.2.4 Protein Synthesis Inhibition

Flavonoids and other polyphenols can attach to bacterial ribosomes, decreasing protein synthesis. These chemicals impede the action of several translation factors, which are important for the initiation, elongation, and termination phases of protein synthesis, impeding the normal assembly and function of the protein synthesis machinery. This process is recognized in chemicals from plants including licorice, garlic, and green tea (Mitropoulou et al. 2023).

3 URI and Antimicrobial Properties

URI, often caused by bacterial or viral pathogens, is among the most common illnesses globally. The increasing resistance to conventional antibiotics has prompted interest in medicinal plants with broad-spectrum antimicrobial activity as potential alternatives for managing URIs (Ho et al. 2024; Campbell 2025). Phytochemicals such as flavonoids, alkaloids, and essential oils have shown inhibitory effects against URI-related pathogens like *Streptococcus pneumoniae* and *Haemophilus influenzae*. Echinacea has demonstrated efficacy against a variety of pathogens, such as *Escherichia coli* and *Staphylococcus aureus* (Shah et al. 2007). Berberine blocks bacterial DNA gyrase, preventing replication of *S. pneumoniae* and *Haemophilus influenzae* (Hernandez-Ruiz et al. 2025). Additionally, they increase bacterial membrane permeability, leading to ion leakage and microbial death (Mutha et al. 2021) and also interfere with bacterial communication, reducing virulence factor production (Hernandez-Ruiz et al. 2025). Eucalyptol and allicin present in eucalyptus and garlic have shown effectiveness on both Gram-positive and Gram-negative bacteria (Pal and Bareth 2023). Quercetin prevents bacterial biofilm formation, reducing *Streptococcus pyogenes* persistence in pharyngitis, while apigenin and luteolin inhibit influenza virus entry into host cells, reducing common cold severity (Oluwole et al. 2022). Flavonoids promote tight junction protein expression such as occluding and claudin, preventing microbial invasion and maintaining barrier integrity as well as blocking bacterial metabolic enzymes limiting microbial survival (Mutha et al.

2021). Terpenoids like thymol and carvacrol from thyme and oregano have significant antibacterial properties. They act in various ways, which included causing bacterial structural instability by weakening the bacterial lipid membranes, blocking the microbial electron transport chains, and preventing virus attachment and replication, reducing influenza virus replication (Mutha et al. 2021). Additionally, they stimulate mucin production, increasing mucus layer thickness and protecting epithelial cells from pathogens. (Nurzynska-Wierdak and Wlasek-Janusz, 2025). Phenolic acids, such as caffeic acid, ferulic acid, gallic acid and chlorogenic acid, had proved to possess antibacterial effects by disrupting bacterial membrane and virulence factor activation, reducing *Streptococcus* infections (Mutha et al. 2021). They also suppress fungal growth by interfering with ergosterol (Oluwole et al. 2022). Anthocyanins are rich in medicinal plants like elderberry and are well known for their antibacterial and antiviral effects. They disrupt the integrity of bacteria cell wall and inhibit bacteria from forming biofilms (Strugala et al. 2018). Tannins block bacterial enzymes essential for metabolism and replication (Pal and Bareth, 2023). They reduce mucus overproduction by alleviating respiratory symptoms. Saponins disrupts cell membrane by interacting with the bacterial membranes, causing leakage of cellular contents and also viral lipid membranes leading to viral inactivation (Banerjee et al. 2022).

4 URI and Anti-inflammatory Action

Inflammation is a key component of upper respiratory infections (URIs), contributing to symptoms such as sore throat, nasal congestion, and coughing. Medicinal plants with anti-inflammatory properties can help alleviate these symptoms by modulating inflammatory mediators like cytokines and prostaglandins (Ullah et al. 2020; Okoduwa et al. 2024). Bioactive compounds such as curcumin, quercetin, and glycyrrhizin have shown potential in reducing inflammation associated with URIs. Several phytochemicals were found to decrease the release of cytokines including TNF-α, IL-1β, and IL-6 by interacting with the Toll-like receptor 4 (TLR4) signaling pathway and downregulation of inflammatory mediators to prevent excessive immune activation (Oluwole et al. 2022). For example, anthocyanins from elderberry and glycyrrhizin from licorice have been found to lower the levels of these cytokines in several inflammatory circumstances (Strugala et al. 2018: Mitropoulou et al. 2023) (Fig. 15.3). Polyphenols and saponins reduce inflammation by modulating the production of pro-inflammatory cytokines and enhancing the activity of macrophages. They improve antigen presentation by stimulating macrophage and dendritic cell activity. These help in alleviating symptoms of inflammatory conditions by reducing tissue damage, enhancing tissue regeneration, mucosal healing in infections, and strengthening respiratory defense (Shukla et al. 2014; Hernandez-Ruiz et al. 2025). Alkaloids suppress cyclooxygenase and lipoxygenase pathways, reducing inflammation in pharyngitis and laryngitis (Mutha et al. 2021). Flavonoids modulate T-cell and B-cell responses, enhancing adaptive immunity. Phenolic compounds from echinacea and anthocyanins from elderberry scavenge free radicals, decreasing oxidative stress and protecting respiratory epithelial cells from damage (Vaou et al. 2023; Oluwole et al. 2022) (Fig. 15.3). This antioxidant activity supports overall immune health and may contribute to its therapeutic effects in managing respiratory infections (Shukla et al. 2014). Elderberry contains substances that can block these pathways, enhance immune response, and assist in pathogens elimination (Strugala et al. 2018) (Fig. 15.3). Terpenoids suppress nitric oxide and reduce airway inflammation and congestion in common cold and prevent fluid accumulation in laryngitis, alleviating vocal strain. As shown in Fig. 15.2, the presence of hydroxyl and other functional groups in the compounds may contribute to their antimicrobial and antioxidant effects. (Fig. 15.2).

5 Safety and Efficacy

It is essential to assess the safety profiles of these plants through scientific research so as to prove that they do not present any health dangers. Studies conducted both in vitro and in vivo are commonly used to evaluate the safety of these plant-based treatments. While in-vivo researches utilize animal models, like Wistar rats, to assess acute and chronic toxicity, in-vitro studies analyze the plant extracts on bacterial cultures to ascertain their antibacterial activity and possible cytotoxicity. Histopathological evaluations, behavioral alterations, and lethality are among the parameters considered. When used in moderation (0.2 mg/kg/day), licorice is usually regarded as safe (Wahab et al. 2021). However, owing to its glycyrrhizin concentration, prolong intake of licorice can have major adverse effects including pseudoaldosteronism (Wang et al. 2020). Short-term usage of echinacea is thought to be safe, but long-term use has not been thoroughly investigated (Hudson et al. 2010). When taken in reasonable proportions, oregano and peppermint are generally harmless for most people, but when used undiluted, oregano oil can irritate skin (Nurzynska-Wierdak and Wlasek-Janusz 2025). In culinary applications, other compounds including elderberry, thyme, ginger, and garlic are generally considered safe for most people; however, large dosages may result in moderate adverse effect (Ali et al. 2020).

The glycyrrhizin in licorice can cause oedema, heart failure, hypokalemia, hypertension, and hormonal instability when consumed in excess. It may intensify corticosteroid effects, raise the risk of hypokalemia, augment diuretic effects, increase the risk of hemorrhage, and reduce the efficacy of certain HIV medications (Ali et al. 2020). Echinacea can upset the gastrointestinal tract and trigger allergic reactions ranging from rash to asthma and anaphylaxis, especially in people who are allergic to plants in the daisy family. Long-term usage might cause implications for the immune system especially with those people at the risk of developing autoimmune diseases. It can also lower caffeine blood concentration, which can lead to side effects such as jitteriness and headaches (Hudson et al. 2010).

Raw elderberries contain cyanogenic glycosides, which can lead to nausea, vomiting, and diarrhea if ingested. It could also inhibit the activity of immunodepressants, interacting with diabetes medication by limiting their ability to control

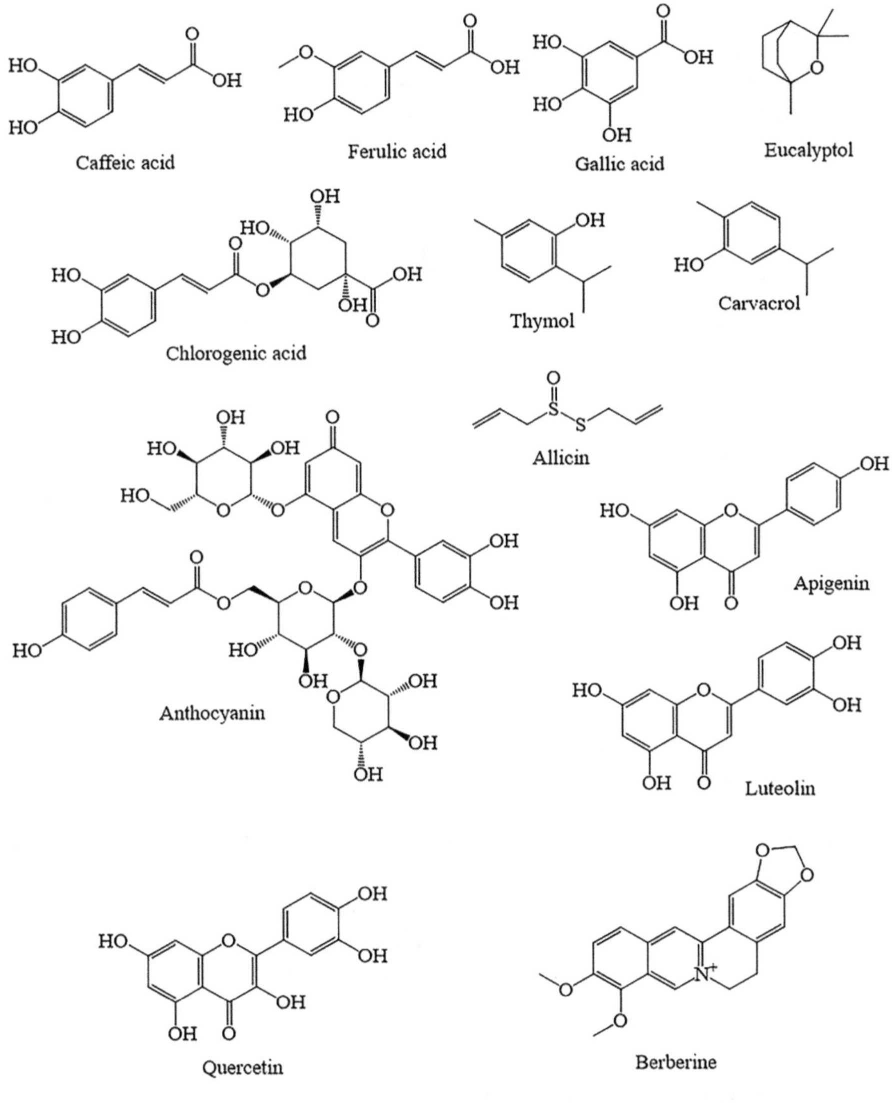

Fig. 15.2 Chemical structure of some phytocompounds used for treatment of upper respiratory infections via antimicrobial or anti-inflammatory action. Chemical structures were drawn using Chem draw Professional 15.0

immune system coupled with reducing the blood sugar level resulting to hypoglycemia (Roschek et al. 2009). Thyme, oregano, and peppermint essential oils can cause skin irritation and allergic reactions including rash and itching if used undiluted, and ingesting large amounts can lead to gastrointestinal issues including acid reflux and heartburn. Thyme may affect hormone levels, so it should be used in patients with hormone-sensitive conditions. It may also lower blood pressure, necessitating

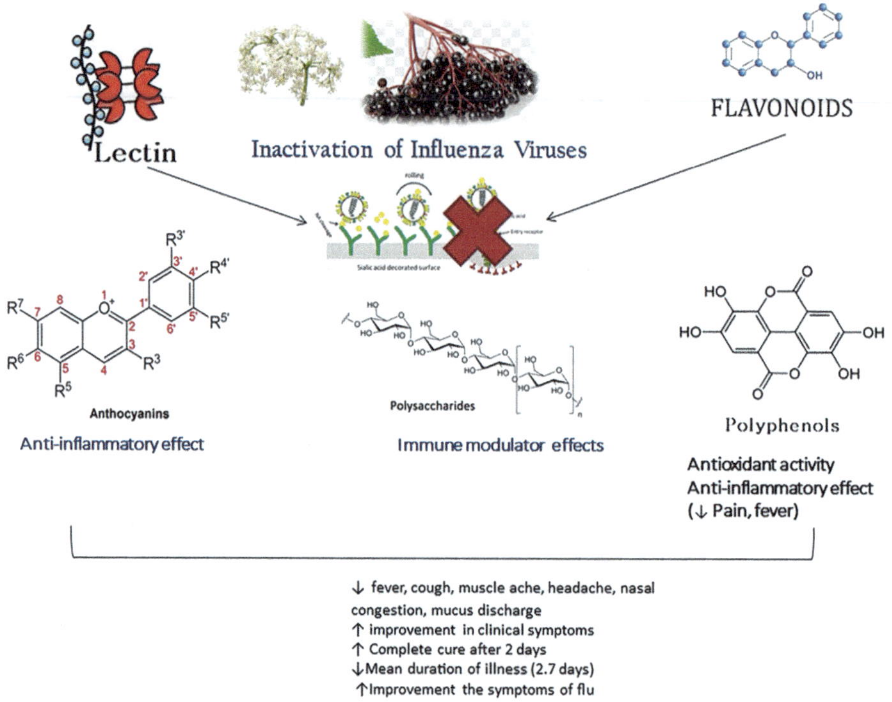

Fig. 15.3 Mechanisms of action for the efficacy of elderberry plant in treatment of URIs (Modified from Mahboubi 2021)

adjustments in hypertension medications (Marchese et al. 2016). Oregano may lower blood sugar levels, requiring adjustments in diabetes medications, and can enhance the effects of blood-thinning medications (Nurzynska-Wierdak and Wlasek-Janusz 2025). Peppermint oil capsules may dissolve too quickly if taken with antacids, leading to heartburn, and may also increase the absorption of cyclosporine, a drug used to prevent organ rejection (Marchese et al. 2016).

6 Dosage and Administration

Licorice root has been used in daily doses ranging from 760 mg to 15 g for conditions like ulcers and gastritis. The acceptable daily intake (ADI) for glycyrrhizin is suggested to be 0.2 mg/kg/day (Wahab et al. 2021). Ginger has been used in clinical trials in dosages of 170 mg to 1 g, taken 3–4 times daily. For respiratory infections, about 1–3 g of powdered ginger or a quarter to half inch slice of fresh ginger is recommended (Harun and Mohamad, 2023). Garlic is typically consumed in doses of 2–5 g of fresh raw garlic, 0.4–1.2 g of dried garlic powder, or 300–1000 mg of garlic extract (Rivlin, 2001). Echinacea is often taken in doses of 300–500 mg of dried extract, 2.5 ml of tincture, or 10 ml of juice, 3 times daily for up to 10 days

(European Medicine Agency (EMA), 2015). Elderberry is commonly used in doses of 15 ml of syrup, 3–4 times daily, or 300 mg of extract, 2–3 times daily (Strugala et al. 2018). Thyme is used in doses of 1–2 g of dried herb, 3 times daily, or 1–2 drops of essential oil diluted in a carrier oil for topical use (EMA 2015). Oregano oil is typically used in doses of 200 mg, 3 times daily, or 1–2 drops of essential oil diluted in a carrier oil for topical use (Nurzynska-Wierdak and Wlasek-Janusz 2025). Peppermint oil is used in doses of 0.2–0.4 ml of essential oil, 3 times daily, or 1–2 drops diluted in a carrier oil for topical use (McKay and Blumberg 2020).

7 Clinical Evidence and Studies

The immunomodulatory activities and effectiveness of Echinacea in managing respiratory diseases were highlighted in a recent study by Ahmadi et al. (2024) that assessed the plant's bioactive components and agronomy. Furthermore, Echinacea can decrease the length of cold symptoms by 1.4 days and lower the probability of getting the flu by 58%, according to a meta-analysis conducted by Shah et al. (2007). Nevertheless, the findings from numerous researches have been inconsistent, suggesting that more study is necessary. Elderberry administration decreased the duration and severity of cold-related symptoms in air travelers, indicating its potential for application in the treatment of respiratory tract infections (Tiralongo et al. 2020). According to a work by Zakay-Rones et al. (2004), elderberry extract also considerably decreased the intensity and length of influenza symptoms. The antiviral characteristics of liquorice root and its efficacy for treating respiratory illnesses were examined by Li et al. (2021). They discovered that glycyrrhizin has strong antiviral and anti-inflammatory qualities, which might make it a useful treatment for diseases like COVID-19. Anti-inflammatory and antioxidant properties of ginger were examined in research work by Marx et al. (2021), which showed promise for treating respiratory infections. The findings were conflicting, though, and more study is required to validate these advantages and establish the ideal dose. Garlic has antibacterial and antiviral qualities that support its applications in treatment of respiratory tract infections, as reported by Banerjee et al. (2022). They discovered that garlic has strong antibacterial and immunomodulatory qualities, especially in its organosulfur compounds, which can lessen the intensity and length of upper respiratory infections. Salehi et al. (2021) investigated the antibacterial and anti-inflammatory qualities of thyme. The essential oils of thyme, especially thymol and carvacrol, were discovered to have strong antibacterial action on a variety of respiratory infection-causing microorganisms and to have potential for lowering inflammation. McKay and Blumberg (2020) examined the bioactivity of peppermint and its health advantages. Its essential oils, especially menthol and menthone, have been shown to have strong antibacterial, antiviral, and anti-inflammatory qualities. The antibacterial, antioxidant, and anti-inflammatory properties of oregano against respiratory disorders were examined in a research work by Leyva-López et al. (2021). They discovered that the chemicals contained in oregano, such as thymol and carvacrol, shown strong antibacterial activity against a range of pathogens and also had the ability to lower inflammation and oxidative stress. In a case study published in 2023, Smith and colleagues described how using echinacea supplements significantly decreased the incidence of recurring respiratory infections in a patient (Smith et al. 2023). The application of elderberry syrup in a family with influenza was detailed in a case study in phytotherapy research conducted by Barak et al. (2021). The family members observed less severe symptoms and quicker recuperation times. The use of licorice root extract in a patient with chronic cold and flu symptoms was documented in a case study by Li et al. (2021). The patient's symptoms and lung function significantly improved. The usage of ginger pills in a patient receiving chemotherapy was detailed in a case report published in Supportive Care in Cancer by Ryan et al. (2022). The patient reported better general health and less nausea.

8 Challenges and Future Directions

Despite the possible advantages, there are a number of difficulties when combining phytotherapy with traditional medicine. The reliability and efficacy of many herbal treatments remain insufficiently tested in clinical research, raising questions about their effectiveness. Additionally, disparities in the composition of herbal products can lead to inconsistent therapeutic outcomes. The use and licensing of herbal medicines are further complicated by differing national regulations, which can create challenges in standardization and accessibility (Raja et al. 2024). To ensure safety, it is essential to closely monitor potential interactions between medications and herbs, as these interactions could lead to adverse effects. Future studies should prioritize several key areas to enhance the use of phytotherapy in treating URIs. Extensive randomized controlled trials are essential for determining the effectiveness and safety of herbal remedies. Research should also explore the biological mechanisms through which herbal treatments exert their therapeutic effects. Additionally, standardization and quality control are vital, requiring the development of uniform guidelines for the dosing and production of herbal medications. Education and training programs are equally important to equip medical professionals with a deeper understanding of the benefits and limitations of phytotherapy.

9 Conclusions

With antibiotic resistance continuing to be a global health concern, the role of phytotherapy becomes more pertinent. These natural remedies, with their antimicrobial, anti-inflammatory, and immunomodulatory properties, offer an important substitute or a supplement to conventional treatments. Moreover, the use of medicinal plants has demonstrated significant potential in reducing symptoms of upper respiratory infections and enhancing the body's immune response. Future research should concentrate on large-scale clinical trials and the investigation of synergistic effects between various plant compounds. Additionally, educating the public about the advantages of phytotherapy and how to effectively employ it safely is a possible and achievable alternative for the comprehensive management of URIs, which may lead to better patient outcomes and reduced medical expenses.

References

Ahmadi F, Kariman K, Mousavi M, Rengel Z (2024) Echinacea: bioactive compounds and agronomy. Plants 13(9):1235. https://doi.org/10.3390/plants13091235

Ali BH, Blunden G, Tanira MO (2020) Ginger: a natural remedy for respiratory infections. Phytother Res 34(5):1003–1016. https://doi.org/10.1002/ptr.6603

Banerjee S, Santra D, Das S (2022) Antimicrobial and antiviral properties of garlic (Allium sativum): a review. J Ethnopharmacol 282:114122. https://doi.org/10.1016/j.jep.2022.114122

Banoon SR, Alrashedi HS, Kareem Al-Ataie SS, Mohamed IA (2021) Potential role of medicinal plants for the treatment of respiratory viruses: a review. Egypt J Chem 64(12):7495–7508. https://doi.org/10.21608/ejchem.2021.97912.4569

Barak V, Halperin T, Kalickman I (2021) The effect of Sambucol, a black elderberry-based, natural product, on the production of human cytokines: I. Inflammatory cytokines. Phytother Res 35(2):290–296. https://doi.org/10.1016/S0944-7113(01)80001-180001-1

Bartlow AW, Stromberg ZR, Gleasner CD, Hu B, Davenport KW et al (2022) Comparing variability in diagnosis of upper respiratory tract infections in patients using syndromic next generation sequencing, and PCR-based methods. PLoS Glob Public Health 2(7):e0000811. https://doi.org/10.1371/journal.pgph.0000811

Campbell L (2025) Traditional herbal plants and their phytoconstituents based remedies for respiratory diseases: a review. Open Respir Med J 19:e18743064341009

Chen J, Li X, Wang W, Jia Y, Lin F, Xu J (2019) The prevalence of respiratory pathogens in adults with community-acquired pneumonia in an outpatient cohort. Infect Drug Resist 12:2335–2341. https://doi.org/10.2147/IDR.S213296

Chouhan AS (2024) Overuse of antibiotics causes serious adverse effects in the body: a literature review. J Clin Med Health Care 12(3):45–57. https://doi.org/10.31579/2688-7517/196

Dutu LE, Popescu ML, Purdel CN, Ilie EI, Luţă EA, Costea L, Gîrd CE (2022) Traditional medicinal plants—a possible source of antibacterial activity on respiratory diseases induced by *chlamydia pneumoniae*, *haemophilus influenzae*, *Klebsiella pneumoniae* and *Moraxella catarrhalis*. Diversity 14:145. https://doi.org/10.3390/d14020145

European Medicine Agency (2015) European Union herbal monograph on *Echinacea purpurea* (L.) Moench, herba recens. Committee on Herbal Medicinal Products. https://www.ema.europa.eu/en/documents/herbal-monograph/final-european-union-herbal-mongraph-echinacea—purpurea-l-moench-herba-recens.

Fiore C, Eisenhut M, Krausse R, Ragazzi E, Pellati D, Armanini D, Bielenberg J (2008) Antiviral effects of *Glycyrrhiza* species. Phytother Res 22(2):141–148. https://doi.org/10.1002/ptr.2295

Fleming-Dutra KE, Shapiro DJ, Hyun DY, Hicks LA (2022) Frequency of first-line antibiotic selection among US ambulatory care visits for otitis media, sinusitis, and pharyngitis. JAMA Netw Open 5(10):e2437409

Garcia ML, Patel RK (2024) Symptomatology and diagnostic criteria for upper respiratory tract infections. Clin Microbiol Infect 29(2):123–130

Geffen L (2006) Common upper respiratory tract problems in the elderly: a guide to clinical diagnosis and prudent prescription. S Afr Fam Pract 48(5):20–23. https://doi.org/10.1080/20786204.2006.10873390

Harun NH, Mohamad MF (2023) *Zingiber officinale* (ginger): a systematic review and meta-analysis on antimicrobial activities. J Appl Pharm Sci 13(3):001–011. https://doi.org/10.7324/JAPS.2023.8330

Hayden FG, Shindo N, Gubareva LV (2023) Limited efficacy of antiviral treatments for upper respiratory infections: a comprehensive review. J Infect Dis 228(4):567–575

Hernandez-Ruiz RG, Olivares-Ochoa XC, Salinas-Varela Y, Guajardo-Esinoza D (2025) Phenolic compounds and anthocyanins in legumes and their impact on inflammation, oxidative stress, and metabolism: comprehensive review. Molecules 30(1):174. https://doi.org/10.3390/molecules30010174

Ho JSS, Ping TL, Paudel KR, El Sherkawi T, De Rubis G, Yeung S, Hansbro PM, Oliver BGG, Chellappan DK, Sin KP, Dua K (2024) Exploring bioactive phytomedicines for advancing pulmonary infection management: insights and future prospects. Phytother Res 38(12):5840–5872

Hu L, Han B, Tong Q, Xiao H, Cao D (2020) Detection of eight respiratory bacterial pathogens based on multiplex real-time PCR with fluorescence melting curve analysis. Can J Infect Dis Med Microbiol 2020:2697230. https://doi.org/10.1155/2020/2697230

Hudson JB, Vimalanathan S, Kang L (2010) Echinacea: antimicrobial and anti-inflammatory properties. Phytomedicine 17(8–9):563–568. https://doi.org/10.1016/j.phymed.2010.01.010

James SL, Abate D, Abate KH, Abay SM, Abbafati C et al (2018) Global, regional, and national incidence, prevalence, and years lived with disability for 354 diseases and injuries for 195 countries and territories, 1990–2017: a systematic analysis for the global burden of disease study. Lancet 392(10159):1789–1858. https://doi.org/10.1016/S0140-6736(18)32279-7

Kha TC, Le LTP (2021) Plant extracts: antimicrobial properties, mechanism of action and applications. In: Inamuddin, Ahamed MI, Prasad R (eds) Advanced antimicrobial materials and applications, Environmental and microbial biotechnology. Springer, Singapore, pp 257–283. https://doi.org/10.1007/978-981-15-7098-8_11

Leyva-López N, Gutiérrez-Grijalva EP, Vazquez-Olivo G, Heredia JB (2021) Essential oils of oregano: biological activity beyond their antimicrobial properties. Molecules 26(7):1906. https://doi.org/10.3390/molecules26071906

Li W, Li Y, Zhang X (2021) *Glycyrrhiza glabra*: a phytochemical and pharmacological review. J Ethnopharmacol 35(2):408–423. https://doi.org/10.1002/ptr.6824

Little P, Stuart B, Francis N et al (2023) Delayed antibiotic prescribing strategies for respiratory tract infections: pragmatic, randomized controlled trial. Lancet Infect Dis 23(1):45–54. https://doi.org/10.1016/S1473-3099(22)00567-8

Lucero M, Wang T, Hernandez P (2025) The role of Brassica juncea, Mellisa officinalis, Taraxacum officinale and Vaccinium vitis-idaea in respiratory infections: a phytochemical perspective. Phytother Res 39(2):112–126

Mahboubi M (2021) *Sambucus nigra* (black elder) as alternative treatment for cold and flu. Adv Trad Med 21:405–414. https://doi.org/10.1007/s13596-020-00469-z

Marchese A, Orhan IE, Daglia M (2016) Thymol and carvacrol: antimicrobial and anti-inflammatory properties. Molecules 21(1):1–15. https://doi.org/10.3390/molecules21010001

Marshall K (2010) Echinacea: a brief history of use and research. NDNR. https://ndnr.com/botanical-medicine/echinacea-a-brief-history-of-use-and-research/.

Marx W, McKavanagh D, McCarthy AL, Bird R, Ried K, D'Cunha NM, Isenring E (2021) The effect of ginger (*Zingiber officinale*) on inflammatory markers in blood: a systematic review of randomized controlled trials. J Nutr Intermed Metab 24:100110. https://doi.org/10.1016/j.jnim.2021.100110

Massingill B, Carter L, Evans D (2021) Matricaria recutita, Rosmamarinus officinalis and Chelidonium majus and their antimicrobial properties in upper respiratory infections. Int J Med 28(1):45–59

McKay DL, Blumberg JB (2020) A review of the bioactivity and potential health benefits of peppermint tea (*Mentha piperita* L.). Phytother Res 34(6):123–132. https://doi.org/10.1002/ptr.6557

Mendicino J, Patel R, Singh A (2022) Phytochemical and therapeutic potential of Aloe Vera, Calendula officinalis, Lavendular angustifolia and Achhillea millefolium in respiratory health. J Herb Med 15(4):289–302

Metz C, Schmid A, Veldhoen S (2023) Increase in complicated upper respiratory tract infection in children during the 2022/2023 winter session: a post-COVID-19 effect? Pediatr Radiol 54:49–57. https://doi.org/10.1007/s00247-023-05808-1

Mitropoulou G, Stavropoulou E, Vaou N, Tsakris Z, Voidarou C, Tsiotsias A, Tsigalou C et al (2023) Insights into antimicrobial and anti-inflammatory applications of plant bioactive compounds. Microorganisms 11(5):1156. https://doi.org/10.3390/microorganisms11051156

Mutha RE, Tatiya AU, Surana SJ (2021) Flavonoids as phenolic compounds and their role in therapeutics: an overview. Futur J Pharm Sci 7(1):25. https://doi.org/10.1186/s43094-020-00161-8

Nazzaro F, Fratianni F, De Martino L et al (2013) Effect of essential oils on pathogenic bacteria. Pharmaceuticals 6(12):1451–1474. https://doi.org/10.3390/ph6121451

Njoku OV, Obi C (2009) Phytochemical constituents of some selected medicinal plants. Afr J Pure Appl Chem 3(11):228–233. https://doi.org/10.5897/AJPAC.9000136

Nurzynska-Wierdak R, Wlasek-Janusz M (2025) Chemical composition, biological activity, and potential uses of oregano (*Origanum vulgare* L.) and oregano essential oil. Pharmaceuticals 18(2):267. https://doi.org/10.3390/ph18020267

Okoduwa SI, Abdulwaliyu I, Igiri BE, Arekemase SO, Okoduwa UJ, Itiat JF, Egbule MN, Mustapha RA (2024) Multi-therapeutic potential of flavonoids as an essential component in nutraceuticals for the treatment and management of human diseases. Phytomedicine Plus 4:100558

Oluwole O, Fernando WB, Lumanlan J, Ademuyiwa O, Jayasena V (2022) Role of phenolic acid, tannins, stilbenes, lignans, and flavonoids in human health—a review. Int J Food Sci Technol 57(10):6326–6335. https://doi.org/10.1111/ijfs.15936

Pal D, Bareth K (2023) Respiratory viral infections and the role of medicinal plants in prevention and treatment. In: Anti-viral metabolites from medicinal plants, pp 397–427. https://doi.org/10.1007/978-3-030-83350-3_10-1

Popova A, Milhaylova D, Spasov A (2021) Plant-based remedies with reference to respiratory diseases: a review. Open Biotechnol J 15(1):46–58. https://doi.org/10.2174/1874434602015010046

Raja M, Cramer H, Lee MS, Wieland LS, Ng JY (2024) Addressing the challenges of traditional, complementary, and integrative medicine research: an international perspective and proposed strategies moving forward. Perspect Integr Med 3(2):86–97. https://doi.org/10.56986/pim.2024.06.004

Rivlin RS (2001) Historical perspective on the use of garlic. J Nutr 131(3):951S–954S. https://doi.org/10.1093/jn/131.3.951S

Roschek B, Fink RC, McMichael MA (2009) Elderberry extracts inhibit influenza virus and modulate inflammatory cytokines. Phytochemistry 70(10):1255–1261. https://doi.org/10.1016/j.phytochem.2009.06.003

Ryan JL, Heckler CE, Roscoe JA, Dakhil SR, Kirshner JJ, Flynn PJ, Morrow GR (2022) Ginger (*Zingiber officinale*) reduces acute chemotherapy-induced nausea: a URCC CCOP study of 576 patients. Support Care Cancer 30(7):1479–1489. https://doi.org/10.1007/s00520-011-1236-3

Salehi B, Mishra AP, Shukla I, Sharifi-Rad M, Contreras MDM, Segura-Carretero A, Sharifi-Rad J (2021) Thymol, carvacrol, and thymol derivatives as antimicrobial agents against respiratory tract infections: a review. Microb Pathog 150:104118. https://doi.org/10.1016/j.micpath.2020.104118

Shah SA, Sander S, White CM, Rinaldi M, Coleman CI (2007) Evaluation of echinacea for the prevention and treatment of the common cold: a meta-analysis. Lancet Infect Dis 7(7):473–480. https://doi.org/10.1016/S1473-3099(07)70160-3

Shukla S, Bajpai VK, Kim M (2014) Plants as potential sources of natural immunomodulators. Rev Environ Sci Biotechnol 13:17–33. https://doi.org/10.1007/s11157-012-9303-x

Smith JA, Brown LM (2024) Global, regional, and national burden of upper respiratory infections. J Global Health 14(1):45–58. https://doi.org/10.1016/S1473-3099(24)00430-4

Smith A, Kardos P, Pfaar O, Randerath W et al (2022) The treatment of mild upper respiratory infections—a position paper with recommendations for best practice. Drugs Context 11:4–2. https://doi.org/10.7573/dic.2023-4-2

Smith T, Kawa K, Eckl V, Morton C, Stredney R (2023) Herbal supplement use for respiratory health: a review of the evidence. J Herbal Med 15(4):245–259. https://doi.org/10.1016/j.hermed.2023.100456

Strugala P, Loi S, Bazanow B et al (2018) A comprehensive study on the biological activity of elderberry extract and cyanidin 3-O-glucoside and their interactions with membranes and human serum albumin. Molecules 23(10):2566. https://doi.org/10.3390/molecules23102566

Thapa S, Gokhale S, Sharma AL, Sapkota LB, Ansari S, Gautam R, Shrestha S, Neopane P (2017) Burden of bacterial upper respiratory tract pathogens in school children of Nepal. BMJ Open Respir Res 4:e000203. https://doi.org/10.1136/bmjresp-2017-000203

Tiralongo E, Wee SS, Lea RA (2020) Elderberry supplementation reduces cold duration and symptoms in air-travelers: a randomized, double-blind placebo-controlled clinical trial. Nutrients 12(4):182. https://doi.org/10.3390/nu12040182

Ullah A, Munir S, Badshah SL, Khan N, Ghani L, Poulson BG, Emwas AH, Jaremko M (2020) Important flavonoids and their role as a therapeutic agent. Molecules 25(22):5243

Vaou N, Tsakris Z, Voidarou C, Tsiotsias A, Tsigalou C, Taban BM, Kourkoutas Y, Bezirtzoglou E (2023) Antimicrobial activity of plant bioactive compounds. Microorganisms 11(5):1156. https://doi.org/10.3390/microorganisms11051156

Ventola CL (2022) The antibiotic resistance crisis: part 1: causes and threats. Pharm Ther 40(4):277–283. https://doi.org/10.1016/j.antibiotres.2022.01.001

Viljoen E, Visser J, Koen N, Musekiwa A (2014) A systematic review and meta-analysis of the effect and safety of ginger in the treatment of pregnancy-associated nausea and vomiting. Nutr J 13:20. https://doi.org/10.1186/1475-2891-13-20

Wadhwa R, Gupta R, Sharma A (2019) Anti-inflammatory and antioxidant properties of medicinal plants used for upper respiratory infections. J Ethnopharmacol 234(3):567–575. https://doi.org/10.13040/IJPSR.0975-8232.10(10).4507-12

Wahab S, Annadurai S, Abullais S et al (2021) Glycyrrhiza glabra (Licorice): a comprehensive review on its phytochemistry, biological activities, clinical evidence and toxicology. Plant 10(12):2751. https://doi.org/10.3389/fphar.2021.680674

Wang Z, Li H, Zhang X (2020) Antimicrobial and anti-inflammatory activities of glycyrrhizin from licorice. J Ethnopharmacol 256:112–120. https://doi.org/10.1016/j.jep.2020.112120

World Health Organization (2023) Antimicrobial resistance. WHO fact sheets. https://www.who.int/news-room/fact-sheets/detail/antimicrobial-resistance

Yap PSX, Yusoff K, Lim SHE et al (2021) Membrane disruption properties of essential oils- a double-edge sword? PRO 9(4):595. https://doi.org/10.3390/pr9040595

Zakay-Rones Z, Thom E, Wollan T, Wadstein J (2004) Randomized study of the efficacy and safety of oral elderberry extract in the treatment of influenza A and B virus infections. J Int Med Res 32(2):132–140. https://doi.org/10.1177/147323000403200205

Plant-Based Treatments for Asthma and Chronic Obstructive Pulmonary Disease (COPD)

16

Shivakumar, Hawa Naafortu Issah, and D. Udayakumar

Abstract

Asthma and chronic obstructive pulmonary disease (COPD) are long-term respiratory disorders characterized by airway inflammation, blockage, and oxidative stress. These conditions compromise lung function and negatively impact overall quality of life. While conventional medications are effective, they can often lead to undesirable side effects, sparking interest in alternative therapeutic approaches. Plant-based therapies, leveraging bioactive compounds found in medicinal herbs, present promising complementary or alternative options. Natural compounds like flavonoids, terpenoids, alkaloids, and phenolic acids possess anti-inflammatory, antioxidant, and bronchodilatory effects, which can help alleviate symptoms of asthma and COPD. This chapter explores the therapeutic potential of plant-based treatments, examining species like *Boswellia serrata*, *Glycyrrhiza glabra*, and *Curcuma longa*, which have demonstrated positive outcomes in both preclinical and clinical research. Topics include the mechanisms by which these phytochemicals exert therapeutic effects, their health benefits, and the challenges in incorporating plant-based remedies into conventional care. Additionally, the chapter discusses current research gaps and future directions to improve the safety, efficacy, and availability of plant-derived treatments for respiratory diseases, aiming to reduce dependence on synthetic medications.

Shivakumar · D. Udayakumar (✉)
Organic and Medicinal Chemistry Laboratory, Department of Chemistry, National Institute of Technology Karnataka, Mangalore, Karnataka, India
e-mail: udayakumar@nitk.edu.in

H. N. Issah
University of Education, Winneba, Ghana

K. Kuca et al. (eds.), *Medicinal Plants and Their Bioactives in Human Diseases*,
https://doi.org/10.1007/978-3-032-01356-9_16

Keywords

Asthma · COPD · Medicinal plants · Phytochemicals · Respiratory disease

1 Introduction

Asthma and Chronic Obstructive Pulmonary Disease (COPD) are among the most prevalent chronic respiratory disorders, affecting millions of people worldwide. Despite medical advancements, these conditions continue to contribute significantly to illness and death, with their prevalence rising globally. While both asthma and COPD involve chronic airway inflammation, airflow restriction, and respiratory symptoms, they differ in underlying mechanisms, triggers, and disease progression. Asthma is a long-term inflammatory condition of the airways, characterized by reversible airflow obstruction and heightened sensitivity to various stimuli. It can affect individuals of all ages but often begins in childhood. Common symptoms include wheezing, shortness of breath, chest tightness, and coughing, which frequently worsen at night or in the early morning. The disease mechanism involves airway inflammation triggered by allergens or irritants, leading to the release of inflammatory substances such as histamine, cytokines, and leukotrienes, resulting in bronchoconstriction (Akinbami et al. 2011). In contrast, COPD is a progressive respiratory condition mainly resulting from prolonged exposure to harmful pollutants, with cigarette smoke being the leading cause. COPD encompasses conditions such as chronic bronchitis and emphysema, both of which are marked by persistent and irreversible airflow obstruction. COPD leads to persistent airway inflammation and the deterioration of lung parenchyma, and airway remodeling, which leads to progressive dyspnea, chronic cough, sputum production, and frequent respiratory infections. Unlike asthma, the inflammation in COPD is often less reversible, and the disease progresses over time, making it a leading cause of death globally (GOLD 2024). Both asthma and COPD are associated with significant healthcare costs, reduced quality of life, and an increased burden on the healthcare system due to frequent hospitalizations and exacerbations. The management of these diseases often requires a combination of pharmacological and non-pharmacological treatments, with a focus on controlling inflammation, improving lung function, and preventing exacerbation. While conventional medications such as bronchodilators, corticosteroids, and other anti-inflammatory drugs have been pivotal in the management of asthma and COPD, they have several limitations such as osteoporosis, cataracts (Hosker et al. 2007), tachycardia, tremors (Marciniuk and Calzetta 2023), and dry mouth, etc., which reduce their long-term effectiveness and increase the risk of adverse effects.

Herbal remedies have been a part of traditional medicine for centuries, used to address various health conditions, including respiratory disorders. In recent years, there has been increasing interest in incorporating plant-based treatments alongside conventional therapies for managing asthma and COPD. This interest is driven by the increasing recognition of the therapeutic potential of plant-derived compounds and the need for alternative options in managing these chronic diseases. Chronic inflammation and oxidative stress are central to the pathophysiology of both asthma and COPD. Numerous plant-derived compounds, including flavonoids, polyphenols, and terpenes, have shown significant anti-inflammatory and antioxidant effects. These properties can help alleviate airway inflammation, prevent oxidative damage, and reduce airway remodeling (Nicikowski and Julita Reguła 2021). For example, curcumin, a polyphenolic compound present in turmeric, has been shown to suppress the release of pro-inflammatory cytokines and reduce oxidative stress, making it a promising adjunct in asthma and COPD management (Safari et al. 2023). Several plants possess natural bronchodilator properties that can help ease symptoms like wheezing and dyspnea. Ginger, for example, contains compounds like gingerol, which have been shown to relax airway smooth muscle and enhance respiratory function (Townsend et al. 2013). Similarly, *Boswellia serrata* has demonstrated significant anti-inflammatory and mucolytic effects, helping to reduce airway obstruction and improve airflow in both asthma and COPD patients (Roy et al. 2019). Many plant-based treatments exhibit antimicrobial, antiviral, and immunomodulatory activities that can help prevent or manage respiratory infections. For instance, thyme and oregano are rich in essential oils with potent antimicrobial properties, which may aid in the prevention of upper respiratory infections and reduce the frequency of exacerbations in asthma and COPD patients (Cushnie and Lamb 2005). One of the primary advantages of plant-based treatments is their generally favorable safety profile when used appropriately. Unlike corticosteroids and long-term bronchodilators, most herbal remedies have fewer side effects and lower risks of dependence. This makes them an attractive option for long-term management, especially in patients who experience adverse reactions to conventional drugs (Rajizadeh et al. 2024). Plant-based treatments are often more affordable and accessible than conventional pharmaceuticals, particularly in low- and middle-income countries where expensive medications may be less available. This makes plant-based therapies a viable option for managing asthma and COPD in resource-constrained settings (Yeh and Horwitz 2017). In this chapter, we explore the therapeutic potential of plant-based treatments, which have demonstrated positive outcomes in both preclinical and clinical research. Topics include the mechanisms by which these phytochemicals exert therapeutic effects, their health benefits, and the challenges in incorporating plant-based remedies into conventional care. Additionally, the chapter discusses current research gaps and future directions to improve the safety, efficacy, and availability of plant-derived treatments for respiratory diseases, aiming to reduce dependence on synthetic medications.

2 Pathophysiology of Asthma and COPD

2.1 Pathophysiology of Asthma

Asthma is a long-term inflammatory condition of the airways that can affect people of all ages, marked by temporary and reversible airflow blockage. The disease is driven by complex interactions between genetic predisposition and environmental factors, leading to hyperresponsiveness of the airways and inflammation. The hallmark of asthma is airway inflammation, which is typically triggered by allergens, respiratory infections, environmental pollutants, or exercise (Mims 2015). The pathophysiology of asthma involves multiple immune and inflammatory cells, including T-helper 2 (Th2) cells, eosinophils, mast cells, and neutrophils. In an asthma attack, the immune system becomes sensitized to environmental triggers, such as pollen or dust mites, which activate Th2 cells. These cells release cytokines like interleukins (IL-4, IL-5, and IL-13), which play a pivotal role in promoting inflammation and recruitment of eosinophils to the airways (Lambrecht and Hammad 2015). Eosinophils release inflammatory substances like leukotrienes, prostaglandins, and cytokines, which play a role in increasing airway sensitivity, causing bronchoconstriction, and promoting mucus production. Additionally, mast cells, which are also activated during an asthma attack, release histamine and other vasoactive mediators that further promote inflammation and constriction of the airways (Méndez-Enríquez and Hallgren 2019). These inflammatory changes result in airway remodeling, which is characterized by thickening of the airway walls, increased mucus production, and smooth muscle hypertrophy, leading to chronic airflow limitation. Asthma is typically classified into different phenotypes, such as allergic asthma, non-allergic asthma, and severe asthma. The treatment of asthma is generally aimed at reducing inflammation and improving air flow using inhaled corticosteroids, bronchodilators, and leukotriene modifiers. However, these treatments may not work for all patients, and some individuals might experience side effects with prolonged use.

2.2 Pathophysiology of COPD

COPD is a chronic lung condition primarily caused by prolonged exposure to harmful particles, especially from smoking. It is marked by ongoing airflow restrictions, chronic inflammation, and gradual lung damage. Unlike asthma, which features reversible airflow obstruction, the airflow limitation in COPD is mostly irreversible and worsens with time (GOLD 2024). COPD includes two primary conditions: chronic bronchitis and emphysema. Chronic bronchitis involves inflammation and thickening of the bronchial walls, causing excessive mucus production and a persistent cough. Emphysema, in contrast, is characterized by the breakdown of alveolar walls and the loss of elastic recoil, leading to compromised gas exchange and decreased lung flexibility. The pathophysiology of COPD involves a complex inflammatory response to inhaled toxins, with neutrophils, macrophages, and T-cells playing central roles in the disease process. Exposure to cigarette smoke or other irritants activates immune cells, prompting the release of pro-inflammatory cytokines like tumor necrosis factor-alpha (TNF-α), IL-8, and IL-1β. These cytokines attract additional neutrophils and macrophages to the lungs, intensifying inflammation, promoting protease release, and leading to tissue damage (Aghasafari et al. 2018). A key characteristic of COPD is the imbalance between proteases (such as matrix metalloproteinases) and their inhibitors, which leads to the destruction of lung tissue. This imbalance is primarily driven by the excessive activation of neutrophils and macrophages, which release proteolytic enzymes that degrade the extracellular matrix of the lung. In emphysema, the breakdown of alveolar walls reduces the surface area available for gas exchange, leading to impaired oxygen absorption. Furthermore, airway remodeling in COPD involves fibrosis and airway thickening, which contributes to increased airflow restriction. The inflammation and mucus hypersecretion in chronic bronchitis contribute to frequent respiratory infections, exacerbations, and a decline in lung function over time (Lloyd and Hawrylowicz 2009). The primary treatment for COPD includes bronchodilators, such as beta-agonists and anticholinergics, with inhaled corticosteroids (ICS) prescribed for certain patients. However, the effectiveness of these treatments is limited, especially in advanced stages of the disease, and they are not sufficient to halt disease progression. As with asthma, there is an increasing interest in alternative and complementary therapies that can address the underlying inflammation and oxidative stress in COPD.

3 Key Plant-Based Compounds and Their Mechanisms of Action

For centuries, plant-derived compounds have been integral to traditional medicine, and modern research continues to highlight their therapeutic potential in managing various conditions, including asthma and COPD. Many of these natural compounds possess significant biological activities, such as anti-inflammatory, antioxidant, bronchodilator, and immunomodulatory properties. Below, we examine key classes of phytochemicals relevant to respiratory health and their mechanisms of action.

3.1 Flavonoids

Flavonoids are a diverse group of polyphenolic compounds found in various fruits, vegetables, herbs, and teas. Known for their antioxidant, anti-inflammatory, and antimicrobial properties, these bioactive molecules play a crucial role in managing respiratory conditions like asthma and COPD. Specific flavonoids, such as quercetin, kaempferol, and catechins, have been shown to inhibit the production of pro-inflammatory cytokines and enzymes involved in inflammation, including cyclooxygenase (COX), lipoxygenase (LOX), and inducible nitric oxide synthase (iNOS) (Maleki et al. 2019). By regulating these inflammatory pathways, flavonoids contribute to reducing airway inflammation, a major factor in asthma and COPD. In addition to their anti-inflammatory effects, flavonoids function as powerful antioxidants, neutralizing free radicals and mitigating oxidative stress in the lungs. Oxidative stress is a key factor in the onset and progression of both asthma and COPD, contributing to airway damage and chronic inflammation. Flavonoids like epicatechins and quercetin help neutralize reactive oxygen species (ROS) while boosting the function of antioxidant enzymes such as superoxide dismutase (SOD) and glutathione peroxidase (GPx) (Tanaka and Takahashi, 2013). Moreover, flavonoids have demonstrated bronchodilator properties, aiding in the relaxation of airway smooth muscles, which helps alleviate bronchoconstriction and improve airflow. This is particularly beneficial for asthma patients who experience wheezing and breathing difficulties due to airway narrowing (Townsend and Emala 2013). Research on plant-derived flavonoids, such as those from green tea (Camellia sinensis) and apples (Malus domestica), suggests they may enhance lung function and reduce inflammation in individuals with asthma and COPD, as evidenced by both preclinical and clinical studies (Townsend and Emala 2013).

3.2 Terpenes and Terpenoids

Terpenes are a broad class of naturally occurring compounds present in numerous plants, contributing to the characteristic scents of various herbs and spices. A specific subset, known as terpenoids, includes oxygenated terpenes that are particularly valued for their medicinal benefits. These compounds possess anti-inflammatory, bronchodilatory, and antimicrobial properties, making them useful in respiratory health. Certain terpenoids, such as eucalyptol (present in eucalyptus oil) and limonene (abundant in citrus fruits), have been shown to promote airway relaxation and improve respiratory airflow. Eucalyptol helps reduce airway constriction, making breathing easier (Muhammad Imran Tousif et al. 2023). Additionally, terpenoids exhibit strong anti-inflammatory effects. For example, caryophyllene, a sesquiterpene found in black pepper, has been reported to suppress pro-inflammatory cytokines like TNF-α and IL-6, aiding in the control of airway inflammation (Gertsch et al. 2008). These compounds also suppress the NF-κB pathway, a critical regulator of inflammation involved in the development of asthma and COPD (Edwards et al. 2009). Beyond their anti-inflammatory properties, many terpenoids act as antioxidants and antimicrobials, helping to reduce oxidative stress and protect against respiratory infections. Menthol, found in peppermint, has demonstrated antimicrobial effects and may help alleviate respiratory symptoms by acting as a mild decongestant and promoting mucus clearance (Eccles 2003). Overall, terpenoids such as eucalyptol, carvacrol, and menthol show significant potential in managing asthma and COPD by reducing inflammation, supporting mucus clearance, and improving lung function.

3.3 Alkaloids

Alkaloids are nitrogen-based compounds present in various plants, recognized for their wide range of pharmacological properties. Several alkaloids, including theophylline, caffeine, and ephedrine, have been widely used in managing respiratory conditions due to their bronchodilator and anti-inflammatory properties. Compounds such as ephedrine (derived from *Ephedra sinica*) and theophylline (found in

Camellia sinensis) help relax bronchial smooth muscles, improving airflow and alleviating bronchoconstriction. Ephedrine works by stimulating the release of norepinephrine, which activates beta-adrenergic receptors in the airways, leading to bronchodilation (Nazar et al. 2023). Theophylline, on the other hand, inhibits phosphodiesterase, increasing cyclic AMP (cAMP) levels in bronchial smooth muscles, which promotes relaxation and enhances ventilation (Boylan et al. 2023). Along with their bronchodilatory effects, alkaloids also influence inflammatory pathways. Caffeine, for instance, suppresses phosphodiesterase activity, which reduces the production of pro-inflammatory cytokines and helps manage lung inflammation. This mechanism may be particularly beneficial in conditions like asthma, where chronic inflammation contributes to airway narrowing (Zuo et al. 2019). Ephedrine has also been shown to modulate immune responses by regulating cytokine production, which can prevent excessive immune activation during respiratory inflammation (Nazar et al. 2023). Overall, alkaloids play a crucial role in respiratory therapy by promoting bronchodilation and reducing inflammation, making them valuable in the treatment of conditions such as asthma and COPD.

3.4 Phenolic Acids

Phenolic acids are plant-derived compounds featuring an aromatic ring with one or more hydroxyl groups. Examples include caffeic acid, ferulic acid, and chlorogenic acid, which are found abundantly in fruits, vegetables, and grains. These compounds are renowned for their powerful antioxidant and anti-inflammatory properties, making them promising for managing respiratory conditions. The primary way phenolic acids exert their effects is through their antioxidant activity. By neutralizing free radicals and reducing oxidative stress, they help protect lung tissue from damage caused by inflammation and environmental pollutants. Caffeic acid, for example, has been shown to inhibit NF-κB activation and reduce the release of pro-inflammatory cytokines in the lungs, offering significant benefits for conditions like asthma and COPD, where inflammation is a key factor in disease progression (Sun et al. 2022). Ferulic acid has also demonstrated therapeutic potential by reducing the release of histamine and other allergic mediators from mast cells, which may help prevent asthma exacerbations triggered by allergic reactions (Zhai et al. 2023). Overall, phenolic acids play a role in modulating inflammatory responses in the lungs and could be effective supplementary treatments for chronic respiratory diseases, particularly when combined with other therapeutic options.

3.5 Other Relevant Phytochemicals

In addition to the previously discussed compound classes, several other phytochemicals also show promise in the treatment of respiratory diseases, including saponins, glycosides, and polysaccharides. Saponins, a group of plant-derived glycosides, possess strong anti-inflammatory and immunomodulatory properties. Compounds extracted from *Panax ginseng* and *Astragalus membranaceus* have been shown to help regulate the immune system and reduce airway inflammation, which may aid in preventing exacerbation in conditions such as asthma and COPD (Kim et al. 2004). Glycosides like baicalin, derived from *Scutellaria baicalensis*, have shown anti-inflammatory properties by reducing the production of pro-inflammatory cytokines and helping to prevent airway remodeling, a critical aspect of chronic respiratory conditions (Liao et al. 2021). Additionally, polysaccharides from medicinal mushrooms like *Ganoderma lucidum* (reishi mushroom) have demonstrated notable immunomodulatory and anti-inflammatory effects, positioning them as promising therapeutic options for managing chronic respiratory diseases (Kozarski et al. 2011).

4 Notable Plants Used in Respiratory Therapy

Plant-derived treatments have been used for centuries in traditional medicine to address respiratory conditions. In recent years, modern science has validated the therapeutic potential of several plants, many of which exhibit anti-inflammatory, bronchodilatory, antioxidant, and mucolytic effects. Below is a detailed discussion of notable plants used in respiratory therapy, their active compounds, and mechanisms of action.

Turmeric (*Curcuma longa*) Turmeric, a spice obtained from the rhizomes of *Curcuma longa*, is renowned for its active compound, curcumin (Fig. 16.1), which possesses strong anti-inflammatory and antioxidant effects. Curcumin blocks nuclear factor-kappa B (NF-κB) and other pro-inflammatory substances, including interleukins (IL-6, IL-8) and tumor necrosis factor-alpha (TNF-α), helping to reduce airway inflammation in asthma and COPD. It also neutralizes reactive oxygen species (ROS), reducing oxidative stress in the lungs (Biswas et al. 2005). Studies have shown that curcumin supplementation improves lung function and reduces inflammatory markers in asthma patients (Quan et al. 2021).

Licorice Root (*Glycyrrhiza glabra*) Licorice root has long been a key element in traditional medicine for treating respiratory issues, thanks to its active compound, glycyrrhizin (Fig. 16.2) and its flavonoids. Glycyrrhizin demonstrates anti-inflammatory effects by suppressing the expression of pro-inflammatory cytokines and decreasing airway hyperresponsiveness. It also has mucolytic properties, helping to clear mucus from the respiratory tract (Nishimoto et al. 2010). Licorice

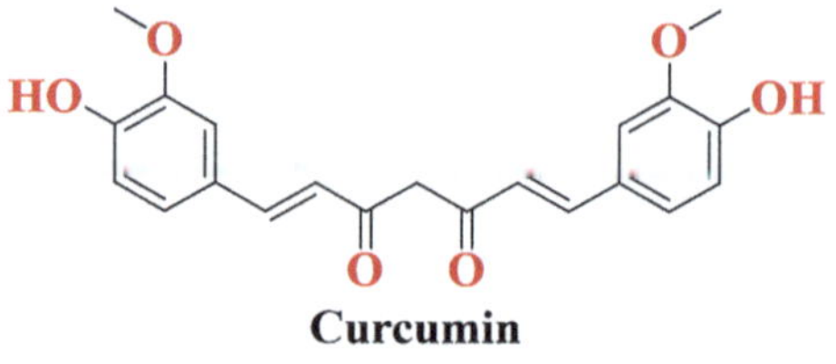

Fig. 16.1 Chemical structure of curcumin

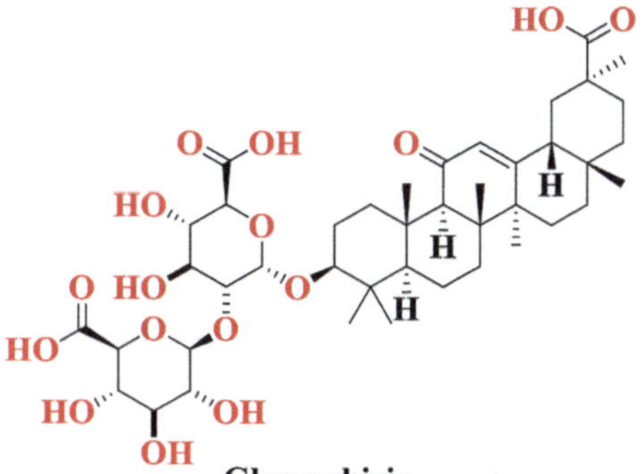

Fig. 16.2 Chemical structure of glycyrrhizin

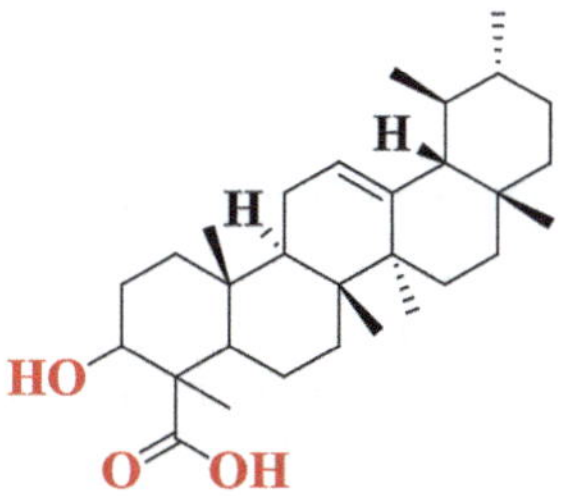

Fig. 16.3 Chemical structure of Boswellic acid

extracts have demonstrated efficacy in reducing coughing and mucus production in patients with chronic bronchitis and asthma (Yang et al. 2016).

Boswellia (*Boswellia serrata*) Boswellia, also known as Indian frankincense, contains boswellic acids (Fig. 16.3), which are celebrated for their anti-inflammatory effects. These acids block 5-lipoxygenase (5-LOX), an enzyme involved in the production of leukotrienes, thereby helping to reduce airway inflammation. This mechanism is particularly beneficial in asthma, where leukotrienes play a major role in airway narrowing (Ammon 2010). Clinical trials have shown that Boswellia extracts improve lung function and reduce asthma symptoms, such as wheezing and shortness of breath (Gupta, et al. 1998).

Ginger (*Zingiber officinale*) Ginger has long been utilized in traditional medicine for its ability to ease respiratory symptoms. Its active compounds, including gingerol and shogaol (Fig. 16.4), play a key role in its therapeutic benefits. Ginger exhibits bronchodilatory effects by relaxing airway smooth muscles. It also inhibits pro-inflammatory pathways, including COX and LOX, reducing inflammation. Additionally, ginger enhances the clearance of mucus from the respiratory tract (Townsend et al. 2013). In asthma, ginger has been shown to improve lung function and reduce inflammation (Elvedin Luković et al. 2021).

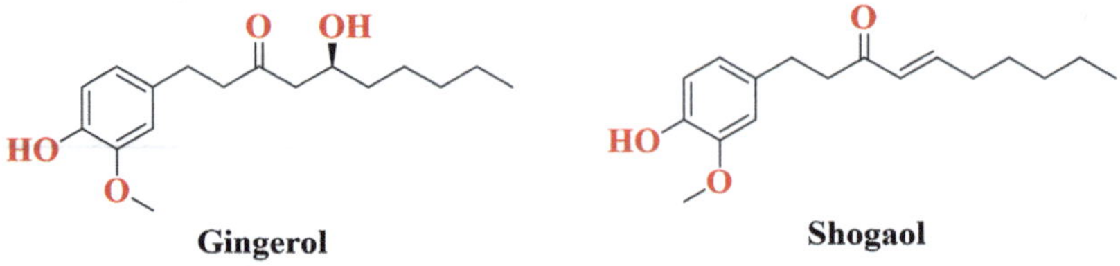

Fig. 16.4 Chemical structure of gingerol and shogaol

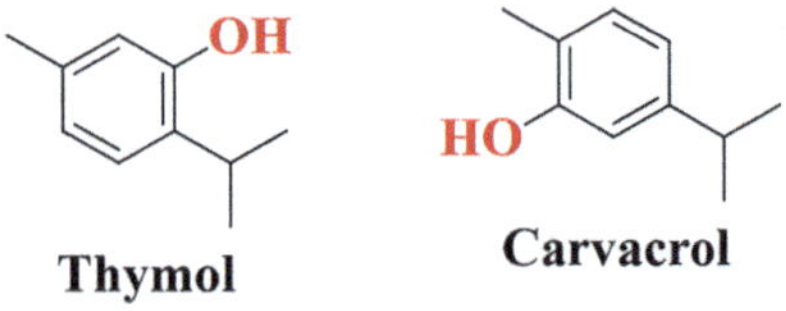

Fig. 16.5 Chemical structure of thymol and carvacrol

Thyme (*Thymus vulgaris*) Thyme is a Mediterranean herb whose essential oils, particularly thymol and carvacrol (Fig. 16.5), have antimicrobial, anti-inflammatory, and antitussive properties. Thymol acts as a bronchodilator and has potent antioxidant effects, reducing oxidative stress in the lungs. It also exhibits antimicrobial activity, making it useful for respiratory infections that exacerbate asthma and COPD (Mohammadi et al. 2018). Thyme extracts have been shown to alleviate coughing and improve mucus clearance in respiratory infections (Lazareva and Karnoukh 2020).

Mullein (*Verbascum thapsus*) Mullein has been traditionally employed to treat respiratory issues because of its anti-inflammatory and expectorant effects. Mullein contains saponin (Fig. 16.6) and flavonoids that help loosen mucus and promote its expulsion. It also has anti-inflammatory effects, reducing irritation and swelling in the airways (Dar et al. 2019). Although largely supported by anecdotal evidence, mullein has shown promise in managing chronic cough and bronchitis symptoms in herbal medicine.

Oregano (*Origanum vulgare*) Oregano, known for its essential oil rich in carvacrol and thymol (Fig. 16.7), exhibits antimicrobial and anti-inflammatory effects. Carvacrol and thymol inhibit pro-inflammatory mediators and have bronchodilatory properties. Their antimicrobial activity also helps combat respiratory infections, which are common in COPD exacerbations (Force et al. 2000). Oregano oil has been shown to reduce symptoms of respiratory infections, such as coughing and congestion, in both preclinical and clinical studies (Leyva-López et al. 2016).

Eucalyptus (*Eucalyptus globulus*) Eucalyptus is widely recognized for its active compound eucalyptol (Fig. 16.8), which has significant respiratory benefits. Eucalyptol acts as a mucolytic, helping to loosen and expel mucus. It also exhibits anti-inflammatory and bronchodilatory effects, improving airflow in asthma and

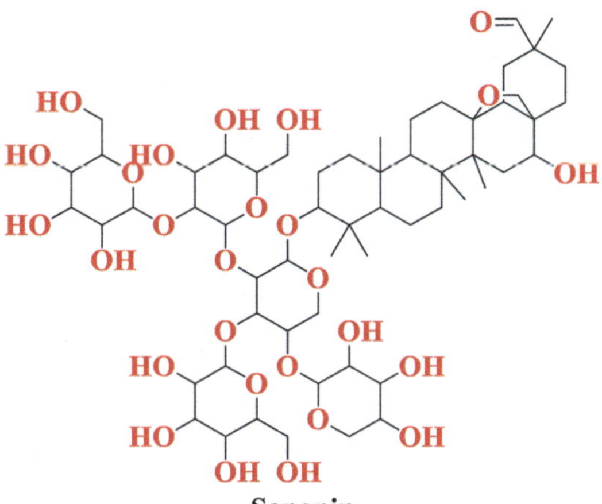

Fig. 16.6 Chemical structure of saponin

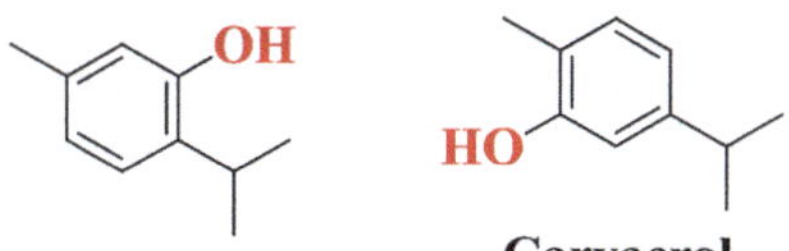

Fig. 16.7 Chemical structure of thymol and carvacrol

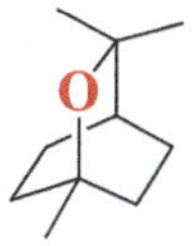

Fig. 16.8 Chemical structure of eucalyptol

COPD patients (JUERGENS et al. 2003). Clinical trials have demonstrated that eucalyptol supplementation improves lung function and reduces exacerbations in COPD patients (Galan et al. 2020).

Ephedra (*Ephedra sinica*) Ephedra, also known as ma huang, has been used in traditional Chinese medicine for its bronchodilatory properties due to its active alkaloids' ephedrine and pseudoephedrine (Fig. 16.9). Ephedrine stimulates beta-adrenergic receptors, causing bronchodilation and reducing airway resistance. It also reduces mucus secretion, alleviating symptoms of asthma and bronchitis (Nazar et al. 2023). Although it is effective, the use of Ephedra has declined due to safety concerns, including cardiovascular side effects (Hollister 2003).

Butterbur (*Petasites hybridus*) Butterbur has gained attention for its anti-inflammatory and antiallergic properties, primarily attributed to its petasin and isopetasin compounds (Fig. 16.10). Petasin inhibits leukotriene synthesis and histamine release, reducing airway inflammation, and bronchial hyperresponsiveness.

Fig. 16.9 Chemical structure of ephedrine

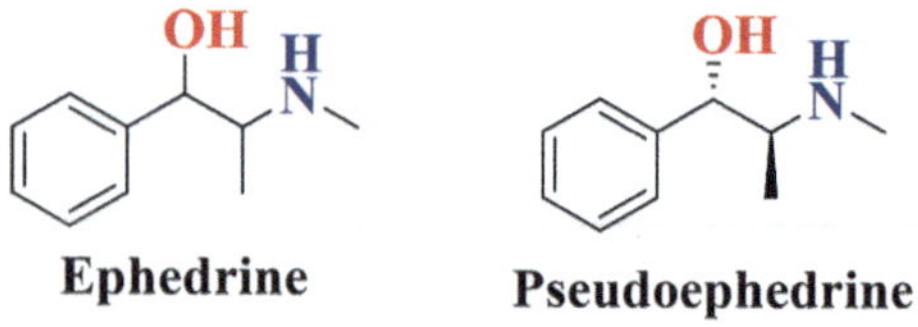

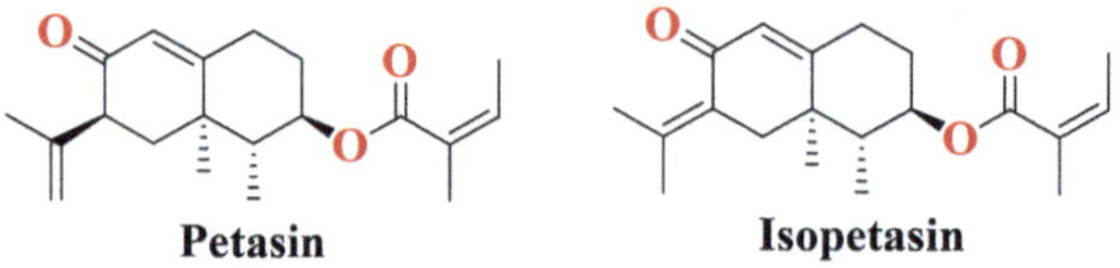

Fig. 16.10 Chemical structure of petasin and isopetasin

This makes it useful in asthma and allergic respiratory conditions (Xin Yi Lim et al. 2024). Butterbur extracts have shown efficacy in reducing allergic rhinitis symptoms, which are often associated with asthma.

5 Efficacy and Safety of Plant-Based Treatments for Asthma and COPD

The increasing interest in plant-based therapies for asthma and COPD is driven by their potential to reduce inflammation, enhance airflow, and combat oxidative stress. While these treatments offer promising benefits, their safety and appropriate dosing require careful evaluation. High doses of certain plant-derived compounds, such as glycyrrhizin from licorice root or ephedrine from *Ephedra sinica*, can lead to adverse effects like hypertension, low potassium levels, or heart rhythm disturbances (Haller and Benowitz 2001; Tang et al. 2023). Additionally, some phytochemicals, including curcumin and boswellic acids, may interact with conventional medications, potentially affecting their metabolism or therapeutic effects. Certain herbal remedies, such as butterbur, may also trigger allergic reactions, especially in individuals sensitive to ragweed (Carstairs et al. 2003). Establishing the correct dosage poses a challenge due to variations in the concentration of active compounds, which depend on factors like plant species, growing conditions, and extraction techniques. To ensure consistent effectiveness and reduce the risk of side effects, standardized formulations are essential. Table 16.1 provides a summary of evidence and dosages for some important plant-based treatments.

Table 16.1 Summary of evidence and dosages for plant-based treatments

Plant	Active compounds	Study type	Key findings	Recommended dosage	Safety considerations	Reference
Turmeric (*Curcuma longa*)	Curcumin, volatile oils	RCT	Reduced airway inflammation and asthma symptoms; improved lung function	500–1000 mg curcumin daily	Generally safe; avoid high doses in individuals with gallstones	Quan et al. (2021)
Licorice root (*Glycyrrhiza glabra*)	Glycyrrhizin, flavonoids	Clinical trial	Improved mucus clearance and reduced chronic bronchitis symptoms in COPD	200–400 mg glycyrrhizin daily	Prolonged use may cause hypokalemia; caution with corticosteroids	Nishimoto et al. (2010)
Boswellia (*Boswellia serrata*)	Boswellic acids	RCT	Improved FEV1 and reduced dyspnea in asthma; lowered leukotriene levels	300–600 mg extract daily	May cause mild gastrointestinal upset; avoid in pregnancy	Ammon (2010)
Ginger (*Zingiber officinale*)	Gingerols, shogaols	Preclinical/ animal study	Reduced pro-inflammatory cytokines; improved airway relaxation and lung function	200–400 mg eucalyptol daily	Safe when used in recommended doses; avoid direct ingestion of essential oil	Townsend et al. (2013)
Butterbur (*Petasites hybridus*)	Petasin, isopetasin	Placebo-controlled study	Reduced airway hyperresponsiveness and improved asthma control	50–100 mg extract twice daily	Must be free of pyrrolizidine alkaloids to avoid liver toxicity	Carstairs et al. (2003)
Ephedra (*Ephedra sinica*)	Ephedrine, pseudoephedrine	Preclinical/ animal study	Rapid bronchodilation and relief from bronchoconstriction in asthma models	Use under medical supervision (variable doses)	Risk of hypertension and tachycardia; avoid in heart conditions	Nazar et al. (2023)
Thyme (*Thymus vulgaris*)	Thymol, carvacrol	Clinical trial	Enhanced mucus clearance; antimicrobial effects against respiratory pathogens	2–3 g dried herb or 300 mg extract daily	Safe in small doses; avoid high doses in pregnancy.	Mohammadi et al. (2018)
Oregano (*Origanum vulgare*)	Carvacrol, thymol	Preclinical/ animal study	Reduced bacterial load in COPD models; antioxidant and anti-inflammatory activity	200 mg oil or 2–4 g dried herb daily	May cause mild allergic reactions in sensitive individuals	Leyva-López et al. (2016)
Mullein (*Verbascum thapsus*)	Flavonoids, saponins	Preclinical/ animal study	Reduced eosinophilic infiltration; eased coughing and mucus clearance	1–4 g dried leaf or 1–2 ml tincture daily	Safe; avoid individuals with known allergies to Verbascum	Dar et al. (2019)

6 Challenges in the Use of Plant-Based Treatments for Respiratory Conditions

Despite the promising potential of plant-based treatments for asthma and COPD, several challenges remain that hinder their widespread adoption and clinical integration. These challenges encompass issues related to the standardization of plant-based products, the limitations of clinical evidence, regulatory barriers, and gaps in cultural knowledge.

6.1 Lack of Standardization

A major challenge in utilizing plant-based treatments is the absence of standardization in the preparation, dosage, and quality of these products. Medicinal plants can vary due to factors like plant species, cultivation techniques, harvesting time, and geographic location. This variability can influence the effectiveness and safety of plant-based therapies, complicating the development of reliable treatment protocols. For instance, the concentration of active compounds in *Boswellia serrata* can differ greatly depending on the extraction method, leading to inconsistent clinical results (Beghelli et al. 2017). Additionally, the preparation of herbal remedies in traditional medicine systems may not adhere to modern pharmaceutical standards, which can lead to contamination, poor bioavailability, or interaction with other medications (Raghavendra et al. 2009). Therefore, there is an urgent need for standardized protocols for the extraction, dosage, and administration of plant-based treatments to ensure safety and efficacy.

6.2 Limited Clinical Evidence and Regulatory Barriers

While preclinical studies and animal models have shown promising results, the clinical evidence supporting the use of plant-based treatments for asthma and COPD is still limited. Many clinical studies lack large sample sizes, proper controls, or long-term follow-up, leading to inconclusive results about the effectiveness and safety of these therapies in humans. In a review of clinical trials on the use of herbal remedies for asthma, it was found that most studies were of low quality, with inadequate reporting of adverse events and standardization issues (Huntley 2000). Furthermore, the regulatory landscape surrounding plant-based treatments remains complex and inconsistent across different countries. In many regions, herbal products are categorized as dietary supplements rather than pharmaceuticals, meaning they are not subjected to the rigorous testing and regulatory oversight required for prescription drugs. This lack of regulation can lead to unsafe practices, such as the use of unverified products that may interact negatively with prescribed medications or cause unexpected side effects (Ekor 2013).

6.3 Cultural and Knowledge Gaps

Another challenge in the widespread use of plant-based treatments is the cultural and knowledge gaps that exist between traditional herbal medicine practitioners and modern medical professionals. Many plant-based remedies are rooted in traditional medical systems that may not be widely understood or accepted by mainstream healthcare providers. In countries where Western medicine predominates, physicians may be hesitant to recommend plant-based treatments due to unfamiliarity with their efficacy or potential interactions with pharmaceutical drugs (Fehér et al. 2020). Additionally, there is a lack of education and training for healthcare providers regarding the use of plant-based therapies. Many clinicians may not be aware of the pharmacological properties of herbal treatments, their potential benefits, or the scientific evidence supporting their use. This gap in knowledge can hinder the integration of plant-based treatments into conventional treatment regimens and discourage patients from using these therapies in conjunction with prescription medications (Fong 2002). Moreover, there are cultural differences in the acceptance of plant-based treatments. In some regions, particularly in the Western world, there is skepticism surrounding the efficacy of herbal remedies, which may be attributed to the rise of pharmaceutical drugs and the perceived "naturalistic fallacy" that all natural substances are inherently safe and effective (Eisenberg et al. 1998). Conversely, in regions where traditional medicine is still widely practiced, there may be insufficient awareness of potential safety issues and interactions with modern medicines. Addressing these challenges requires collaborative efforts from researchers, healthcare providers, and policymakers to ensure that plant-based treatments are rigorously tested, standardized, and integrated into mainstream respiratory care. Developing evidence-based guidelines for their use, improving regulatory frameworks, and enhancing education for both healthcare professionals and patients will be key in overcoming these obstacles.

7 Future Directions in Plant-Based Treatments for Asthma and COPD

The future of plant-based treatments for asthma and COPD lies in overcoming current challenges and exploring new avenues in research, therapy, and integration with conventional treatments. As scientific understanding of plant compounds and their therapeutic potential grows, innovative approaches are being developed to maximize their benefits in managing respiratory conditions.

7.1 Innovative Approaches in Plant-Based Medicine

In the field of plant-based medicine, research is increasingly focusing on new methods of extraction and formulation that enhance the bioavailability and efficacy of active compounds. Traditional herbal remedies, although beneficial, often suffer from issues like poor absorption or inconsistent dosages. Therefore, innovative approaches in biotechnological advancements such as nano-formulations, encapsulation techniques, and enhanced delivery systems are being explored to optimize the therapeutic potential of plant-based treatments. For example, the use of nanoparticle carriers for herbal extracts, like curcumin from turmeric, is a promising strategy to improve its solubility and bioavailability, especially in treating conditions such as asthma and COPD. Nano-curcumin has shown improved anti-inflammatory and antioxidant effects in preclinical studies, suggesting its potential as an effective adjunct therapy for respiratory diseases (Rajizadeh et al. 2024) Additionally, plant-based compounds are being investigated for their synergistic effects when combined with other complementary natural substances. This can potentially lead to more comprehensive treatments that not only address inflammation but also tackle multiple aspects of respiratory conditions, including oxidative stress, mucous secretion, and airway remodeling (Gupta, et al. 1998).

7.2 Personalized Medicine in Plant-Based Respiratory Treatments

As precision medicine continues to evolve, personalized approaches to treatment are gaining more attention. Personalized medicine involves tailoring treatment to individual patients based on genetic, environmental, and lifestyle factors. In the context of asthma and COPD, personalized plant-based therapies could be developed by identifying specific biomarkers and genetic profiles that predict how patients will respond to a particular plant compound. For instance, the use of genetic testing to identify polymorphisms in genes related to inflammatory pathways (such as TNF-α, IL-6, and other cytokine genes) could guide the selection of plant-based compounds that target these pathways more effectively. Personalized strategies could also consider factors like the patient's microbiome, which may influence the therapeutic response to plant-based treatments (Kuntz and Gilbert 2017). Furthermore, research into the pharmacogenomics of plant compounds, including how they are metabolized in different individuals, could improve the personalization of plant-based therapies, ensuring that patients receive the most effective dosages and formulations.

7.3 Integration with Conventional Therapies

Another important future direction is the integration of plant-based treatments with conventional pharmaceutical therapies. While pharmaceutical treatments for asthma and COPD, such as bronchodilators and corticosteroids, remain the standard of care, there is growing interest in combining these with plant-based remedies to enhance therapeutic outcomes and reduce side effects. For instance, combining turmeric-derived curcumin with corticosteroids may reduce the steroid dose required to achieve effective anti-inflammatory effects, potentially decreasing the side effects associated with long-term steroid use, such as weight gain, osteoporosis, and increased risk of infections (Razavi et al. 2021). Furthermore, plant-based treatments such as *Boswellia serrata* could complement the action of bronchodilators by reducing inflammation and bronchoconstriction, which would improve airflow and reduce symptoms (Marogna et al. 2015). Integrating plant-based treatments could also offer patients more holistic care, addressing the root causes of inflammation and oxidative stress while providing symptomatic relief, reducing the reliance on pharmaceutical drugs, and ultimately improving quality of life.

7.4 Potential for Policy and Regulatory Evolution

For plant-based treatments to gain wider acceptance in clinical practice, significant improvements in policy and regulation are essential. Currently, the regulatory landscape for herbal products is inconsistent, with varying standards across different regions. In many countries, these products are categorized as dietary supplements, meaning they are not subjected to the same rigorous testing and approval processes as conventional pharmaceuticals. Strengthening regulatory frameworks can help ensure the safety, efficacy, and quality of plant-based treatments while also fostering innovation. One effective strategy is the implementation of stricter quality control measures, such as enforcing Good Manufacturing Practices (GMP) for herbal medicines. Additionally, there is a growing need for well-designed clinical trials to provide scientific evidence supporting the effectiveness of these treatments (Klein-Junior et al. 2021). Moving forward, greater collaboration between traditional herbal medicine practitioners and modern healthcare professionals could be facilitated through policies that promote evidence-based integrative medicine. A regulatory system that prioritizes thorough clinical validation while ensuring patient safety will be crucial in bridging the gap between alternative and conventional healthcare. The future of plant-based treatments for asthma and COPD is promising. Advances in biotechnology, personalized medicine, and regulatory improvements could enable these therapies to become a more integral part of respiratory care. By addressing current challenges and leveraging the synergistic potential of plant-based treatments, researchers, healthcare professionals, and policymakers can contribute to the development of more effective, accessible, and holistic options for managing asthma and COPD.

8 Conclusions

Plant-based therapies are increasingly recognized as valuable adjuncts to conventional treatments for asthma and chronic obstructive pulmonary disease (COPD). A variety of plant-derived compounds, including flavonoids, terpenes, alkaloids, and phenolic acids, have demonstrated therapeutic promise in both preclinical and clinical studies. These compounds are known for their anti-inflammatory, antioxidant, and immune-modulating properties, which can help alleviate respiratory symptoms and enhance lung function. Prominent medicinal plants like turmeric, licorice root, Boswellia, and ginger have shown potential in managing these conditions. However, the widespread use of plant-based treatments faces several challenges. Key issues include a lack of standardization, limited high-quality clinical evidence, regulatory barriers, and insufficient awareness in various medical and cultural communities. To fully realize the benefits of plant-based therapies, further research is needed. Future clinical trials should focus on large-scale, rigorous studies to confirm the efficacy and safety of these treatments across diverse populations. Additionally, optimizing dosages and ensuring consistency in herbal formulations will improve their reliability and therapeutic effectiveness. Investigating the combined effects of plant-based treatments with conventional drugs could also lead to more effective, integrated approaches for managing asthma and COPD.

References

Aghasafari P, George U et al (2018) A review of inflammatory mechanism in airway diseases. Inflamm Res 68(1):59–74

Akinbami LJ, Moorman JE et al (2011) Asthma prevalence, health care use, and mortality: United States, 2005-2009. Natl Health Stat Report 32:1–14

Ammon HPT (2010) Modulation of the immune system by Boswellia serrata extracts and boswellic acids. Phytomedicine 17(11):862–867

Beghelli D, Isani G et al (2017) Antioxidant and ex vivo immune system regulatory properties of Boswellia serrata extracts. Oxidative Med Cell Longev 2017:1–10

Biswas SK, McClure D et al (2005) Curcumin induces glutathione biosynthesis and inhibits NF-κB activation and Interleukin-8 release in alveolar epithelial cells: mechanism of free radical scavenging activity. Antioxid Redox Signal 7(1):32–41

Boylan PM, Abdalla M et al (2023) Theophylline for the management of respiratory disorders in adults in the 21st century: a scoping review from the American College of Clinical Pharmacy Pulmonary Practice and Research Network. Pharmacotherapy 43(9):963–990

Carstairs IJ, Lee DKC et al (2003) Butterbur, a herbal remedy, protects against allergen-sensitized mast cell priming in seasonal allergic rhinitis. J Allergy Clin Immunol 111(2):76–76

Cushnie TPT, Lamb AJ (2005) Antimicrobial activity of flavonoids. Int J Antimicrob Agents 26(5):343–356

Dar MA, Bhat MF et al (2019) Extensive phytochemistry, comprehensive traditional uses, and critical pharmacological profile of the great mullein: *Verbascum thapsus* L. Nat Prod J 9(3):158–171

Eccles R (2003) Menthol: effects on nasal sensation of airflow and the drive to breathe. Curr Allergy Asthma Rep 3(3):210–214

Edwards MR, Bartlett NW et al (2009) Targeting the NF-κB pathway in asthma and chronic obstructive pulmonary disease. Pharmacol Ther 121(1):1–13

Eisenberg DM, Davis RB et al (1998) Trends in alternative medicine use in the United States, 1990-1997: results of a follow-up national survey. JAMA 280(18):1569–1575

Ekor M (2013) The growing use of herbal medicines: issues relating to adverse reactions and challenges in monitoring safety. Front Pharmacol 4:177

Fehér A, Gazdecki M et al (2020) A comprehensive review of the benefits of and the barriers to the switch to a plant-based diet. Sustainability 12(10):4136. https://doi.org/10.3390/su12104136

Fong HHS (2002) Integration of herbal medicine into modern medical practices: issues and prospects. Integr Cancer Ther 1(3):287–293

Force M, Spark WS et al (2000) Inhibition of enteric parasites by emulsified oil of oregano in vivo. Phytother Res 14(3):213–214

Galan D, Ezeudu NE et al (2020) Eucalyptol (1,8-cineole): an underutilized ally in respiratory disorders? J Essent Oil Res 32(2):103–110

Gertsch J, Leonti M et al (2008) Beta-caryophyllene is a dietary cannabinoid. Proc Natl Acad Sci USA 105(26):9099–9104

GOLD (2024) Global strategy for prevention, diagnosis and management of COPD: 2024 report. https://goldcopd.org/2024-gold-report

Gupta IG, Parihar A et al (1998) Effects of Boswellia serrata gum resin in patients with bronchial asthma: results of a double-blind, placebo-controlled, 6-week clinical study. Eur J Med Res3(11):511–514

Haller CA, Benowitz NL (2001) Adverse cardiovascular and central nervous system events associated with dietary supplements containing ephedra alkaloids. ACC Curr J Rev 10(3):25–26

Hollister A (2003) American Society for Clinical Pharmacology and Therapeutics position statement on the public health risks of ephedra. Clin Pharmacol Ther 74(5):403–405

Hosker H, Anstey K et al (2007) Variability in the organization and management of hospital care for COPD exacerbations in the UK. Respir Med 101(4):754–761

Huntley A (2000) Herbal medicines for asthma: a systematic review. Thorax 55(11):925–929

Juergens UR, Dethlefsen U et al (2003) Anti-inflammatory activity of 1.8-cineol (eucalyptol) in bronchial asthma: a double-blind placebo-controlled trial. Respir Med 97(3):250–256

Kim HP, Son KH et al (2004) Anti-inflammatory plant flavonoids and cellular action mechanisms. J Pharmacol Sci 96(3):229–245

Klein-Junior LC, de Souza MR et al (2021) Quality control of herbal medicines: from traditional techniques to state-of-the-art approaches. Planta Med 87(12):964–988

Kozarski M, Klaus A et al (2011) Antioxidative and immunomodulating activities of polysaccharide extracts of the medicinal mushrooms *Agaricus bisporus*, *Agaricus brasiliensis*, *Ganoderma lucidum* and *Phellinus linteus*. Food Chem 129(4):1667–1675

Kuntz TM, Gilbert JA (2017) Introducing the microbiome into precision medicine. Trends Pharmacol Sci 38(1):81–91

Lambrecht BN, Hammad H (2015) The immunology of asthma. Nat Immunol 16(1):45–56

Lazareva NB, Karnoukh KI (2020) Herbal medicines: current use in the treatment of respiratory infections. Meditsinskiy Sovet Medical Council 17:114–122

Leyva-López N, Nair V et al (2016) Protective role of terpenes and polyphenols from three species of oregano (Lippia graveolens, Lippia palmeri and Hedeoma patens) on the suppression of lipopolysaccharide-induced inflammation in RAW 264.7 macrophage cells. J Ethnopharmacol 187:302–312

Liao H, Ye J et al (2021) The main bioactive compounds of Scutellaria baicalensis Georgi. For alleviation of inflammatory cytokines: a comprehensive review. Biomed Pharmacother 133:110917

Lim XY, Lau MS et al (2024) Medicinal plants for allergic rhinitis: a systematic review and meta-analysis. PLoS One 19(4):e0297839

Lloyd CM, Hawrylowicz CM (2009) Regulatory T cells in asthma. Immunity 31(3):438–449. https://doi.org/10.1016/j.immuni.2009.08.007

Luković E, Perez-Zoghbi JF et al (2021) Ginger metabolites and metabolite-inspired synthetic products modulate intracellular calcium and relax airway smooth muscle. Am J Phys Lung Cell Mol Phys 321(5):912–924

Maleki SJ, Crespo JF et al (2019) Anti-inflammatory effects of flavonoids. Food Chem 299:125124

Marciniuk DD, Calzetta L (2023) Inhaled corticosteroids and COPD. Chest 163(1):8–9

Marogna M, Braidi C et al (2015) A randomized controlled trial of a phytotherapic compound containing boswellia serrata and bromeline for seasonal allergic rhinitis complicated by upper airways recurrent respiratory infections. J Allergy Clin Immunol 135(2):271

Méndez-Enríquez E, Hallgren J (2019) Mast cells and their progenitors in allergic asthma. Front Immunol 10:821

Mims JW (2015) Asthma: definitions and pathophysiology. Int Forum Allergy Rhinol 5(S1):S2–S6

Mohammadi A, Mahjoub S et al (2018) Immunomodulatory effects of thymol through modulation of redox status and trace element content in experimental models of asthma. Biomed Pharmacother 105:856–861
Nazar H, Amar AL et al (2023) Review of pharmacological activity of ephedra gerardiana. Sci Inq Rev 7(2):81–100. https://doi.org/10.32350/sir.72.05
Nicikowski J, Reguła J (2021) Selected bioactive compounds in food of plant origin as natural immunomodulators in asthma and chronic obstructive pulmonary disease. Acta Scientiarum Polonorum 20(4):383–397
Nishimoto Y, Hisatsune A et al (2010) Glycyrrhizin attenuates mucus production by inhibition of MUC5AC mRNA expression in vivo and in vitro. J Pharmacol Sci 113(1):76–83. https://doi.org/10.1254/jphs.09344FP
Quan M, Alismail A et al (2021) Randomized, placebo controlled, double blinded pilot superiority phase 2 trial to evaluate the effect of curcumin in moderate to severe asthmatics. BMC Pulm Med 21(1):268
Raghavendra H, Yogesh H et al (2009) An overview of herbal medicine. Pharm Sci 1(1):1–20
Rajizadeh MA, Najafipour H et al (2024) An updated comprehensive review of plants and herbal compounds with Antiasthmatic effect. Evid Based Complement Alternat Med 2024:1–36
Razavi BM, Ghasemzadeh Rahbardar M et al (2021) A review of therapeutic potentials of turmeric (Curcuma longa) and its active constituent, curcumin on inflammatory disorders, pain, and their related patents. Phytother Res 35(12):6489–6513
Roy NK, Parama D et al (2019) An update on pharmacological potential of Boswellic acids against chronic diseases. Int J Mol Sci 20(17):4101
Safari S, Davoodi P et al (2023) Curcumin effects on chronic obstructive pulmonary disease: a systematic review. Health Sci Rep 6(3):e1145
Sun W, Xie W et al (2022) Caffeic acid phenethyl ester attenuates osteoarthritis progression by activating NRF2/HO-1 and inhibiting the NF-κB signaling pathway. Int J Mol Med 50(5):134
Tanaka T, Takahashi R (2013) Flavonoids and asthma. Nutrients 5(6):2128–2143
Tang S, Ren J et al (2023) Ephedrae herba: a review of its phytochemistry, pharmacology, clinical application, and alkaloid toxicity. Molecules 28(2):663
Tousif MI, Nazir M et al (2023) Terpenoids as human neutrophil elastase (HNE) inhibitors: a comprehensive review of natural anti-inflammatory isoprenoids. Chembiochem 24(22):e202300346
Townsend EA, Emala CW (2013) Quercetin acutely relaxes airway smooth muscle and potentiates β-agonist-induced relaxation via dual phosphodiesterase inhibition of PLCβ and PDE4. Am J Phys Lung Cell Mol Phys 305(5):L396–L403
Townsend EA, Siviski ME et al (2013) Effects of ginger and its constituents on airway smooth muscle relaxation and calcium regulation. Am J Respir Cell Mol Biol 48(2):157–163
Yang R, Yuan BC (2016) The anti-inflammatory activity of licorice, a widely used Chinese herb. Pharm Biol 55(1):5–18
Yeh GY, Horwitz R (2017) Integrative medicine for respiratory conditions: asthma and COPD. Med Clin North Am 101(5):925–941
Zhai Y, Wang T et al (2023) Ferulic acid: a review of pharmacology, toxicology, and therapeutic effects on pulmonary diseases. Int J Mol Sci 24(9):8011
Zuo H, Cattani-Cavalieri I et al (2019) Phosphodiesterases as therapeutic targets for respiratory diseases. Pharmacol Ther 197:225–242

Antiviral and Immunomodulatory Effects of Medicinal Plants in Respiratory Diseases

17

Adil Husain and Firoz Ahmad

Abstract

Respiratory diseases caused by viral pathogens represent a significant global health burden. After facing a severe pandemic like COVID-19, respiratory diseases arise as a major area of public health and research. The persistent and ubiquitous threat that viral infections pose to global health highlights the urgent need for resourceful approaches to mitigate the effects of viruses. Viral infections continue to be a major cause of morbidity and mortality even though conventional antiviral drugs like interferon and ribavirin exhibit some level of effectiveness against some viruses in testing environments. As resistance to conventional antiviral drugs increases and concerns about side effects remain, medicinal plants have gained importance as alternative therapeutics due to their antiviral and immunomodulatory properties. This chapter explores the dual role of medicinal plants in combating respiratory viruses and modulating immune responses. We address the molecular mechanisms underlying the antiviral effects of important phytochemicals such as flavonoids, alkaloids, and terpenoids, which inhibit virus replication, attachment, and entry into host cells. In addition, we discuss how these compounds regulate immune functions, reduce inflammation and cytokine storms and thus prevent the exacerbation of respiratory infections. Highlighting current preclinical and clinical evidence, we highlight the potential of medicinal

Authors Adil Husain and Firoz Ahmad have equally contributed to this chapter.

A. Husain
Department of Biochemistry, Babu Banarasi Das College of Dental Sciences, Babu Banarasi Das University, Lucknow, Uttar Pradesh, India

F. Ahmad (✉)
Department of Physiological Sciences, Oklahoma Center for Respiratory and Infectious Diseases, Oklahoma State University, Stillwater, OK, USA
e-mail: firoz.ahmad@okstate.edu

K. Kuca et al. (eds.), *Medicinal Plants and Their Bioactives in Human Diseases*,
https://doi.org/10.1007/978-3-032-01356-9_17

plants to complement existing antiviral therapies and advocate for their inclusion in integrative approaches to the treatment of respiratory diseases.

Keywords

Antimicrobial · Immunomodulatory · COPD · Antiviral compounds · Respiratory disorders

1 Introduction

Respiratory diseases continue to be among the most common and deadly illnesses worldwide, contributing significantly to morbidity and mortality. They can be classified into two primary categories: infectious respiratory diseases, including viral infections like influenza, RSV, rhinoviruses, and more recently, coronaviruses (e.g., SARS-CoV-2), and chronic respiratory conditions such as asthma, chronic obstructive pulmonary disease (COPD), and pulmonary fibrosis (James et al. 2018; Labaki and Han 2020). Both types of diseases place enormous strain on healthcare systems, particularly as the global population ages and environmental factors, such as air pollution and smoking, contribute to their increasing prevalence. Although modern medicine has provided important advances in both the prevention and treatment of respiratory diseases through the development of antiviral drugs, vaccines, and other interventions, these treatments are not without limitations. Antiviral drugs, for example, often have narrow therapeutic windows, limited effectiveness, and can contribute to the development of resistance. Moreover, certain populations, such as those in low-resource settings, may lack access to these modern therapies. Furthermore, chronic inflammatory diseases like asthma and COPD require long-term management strategies to mitigate inflammation and improve lung function (Jasemi et al. 2022; Mitra et al. 2022; Wang et al. 2021). As a result, the demand for complementary and alternative treatment strategies remains high. In this context, medicinal plants have re-emerged as important therapeutic agents. Phytotherapy, or plant-based medicine, has been a staple of human health for centuries, particularly in the treatment of respiratory diseases. Many plants have been shown to possess antiviral, anti-inflammatory, and immunomodulatory properties, making them attractive options for addressing both viral infections and chronic inflammatory conditions of the lungs. These properties allow medicinal plants to not only directly target respiratory viruses but also to modulate immune responses, thereby improving immune function and reducing chronic inflammation (Mustra Rakic and Wang 2020; Rahman et al. 2022; Zhai et al. 2018). This chapter explores the antiviral and immunomodulatory effects of medicinal plants, focusing on their potential to treat and prevent respiratory diseases. We examined key plants with a history of use in traditional medicine and review modern scientific evidence supporting their antiviral and immune-regulating effects. We also discussed the mechanisms through which these plants exert their beneficial actions and explore the challenges and opportunities for integrating these natural remedies into mainstream medical practice (Bibi et al. 2021; Dhamodharan et al. 2024).

Respiratory diseases, both acute and chronic, affect a vast portion of the global population. According to the World Health Organization (WHO), respiratory diseases are responsible for millions of deaths each year, particularly in low- and middle-income countries. Chronic obstructive pulmonary disease (COPD) and asthma contribute heavily to this burden, but viral infections like influenza, COVID-19, and RSV also cause significant morbidity, particularly during seasonal epidemics or pandemics. These conditions put immense pressure on healthcare systems, especially during viral outbreaks, leading to overwhelmed hospital systems and substantial mortality rates (Rahman et al. 2022; Zhai et al. 2018).

Chronic respiratory diseases (CRDs) like asthma, COPD, and interstitial lung disease are a growing public health challenge. These diseases are characterized by long-term inflammation, impaired lung function, and immune system dysregulation. They often involve both viral infections (which exacerbate symptoms) and chronic inflammation, creating a complex interplay between pathogens, immune cells, and airway tissues (Jasemi et al. 2022; Rahman et al. 2022).

Acute respiratory infections (ARIs), particularly those caused by viruses, are among the most common causes of morbidity and mortality worldwide. The global burden is exacerbated by the emergence of new viral pathogens, the rise of antibiotic resistance in bacterial co-infections, and the increasing prevalence of viral resistance. The COVID-19 pandemic caused by SARS-CoV-2 demonstrated the profound impact that respiratory viral infections can have on global health, with millions of cases and deaths worldwide, along with long-term health consequences for survivors (i.e., "long COVID") (James et al. 2018; Labaki and Han 2020).

2 Medicinal Plants and Their Phytocompounds in Respiratory Health

Medicinal plants have long been recognized for their therapeutic potential, owing to the vast array of bioactive compounds they contain. These compounds have significant effects on human health and are often the key to the plants' therapeutic properties. Among the most studied groups of bioactive compounds are flavonoids, alkaloids, terpenoids, glycosides, and polyphenols (Bibi et al. 2021; Dhamodharan et al. 2024; Zhai et al. 2018). Many of these compounds possess antiviral, anti-inflammatory, antimicrobial, and immunomodulatory properties, making them valuable in the prevention and treatment of a variety of diseases, including respiratory conditions. This section provides a deeper exploration of the various classes of active compounds found in medicinal plants and their specific actions, particularly in relation to respiratory health (Fig. 17.1).

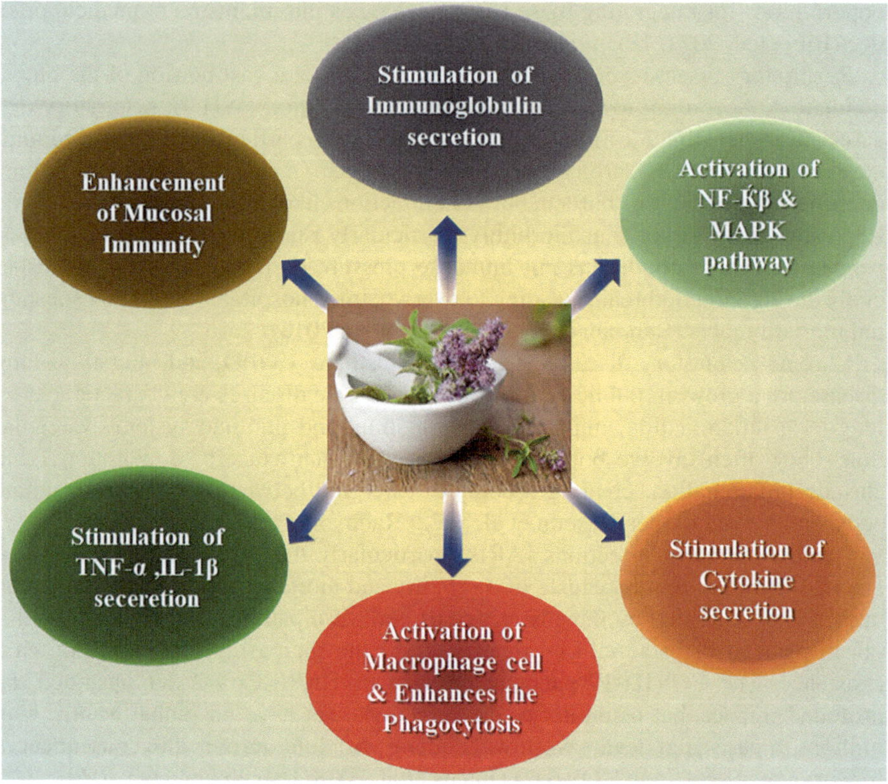

Fig. 17.1 Biological functions of phytocompound against the respiratory diseases

2.1 Flavonoids

Flavonoids are a large group of polyphenolic compounds widely distributed in the plant kingdom. These compounds are known for their potent antioxidant and anti-inflammatory effects, making them important in the management of diseases associated with oxidative stress and inflammation, including respiratory conditions like asthma, chronic obstructive pulmonary disease (COPD), and viral infections (Jasemi et al. 2020).

2.2 Alkaloids

Alkaloids are a diverse group of nitrogen-containing compounds found in many medicinal plants. They are known for their antimicrobial, anti-inflammatory, and immunomodulatory activities. Several alkaloids have been shown to be effective in managing respiratory infections and chronic conditions by targeting inflammation, immune function, and pathogen defense mechanisms (Bibi et al. 2021; Jasemi et al. 2020).

2.3 Terpenoids

Terpenoids are a large and diverse class of naturally occurring compounds found in the essential oils of plants. They are known for their anti-inflammatory, antimicrobial, and immune-modulating properties. In the context of respiratory diseases, terpenoids can play a crucial role in managing inflammation, promoting expectoration, and enhancing immune defense (Jasemi et al. 2020; Bibi et al. 2021; Dhamodharan et al. 2024; Zhang et al. 2024).

2.4 Glycosides

Glycosides are compounds consisting of a sugar molecule bound to a non-sugar molecule, which can be a steroid, flavonoid, or another compound. Glycosides have diverse pharmacological effects, including anti-inflammatory, antimicrobial, and expectorant actions, which make them valuable in treating respiratory conditions (Zhang et al. 2024).

2.5 Saponins

A type of glycoside found in plants like ginseng (*Panax ginseng*) and licorice (*Glycyrrhiza glabra*), saponins are particularly known for their ability to enhance immune function. Saponins stimulate the production of immune cells such as macrophages and natural killer (NK) cells, which are crucial for fighting infections. They also have anti-inflammatory properties, which can help manage the chronic inflammation seen in diseases like asthma and COPD. In addition, saponins have expectorant effects, helping to clear mucus from the respiratory tract, making them valuable in the treatment of bronchitis and other respiratory infections (Lin et al. 2022; Wang et al. 2022a, b).

2.6 Polyphenols

Polyphenols, including flavonoids, tannins, lignans, and stilbenes, are a large group of plant-derived compounds that possess potent antioxidant, anti-inflammatory, and antiviral properties. Polyphenols are particularly effective at reducing oxidative stress, a major contributor to respiratory diseases such as COPD, asthma, and lung fibrosis (Fu et al. 2022).

3 Antiviral Effects of Medicinal Plants Against the Respiratory Diseases

With their strong antiviral qualities, medicinal plants have great therapeutic potential for treating respiratory conditions. These plants contain flavonoids, alkaloids, terpenoids, and polyphenols, which are phytochemicals that can block viral entry, inhibit viral replication, and alter host immune responses. There is evidence that plants can effectively combat viruses that cause coronavirus infections, influenza, and respiratory syncytial virus (RSV) infections. These organic substances improve respiratory health by reducing inflammation and oxidative stress in addition to fighting infections. For safe and efficient respiratory treatments, their full antiviral potential must be realized through further study and clinical validation.

3.1 Echinacea

Echinacea purpurea is one of the most well-studied plants for its antiviral properties. The plant contains bioactive compounds such as alkylamides and phenolic compounds, which have been shown to inhibit the replication of viruses such as influenza and rhinoviruses. Echinacea enhances immune function by increasing the activity of phagocytes and stimulating the production of cytokines, which help modulate the immune response. Studies have shown that Echinacea can reduce the severity and duration of cold symptoms, particularly when used at the first sign of illness. Echinacea stimulates the activity of white blood cells and interferons, boosting the body's antiviral defense. It also has direct antiviral effects, preventing viruses from entering cells and inhibiting viral replication (Ogal et al. 2021; Nicolussi et al. 2022; Guo et al. 2023).

3.2 Elderberry

Sambucus nigra, commonly known as elderberry, has been used in traditional medicine for centuries to treat colds, flu, and respiratory infections. Elderberry extract contains flavonoids like anthocyanins, which have been shown to inhibit the replication of influenza viruses. Clinical trials suggest that elderberry extract can reduce the duration of flu symptoms by up to 50% and is effective in treating respiratory infections. Elderberry blocks viral attachment to cells, preventing viral entry and replication. It also modulates the immune response, enhancing cytokine production to boost the body's defense (Asgary and Pouramini 2022; Liu et al. 2022).

3.3 Garlic (*Allium sativum*)

Garlic has long been celebrated for its antimicrobial and immune-boosting properties. The active compound Allicin found in garlic has demonstrated antiviral activity

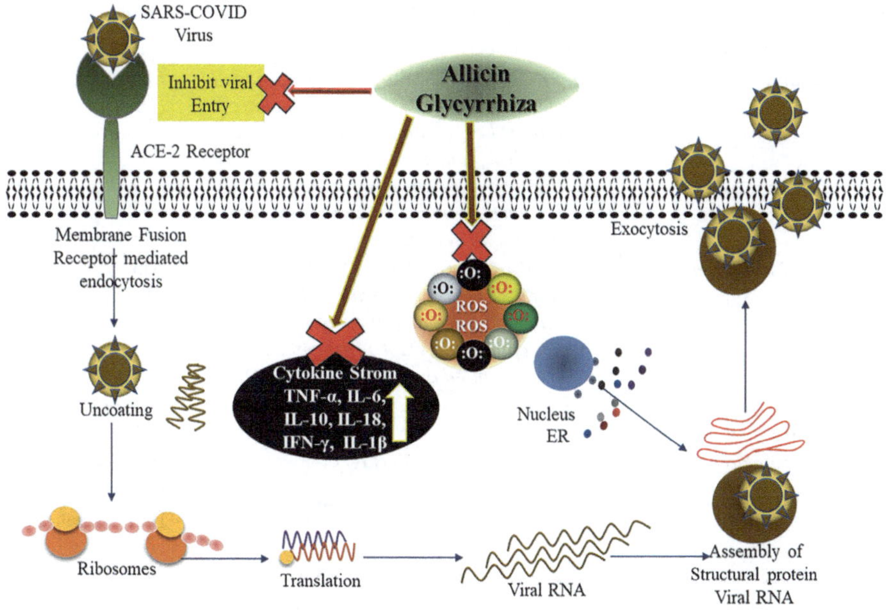

Fig. 17.2 Antiviral & immunomodulatory effect of allicin and glycyrrhizin against SARC-COVID virus infection

against a variety of viruses, including influenza, rhinovirus, and coronaviruses. Research indicates that garlic supplementation can reduce the frequency and duration of colds and flu and may reduce the severity of respiratory infections. Allicin acts by disrupting the viral envelope, preventing viral entry into host cells (Fig. 17.2). Garlic also enhances immune function by stimulating the production of cytokines and interferons (Abd El-Ghany 2024; Adjei-Mensah et al. 2023) (Table 17.1).

4 Immunomodulatory Effects of Medicinal Plants in Respiratory Diseases

Immunomodulation refers to the adjustment or regulation of the immune system, either to enhance or suppress immune responses. In respiratory diseases, particularly chronic conditions like asthma and COPD, dysregulated immune responses lead to chronic inflammation and airway remodeling. Effective immunomodulatory agents can help regulate this inflammation, reduce airway hyperreactivity, and improve lung function (Paulovičová et al. 2022; Zrig 2022). Some of the important plants with phytochemical having immunomodulatory effects are discussed below.

Table 17.1 List of medicinal plants against the respiratory disease

S.N.	Medicinal plants	Antiviral property	References
1.	*Allium sativum* L.	*Influenza B, human rhinovirus type 2, human cytomegalovirus (HCMV), Para influenza virus type 3, herpes simplex type 1 and 2, vaccinia virus, and vesicular stomatitis virus; Amerliorate conditions associated with HIV infection such as fungal infections (thrush) and parasitic infections (cryptosporidium)*	Mukhtar et al. (2008)
2.	*Azadirachta indica* Juss	*Dengue virus type-2 (DEN-2), HSV1, Polio virus, Influenza, HIV, Coxsackie B group virus, and Dengue virus at early step of viral genome replication, Duck viral enteritis (DEV), also called duck plague virus (DPV)*	Ahmad et al. (2016), Parida et al. (2002)
3.	*Glycyrrhiza uralensis* Fisch	*SARS-CoV*	Ben-Shabat et al. (2020)
4.	*Globularia arabica*	*Poliomyelitis-1 virus (POLIO)*	Ben-Shabat et al. (2020)
5.	*Ficus carica*	*HSV-1, ECV-11, ADV, influenza irus*	Ben-Shabat et al. (2020)
6.	*Diospyros kaki* L.	*Influenza virus H3N2, H5N3, HSV-1, VSV, Sendai virus, PV, coxsachievirus, adenovirus, rotavirus, feline calicivirus, mouse norovirus, NDV*	Ben-Shabat et al. (2020), Khouchlaa et al. (2021) and Ueda et al. (2013)
7.	*Glycine max* (L.) Merr	*Human adenovirus type 1, coxsackievirus B1*	Yamai et al. (2003)
8.	*Euphorbia hirta*	*HIV-1, HIV-2, SIV mac 251*	Serkedjieva (2003)
9.	*Geranium sanguineum* L.	*Influenza virus*	Zakay-Rones et al. (2004)
10.	*Myristica fragrans*	*Human rotavirus*	Ben-Shabat et al. (2020)
11.	*Olea europaea* L.	*Influenza virus subtype H9N2, Viral haemorrhagic septicaemia virus (VHSV); HIV*	Ben-Shabat et al. (2020)
12.	*Sambucus nigra* L.	*Influenza virus*	Zakay-Rones et al. (2004)
13.	*Sanguisorba minor*	*VSV, HSV-1 HIV*	Ben-Shabat et al. (2020)
14.	*Prunella vulgaris*	*HIV-1, Ebola virus*	Ben-Shabat et al. (2020)
15.	*Thymus caramanicus* Thymus daenensis Thymus kotschyanus	*HIV-1* *HIV-1* *HIV-1*	Ben-Shabat et al. (2020)
16.	*Trifollium species* Secomet-V	*HPV, Marburg, influenza, HIV*	Kotwal et al. (2005)
17.	*Trachyspermum ammi* L.	*HCV*	Hussein et al. (2000)

(continued)

Table 17.1 (continued)

S.N.	Medicinal plants	Antiviral property	References
18.	*Stevia rebaudiana* L.	*Human Rhinoviruses (HRV)*	Zakay-Rones et al. (2004)
19.	*Salacia reticulata*	*H1N1 influenza*	Ben-Shabat et al. (2020)
20.	*Tuberaria lignosa*	*HIV*	Ben-Shabat et al. (2020)
21.	*Vitis labrusca* *Vitis macrocarpon*	*(SA-11) and human (HCR3) rotaviruses*	Ben-Shabat et al. (2020)

4.1 Astragalus (*Astragalus membranaceus*)

Astragalus is a well-known herb in Traditional Chinese Medicine that has been used to strengthen the immune system and improve lung health. Its active compounds, such as polysaccharides and flavonoids, have been shown to enhance the activity of macrophages and natural killer (NK) cells, which play vital roles in immune defense. Studies have shown that astragalus extract can reduce symptoms of asthma, prevent viral infections, and enhance immune responses in individuals with weakened immunity. Astragalus stimulates the production of T-cells, enhances phagocytosis, and increases the production of interleukins, improving overall immune function (Li et al. 2020; Zhang et al. 2023).

4.2 Cat's Claw (*Uncaria tomentosa*)

Cat's Claw, a vine native to the Amazon rainforest, has powerful anti-inflammatory and immune-enhancing effects. The plant's active compounds, including pentacyclic alkaloids, are believed to modulate immune responses, enhance phagocytosis, and protect against viral infections. Research indicates that Cat's Claw may help manage inflammatory respiratory diseases like COPD and asthma by reducing airway inflammation and improving immune function. Cat's Claw promotes the production of immune cells and enhances the activity of macrophages, which are critical in defending the body against pathogens (Brito et al. 2023).

4.3 Echinacea (*Echinacea purpurea*)

Echinacea, one of the most well-known herbs for respiratory health, has been traditionally used to prevent and treat colds and respiratory infections. Research indicates that Echinacea can stimulate the immune system by increasing the activity of macrophages, natural killer (NK) cells, and the production of cytokines like interleukin-1 (IL-1), interleukin-6 (IL-6), and tumor necrosis factor-alpha (TNF-α). In clinical studies, Echinacea has been found to reduce the duration and severity of cold symptoms, likely due to its immune-boosting and antiviral properties.

Additionally, it has anti-inflammatory effects, which are beneficial in managing respiratory tract infections (Sharifi-Rad et al. 2018; Vieira et al. 2023).

4.4 Andrographis (*Andrographis paniculata*)

Andrographis is a well-researched plant in traditional medicine for respiratory ailments, particularly in Asia. It is known for its immunomodulatory, anti-inflammatory, and antimicrobial properties. Andrographis enhances the production of pro-inflammatory cytokines, such as IL-1β and TNF-α, while also increasing the activity of T lymphocytes and NK cells, which play key roles in fighting respiratory infections. Studies suggest that Andrographis can help reduce the severity and duration of cold symptoms, bronchitis, and asthma by modulating immune responses. It also exhibits antiviral properties, which may be beneficial in treating viral respiratory infections like the common cold and influenza (Hossain et al. 2021; Veldman et al. 2023).

4.5 Astragalus (*Astragalus membranaceus*)

Astragalus is a popular herb in traditional Chinese medicine and is renowned for its ability to enhance immune function and support lung health. It has been shown to stimulate the production of interferons, increase the activity of macrophages and NK cells, and enhance the synthesis of immune proteins like cytokines and antibodies. In respiratory diseases, particularly chronic conditions like asthma and COPD, Astragalus can help improve lung function by reducing inflammation and promoting airway relaxation. It may also have a protective effect against oxidative stress, which plays a key role in the progression of chronic respiratory diseases (Li et al. 2022; Zhang et al. 2020).

4.6 Turmeric (*Curcuma longa*)

The active compound in turmeric, curcumin, is known for its anti-inflammatory, antioxidant, and immunomodulatory properties. Curcumin can inhibit the production of pro-inflammatory cytokines and reduce oxidative stress, both of which are central to the pathogenesis of diseases like asthma, COPD, and pulmonary fibrosis. Curcumin has been found to suppress the activation of NF-kB (a key regulator of inflammation) and modulate T-cell activity, making it a potential therapeutic agent in managing respiratory inflammation and enhancing overall lung function. Its ability to regulate both innate and adaptive immune responses makes it a promising candidate for the management of respiratory diseases (Daily et al. 2016; Kotha and Luthria 2019).

4.7 Ginseng (*Panax ginseng*)

Ginseng is a well-known adaptogen with proven effects on immune system modulation. It is particularly beneficial in boosting the immune response in individuals suffering from respiratory infections. Ginseng has been shown to enhance the production of cytokines, particularly IL-2, and improve the function of NK cells, which are crucial for combating respiratory pathogens. Research suggests that ginseng may be effective in preventing upper respiratory infections, reducing the duration of colds, and improving lung function in individuals with asthma and COPD. Its immunomodulatory properties make it a valuable plant in managing both acute and chronic respiratory diseases (Im 2020; Mancuso and Santangelo 2017).

4.8 Licorice (*Glycyrrhiza glabra*)

Licorice root has a long history of use in traditional medicine for treating respiratory issues, including cough, bronchitis, and asthma. Glycyrrhizin, the primary active compound in licorice, has been shown to possess anti-inflammatory, antiviral, and immunomodulatory effects. It can stimulate the production of certain cytokines, such as interferon-α and IL-1β, which play a role in the body's immune response to respiratory infections. Licorice also has the ability to inhibit the activity of inflammatory enzymes like cyclooxygenase (COX) and lipoxygenase (LOX), which are involved in the inflammatory processes seen in respiratory diseases like asthma and COPD. Additionally, licorice's expectorant properties make it useful for loosening mucus in the airways and alleviating symptoms of bronchitis and cough (Tuli et al. 2022; Wang et al. 2022a, b).

4.9 Holy Basil (*Ocimum sanctum*)

Holy basil, also known as Tulsi, is highly regarded in Ayurvedic medicine for its adaptogenic, anti-inflammatory, and immune-boosting properties. Studies have shown that Tulsi has the ability to modulate immune cell activity by enhancing the function of macrophages and NK cells, which are vital in fighting off respiratory pathogens. The herb also possesses anti-inflammatory properties, which may help reduce airway inflammation and prevent exacerbations in conditions like asthma and COPD. Additionally, Tulsi has antimicrobial effects against a range of respiratory pathogens, including bacteria, viruses, and fungi. This makes it a potent herbal option for preventing or managing respiratory infections (Bharath et al. 2023; Kustiati et al. 2022).

4.10 Ginger (*Zingiber officinale*)

Ginger is widely known for its anti-inflammatory and antioxidant properties. The active compounds in ginger, such as gingerol and shogaol, can modulate the immune response by reducing the release of pro-inflammatory cytokines like TNF-α and IL-6. This can be particularly beneficial in managing conditions characterized by chronic inflammation, such as asthma and COPD. In addition to its immunomodulatory effects, ginger has antitussive (cough-relieving) and expectorant properties, making it useful for managing coughs and mucus production in respiratory infections. Ginger also has antimicrobial properties, which may contribute to its ability to help the body fight respiratory infections like colds and influenza (Ayustaningwarno et al. 2024; Zhang et al. 2022).

4.11 Mullein (*Verbascum thapsus*)

Mullein is a plant commonly used in traditional herbal medicine for treating respiratory ailments, including asthma, bronchitis, and pneumonia. Mullein's active compounds, including flavonoids and saponins, have been found to have anti-inflammatory and immunomodulatory effects. Mullein works by promoting the activity of immune cells such as macrophages and lymphocytes, which helps strengthen the body's defense against respiratory infections. Additionally, it has expectorant and antispasmodic properties, which help alleviate coughing and improve mucus clearance in the airways. Mullein is also used as a natural remedy for soothing inflamed respiratory tissues, making it helpful in conditions like bronchitis and sore throat (Chen et al. 2022; Soto et al. 2022).

4.12 Black Seed (*Nigella sativa*)

Black seed, or Nigella sativa, is an herb known for its potent medicinal properties, including its immunomodulatory, anti-inflammatory, and antimicrobial effects. The active compound thymoquinone has been extensively studied for its ability to modulate immune function by increasing the activity of macrophages, NK cells, and T lymphocytes. In respiratory diseases like asthma, Black Seed has shown promise in reducing inflammation and improving airflow by regulating immune responses that trigger bronchoconstriction. It also has antiviral properties, particularly against respiratory viruses like influenza and rhinoviruses, making it beneficial in preventing and managing respiratory infections. Some studies have indicated that Black Seed oil may improve lung function and reduce exacerbations in individuals with asthma (Mahboubi 2018; Rahim et al. 2022).

4.13 Peppermint (*Mentha piperita*)

Peppermint, particularly its essential oil, is commonly used to relieve symptoms of respiratory distress, including cough, congestion, and sore throat. Menthol, the primary active component in peppermint, has bronchodilator effects, which can help open up the airways, improving airflow and reducing symptoms of conditions like asthma and bronchitis. Peppermint oil has also been found to modulate the immune response by enhancing the activity of specific immune cells, such as macrophages, which play a central role in immune surveillance and pathogen clearance in the respiratory tract. Additionally, peppermint has antimicrobial properties and can help clear mucus, making it useful for managing infections like the common cold and sinusitis (Hudz et al. 2023; Zhao et al. 2022).

4.14 Elderberry (*Sambucus nigra*)

Elderberry has been shown to be a potent immunomodulator, particularly for its ability to stimulate immune cell activity and enhance the body's ability to fight viral infections. The anthocyanins and flavonoids found in elderberries have demonstrated antiviral effects, particularly against respiratory viruses such as influenza, by inhibiting the virus's ability to enter human cells. Elderberry also has anti-inflammatory effects that may help reduce the severity of symptoms in respiratory conditions like the common cold, flu, and upper respiratory tract infections (URIs). Research has indicated that elderberry can shorten the duration of cold symptoms and reduce the need for medication, making it an effective natural alternative or adjunct to traditional treatments (Ferreira et al. 2022; Uhl and Mitchell 2024).

4.15 Cinnamon (*Cinnamomum verum*)

Cinnamon, particularly Cinnamomum verum, is not only a popular spice but also a powerful medicinal plant with a variety of pharmacological effects, including immune modulation. Cinnamon's active compound, cinnamaldehyde, has demonstrated strong anti-inflammatory and antioxidant effects, which can help alleviate symptoms of respiratory diseases like asthma and COPD. Cinnamon also has antimicrobial properties, making it effective in preventing and managing respiratory infections caused by bacteria and viruses. It has been shown to improve respiratory function, reduce inflammation in the airways, and relieve cough and congestion. Additionally, cinnamon may help in reducing the risk of respiratory infections by enhancing the body's overall immune response (Silva et al. 2022; Wijesinghe et al. 2021).

4.16 Thyme (*Thymus vulgaris*)

Thyme is a well-known herb used in the treatment of respiratory infections, coughs, and bronchitis. The active compounds thymol and carvacrol in thyme have both antimicrobial and immunomodulatory effects. Thyme has been found to stimulate the immune system by increasing the activity of immune cells, such as macrophages, which are essential for fighting respiratory pathogens. Thyme's expectorant and antispasmodic properties help clear mucus from the airways and reduce coughing, making it particularly useful in managing conditions like bronchitis, asthma, and chronic cough. Additionally, thyme has anti-inflammatory effects, which can help reduce airway inflammation in diseases like asthma and COPD (Alotaibi et al. 2023; Klimiuk et al. 2023).

5 Conclusions and Future Perspectives

In conclusion, medicinal plant products are a valuable source of bioactive substances with important antiviral and immunomodulatory potential in respiratory diseases. Their ability to inhibit viral replication, block viral entry, and modulate the immune response makes them promising candidates for alternative or complementary therapies. Key phytochemicals such as flavonoids, terpenoids, and alkaloids offer a multifunctional benefit, acting as direct antiviral agents as well as modulators of immune pathways involved in inflammatory and immune responses during a viral infection. These properties are particularly important for the treatment of acute and chronic respiratory diseases aggravated by viral pathogens.

In addition, medicinal plants offer the benefits of lower toxicity and better patient adherence than traditional medicines. Recent studies, including preclinical and limited clinical studies, provide supportive evidence of their antiviral efficacy against influenza, respiratory syncytial virus (RSV), and coronaviruses. However, pharmacokinetic, standardization and interaction profiles of plant derived compounds are still largely unexplored. Rigorous scientific validation, including large-scale randomized clinical trials, is urgently needed to assess their safety, dose-proportional optimization, and therapeutic potential in real-world settings. Integrating traditional knowledge with modern pharmacological research will be crucial to identify novel lead compounds and understand their mechanisms of action.

As global health challenges evolve, medicinal plants represent a promising area for future research. Emerging fields like phytomedicine, molecular pharmacology, and genomics will likely provide more insights into how medicinal plants work at the molecular level and how they can be effectively integrated into therapeutic regimens for respiratory diseases. Looking ahead, the future of respiratory antiviral therapy lies in a holistic and integrated approach combining traditional medicines with evidence-based herbal preparations. Biotechnological advances such as plant-derived nanoparticles and targeted delivery systems can increase the bioavailability and efficacy of plant chemicals. Collaborative research efforts involving ethnobotanists, pharmacologists, clinicians, and policymakers are crucial for translating laboratory findings into effective, accessible, and affordable treatments. Support for sustainable cultivation and preservation of medicinal plants will also be crucial to ensure a sustainable supply of these valuable resources in the global fight against respiratory diseases.

References

Abd El-Ghany WA (2024) Potential effects of garlic (Allium sativum L.) on the performance, immunity, gut health, antioxidant status, blood parameters, and intestinal microbiota of poultry: an updated comprehensive review. Animals 14(3):498

Adjei-Mensah B, Quaye B, Opoku O, Atuahene CC (2023) Antiviral potentials of garlic (Allium sativum) in poultry production: a mini review. Vet Med Sci 9(6):2711–2718

Ahmad A, Javed MR, Rao AQ, Husnain T (2016) Designing and screening of universal drug from neem (Azadirachta indica) and standard drug chemicals against influenza virus nucleoprotein. BMC Complement Altern Med 16(1):519

Alotaibi NM, Alotaibi MO, Alshammari N, Adnan M, Patel M (2023) Network pharmacology combined with molecular docking, molecular dynamics, and in vitro experimental validation reveals the therapeutic potential of Thymus vulgaris L. essential oil (thyme oil) against human breast cancer. ACS Omega 8(50):48344–48359

Asgary S, Pouramini A (2022) The pros and cons of using elderberry (Sambucus nigra) for prevention and treatment of COVID-19. Adv Biomed Res 11(1):96

Ayustaningwarno F, Anjani G, Ayu AM, Fogliano V (2024) A critical review of ginger's (Zingiber officinale) antioxidant, anti-inflammatory, and immunomodulatory activities. Front Nutr 11:1364836

Ben-Shabat S, Yarmolinsky L, Porat D, Dahan A (2020) Antiviral effect of phytochemicals from medicinal plants: applications and drug delivery strategies. Drug Deliv Transl Res 10(2):354–367

Bharath MN, Gupta S, Vashistha G, Ahmad S, Singh SV (2023) Bioprospective role of Ocimum sanctum and Solanum xanthocarpum against emerging pathogen: Mycobacterium avium subspecies paratuberculosis: a review. Molecules 28(8):3490

Bibi S, Sarfraz A, Mustafa G, Ahmad Z, Zeb MA, Wang YB, Khan T, Khan MS, Kamal MA, Yu H (2021) Impact of traditional plants and their secondary metabolites in the discovery of COVID-19 treatment. Curr Pharm Des 27(9):1123–1143

Brito ASD, Souza JG, Moraes Filho AV, Oliveira PG, Oliveira MAP, Barbosa MS, Sousa JMG, Pinto RM, Conceição EC, Silva CA (2023) Immunostimulating activity of Uncaria tomentosa in RAW 264.7 macrophages. Braz J Biol 83:e271339

Chen S, Liu H, Wang S, Jiang H, Gao L, Wang L, Teng L, Wang C, Wang D (2022) The neuroprotection of verbascoside in Alzheimer's disease mediated through mitigation of neuroinflammation via blocking NF-κB-p65 signaling. Nutrients 14(7):1417

Daily JW, Yang M, Park S (2016) Efficacy of turmeric extracts and curcumin for alleviating the symptoms of joint arthritis: a systematic review and meta-analysis of randomized clinical trials. J Med Food 19(8):717–729

Dhamodharan P, Pillai GG, Arumugam M (2024) Identification of potential phytochemical inhibitors for the proven six chronic obstructive pulmonary disease biomarkers – a molecular dynamics study. J Biomol Struct Dyn 42(20):10994–11019

Ferreira SS, Martins-Gomes C, Nunes FM, Silva AM (2022) Elderberry (Sambucus nigra L.) extracts promote anti-inflammatory and cellular antioxidant activity. Food Chem X 15:100437

Fu YS, Kang N, Yu Y, Mi Y, Guo J, Wu J, Weng CF (2022) Polyphenols, flavonoids and inflammasomes: the role of cigarette smoke in COPD. Eur Respir Rev 31(164):220028

Guo J, Luo Y, Zuo J, Teng J, Shen B, Liu X (2023) Echinacea polyphenols inhibit NLRP3-dependent pyroptosis, apoptosis, and necroptosis via suppressing NO production during lipopolysaccharide-induced acute lung injury. J Agric Food Chem 71(19):7289–7298
Hossain S, Urbi Z, Karuniawati H, Mohiuddin RB, Moh Qrimida A, Allzrag AMM, Ming LC, Pagano E, Capasso R (2021) Andrographis paniculata (Burm. f.) Wall. ex Nees: an updated review of phytochemistry, antimicrobial pharmacology, and clinical safety and efficacy. Life 11(4):348
Hudz N, Kobylinska L, Pokajewicz K, Horčinová Sedláčková V, Fedin R, Voloshyn M, Myskiv I, Brindza J, Wieczorek PP, Lipok J (2023) Mentha piperita: essential oil and extracts, their biological activities, and perspectives on the development of new medicinal and cosmetic products. Molecules 28(21):7444
Hussein G, Miyashiro H, Nakamura N, Hattori M, Kakiuchi N, Shimotohno K (2000) Inhibitory effects of Sudanese medicinal plant extracts on hepatitis C virus (HCV) protease. Phytother Res 14(7):510–516
Im DS (2020) Pro-resolving effect of ginsenosides as an anti-inflammatory mechanism of Panax ginseng. Biomolecules 10(3):444
James SL, Abate D, Abate KH, Abay SM, Abbafati C, Abbasi N, Abbastabar H, Abd-Allah F, Abdela J, Abdelalim A et al (2018) Global, regional, and national incidence, prevalence, and years lived with disability for 354 diseases and injuries for 195 countries and territories, 1990–2017: a systematic analysis for the Global Burden of Disease Study 2017. Lancet 392(10159):1789–1858
Jasemi SV, Khazaei H, Aneva IY, Farzaei MH, Echeverría J (2020) Medicinal plants and phytochemicals for the treatment of pulmonary hypertension. Front Pharmacol 11:145
Jasemi SV, Khazaei H, Momtaz S, Farzaei MH, Echeverría J (2022) Natural products in the treatment of pulmonary emphysema: therapeutic effects and mechanisms of action. Phytomedicine 99:153988
Khouchlaa A, El Menyiy N, Guaouguaou F-E, El Baaboua A, Charfi S, Lakhdar F, El Omari N, Taha D, Shariati MA, Rebezov M, El-Shazly M, Bouyahya A (2021) Ethnomedicinal use, phytochemistry, pharmacology, and toxicology of Daphne gnidium: a review. J Ethnopharmacol 275:114124
Klimiuk K, Sembratowicz I, Tutaj K, Czech A (2023) Effect of Thyme (Thymus vulgaris L.) used in diets with extruded flaxseed on the antioxidant and lipid profile of the blood and tissues of fattening pigs. Antioxidants 12(5):1045
Kotha RR, Luthria DL (2019) Curcumin: biological, pharmaceutical, nutraceutical, and analytical aspects. Molecules 24(16):2930
Kotwal GJ, Kaczmarek JN, Leivers S, Ghebremariam YT, Kulkarni AP, Bauer G, De Beer C, Preiser W, Mohamed AR (2005) Anti-HIV, anti-poxvirus, and anti-SARS activity of a non-toxic, acidic plant extract from the Trifollium species Secomet-V/anti-Vac suggests that it contains a novel broad-spectrum antiviral. Ann N Y Acad Sci 1056(1):293–302
Kustiati U, Wihadmadyatami H, Kusindarta DL (2022) Dataset of phytochemical and secondary metabolite profiling of holy basil leaf (Ocimum sanctum Linn) ethanolic extract using spectrophotometry, thin layer chromatography, Fourier transform infrared spectroscopy, and nuclear magnetic resonance. Data Brief 40:107774
Labaki WW, Han MK (2020) Chronic respiratory diseases: a global view. Lancet Respir Med 8(6):531–533
Li W, Hu X, Wang S, Jiao Z, Sun T, Liu T, Song K (2020) Characterization and anti-tumor bioactivity of Astragalus polysaccharides by immunomodulation. Int J Biol Macromol 145:985–997
Li C, Liu Y, Zhang Y, Li J, Lai J (2022) Astragalus polysaccharide: a review of its immunomodulatory effect. Arch Pharm Res 45(6):367–389
Lin H, Wang C, Yu H, Liu Y, Tan L, He S, Li Z, Wang C, Wang F, Li P, Liu J (2022) Protective effect of total saponins from American ginseng against cigarette smoke-induced COPD in mice based on integrated metabolomics and network pharmacology. Biomed Pharmacother 149:112823

Liu D, He X-Q, Wu DT, Li H-B, Feng Y-B, Zou L, Gan R-Y (2022) Elderberry (Sambucus nigra L.): bioactive compounds, health functions, and applications. J Agric Food Chem 70(14):4202–4220

Mahboubi M (2018) Natural therapeutic approach of Nigella sativa (black seed) fixed oil in management of sinusitis. Integr Med Res 7(1):27–32

Mancuso C, Santangelo R (2017) Panax ginseng and Panax quinquefolius: from pharmacology to toxicology. Food Chem Toxicol 107:362–372

Mitra S, Anand U, Ghorai M, Vellingiri B, Jha NK, Behl T, Kumar M, Radha SMS, Proćków J, Dey A (2022) Unravelling the therapeutic potential of botanicals against chronic obstructive pulmonary disease (COPD): molecular insights and future perspectives. Front Pharmacol 13:824132

Mukhtar M, Arshad M, Ahmad M, Pomerantz RJ, Wigdahl B, Parveen Z (2008) Antiviral potentials of medicinal plants. Virus Res 131(2):111–120

Mustra Rakic J, Wang X-D (2020) Role of lycopene in smoke-promoted chronic obstructive pulmonary disease and lung carcinogenesis. Arch Biochem Biophys 689:108439

Nicolussi S, Ardjomand-Woelkart K, Stange R, Gancitano G, Klein P, Ogal M (2022) Echinacea as a potential force against coronavirus infections? A mini-review of randomized controlled trials in adults and children. Microorganisms 10(2):211

Ogal M, Johnston SL, Klein P, Schoop R (2021) Echinacea reduces antibiotic usage in children through respiratory tract infection prevention: a randomized, blinded, controlled clinical trial. Eur J Med Res 26(1):33

Parida MM, Upadhyay C, Pandya G, Jana AM (2002) Inhibitory potential of neem (Azadirachta indica Juss) leaves on Dengue virus type-2 replication. J Ethnopharmacol 79(2):273–278

Paulovičová E, Paulovičová L, Pawlaczyk-Graja I, Gancarz R, Kopáčová M, Capek P (2022) Effectivity of polyphenolic polysaccharide-proteins isolated from medicinal plants as potential cellular immune response modulators. Biologia 77(12):3581–3593

Rahim MA, Shoukat A, Khalid W, Ejaz A, Itrat N, Majeed I, Koraqi H, Imran M, Nisa MU, Nazir A, Alansari WS, Eskandrani AA, Shamlan G, Al-Farga A (2022) A narrative review on various oil extraction methods, encapsulation processes, fatty acid profiles, oxidative stability, and medicinal properties of black seed (Nigella sativa). Foods 11(18):2826

Rahman MM, Bibi S, Rahaman MS, Rahman F, Islam F, Khan MS, Hasan MM, Parvez A, Hossain MA, Maeesa SK, Islam MR, Najda A, Al-malky HS, Mohamed HRH, AlGwaiz HIM, Awaji AA, Germoush MO, Kensara OA, Abdel-Daim MM, Kamal MA (2022) Natural therapeutics and nutraceuticals for lung diseases: traditional significance, phytochemistry, and pharmacology. Biomed Pharmacother 150:113041

Serkedjieva J (2003) Influenza virus variants with reduced susceptibility to inhibition by a polyphenol extract from Geranium sanguineum L. Pharmazie 58(1):53–57

Sharifi-Rad M, Mnayer D, Morais-Braga MFB, Carneiro JNP, Bezerra CF, Coutinho HDM, Salehi B, Martorell M, Del Mar Contreras M, Soltani-Nejad A, Uribe YAH, Yousaf Z, Iriti M, Sharifi-Rad J (2018) Echinacea plants as antioxidant and antibacterial agents: from traditional medicine to biotechnological applications. Phytother Res 32(9):1653–1663

Silva ML, Bernardo MA, Singh J, De Mesquita MF (2022) Cinnamon as a complementary therapeutic approach for dysglycemia and dyslipidemia control in type 2 diabetes mellitus and its molecular mechanism of action: a review. Nutrients 14(13):2773

Soto KM, Luzardo-Ocampo I, López-Romero JM, Mendoza S, Loarca-Piña G, Rivera-Muñoz EM, Manzano-Ramírez A (2022) Gold nanoparticles synthesized with common mullein (Verbascum thapsus) and castor bean (Ricinus communis) ethanolic extracts displayed antiproliferative effects and induced caspase 3 activity in human HT29 and SW480 cancer cells. Pharmaceutics 14(10):2069

Tuli HS, Garg VK, Mehta JK, Kaur G, Mohapatra RK, Dhama K, Sak K, Kumar A, Varol M, Aggarwal D, Anand U, Kaur J, Gillan R, Sethi G, Bishayee A (2022) Licorice (Glycyrrhiza glabra L.)-derived phytochemicals target multiple signaling pathways to confer oncopreventive and oncotherapeutic effects. Onco Targets Ther 15:1419–1448

Ueda K, Kawabata R, Irie T, Nakai Y, Tohya Y, Sakaguchi T (2013) Inactivation of pathogenic viruses by plant-derived tannins: strong effects of extracts from persimmon (Diospyros kaki) on a broad range of viruses. PLoS One 8(1):e55343

Uhl K, Mitchell AE (2024) Elderberry, an ancient remedy: a comprehensive study of the bioactive compounds in three Sambucus nigra L. subspecies. Annu Rev Food Sci Technol 15(1):27–51

Veldman LBM, Belt-Van Zoen E, Baars EW (2023) Mechanistic evidence of Andrographis paniculata, Pelargonium sidoides, Echinacea species and a combination of Hedera helix, Primula veris/Primula elatior and Thymus vulgaris/Thymus zygis in the treatment of acute, uncomplicated respiratory tract infections: a systematic literature review and expert interviews. Pharmaceuticals 16(9):1206

Vieira SF, Gonçalves SM, Gonçalves VMF, Llaguno CP, Macías F, Tiritan ME, Cunha C, Carvalho A, Reis RL, Ferreira H, Neves NM (2023) Echinacea purpurea fractions represent promising plant-based anti-inflammatory formulations. Antioxidants 12(2):425

Wang J, Wu Q, Ding L, Song S, Li Y, Shi L, Wang T, Zhao D, Wang Z, Li X (2021) Therapeutic effects and molecular mechanisms of bioactive compounds against respiratory diseases: traditional Chinese medicine theory and high-frequency use. Front Pharmacol 12:734450

Wang X, Wang Y, Huang D, Shi S, Pei C, Wu Y, Shen Z, Wang F, Wang Z (2022a) Astragaloside IV regulates the ferroptosis signaling pathway via the Nrf2/SLC7A11/GPX4 axis to inhibit PM2.5-mediated lung injury in mice. Int Immunopharmacol 112:109186

Wang KL, Yu YC, Chen HY, Chiang YF, Ali M, Shieh TM, Hsia SM (2022b) Recent advances in Glycyrrhiza glabra (Licorice)-containing herbs alleviating radiotherapy- and chemotherapy-induced adverse reactions in cancer treatment. Meta 12(6):535

Wijesinghe G, Oliveira T, Maia F, de Feiria S, Joia F, Barbosa J, Boni G, Höfling J (2021) Efficacy of true cinnamon (Cinnamomum verum) leaf essential oil as a therapeutic alternative for Candida biofilm infections. Iran J Basic Med Sci. https://doi.org/10.22038/ijbms.2021.53981.12138

Yamai M, Tsumura K, Kimura M, Fukuda S, Murakami T, Kimura Y (2003) Antiviral activity of a hot water extract of black soybean against a human respiratory illness virus. Biosci Biotechnol Biochem 67(5):1071–1079

Zakay-Rones Z, Thom E, Wollan T, Wadstein J (2004) Randomized study of the efficacy and safety of oral elderberry extract in the treatment of influenza A and B virus infections. J Int Med Res 32(2):132–140

Zhai T, Li S, Hu W, Li D, Leng S (2018) Potential micronutrients and phytochemicals against the pathogenesis of chronic obstructive pulmonary disease and lung cancer. Nutrients 10(7):813

Zhang J, Wu C, Gao L, Du G, Qin X (2020) Astragaloside IV derived from Astragalus membranaceus: a research review on the pharmacological effects. Adv Pharmacol 87:89–112

Zhang S, Kou X, Zhao H, Mak KK, Balijepalli MK, Pichika MR (2022) Zingiber officinale var. rubrum: red ginger's medicinal uses. Molecules 27(3):775

Zhang X, Qu X, Zou Y (2023) The effect of Astragalus on humoral and cellular immune response: a systematic review and meta-analysis of human studies. Complement Med Res 30(6):535–543

Zhang J, Shao MM, Wang M (2024) Therapeutic potential of natural flavonoids in pulmonary arterial hypertension: a review. Phytomedicine 128:155535

Zhao H, Ren S, Yang H, Tang S, Guo C, Liu M, Tao Q, Ming T, Xu H (2022) Peppermint essential oil: its phytochemistry, biological activity, pharmacological effect and application. Biomed Pharmacother 154:113559

Zrig A (2022) The effect of phytocompounds of medicinal plants on coronavirus (2019-NCOV) infection. Pharm Chem J 55(10):1080–1084

Comparative Efficacy of Phytochemicals in Respiratory Health

18

Nidhi Gupta, Nitin Sharma, Kamil Kuca, Jiri Patocka, Rajni Dhalaria, and Vikas Kumar

Abstract

Respiratory diseases, including asthma, chronic obstructive pulmonary disease (COPD), pulmonary fibrosis and lung cancer, impose a significant global health burden. Conventional treatments, such as corticosteroids, bronchodilators and antibiotics, often have limitations, including drug resistance, side effects and high costs. Phytochemicals derived from medicinal plants have gained attention as potential therapeutic agents due to their anti-inflammatory, antioxidant, bronchodilatory, mucolytic and immunomodulatory properties. This chapter explores the comparative efficacy of key phytochemicals, including polyphenols, terpenoids, alkaloids and saponins, in the management of respiratory diseases. Specific compounds such as quercetin, curcumin, eucalyptol and berberine have shown promise in mitigating airway inflammation, reducing oxidative stress and enhancing immune responses. Furthermore, the integration of phytochemicals with nanotechnology and conventional therapies is discussed to optimize their

N. Gupta · N. Sharma
Department of Biotechnology, Chandigarh Group of Colleges,
Landran, Mohali, Punjab, India

K. Kuca
Department of Chemistry, University of Hradec Kralove, Hradec Kralove, Czech Republic
e-mail: kamil.kuca@uhk.cz

J. Patocka
Faculty of Science, University of Hradec Kralove, Kralove, Czech Republic
e-mail: toxicology@toxicology.cz

R. Dhalaria
Department of Biosciences, Minerva P.G. College of Arts, Science and Commerce,
Indora, Kangra, India

V. Kumar (✉)
University Institute of Biotechnology, Chandigarh University, Gharuan, Mohali, Punjab, India

K. Kuca et al. (eds.), *Medicinal Plants and Their Bioactives in Human Diseases*,
https://doi.org/10.1007/978-3-032-01356-9_18

therapeutic potential. Challenges related to bioavailability, standardization and clinical validation are also highlighted. Advancing research on phytochemicals may lead to more effective and safer alternatives for respiratory health management.

Keywords

Phytochemicals · Respiratory diseases · Antioxidants · Medicinal plants · Nanotechnology · Integrative therapy

1 Introduction

Respiratory health is a critical aspect of overall well-being, as the respiratory system plays a fundamental role in oxygen exchange and immune defence in humans. Respiratory diseases include a variety of conditions, ranging from mild and temporary illnesses to severe and long-lasting diseases that can be life-threatening (Quecchia and Vianello 2025). Respiratory disorders, including asthma, chronic obstructive pulmonary disease (COPD), pulmonary fibrosis, pneumoconiosis, lung cancer, bronchiectasis, occupational lung diseases, interstitial lung disease and pulmonary sarcoidosis, pose significant health risks and challenges to humans globally. In 2019, chronic respiratory diseases (CRDs) were the third leading cause of death worldwide, claiming around four million lives. Globally, nearly 454.6 million people were affected with CRD (Momtazmanesh et al. 2023). Asthma and COPD are the most prevalent chronic respiratory diseases, affecting nearly half a billion people and causing around four million deaths annually. A significant portion, i.e., 90%, of these fatalities occur in people aged under 70 years in low- and middle-income countries (WHO 2023a, b). COPD alone claims 3.2 million lives annually, making up 81.7% of all chronic respiratory disease-related deaths, and is ranked as the third-leading cause of death worldwide in 2019 (Levine and Marciniuk 2022). As per WHO, the count of COPD deaths was increased to 3.5 million deaths in 2021, accounting for about 5% of all deaths worldwide. As per WHO report (2023a, b), 262 million people were affected by asthma in 2019 and it caused 0.455 million deaths. The main risk factors for lung diseases include smoking, air pollution (both indoors and outdoors), workplace hazards and poverty (Soriano et al. 2020).

Several methods of conventional treatments are available for respiratory disorders including corticosteroids, bronchodilators like beta2-agonists and antimuscarinic drugs and antibiotics like aminopenicillin with clavulanic acid, tetracycline, macrolide, which are commonly used to manage these conditions but are often associated with minor to severe side effects, drug resistance and high costs (Fan and Xu 2014; Melani et al. 2023). This highlights the need for alternative and complementary approaches. Phytochemicals that are derived from medicinal plants have gained increasing attention for their therapeutic potential in respiratory health (Zhai et al. 2018; Mehta and Pradhan 2024). These compounds, including polyphenols, terpenoids, alkaloids and saponins, exhibit a wide range of pharmacological activities such as anti-inflammatory, antiviral, bronchodilatory, mucolytic and immunomodulatory effects (Juergens et al. 2020). For instance, monoterpene 1,8-cineole (eucalyptol) has demonstrated potent anti-inflammatory, antioxidant, bronchodilatory, antiviral and antimicrobial effects, while thymol provides bronchodilatory and expectorant benefits (Nagoor Meeran et al. 2017).

Despite the promising benefits of phytochemicals, their comparative efficacy in different respiratory conditions remains an area of active research. Some phytochemicals show superior effects in reducing airway inflammation, while others excel in modulating immune responses or enhancing lung function. Understanding these differences is essential for optimizing their therapeutic use and integrating them into conventional treatment strategies. Integrating phytochemicals with nanotechnology could optimize lung disease therapy. Nanoformulations can enhance delivery of phytochemicals, especially via the pulmonary route, thereby improving effectiveness of phytochemicals. While preclinical studies are promising, further research is needed in this field (Supramaniam et al. 2024). This chapter provides a comprehensive analysis of the efficacy of various phytochemicals in respiratory health, examining their mechanisms of action, clinical relevance and potential synergies with conventional therapies. It also highlights the challenges associated with their use, such as standardization, bioavailability and potential drug interactions. This chapter aims to contribute to the growing body of knowledge on plant-based therapies for respiratory diseases and their potential role in modern medicine.

2 Common Respiratory Diseases

With the growing population, pollution and modern lifestyle, there is rise of various diseases globally, with respiratory disorders being among the most prevalent, mostly in middle- and lower-income countries. The human lungs when constantly exposed to environmental pollutants from both natural and human-made sources become highly susceptible to a range of respiratory conditions. These disorders can significantly impact quality of life and, in severe cases, be life-threatening. Cigarette smoking (CS) and tobacco addiction are one of the major reasons of respiratory disorders such as Chronic obstructive pulmonary disease (COPD) as well as lung cancer. Although genetic predisposition can be there for some respiratory disorders, some of the major respiratory diseases affecting millions worldwide are as listed below.

2.1 Asthma

Asthma is an obstructive inflammatory disorder in the lower airways of humans that can occur at any age. It is a complex disease with no single cause, influenced by various factors like indoor and outdoor pollutants, allergens, tobacco smoke, social deprivation, climate-induced extreme weather events, genetic history and occupational exposures. It can be periodic, with symptoms appearing and resolving, or it can be persistent, with continuing symptoms like breathing difficulties. Major symptoms of asthma include wheezing, breathlessness, coughing and shortness of breath, which may worsen due to allergens, respiratory tract infections, exercise, or weather changes (Vicedo-Cabrera et al. 2023; Melén et al. 2024). Most common medications used for treatment of asthma are glucocorticoids, β_2-agonist, short-acting β-agonists, epinephrine, magnesium sulfate, non-invasive positive pressure ventilation (NPPV), azithromycin and monoclonal antibodies like anti-IL-5 (mepolizumab) and anti-IgE mAB.

2.2 Chronic Obstructive Pulmonary Disease (COPD)

COPD is a progressive, irreversible lung disease, which includes emphysema, chronic bronchitis and small airway damage. Symptoms include coughing, wheezing, tiredness, sputum production, shortness of breath and chest tightness, often leading to exercise intolerance and other respiratory complications (Shukla et al. 2020). In COPD, a condition called emphysema arises where lung function declines due to damage of alveoli and airway support, thereby reducing gas exchange and causing trapped air which leads to breathing difficulties (Shah et al. 2017). Another condition called as chronic bronchitis, mucus buildup and inflammation in the airways of the lungs occurs, especially in people who smoke or have smoking history. This reduces oxygen exchange, leading to breathing difficulties, low blood oxygen levels (hypoxemia) and potential organ failure. The basic treatments given for COPD to relax and increase the airflow in lungs are corticosteroids, glucocorticosteroids, bronchodilators, antibiotics and theophylline. Lifestyle modifications can also lead to improvement in patients suffering with COPD (Putcha and Wise 2018).

2.3 Interstitial Lung Diseases/Diffused Parenchymal Diseases

Interstitial lung diseases (ILD) include group of diseases (more than 100) such as acute eosinophilic pneumonia, bronchiolitis obliterans organizing pneumonia (BOOP), lymphocytic interstitial pneumonitis, cellular interstitial pneumonia, hypersensitivity pneumonitis (HP), idiopathic interstitial pneumonia, alveolar proteinosis and pulmonary capillaritis, all of them is characterized by inflammation and scarring or fibrosis of the lungs (Rivera-Ortega and Molina-Molina 2019). The main symptoms include shortness of breath, discomfort in chest, dry cough and fatigue. The reason for ILD can be occupational and environmental exposure to chemicals, mineral dusts, toxic gases, organic dust, drug-induced, autoimmune disorders and Idiopathic. Like other type of lung diseases, corticosteroids are choice of medications. Immunosuppressant therapy is also another method to treat ILD. Some new drugs tyrosine kinase inhibitor, pyridines and prostacyclin vasodilator have been approved by FDA for management of ILD (Bang et al. 2023).

2.4 Pulmonary Tuberculosis

Tuberculosis (TB) is one of leading killer in the world, which is caused by bacteria *Mycobacterium tuberculosis*, which mainly affect the lungs but can affect other organs of body called as extrapulmonary tuberculosis (Chai et al. 2018). Lung TB spreads through air when the patient sneezes, coughs or spits and susceptible individual when comes in contact and inhale those droplets cause inhalation of the bacteria into the alveoli of the lower airway. TB causes lung remodelling and structural lung damage includes fibrosis, cavitation, bronchiectasis, bronchostenosis and parenchymal destruction, leading to varying degrees of pulmonary dysfunction, which can increase mortality from respiratory causes (Ravimohan et al. 2018). Isoniazid, pyrazinamide, rifampicin and ethambutol are first line of antibiotics for treatment of TB. Second-line anti-TB drugs are given in case of multidrug-resistant (MDR) TB.

2.5 Cystic Fibrosis

Cystic Fibrosis (CF) is an autosomal recessive genetic disorder caused by mutations in Cystic fibrosis transmembrane conductance regulator (CFTR) gene encoding chloride-conducting transmembrane channel on chromosome 7 at position q31.2 (Shteinberg et al. 2021). This results in mucus congestion, chronic infections and inflammation in lung. This disorder primarily depresses airway cells, alongside producing effects on the immune system. Lung damage from bronchiectasis alongside airway obstruction remains the main reason that leads to illness and death in CF patients (Elborn 2016). The one of the approaches used for therapeutic treatment for CF is inhalation of mucoactive agents such as hypertonic saline, mannitol powder and rhDNase, which thins lung mucus that enables easier coughing to remove it from the body. Other treatment includes bronchodilators, corticosteroids and antibiotics. Molecular therapies can also be used in single or double combination of drugs Ivacaftor, Lumacaftor and Tezacaftor. Triple combination molecular therapy is also used consisting of Tezacaftor/Elexacaftor/Ivacaftor for CF treatment (Jaques et al. 2020).

2.6 Lung Cancer

Lung cancer, one of the leading causes of cancer-related deaths, is a highly aggressive and invasive disease, a high susceptibility to infiltrate different locations of the human body and there are two subtypes- small-cell lung carcinoma and non-small-cell lung carcinoma (NSCLC), where latter is 80% of all the total cases. The universal occurrence of this disease among men and women leads to advanced-stage diagnoses because of its fast-spreading capabilities. Smoking cigarettes/tobacco addiction is one of the major causes of lung cancer in addition to genetic risk and environmental factors (Zappa and Mousa 2016). Treatment involves surgery, chemotherapy, radiotherapy and immunotherapy in combinations depending upon cancer stage. The chemotherapy treatment includes the -use of proapoptotic agents, angiogenesis inhibitors, insulin-like growth factor inhibitors and histone deacetylase inhibitors. Smoking cessation and lifestyle changes can prevent lung cancer (Meyer et al. 2024).

3 Key Phytochemicals and Their Effects in Respiratory Health

Phytocompounds produced by plants are known to have a beneficial effect on respiratory health. COPD and lung cancer are linked, and root cause is smoking for both. Cigarette smoking can induce COPD and lung cancer. The causes for COPD alongside lung cancer development may stem from genetic susceptibility alongside telomere reduction, faulty mitochondria function and accelerated biological aging. COPD serves as a lung cancer promoter because it magnifies oxidative stress by accumulating reactive oxygen species (ROS) and reactive nitrogen species (RNS) produced by smoking together with chronic inflammation, while damaging DNA, driving cellular proliferation and blocking DNA repair mechanisms (Durham and Adcock 2015; Padmavathi et al. 2018). Epidemiological and cross-sectional studies demonstrated the significance of dietary intake in asthma and COPD development and management (Berthon and Wood 2015; Joshi et al. 2015). As established in a study, fruits and tomatoes have shown potential to slow down lung function deterioration, specifically in former smokers based on the antioxidants including vitamin C and flavonoids and carotenoids they contain. Smoking-inflicted lung damage and inflammation can be minimized by these dietary compounds, which also fight oxidative stress through their protective mechanisms for lung health preservation and maintain ventilatory function (Garcia-Larsen et al. 2017, 2018). The phytocompounds present in plant have anti-fibrogenic, anti-inflammatory, antitussive, bronchodilatory effects and modulate many signalling pathways like mitogen-activated protein kinase (MAPK)/NF-κB signalling pathways. Many of these phytocompounds have antibacterial activity against the tuberculosis bacteria (Sasikumar et al. 2018). Many of the phytocompounds isolated from medicinal plants have been found to have beneficial effects on respiratory health along with their mode of action is listed in Table 18.1.

4 Integration of Phytochemicals with Conventional Therapies

4.1 Integration of Phytochemicals with Anti-TB Drugs

Anti-TB drugs are found to damage liver during the course of Directly Observed Therapy Short (DOTs) course. Adjunct therapy with phytochemicals or ayurvedic medicines can exert hepatoprotective activities, thus prevent hepatic damage in TB

Table 18.1 Phytocompounds from medicinal plants and their beneficial effects in respiratory health

Source/medicinal plants	Phytocompounds/plant extracts	Respiratory disease	Mode of action	References
Citrus fruits, apples, onions, parsley, sage, tea, red wine, olive oil, grapes, dark cherries, dark berries and other plant sources	Quercetin	Rheumatoid arthritis (RA)-associated ILD	Regulate IL-17 signalling and reduce inflammation	Wang et al. (2024a)
		Idiopathic pulmonary fibrosis	Anti-fibrogenic, anti-inflammatory effects, modulate redox balance by inducing Nrf2	Boots et al. (2020)
		Isoniazid (INH)-induced hepatotoxicity in TB	Inhibit liver injury, regulate NLRP3 inflammasome via SIRT1	Zhang et al. (2020b)
		TB	Antibacterial effect	Sasikumar et al. (2018)
Olives and olive oil	Uvaol	Allergic asthma	Inhibit eosinophil influx, IL-5 production	Agra et al. (2016)
		Degenerative fibrotic disorders and cancer	Inhibit TGF-β1-induced epithelial-mesenchymal transition	Patrícia Gonçalves Tenório et al. (2025)
Eugenia jambolana, Rosemarinus officinalis, Calluna vulgaris, Eriobotrya japonica and Ocimum sanctum	Ursolic acid	TB	Induce autophagy in TB infected macrophages, downregulate Akt/mTOR and TNF-α/TNFR1 pathways	Shen et al. (2023)
Matricaria chamomilla, herbs, fruits, vegetables, nuts and honey	Apigenin	Chronic obese asthma	Inhibit apoptosis, modulate ROS-ASK1-MAPK pathway	Yu et al. (2023)
		COPD	Remove cellular senescence via regulating SIRT1-NAD++-CD38 axis	Li et al. (2021a)
Myrtus communis	Myrtenol	Allergic asthma	Prevent airway remodelling, decrease inflammatory indices and restore IL levels	Bejeshk et al. (2019)

(continued)

Table 18.1 (continued)

Source/medicinal plants	Phytocompounds/plant extracts	Respiratory disease	Mode of action	References
Rudbeckia fulgida	α-L-arabino(4-*O*-methyl-α-D-glucurono)-β-D-xylan (AGX)	Allergic asthma	Antitussive, bronchodilatory effect	Capek et al. (2020)
Curcuma longa	Curcumin	Allergic asthma	Modulate mitogen-activated protein kinase (MAPK)/NF-κB signalling pathways	Chauhan et al. (2018)
		Non-small-cell lung cancer	Inhibit PI3K/Akt/mTOR pathway, modulate apoptosis and autophagy	Liu et al. (2018)
			Alleviate chemo drug resistance and induce autophagy	Chen et al. (2019)
Panax ginseng	Ginsenoside Rg1	High-altitude hypoxia-induced acute lung injury (ALI)	Improve oxygen transport and use, minimize inflammation, support cellular metabolism and vascular health	Chen et al. (2024)
		COPD	Improve Epithelial-mesenchymal transition (EMT) by inhibiting TGF-*β*1/Smad pathway	Guan et al. (2017)
		COPD	Enhance lung function, prevent airway remodelling, down-regulate TGF-β1/Smad3 pathway	Guan et al. (2020)
		Allergic asthma	Reduce total eosinophil counts in BALF, airway hyperresponsiveness (AHR) and IgE serum levels	Lim et al. (2018)
Panax ginseng	Ginsenoside Rd. (GS-Rd)	Non-small-cell lung cancer (NSCLC)	Alleviate cisplatin drug resistance and downregulate NRF2 pathway	Chian et al. (2019)
Forsythia suspensa	Forsythiaside A	ALI and acute respiratory distress syndrome (ARDS)	Inhibit inflammation, regulate PPAR-γ/RXR-α complex	Wang et al. (2024b)

Polygonum cuspidatum	Polydatin (PD) (3,40,5-trihydroxystilbene-3-b-D-glucoside)	Bronchial asthma	Decrease inflammatory mediators, restore oxidative and nitrosative stress biomarkers	Hanna et al. (2019)
Herba Epimedii	Epimedin C	Allergic asthma	Regulate noncanonical NF-κB p52/RelB pathway, MAPKs activation	Huang et al. (2020)
Angelica sinensis	Vanillic acid	Allergic asthma	Prevent lung airway inflammation	Bai et al. (2019)
Eucalyptus species	Eucalyptol	Cigarette smoke (CS)-induced emphysema	Improve lung repair, SOD, elastin, TIMP-1 and decrease neutrophil elastase (NE) levels	Kennedy-Feitosa et al. (2019)
Arcangelisia flava, Berberis aquifolium, Rhizoma coptidis, Berberis vulgaris, Coptis chinensis, Berberis aristata, Hydrastis Canadensis, Phellodendron chinense, Phellodendron amurense	Berberine	Chronic respiratory diseases	Reduce pro-inflammatory cytokine levels, regulate NF-kB, ERK1/2 and p38 MAPK pathways	Tew et al. (2020)
		Lung cancer	Induce apoptosis, regulate miR-19a and TF expression	Chen et al. (2019)
Coptis chinensis	Berberine hydrochloride	Pulmonary fibrosis	Activate PPARγ signalling and prevent lung injury	Zhao et al. (2023)
	Berberine	COPD	Regulate TGF-β1/Smads signalling pathway	Wang et al. (2019)
Moraceae family	Morin	Allergic asthma	Alter SUMF2/IL-13 and BLT2/NF-kB signalling pathway	Kandhare et al. (2019)
Olive oil and white wine	Tyrosol	Allergic asthma	Regulate NF-κB, TNF-α, IL-4, IL-5, IL-13, IFN-γ and IgE levels	Cellat et al. (2021)
Ligusticum wallichii	Ligustrazine	Idiopathic pulmonary fibrosis (IPF)	Improve lung function, arterial blood gas analysis, pulmonary diffusion function	Wu et al. (2020)

(continued)

Table 18.1 (continued)

Source/medicinal plants	Phytocompounds/plant extracts	Respiratory disease	Mode of action	References
Citrus reticulata	Citrus alkaline extract (CAE) of peel	Pulmonary fibrosis	Produce collagen crosslink pyridinoline, lysyl oxidase (LOX), lysyl oxidase-like protein 1 (LOXL1) crosslinking-related enzymes, downregulate TGF-β1 and Smad3 levels	Li et al. (2021b)
Salvia miltiorrhiza	Cryptotanshinone	IPF	Reverse epithelial-mesenchymal transformation process	Zhang et al. (2020a)
Sophora japonica	Sophoricoside	Allergic asthma	Inhibit mast cell activation, $CD4^+$ T cell differentiation	Kim and Lee (2021)
Fruits like blueberries, grapes, plums, apples, peanut and other plant sources	Resveratrol	Allergic asthma	Reduce asthma by inducing beneficial microbiota in gut-lung axis, promotion of normal barrier functions of the lung	Alharris et al. (2022)
		Cystic fibrosis	Reinstates intracellular transport in epithelial cells	Lu et al. (2020)
		Pulmonary fibrosis	Regulate miR-21 through MAPK/AP-1 pathways	Wang et al. (2018a)
		Interstitial lung disease	Regulate TGF-β/Smad2/3/4, NF-κB and JAK/STAT pathways and overactivation of immune cells	Huo et al. (2023)
Artocarpus lakoocha, peanuts, grapes, mulberries and other plant sources	Oxyresveratrol	Allergic asthma	Anti-inflammatory effects, ameliorates allergic airway inflammation via attenuation of IL-4, IL-5 and IL-13 expression levels	Ashraf et al. (2015)

Traditional Chinese medicine (TCM)	Kang Ai injection, herbal decoction, Zhenqifuzheng capsules	Non-small cell lung cancer	Improve quality of life (QOL), reduce symptoms and reduce adverse events (AEs) of chemotherapy	Xiao et al. (2021)
Grape, strawberries, apple, persimmon, onion, cucumber and other plant sources	Carnosic acid and Fisetin (3,3′,4′,7-tetrahydroxyflavone)	Lung cancer	Inhibit lung cancer growth by caspase-3 activation	Shi et al. (2017)
Green tea	Epigallocatechin-3-gallate (EGCG)	Non-small cell lung cancer	Regulates CTR1 expression, enhance uptake of drug cisplatin	Chen et al. (2020)
Citrus fruits, Brussels sprouts, broccoli, apples and other plant sources	Kaempferol-3-O-rhamnoside and kaempferol	Allergic asthma	Reduce inflammatory cells in bronchoalveolar lavage fluid (BALF) and lung tissues, inhibit Th2 cytokines and IgE production	Chung et al. (2015)
	Kaempferol	Lung ischemia-reperfusion injury	Anti-inflammation and antioxidative stress involving the SIRT1/HMGB1/NF-κB axis	Yang et al. (2020)
		Cystic fibrosis	Activates cl^- secretion activating apical Cl^- current and basolateral K^+ current, regulate cAMP/PKA pathway and CFTR expression, increase fluid secretion	Thaweewattanodom et al. (2024)
Scutellaria baicalensis, *Oroxylum indicum*	Baicalin and its aglycone, baicalein	Respiratory diseases	Inhibit NF-κB activity, inflammatory cytokines and chemokines	Dinda et al. (2017)

patients (Mangwani et al. 2020). *Maytenus royleanus* leaves' extract has been found to ameliorate oxidative damage caused by anti-TB drug by enhancing antioxidants level (Shabbir et al. 2020). Similarly, *Warburgia ugandensis*, *Acacia hockii*, *Combretum molle*, *Albizia coriaria* and *Zanthoxylum leprieurii* have also shown anti-mycobacterial effect (Oloya et al. 2022). Phytocompound bergenin can be used as a prospective adjunct agent in TB therapy as they reduce immune impairment, reduces bacterial loads when used in combination as well as length of treatment of TB (Kumar et al. 2019). Essential oil from the leaves of *Ocimum sanctum* have also shown anti-mycobacterial effect (Jayapal et al. 2021). Luteolin is a hepatoprotective immunomodulator and can be used in combination therapy with conventional TB drugs (INH) by enhancing memory T cell responses, reducing treatment duration and preventing relapse of TB in patients (Singh et al. 2021).

4.2 Integration of Phytochemicals with Anticancer Drugs

Traditional Chinese medicines (TCM) based on phytocompounds are used as an adjuvant therapy for stage IB-IIIA non-small-cell lung cancer (NSCLC) patients by studying its effects on survival rates, recurrence risks and quality of life outcomes (Jiang et al. 2019; Singh et al. 2020). Many studies have demonstrated that phytocompounds alone or with chemotherapy treatment break through drug resistance to boost chemotherapy effectiveness. The natural compounds possess the ability to enhance cancer cell treatment sensitivity through their drug-resistant pathway targeting while decreasing side effects and enhancing therapeutic effectiveness (Chen et al. 2019). Combination therapy of anticancer drugs and phytochemicals can enhance the cytotoxic effect in patients suffering with lung cancer (Li et al. 2024). Cisplatin is an anticancer chemotherapeutic agent used in wide varieties of cancer treatment, which can become ineffective over the time as the cancer cells become resistance, leading to therapeutic failure, which can cause reoccurrence of tumours as well as progression in cancer (Sosa Iglesias et al. 2018). Ginsenosides Rg3, Rb1 and Rd. reverse cisplatin resistance in lung cancer through various mechanisms, including PD-L1 downregulation, NF-κB inhibition, suppression of epithelial–mesenchymal transition, modulation of drug efflux pumps (ABCB1 and PTCH1), and downregulating NRF2 signalling pathway (Jiang et al. 2017; Wang et al. 2018b; Chian et al. 2019; Niu et al. 2023). Emodin can exert a similar effect by inhibiting P-glycoprotein (Pgp) expression (Peng et al. 2021). An Epigallocatechin gallate derivative, EGCGD (2R,3R-6-methoxycarbonylgallocatechin-3-*O*-gallate) isolated from Anhua dark tea, was found efficient against gefitinib-resistant via inhibition of PI3K/mTOR, and EMT can be used as a combination therapy to combat NSCLC (Liu et al. 2020).

4.3 Integration of Phytochemicals with Drugs for Other Respiratory Disorders

Glycyrrhizin and β2-adrenoceptor (β2-AR) agonists drug like salbutamol have been found to work in synergy for treatment of asthma and COPD by increasing its mRNA levels, I-κB degradation, reducing NF-κB and IL-8 production thus improving symptoms in an in vivo study and supported by simulation studies as well (Yang et al. 2010; Qi et al. 2023). A cross-sectional study has revealed that a high intake of phytochemicals may support better respiratory and muscle function during COPD therapy in middle-aged COPD patients (Ranogajec et al. 2024). The nutritional intake of polyphenols shows protective benefits against cigarette smoke-induced oxidative damage combined (Rudrapal et al. 2022). Phase 2 trials of curcumin in reducing lung inflammation in asthmatic patients are conducted based on previous studies (Quan et al. 2021). In a randomized, triple-blind controlled human trial, Squill Oxymel from *Drimia Maritima* showed significant improvement in spirometry parameters when given to asthmatic patients as an add-on treatment (Nejatbakhsh et al. 2017).

5 Challenges and Future Perspectives

Phytochemicals have a very promising role in the respiratory health, but there are certain challenges associated with it. Phytochemical composition varies according to plant species, environmental conditions like temperature, rainfall, humidity and cultivation practices (Mwamatope et al. 2021; Agidew 2022). This poses a significant challenge for the extraction and standardization of the phytochemicals in a consistent manner. Also, many countries lack modern technology and infrastructure for making of new drug product and drug leading compounds from different parts of the medicinal plants (Agidew 2022). Another challenge associated with the phytochemicals is their poor bioavailability, low solubility and hydrophobic nature, which limit their absorption and distribution in the body. One of the possible solutions to this problem can be nanoformulations based on liposomes, nanostructure, nanoparticles or nanoemulsions (Hu et al. 2023). Apart from that, although laboratory and preclinical studies have seen positive results, large-scale clinical trials are insufficient. Without rigorous human studies, dosages, safety profiles and long-term consequences cannot be established precisely. Moreover, the regulatory barriers are an obstacle for acquiring approvals, because obtaining phytochemical-based treatments requires scientific validation and follows strict health regulation. The synergistic role of phytocompounds with the conventional therapies should be thoroughly investigated and incorporation of phytocompounds in the functional foods and nutraceuticals should be considered for the respiratory health and prevention of other lung complications.

6 Conclusions

Respiratory disorders like asthma, COPD, pulmonary fibrosis, CF, tuberculosis and lung cancer pose a significant global health challenge. Conventional therapies though widely used have limitations such as adverse effects, high costs and drug resistance. Phytochemicals derived from medicinal plants have emerged as promising alternative or complementary therapies due to their diverse pharmacological properties, including anti-inflammatory, antioxidant, bronchodilatory, anticholinergic, antimicrobial and immunomodulatory effects. This chapter highlights the therapeutic potential of key phytochemicals such as quercetin, curcumin, berberine and eucalyptol in managing respiratory diseases. Their mechanism of action includes modulation of inflammatory pathways, reduction of oxidative stress and enhancement of immune responses. Furthermore, integrating phytochemicals with nanotechnology and conventional drugs may improve their bioavailability and therapeutic efficacy. However, several challenges remain associated with the use of phytochemicals such as variability in phytochemical composition, standardization issues and limited clinical validation. Future research should focus on large-scale clinical trials, advanced drug delivery systems and regulatory frameworks to facilitate the integration of phytochemicals into conventional respiratory disease management. Strengthening the scientific foundation of phytochemical-based therapies could lead to more effective, safer and affordable treatment options for respiratory health.

References

Agidew MG (2022) Phytochemical analysis of some selected traditional medicinal plants in Ethiopia. Bull Natl Res Cent 46(1):87

Agra LC, Lins MP, da Silva MP, Smaniotto S, De Melo CB, Lagente V, Barreto E (2016) Uvaol attenuates pleuritis and eosinophilic inflammation in ovalbumin-induced allergy in mice. Eur J Pharmacol 780:232–242

Alharris E, Mohammed A, Alghetaa H, Zhou J, Nagarkatti M, Nagarkatti P (2022) The ability of resveratrol to attenuate ovalbumin-mediated allergic asthma is associated with changes in microbiota involving the gut-lung axis, enhanced barrier function and decreased inflammation in the lungs. Front Immunol 13:805770

Ashraf MI, Shahzad M, Shabbir A (2015) Oxyresveratrol ameliorates allergic airway inflammation via attenuation of IL-4, IL-5, and IL-13 expression levels. Cytokine 76(2):375–381

Bai FB, Fang LY, Hu HH, Yang YY, Feng XX, Sun DQ (2019) Vanillic acid mitigates the ovalbumin (OVA)-induced asthma in rat model through prevention of airway inflammation. Biosci Biotechnol Biochem 83(3):531–537

Bang AA, Bang S, Bang A, Acharya S, Shukla S (2023) Recent advances in the treatment of interstitial lung diseases. Cureus 15(10):e48016

Bejeshk MA, Fekri MS, Najafipour H, Rostamzadeh F, Jafari E, Rajizadeh MA, Masoumi-Ardakani Y (2019) Anti-inflammatory and anti-remodeling effects of myrtenol in the lungs of asthmatic rats: histopathological and biochemical findings. Allergol Immunopathol 47(2):185–193

Berthon BS, Wood LG (2015) Nutrition and respiratory health—feature review. Nutrients 7(3):1618–1643

Boots AW, Veith C, Albrecht C, Bartholome R, Drittij MJ, Claessen SM, Bast A, Rosenbruch M, Jonkers L, van Schooten FJ, Schins RP (2020) The dietary antioxidant quercetin reduces hallmarks of bleomycin-induced lung fibrogenesis in mice. BMC Pulm Med 20:6

Capek P, Šutovská M, Barboríková J, Kazimierová I, Fraňová S, Kopačová M (2020) Structural characterization and anti-asthmatic effect of α-l-arabino (4-O-methyl-α-d-glucurono)-β-d-xylan from the roots of *Rudbeckia fulgida*. Int J Biol Macromol 165:842–848

Cellat M, Kuzu M, İşler CT, Etyemez M, Dikmen N, Uyar A, Gökçek İ, Türk E, Güvenç M (2021) Tyrosol improves ovalbumin (OVA)-induced asthma in rat model through prevention of airway inflammation. Naunyn Schmiedeberg's Arch Pharmacol 394(10):2061–2075

Chai Q, Zhang Y, Liu CH (2018) *Mycobacterium tuberculosis*: an adaptable pathogen associated with multiple human diseases. Front Cell Infect Microbiol 8:158

Chauhan PS, Singh DK, Dash D, Singh R (2018) Intranasal curcumin regulates chronic asthma in mice by modulating NF-κB activation and MAPK signaling. Phytomedicine 51:29–38

Chen P, Huang HP, Wang Y, Jin J, Long WG, Chen K, Zhao XH, Chen CG, Li J (2019) Curcumin overcomes primary gefitinib resistance in non-small-cell lung cancer cells through inducing autophagy-related cell death. J Exp Clin Cancer Res 38:1–7

Chen A, Jiang P, Zeb F, Wu X, Xu C, Chen L, Feng Q (2020) EGCG regulates CTR1 expression through its pro-oxidative property in non-small-cell lung cancer cells. J Cell Physiol 235(11):7970–7981

Chen J, Zhang Z, Huang M, Yan J, Gao R, Cui J, Gao Y, Ma Z (2024) Ginsenoside Rg1 prevents and treats acute pulmonary injury induced by high-altitude hypoxia. Int J Mol Sci 25(22):12051

Chian S, Zhao Y, Xu M, Yu X, Ke X, Gao R, Yin L (2019) Ginsenoside Rd reverses cisplatin resistance in non-small-cell lung cancer A549 cells by downregulating the nuclear factor erythroid 2-related factor 2 pathway. Anti-Cancer Drugs 30(8):838–845

Chung MJ, Pandey RP, Choi JW, Sohng JK, Choi DJ, Park YI (2015) Inhibitory effects of kaempferol-3-O-rhamnoside on ovalbumin-induced lung inflammation in a mouse model of allergic asthma. Int Immunopharmacol 25(2):302–310

Dinda B, Dinda S, DasSharma S, Banik R, Chakraborty A, Dinda M (2017) Therapeutic potentials of baicalin and its aglycone, baicalein against inflammatory disorders. Eur J Med Chem 131:68–80

Durham AL, Adcock IM (2015) The relationship between COPD and lung cancer. Lung Cancer 90(2):121–127

Elborn JS (2016) Cystic fibrosis. Lancet 388(10059):2519–2531

Fan LC, Xu JF (2014) Advantages and drawbacks of long-term macrolide use in the treatment of non-cystic fibrosis bronchiectasis. J Thorac Dis 6(7):867

Garcia-Larsen V, Potts JF, Omenaas E, Heinrich J, Svanes C, Garcia-Aymerich J, Burney PG, Jarvis DL (2017) Dietary antioxidants and 10-year lung function decline in adults from the ECRHS survey. Eur Respir J 50(6):1702286

Garcia-Larsen V, Thawer N, Charles D, Cassidy A, Van Zele T, Thilsing T, Ahlström M, Haahtela T, Keil T, Matricardi PM, Brożek G (2018) Dietary intake of flavonoids and ventilatory function in European adults: a GA2LEN study. Nutrients 10(1):95

Guan S, Xu W, Han F, Gu W, Song L, Ye W, Liu Q, Guo X (2017) Ginsenoside Rg1 attenuates cigarette smoke-induced pulmonary epithelial-mesenchymal transition via inhibition of the TGF-β1/Smad pathway. Biomed Res Int 2017:7171404

Guan S, Yu P, Cao J, Xi X, Zhang Q, Zhu C, Hu H, Gong X, Fan H (2020) Ginsenoside Rg1 protects against cigarette smoke-induced airway remodeling by suppressing the TGF-β1/Smad3 signaling pathway. Am J Transl Res 12(2):493

Hanna DA, Khalaf MM, Abo-Saif AA (2019) Polydatin protects against ovalbumin-induced bronchial asthma in rats; involvement of urocortin and surfactant-D expression. Immunopharmacol Immunotoxicol 41(3):403–412

Hu Y, Lin Q, Zhao H, Li X, Sang S, McClements DJ, Long J, Jin Z, Wang J, Qiu C (2023) Bioaccessibility and bioavailability of phytochemicals: influencing factors, improvements, and evaluations. Food Hydrocoll 135:108165

Huang M, Wei Y, Dong J (2020) Epimedin C modulates the balance between Th9 cells and Treg cells through negative regulation of noncanonical NF-κB pathway and MAPKs activation to inhibit airway inflammation in the ovalbumin–induced murine asthma model. Pulm Pharmacol Ther 65:102005

Huo R, Huang X, Yang Y, Yang Y, Lin J (2023) Potential of resveratrol in the treatment of interstitial lung disease. Front Pharmacol 14:1139460

Jaques R, Shakeel A, Hoyle C (2020) Novel therapeutic approaches for the management of cystic fibrosis. Multidiscip Respir Med 15(1):690

Jayapal V, Raj CV, Muthaiah M, Chadha VK, Brammacharry U, Selvaraj S, Easow JM (2021) In-vitro anti-Mycobacterium tuberculosis effect of essential oil of *Ocimum sanctum* L. (Tulsi/Basil) leaves. Indian J Tuberc 68(4):470–473

Jiang Z, Yang Y, Yang Y, Zhang Y, Yue Z, Pan Z, Ren X (2017) Ginsenoside Rg3 attenuates cisplatin resistance in lung cancer by downregulating PD-L1 and resuming immune. Biomed Pharmacother 96:378–383

Jiang Y, Liu LS, Shen LP, Liu JX, Jiang GN, Gu AQ, Li HC, Li Q, Li HG, Huang PX (2019) Traditional Chinese medicine treatment as adjuvant therapy in completely resected stage IB–IIIA non–small-cell lung cancer: study protocol for a multicenter, double-blind, randomized, Placebo-Controlled Trial. Clin Lung Cancer 20(5):e541–e547

Joshi P, Kim WJ, Lee SA (2015) The effect of dietary antioxidant on the COPD risk: the community-based KoGES (Ansan–Anseong) cohort. Int J Chron Obstruct Pulmon Dis 10:2159–2168

Juergens LJ, Worth H, Juergens UR (2020) New perspectives for mucolytic, anti-inflammatory and adjunctive therapy with 1,8-cineole in COPD and asthma: review on the new therapeutic approach. Adv Ther 37(5):1737–1753

Kandhare AD, Liu Z, Mukherjee AA, Bodhankar SL (2019) Therapeutic potential of morin in ovalbumin-induced allergic asthma via modulation of SUMF2/IL-13 and BLT2/NF-kB signaling pathway. Curr Mol Pharmacol 12(2):122–138

Kennedy-Feitosa E, Cattani-Cavalieri I, Barroso MV, Romana-Souza B, Brito-Gitirana L, Valenca SS (2019) Eucalyptol promotes lung repair in mice following cigarette smoke-induced emphysema. Phytomedicine 55:70–79

Kim BH, Lee S (2021) Sophoricoside from *Sophora japonica* ameliorates allergic asthma by preventing mast cell activation and CD4+ T cell differentiation in ovalbumin-induced mice. Biomed Pharmacother 133:111029

Kumar S, Sharma C, Kaushik SR, Kulshreshtha A, Chaturvedi S, Nanda RK, Bhaskar A, Chattopadhyay D, Das G, Dwivedi VP (2019) The phytochemical bergenin as an adjunct immunotherapy for tuberculosis in mice. J Biol Chem 294(21):8555–8563

Levine SM, Marciniuk DD (2022) Global impact of respiratory disease: what can we do, together, to make a difference? Chest 161(5):1153–1154

Li RF, Chen XY, Xu Y, Feng FC, He HL, Zhou XM (2021a) Inhibitory effects of alkaline extract from the pericarp of *Citrus reticulata* Blanco on collagen behavior in bleomycin-induced pulmonary fibrosis. J Ethnopharmacol 269:113761

Li BS, Zhu RZ, Lim SH, Seo JH, Choi BM (2021b) Apigenin alleviates oxidative stress-induced cellular senescence via modulation of the SIRT1–NAD+–CD38 axis. Am J Chin Med 49(5):1235–1250

Li K, Li J, Li Z, Men L, Zuo H, Gong X (2024) Cisplatin-based combination therapies: their efficacy with a focus on ginsenosides co-administration. Pharmacol Res 203:107175

Lim C, Lim S, Lee B, Cho S (2018) Ginsenoside Rg1 exhibits anti-asthmatic activity in an *Aspergillus* protease-induced asthma model in mice. Nat Prod Commun 13(4):1934578X1801300409

Liu F, Gao S, Yang Y, Zhao X, Fan Y, Ma W, Yang D, Yang A, Yu Y (2018) Antitumor activity of curcumin by modulation of apoptosis and autophagy in human lung cancer A549 cells through inhibiting PI3K/Akt/mTOR pathway. Oncol Rep 39(3):1523–1531

Liu J, Zhong T, Yi P, Fan C, Zhang Z, Liang G, Xu Y, Fan Y (2020) A new epigallocatechin gallate derivative isolated from Anhua dark tea sensitizes the chemosensitivity of gefitinib via the suppression of PI3K/mTOR and epithelial-mesenchymal transition. Fitoterapia 143:104590

Lu B, Corey DA, Kelley TJ (2020) Resveratrol restores intracellular transport in cystic fibrosis epithelial cells. Am J Physiol Lung Cell Mol Physiol 318(6):L1145–L1157

Mangwani N, Singh PK, Kumar V (2020) Medicinal plants: adjunct treatment to tuberculosis chemotherapy to prevent hepatic damage. J Ayurveda Integr Med 11(4):522–528

Mehta SK, Pradhan RB (2024) Phytochemicals in antiviral drug development against human respiratory viruses. Drug Discov Today 29:104107
Melani AS, Croce S, Cassai L, Montuori G, Fabbri G, Messina M, Viani M, Bargagli E (2023) Systemic corticosteroids for treating respiratory diseases: less is better, but… when and how is it possible in real life? Pulm Ther 9(3):329–344
Meleń E, Zar HJ, Siroux V, Shaw D, Saglani S, Koppelman GH, Hartert T, Gern JE, Gaston B, Bush A, Zein J (2024) Asthma inception: epidemiologic risk factors and natural history across the life course. Am J Respir Crit Care Med 210(6):737–754
Meyer ML, Peters S, Mok TS, Lam S, Yang PC, Aggarwal C, Brahmer J, Dziadziuszko R, Felip E, Ferris A, Forde PM (2024) Lung cancer research and treatment: global perspectives and strategic calls to action. Ann Oncol 35:1088
Momtazmanesh S, Moghaddam SS, Ghamari SH, Rad EM, Rezaei N, Shobeiri P, Aali A, Abbasi-Kangevari M, Abbasi-Kangevari Z, Abdelmasseh M, Abdoun M (2023) Global burden of chronic respiratory diseases and risk factors, 1990–2019: an update from the global burden of disease study 2019. EClinicalMedicine 59:101936
Mwamatope B, Tembo D, Kampira E, Maliwichi-Nyirenda C, Ndolo V (2021) Seasonal variation of phytochemicals in four selected medicinal plants. Pharm Res 13(4)
Nagoor Meeran MF, Javed H, Al Taee H, Azimullah S, Ojha SK (2017) Pharmacological properties and molecular mechanisms of thymol: prospects for its therapeutic potential and pharmaceutical development. Front Pharmacol 8:380
Nejatbakhsh F, Karegar-Borzi H, Amin G, Eslaminejad A, Hosseini M, Bozorgi M, Gharabaghi MA (2017) Squill Oxymel, a traditional formulation from *Drimia maritima* (L.) Stearn, as an add-on treatment in patients with moderate to severe persistent asthma: a pilot, triple-blind, randomized clinical trial. J Ethnopharmacol 196:186–192
Niu X, Shi Y, Li Q, Chen H, Fan X, Yu Y, Lv C, Lu J (2023) Ginsenoside Rb1 for overcoming cisplatin-insensitivity of A549/DDP cells in vitro and vivo through the dual-inhibition on two efflux pumps of ABCB1 and PTCH1. Phytomedicine 115:154776
Oloya B, Namukobe J, Ssengooba W, Afayoa M, Byamukama R (2022) Phytochemical screening, antimycobacterial activity and acute toxicity of crude extracts of selected medicinal plant species used locally in the treatment of tuberculosis in Uganda. Trop Med Health 50(1):16
Padmavathi P, Raghu PS, Reddy VD, Bulle S, Marthadu SB, Maturu P, Varadacharyulu NC (2018) Chronic cigarette smoking-induced oxidative/nitrosative stress in human erythrocytes and platelets. Mol Cell Toxicol 14:27–34
Patrícia Gonçalves Tenório L, Xavier FH, Silveira Wagner M, Moreira Bagri K, Alves Ferreira EG, Galvani R, Mermelstein C, Bonomo AC, Savino W, Barreto E (2025) Uvaol attenuates TGF-β1-induced epithelial-mesenchymal transition in human alveolar epithelial cells by modulating expression and membrane localization of β-catenin. Front Pharmacol 15:1504556
Peng S, Wang J, Lu C, Xu Z, Chai JJ, Ke Q, Deng XZ (2021) Emodin enhances cisplatin sensitivity in non-small cell lung cancer through Pgp downregulation. Oncol Lett 21(3):230
Putcha N, Wise RA (2018) Medication regimens for managing COPD exacerbations. Respir Care 63(6):773–782
Qi JH, Xu DC, Wang XL, Cai DY, Wang Y, Zhou W (2023) Micro-simulation insights into the functional and mechanistic understanding of glycyrrhizin against asthma. Front Pharmacol 14:1220368
Quan M, Alismail A, Daher N, Cleland D, Chavan S, Tan LD (2021) Randomized, placebo controlled, double blinded pilot superiority phase 2 trial to evaluate the effect of curcumin in moderate to severe asthmatics. BMC Pulm Med 21:1–8
Quecchia C, Vianello A (2025) The therapeutic potential of Myo-inositol in managing patients with respiratory diseases. Int J Mol Sci 26(5):2185
Ranogajec A, Ilić A, Benko Meštrović S, Rumbak I (2024) Effect of phytochemical-rich food intake on respiratory and muscle function in middle-aged patients with COPD: a cross-sectional study. Nutrients 16(22):3962
Ravimohan S, Kornfeld H, Weissman D, Bisson GP (2018) Tuberculosis and lung damage: from epidemiology to pathophysiology. Eur Respir Rev 27(147):170077

Rivera-Ortega P, Molina-Molina M (2019) Interstitial lung diseases in developing countries. Ann Glob Health 85(1):4

Rudrapal M, Maji S, Prajapati SK, Kesharwani P, Deb PK, Khan J, Mohamed Ismail R, Kankate RS, Sahoo RK, Khairnar SJ, Bendale AR (2022) Protective effects of diets rich in polyphenols in cigarette smoke (CS)-induced oxidative damages and associated health implications. Antioxidants 11(7):1217

Sasikumar K, Ghosh AR, Dusthackeer A (2018) Antimycobacterial potentials of quercetin and rutin against *Mycobacterium tuberculosis* H37Rv. 3 Biotech 8(10):427

Shabbir M, Afsar T, Razak S, Almajwal A, Khan MR (2020) Phytochemical analysis and evaluation of hepatoprotective effect of Maytenus royleanus leaves extract against anti-tuberculosis drug induced liver injury in mice. Lipids Health Dis 19:1–5

Shah PL, Herth FJ, van Geffen WH, Deslee G, Slebos DJ (2017) Lung volume reduction for emphysema. Lancet Respir Med 5(2):147–156

Shen J, Fu Y, Liu F, Ning B, Jiang X (2023) Ursolic acid promotes autophagy by inhibiting akt/mTOR and TNF-α/TNFR1 signaling pathways to alleviate pyroptosis and necroptosis in mycobacterium tuberculosis-infected macrophages. Inflammation 46(5):1749–1763

Shi B, Wang LF, Meng WS, Chen L, Meng ZL (2017) Carnosic acid and fisetin combination therapy enhances inhibition of lung cancer through apoptosis induction. Int J Oncol 50(6):2123–2135

Shteinberg M, Haq IJ, Polineni D, Davies JC (2021) Cystic fibrosis. Lancet 397(10290):2195–2211

Shukla SD, Vanka KS, Chavelier A, Shastri MD, Tambuwala MM, Bakshi HA, Pabreja K, Mahmood MQ, O'toole RF (2020) Chronic respiratory diseases: an introduction and need for novel drug delivery approaches. In: Targeting chronic inflammatory lung diseases using advanced drug delivery systems. Academic, pp 1–31

Singh J, Luqman S, Meena A (2020) Emerging role of phytochemicals in targeting predictive, prognostic, and diagnostic biomarkers of lung cancer. Food Chem Toxicol 144:111592

Singh DK, Tousif S, Bhaskar A, Devi A, Negi K, Moitra B, Ranganathan A, Dwivedi VP, Das G (2021) Luteolin as a potential host-directed immunotherapy adjunct to isoniazid treatment of tuberculosis. PLoS Pathog 17(8):e1009805

Soriano JB, Kendrick PJ, Paulson KR, Gupta V, Abrams EM, Adedoyin RA, Adhikari TB, Advani SM, Agrawal A, Ahmadian E, Alahdab F (2020) Prevalence and attributable health burden of chronic respiratory diseases, 1990–2017: a systematic analysis for the Global Burden of Disease Study 2017. Lancet Respir Med 8(6):585–596

Sosa Iglesias V, Giuranno L, Dubois LJ, Theys J, Vooijs M (2018) Drug resistance in non-small cell lung cancer: a potential for NOTCH targeting? Front Oncol 8:267

Supramaniam K, Gazzali AM, Noor NM, Abdul-Aziz A, Nimmano N, Mohtar N (2024) A review on supporting roles of phytochemicals and advances in their Nanoformulations for lung diseases. Malays J Med Health Sci 20

Tew XN, Lau NJ, Chellappan DK, Madheswaran T, Zeeshan F, Tambuwala MM, Aljabali AA, Balusamy SR, Perumalsamy H, Gupta G, Oliver BG (2020) Immunological axis of berberine in managing inflammation underlying chronic respiratory inflammatory diseases. Chem Biol Interact 317:108947

Thaweewattanodom J, Deachapunya C, Poonyachoti S (2024) Kaempferol activates chloride secretion via the cAMP/PKA signaling pathway and expression of CFTR in T84 cells. Front Pharmacol 15:1401273

Vicedo-Cabrera AM, Melén E, Forastiere F, Gehring U, Katsouyanni K, Yorgancioglu A, Ulrik CS, Hansen K, Powell P, Ward B, Hoffmann B (2023) Climate change and respiratory health: a European Respiratory Society position statement. Eur Respir J 62(2):2201960

Wang J, He F, Chen L, Li Q, Jin S, Zheng H, Lin J, Zhang H, Ma S, Mei J, Yu J (2018a) Resveratrol inhibits pulmonary fibrosis by regulating miR-21 through MAPK/AP-1 pathways. Biomed Pharmacother 105:37–44

Wang J, Tian L, Khan MN, Zhang L, Chen Q, Zhao Y, Yan Q, Fu L, Liu J (2018b) Ginsenoside Rg3 sensitizes hypoxic lung cancer cells to cisplatin via blocking of NF-κB mediated epithelial–mesenchymal transition and stemness. Cancer Lett 415:73–85

Wang W, Zha G, Zou JJ, Wang X, Li CN, Wu XJ (2019) Berberine attenuates cigarette smoke extract-induced airway inflammation in mice: involvement of TGF-β1/Smads signaling pathway. Curr Med Sci 39(5):748–753

Wang J, Wang Z, Zhao Y, Bai L, Wei Y, Huang T, Xu Y, Zhou X (2024a) Molecular mechanism of quercetin in treating RA-ILD based on network pharmacology, molecular docking, and experimental validation. Naunyn Schmiedeberg's Arch Pharmacol 397(5):3077–3092

Wang J, Xue X, Zhao X, Luo L, Liu J, Dai S, Zhang F, Wu R, Liu Y, Peng C, Li Y (2024b) Forsythiaside A alleviates acute lung injury by inhibiting inflammation and epithelial barrier damages in lung and colon through PPAR-γ/RXR-α complex. J Adv Res 60:183–200

World Health Organization (2019) Global health estimates 2019. WHO

World Health Organization (2023a) Asthma. https://www.who.int/news-room/fact-sheets/detail/asthma

World Health Organization (2023b) Chronic obstructive pulmonary disease (COPD). https://www.who.int/news-room/fact-sheets/detail/chronic-obstructive-pulmonary-disease-(copd)

Wu X, Li W, Luo Z, Chen Y (2020) Meta-analysis of clinical efficacy and safety of ligustrazine in the treatment of idiopathic pulmonary fibrosis. Evid Based Complement Alternat Med 2020:2416132

Xiao Z, Chen Z, Han R, Lu L, Li Z, Lin J, Hu L, Huang X, Lin L (2021) Comprehensive TCM treatments combined with chemotherapy for advanced non-small cell lung cancer: a randomized, controlled trial. Medicine 100(18):e25690

Yang Y, Shi Q, Liu Z, Li RJ, Pan PW, Hou YY, Lu WG, Bai G (2010) The synergistic anti-asthmatic effects of glycyrrhizin and salbutamol. Acta Pharmacol Sin 31(4):443–449

Yang C, Yang W, He Z, He H, Yang X, Lu Y, Li H (2020) Kaempferol improves lung ischemia-reperfusion injury via anti-inflammation and antioxidative stress regulated by SIRT1/HMGB1/NF-κB axis. Front Pharmacol 10:1635

Yu H, Huang X, Zhu HH, Wang N, Xie C, Zhou YL, Shi HL, Chen MM, Wu YR, Ruan ZH, Lyu YB (2023) Apigenin ameliorates non-eosinophilic inflammation, dysregulated immune homeostasis and mitochondria-mediated airway epithelial cell apoptosis in chronic obese asthma via the ROS-ASK1-MAPK pathway. Phytomedicine 111:154646

Zappa C, Mousa SA (2016) Non-small cell lung cancer: current treatment and future advances. Transl Lung Cancer Res 5(3):288

Zhai T, Li S, Hu W, Li D, Leng S (2018) Potential micronutrients and phytochemicals against the pathogenesis of chronic obstructive pulmonary disease and lung cancer. Nutrients 10(7):813

Zhang Q, Gan C, Liu H, Wang L, Li Y, Tan Z, You J, Yao Y, Xie Y, Yin W, Ye T (2020a) Cryptotanshinone reverses the epithelial–mesenchymal transformation process and attenuates bleomycin-induced pulmonary fibrosis. Phytother Res 34(10):2685–2696

Zhang Y, Qu X, Gao H, Zhai J, Tao L, Sun J, Song Y, Zhang J (2020b) Quercetin attenuates NLRP3 inflammasome activation and apoptosis to protect INH-induced liver injury via regulating SIRT1 pathway. Int Immunopharmacol 85:106634

Zhao J, Ma X, Li S, Liu C, Liu Y, Tan J, Yu L, Li X, Li W (2023) Berberine hydrochloride ameliorates PM2.5-induced pulmonary fibrosis in mice through inhibiting oxidative stress and inflammatory. Chem Biol Interact 386:110731

Mechanisms of Neuroprotection by Medicinal Plants and Their Phytochemicals

19

Kandasamy Saravanan and Smita P. Pilla

Abstract

Neurodegenerative conditions like Alzheimer's and Parkinson's lead to nerve cell decline and impaired brain function over time. Current medications manage symptoms but do not effectively halt or reverse these diseases. Recently, interest has grown in using plants and their compounds to protect the brain, potentially with fewer side effects. Many plants contain healing substances that may safeguard brain health by enhancing the body's defense mechanisms against disease and injury. Key compounds include flavonoids (quercetin, rutin), alkaloids (berberine, huperzine A), terpenoids (ginkgolides, bilobalide), and polyphenols (curcumin, resveratrol). These substances combat harmful particles and activate the Nrf2/ARE pathway, boosting internal antioxidants. They also help control brain inflammation, regulate chemicals like acetylcholine and dopamine, balance calcium, and support cellular energy production. Additionally, phytochemicals promote brain flexibility by activating the BDNF/TrkB pathway, preventing protein clumping, and managing cell death. Emerging research indicates they may also protect the blood–brain barrier and improve cerebral blood flow. Understanding these mechanisms is crucial for developing natural therapies for neurodegenerative diseases, strokes, and brain injuries, ultimately advancing phytochemical-based drug development and improving outcomes in neurological disorders.

Keywords

Neuroprotective herbs · Phytomedicine · Curcumin · Alzheimer Disease · Neurotransmitter

K. Saravanan (✉) · S. P. Pilla
Laboratory of Computational Biology, Biological and Chemical Research Centre, Faculty of Chemistry, University of Warsaw, Warszawa, Poland

K. Kuca et al. (eds.), *Medicinal Plants and Their Bioactives in Human Diseases*,
https://doi.org/10.1007/978-3-032-01356-9_19

1 Introduction

Due to changes in lifestyle and an aging population, chronic neurodegenerative disease has evolved into a major public health dilemma. The human brain is a feat of fantastic complexity, but it is also afflicted by Alzheimer's disease, Parkinson's disease, Huntington's disease, and amyotrophic lateral sclerosis (Kesidou et al. 2023). Unfortunately, this has a tangible effect on human health. These diseases are characterized by progressive neuronal loss, which collapses your intellectual abilities as well as remote mechanical functions (Saravanan et al. 2017, 2019, 2020, 2021). This significantly diminishes quality of life. Despite significant advances in the field, today, only symptomatic treatments exist for these conditions. Given these circumstances, it is essential to consider other alternative strategies or combinations of treatments. The medicinal plants found in traditional medicine have long been in use, and research done to date has confirmed the therapeutic effect of bioactive compounds like flavonoids alkaloids as well as phenolic acids. These compounds confer a wide range of beneficial pharmacological effects, including the power to attenuate oxidative stress, mitochondrial damage, misfolding of proteins, and neuroinflammation, which are fundamental pathogenic mechanisms encountered in diseases of brain death or senility. However, for this reason, phytochemicals like curcumin, resveratrol, and ginsenosides are indeed antioxidants and anti-inflammatory agents while being effective agonists of neuronal differentiation (Elufioye et al. 2017; Rehman et al. 2019). The phytochemicals are responsible for protection against oxidative damage, mainly to prevent the imbalance between the production of many reactive oxide substances (ROS) and defense against these and ROS from materializing, so that they can be neutralized or prevented from causing damage. Plant-derived antioxidants, such as polyphenols taken from *Laurus nobilis* and *Aronia melanocastorpa*, intercept and metabolize free radicals and also can raise the activity of endogenous antioxidant enzymes (Pilipović et al. 2023). In addition, they may also set in motion signaling pathways like Nrf2, which increase gene expression for antioxidants. Neuroinflammation-producing microglia and astrocytes lead to a worse state of neuronal injury as these reactive cells go ahead and release pro-inflammatory cytokines. Phytochemicals, such as astragaloside IV from *Astragalus membranaceus*, can inhibit the mediators of inflammation like NF-κB and reduce cytokine production to alleviate inflammation (Lobine et al. 2021). Furthermore, they change the status of neurotransmitter systems (neurotransmitter modulation) to prevent excitotoxic effects. This is the situation in which over-excretion of glutamate sends nerve cells into an early grave. For example, ginsenosides from *Panax ginseng* and *rosmarinic acid* taken from *Rosmarinus officinalis* stabilize the calcium level within cells and prevent glutamate-induced excitotoxicity (Pandey and Chakrabarty 2024). But despite the fact of their initial promise, phytochemical treatment in a clinical setting has its problems, with molecules that aren't easily absorbed and can't enter the blood–brain barrier (BBB). There is some potential in early research for those techniques that enhance the solubility of these molecules and aim to deliver them, such as using phytosome systems or nanoemulsions (Gunasekaran et al. 2014; Rana et al. 2025). Medicinal plants and their phytochemicals may hold the key to finding remedies for neurodegenerative diseases. This natural line of compounds strikes against many of the pathologic processes contributing to neurodegeneration, offering a systematic and therapeutic approach for reversing neurodegenerative damage. Nonetheless, it remains to be confirmed in practice how effectively these drugs work and what their bioavailability is. Also, of interest is to development of these findings into combination therapy mode based on an understanding of present pattern and character class. Given their therapeutic potential for neurodegenerative diseases, these phytochemicals require further research, to overcome the challenges ahead and maximize treatment effect.

1.1 Neurodegenerative Disease

The main parts of the nervous system are the neurons residing in the brain and spinal cord. These specific cells enable the communication of information to other nerve cells, glands, or muscles. In contrast to various other cell types, neurons lack the ability to replicate; therefore, when they are harmed or perish, the body cannot replace them. Neurodegeneration denotes a state exhibited through gradual decline and impairment of neurons, leading to their eventual demise. This occurrence is linked to neurodegenerative disorders, which stem from multiple factors, including the unusual buildup of proteins in neurons, environmental factors, protein breakdown, genetic susceptibility, mitochondrial impairment, and the aging process. Studies show that the dominant pathological quality identified in these ailments is the clustering of protein substances, along with shifts observed in the physicochemical properties of the human brain (Dugger and Dickson 2017; Daniele et al. 2018; Madhu et al. 2020; Behl et al. 2021).

1.1.1 Alzheimer Disease (AD)

In a neurology conference in 1907, German psychiatrist *Alois Alzheimer* introduced a newly identified condition diagnosed in an elderly woman. At the time, this was considered a rare form of pre-senile dementia, recognized as the most common form of neurodegenerative dementia, later named "Alzheimer's disease," primarily impacts an individual's capacity to perform everyday tasks and ultimately results in death (Hippius and Neundörfer 2003; Nitrini 2023). An individual with Alzheimer's disease exhibits cognitive deficits, memory challenges, and erratic behavior (Félix Bermejo-Pareja and Teodoro del Ser 2024). AD is an irreversible, complex, multifactorial, progressive deterioration, chronic, common cause of dementia and neurodegenerative disorder (Breijyeh and Karaman 2020; Kamatham et al. 2024). Current therapies utilizing cholinesterase inhibitors can provide temporary relief from the cognitive symptoms associated with the condition and may also decelerate the process of neuronal cell death (Chen et al. 2022; Zhang et al. 2024). In order to provide exact or suitable therapy for AD, many scientists and pharmaceutical companies are optimistically working to identify the exact disease mechanism. To date, no medical treatment available to cure the disease completely (Golde 2022; Boxer and Sperling 2023). Globally, ~55 million people are living with AD, and every year ~8 million people are suffering newly [Alzheimer's Association 2024]. Due to that, the CEO of Alzheimer's Disease International (ADI) addressed a global plan (2017–2025) on dementia at the 70th session of the World Health Assembly in Geneva on 29 May 2017 and mentioned that *"the most devastating disease of the 21st century is Dementia, and the future treatment depends on research. Therefore, the ADI proposed a global plan that nationally 1% of the cost should be devoted to dementia research"* (Livingston et al. 2020). Various factors lead to neuronal cell death, such as lower levels of acetylcholine, the buildup of amyloid beta, tau protein aggregation, oxidative imbalance and inflammatory response in the brain, excitotoxicity, and both genetic and hereditary influences. Recent studies (Nasb et al. 2024) have explored Alzheimer's disease (AD) through the evaluation of different hypotheses, including the cholinergic hypothesis, amyloid hypothesis, and tau hypothesis, along with other contributing factors like oxidative stress and genetic influences (Fišar 2022).

1.1.2 Parkinson's Disease (PD)

This ranks as the second leading neurodegenerative disorder; nearly 1.5 million people are affected worldwide (Bloem et al. 2021; Georgios Papagiouvannis et al. 2024). It is a chronic disorder that primarily affects the elderly. The PD affected people could not be able to perform their daily activities. It causes tremors (in hands, legs, jaw, and head), stiffness, and difficulty in walking etc. PD is also characterized as a chronic and progressive condition; the symptoms of PD gradually increase and lead to mental illness, changes in behavior, sleep disruptions, depression, urinary problems, and difficulties with memory. Presently, there is no laboratory test available to diagnose the PD, and no potential drug exists to cure the disease. The main cause of Parkinson's disease is the reduced levels pertaining to dopamine in the substantia nigra, a part of the midbrain that plays a crucial role in controlling motor functions in the body (Dauer and Przedborski 2003; Davie 2008). There are various factors involved for the decrease associated with dopamine levels within the brain, such as dopaminergic neuron cell death, low level of dopamine secretion, aggregation of cleaved alpha-synuclein fragments, affection of central serotonergic systems, etc. In which, dopamine is metabolized in the pathway of dopamine secretions by a set of enzymes combined and degrades it into an inactive metabolite state (Juárez Olguín et al. 2015; Yang et al. 2022; Zhu et al. 2024).

1.1.3 Huntington's Disease (HD)

It is a hereditary condition stemming due to a mutation in the huntingtin gene, which results in the synthesis of a dysfunctional protein that builds up in the brain. This pathological accumulation contributes to the gradual degeneration of neurons, especially within the basal ganglia, leading to impairments in motor function, cognitive abilities, and psychological health. The disorder is marked by persistent inflammation and oxidative stress, both of which intensify neuronal injury (Walker 2007; Ross and Tabrizi 2011; Jimenez-Sanchez et al. 2016; Jurcau 2022).

Medicinal plants have long been regarded as being used for centuries across diverse cultures to promote health and combat diseases, including neurological disorders. Modern pharmacology has begun to validate these traditional practices, uncovering a wealth of bioactive ingredients sourced from medicinal plants and their phytochemicals that possess potent neuroprotective properties. These compounds, such as flavonoids, alkaloids, terpene derivatives, and polyphenolic substances, target pathological mechanisms including oxidative stress, inflammation, and impaired neurotransmitter signaling, offering a multifaceted approach to neuroprotection. Medicinal plants are important in treating different health problems including neurodegenerative disorder. The therapeutic properties of the plants are due to their bioactive compounds and therapeutic activity. Reports (Perry and Howes 2010; Shikha Kushwah et al. 2023; Nagori et al. 2023) confirm that many herbal medicines are still recommended worldwide to improve memory and treat dementia-related conditions. A variety of medicinal plants have been proven to possess acetylcholinesterase inhibitory properties, which may contribute significantly to the treatment of neurodegenerative conditions like Alzheimer's (Patel et al. 2018). *Bacopa monnieri* and *Ginkgo biloba* are widely acknowledged for their strong cognitive enhancement properties in both traditional Indian Ayurveda and Chinese medical systems. The memory-enhancing effects of *G. biloba* leaves and the extracts of *Lycium barbarum* fruit are further supported by their impressive properties that combat oxidative stress and inflammation (Aguiar and Borowski 2013; Singh et al. 2019; Malík and Tlustoš 2023; Yang et al. 2023).

2 Major Classes of Neuroprotective Phytochemicals

The increasing prevalence of neurodegenerative diseases among the aging population necessitates the exploration of neuroprotective agents that can mitigate cognitive decline and other neurological impairments (Kumar and Khanum 2012; Velmurugan et al. 2018; Bansal et al. 2023; Ayaz et al. 2024). Phytochemicals, particularly those derived from plants, have garnered significant attention for their potential neuroprotective effects. This chapter focuses on findings from recent studies to highlight major classes of neuroprotective phytochemicals and therapeutic implications (Fig. 19.1). The scientific reports outline (Grotewold 2006; Agrawal 2011; Kumar and Pandey 2013; Gajender et al. 2023) that the natural products, including polyphenols, vitamins, flavonoids, carbamates, lycopenes, carotenes, and alkaloids, demonstrate a variety of biological properties and confer numerous health benefits (Ramasamy et al. 2012; Rajendran and Ponnuraj 2024). In which the flavonoids are present naturally in all kinds of plants, and it has huge structural and functional diversity based on the C-rings and position of OH groups. In addition, the biological properties and flavonoid–protein interactions are investigated to identify their important health effects and suggest that flavonoids are more essential in the human diet. Furthermore, reports (Kim et al. 2014; Khan et al. 2018) also strongly support that the flavonoids can be used as a therapeutic for cancer, cardiovascular, and Alzheimer's diseases. Although the flavonoid molecules play a vital role in the cognitive deficit, particularly, it increase the acetylcholine level by controlling the AChE activity. From the reports (Wang et al. 2016; Naoi et al. 2019), all the

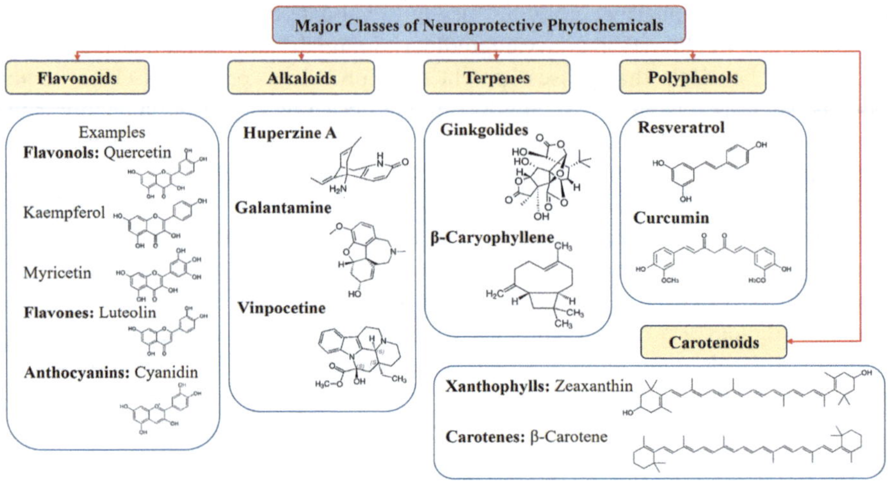

Fig. 19.1 Chemical structure of major classes of neuroprotective phytochemicals

neuroprotective phytochemicals are classified into flavonoids such as quercetin, kaempferol, and hesperidin and non-flavonoids such as curcumin, galantamine, huperzine-A, piperine, and resveratrol. Flavonoids represent a category of polyphenolic compounds characterized by their varied structures, prevalent throughout the plant kingdom. With their rich colors and different structures, flavonoids are mainly responsible for the pigmentation of fruits, vegetables, and flowers. They fall into several subgroups, including flavonols, flavones, flavanones, isoflavones, and anthocyanins (Grotewold 2006). Apart from their aesthetic contribution to our living world, however, these compounds have many human applications, most notably in protection against neurodegenerative diseases. The neuroprotective effect of flavonoids can be attributed to antioxidant, anti-inflammatory, neurotrophic factors, modulation of neurotransmitter systems, and neuroprotection against excitotoxicity (Bellavite 2023; Nezhad et al. 2024). Flavonoids provide neuroprotective advantages by neutralizing free radicals, lowering oxidative stress, and inhibiting the production of pro-inflammatory cytokines, which in turn reduces neuroinflammation (Kumar et al. 2010). They also promote the expression of neurotrophic factors such as brain-derived neurotrophic factor (BDNF), which is essential for neuronal survival and cognitive function (Xu et al. 2013). Additionally, flavonoids influence neurotransmitter systems, including serotonin and dopamine, potentially improving mood and cognitive abilities (Pannu et al. 2021). They also help regulate glutamate levels, protecting against excitotoxicity and neuronal death. Flavonoids, alkaloids, and terpenoids are key neuroprotective phytochemicals found in various dietary sources that can enhance brain health and potentially reduce the risk of neurodegenerative diseases (Balakrishnan et al. 2021). Flavonoids are abundant in fruits like berries, apples, and citrus, as well as in vegetables, teas, and herbs, contributing to antioxidant and anti-inflammatory effects while promoting neurotrophic factors essential for cognitive function (Ullah et al. 2020). Alkaloids, including caffeine and morphine, also offer neuroprotective benefits by modulating neurotransmitter systems and reducing oxidative stress (Hussain et al. 2018). Terpenoids, found in aromatic herbs and citrus fruits, possess antioxidant and anti-inflammatory properties, supporting neuronal health and growth (Câmara et al. 2024). Incorporating a variety of these phytochemicals into a balanced diet is vital for maintaining cognitive function and preventing neurodegeneration. As research advances, understanding the mechanisms of these compounds underscores their importance in dietary strategies for brain health.

3 Important Medicinal Plants with Neuroprotective Properties

Medicinal plants have gained substantial attention for their neuroprotective benefits, particularly in the context of treating neurodegenerative diseases. These plants act as sources of a variety of bioactive ingredients, including flavonoids, alkaloids, and polyphenols. Through diverse molecular pathways, they mitigate oxidative stress, suppress neuroinflammation, and stave off neural damage (Khazdair et al. 2019; Lobine et al. 2021; Moreira et al. 2023; Ghosh et al. 2023; Moise et al. 2024). The WHO predicts that neurodegenerative diseases are likely to become the most important cause of death in the coming decades, affecting billions of people around the world. It is anticipated that millions will suffer from major diseases like Alzheimer's and Parkinson's disease (Steinmetz et al. 2024). Various experiments have indicated that medicinal plants hold promise for combatting the diseases in terms of both cellular and molecular function, for example through blocking off inflammation, suppressing pro-inflammatory cytokines, promoting anti-oxidative properties, or otherwise interfering with these influences. For neurodegeneration, it is one of the most important destructive factors, among which protein depletion, oxidative stress, and aging are considered to play considerable roles. Several traditional herbal remedies such as Ginseng, *Withania somnifera, Bacopa monnieri*, and *Ginkgo biloba* are said to have great potential for alleviating neurological problems. Dietary flavonoids have been shown to possess a variety of therapeutic advantages by numerous reports. They can prohibit amyloid formation, offer antioxidant effects, and act as potent anti-inflammatory substances (Uddin et al. 2020). This piece also maintains that flavonoids could help protect against Alzheimer's by interrupting the assembly of neurotoxic amyloid-β aggregates and inhibiting tau aggregation (Baptista et al. 2014; Saravanan et al. 2022). It also engages with essential signaling pathways in the brain, where the PI3K/Akt and MAPK signaling pathways are critical for regulating cell maintenance and gene function. It also results in the interruption of amyloid-β aggregation and alters the modulation of amyloid precursor protein through the inhibition of β-secretase and activation of α-secretase. Additionally, the suppression of CDK5 and GSK-3β helps to avert abnormal phosphorylation of tau (Baptista et al. 2014; Saravanan et al. 2020). Emodin, a bioactive phytochemical, exhibits significant properties that reduce inflammation, combat oxidative stress, and prevent fibril formation which may also address the pathogenic mechanisms of

AD. Findings indicate that emodin provides robust neuroprotection by preventing the formation of tau and amyloid-beta (Aβ) oligomers and reducing their neurotoxicity through the attenuation of neuroinflammation, oxidative stress, and bioenergetic defects. Additionally, Emodin has shown a correlation with better memory function, learning, and cognition (Saha and Ahmad 2024). The study (Qiao et al. 2015) investigates the neuroprotective benefits of artoindonesianin O (AIO), a phenolic compound sourced from mulberry, on rat neurons in the context of (AD). Mulberries, recognized for their health advantages, include bioactive compounds like AIO that may enhance neuronal function and protect against chronic diseases. AIO was found to protect against neurotoxicity induced by oligomer Aβ42, NMDA, and okadaic acid, while also improving synaptic plasticity. The mechanism involves interfering with the expression of the kinase p-ERK1/2, which is linked to tau protein hyperphosphorylation. Ferreres et al. (2010) investigate the flavonoid content and cytotoxic effects of tomato (*Lycopersicon esculentum*) seeds, and a total of 14 flavonoids were recognized, comprising derivatives of quercetin, kaempferol, and isorhamnetin. Notable compounds among these featured isorhamnetin-3-O-sophoroside, quercetin-3-O-sophoroside, and kaempferol-3-O-sophoroside. The extract from the seeds demonstrated a considerable inhibition of cell proliferation, exceeding 80%, in the rat basophile leukemia (RBL-2H3) cell line, with an IC_{50} value of 5980 μg/mL. It also demonstrated concentration-dependent acetylcholinesterase inhibitory activity (IC_{20} = 2400 μg/mL) and antioxidant mechanism towards DPPH, and superoxide free radicals. Strawberries, known for their high concentrations of phytochemicals and antioxidant capacity, were tested for neuroprotective impacts on PC12 cells subjected to hydrogen peroxide (H_2O_2). Compared to bananas and oranges, strawberries exhibited the most significant reduction in oxidative stress-induced neurotoxicity, as shown by cell viability tests and assays measuring lactate dehydrogenase (LDH) release. The protective effects are attributed to the high phenolic content, particularly anthocyanins, in strawberries (Heo and Lee 2005). Research highlights the potential of crocin, a water-soluble antioxidant derived from saffron (*Crocus sativus* L.), in improving memory and protecting brain cells. Studies suggest crocin may offer multifunctional protective effects against AD, leading to a potential compound for the purpose of prevention or therapeutic intervention (Finley and Gao 2017). Lee et al. (2022) reported the neuroprotective efficacy of compounds extracted from the twigs of *Rosa multiflora,* commonly known as Multiflora Rose, which is traditionally used in herbal medicine for various ailments. They discovered procyanidins, phlobatannins, flavan-3,4-diol glycoside, and neolignan glycoside, in addition to three previously identified compounds. The isolated compounds underwent assessment for their ability to reduce neuroinflammation and promote neurotrophic activity, with a specific focus in relation to nitric oxide production in LPS-stimulated murine microglial cells. Figure 19.2 shows some neuroprotective plants and their possible role.

Berries are a wholesome food source that may help in the treatment of diseases and improve body regulation due to their rich antioxidant properties. Research shows that a diet including blueberry or strawberry extract can prevent radiation-induced brain damage. Strawberries enhance spatial placement, while blueberries

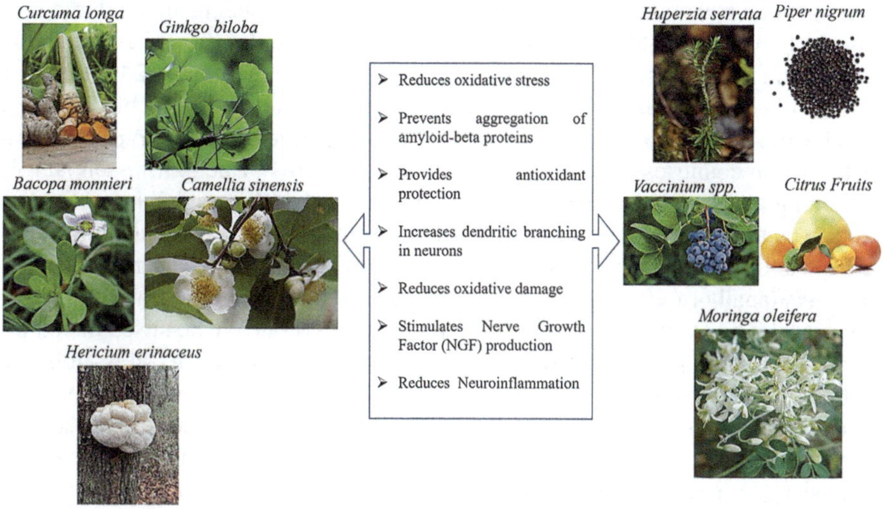

Fig. 19.2 Selected neuroprotective plants and their mechanism of action

improve learning, suggesting different berries benefit various brain regions. A variety of natural antioxidants and anti-inflammatory substances present in berries like strawberries, blueberries, and blackcurrants deliver neuroprotective effects (Subash et al. 2014; Joseph et al. 2014; Dhalaria et al. 2020; Bilawal et al. 2021). Avocado is unique in providing significant monounsaturated fat. It contains various phytochemicals, including antioxidants like flavonoids and coumarins, which may protect against neuronal damage from oxidative stress and slow degenerative diseases. These compounds could serve as safer alternatives to pharmaceuticals for treating neurodegenerative conditions. While there are limitations to their immediate use, polyphenolic phytochemicals from avocados show promise as accessible preventive agents for Alzheimer's Disease (AD) and other related disorders. Continued clinical trials are essential to explore their potential benefits fully.

A majority of Indian spices can help alleviate oxidative stress in the brain, but clinical outcomes for Alzheimer's disease (AD) have been limited due to their low solubility and bioavailability. Higher doses are required for efficacy, making them less palatable despite their safety. Current research focuses on enhancing their bioavailability through methods like nano-formulation. Wide-ranging clinical trials are needed to assess their potential such as disease treatments, supplemental strategies to medication therapy, or preventive measures. Spices, traditionally used in cooking, show therapeutic promise for CNS disorders, but continued research is needed to elucidate their mechanisms and boost clinical efficacy. A primary constituent of turmeric (*Curcuma longa*) is curcumin (Curcuminoid) (Esatbeyoglu et al. 2012). Usually added as coloring agent and given a spicy taste to food. Being grown for numerous generations of Indian and neighboring countries' traditions, it is widely used as a medicinal plant owing to its several medicinal properties (Saravanan et al. 2017, 2021; Esatbeyoglu et al. 2012). Over recent years, a variety of research investigations have taken place to investigate its physical, chemical, and biological nature of curcumin (Mishra and Palanivelu 2008; Pandey et al. 2011; Indira Priyadarsini 2013; Yandra Cervelim Nunes et al. 2024; Jafari et al. 2024; Manolova et al. 2014). Piperine is a bioactive alkaloid, a natural constituent of black pepper (Black Gold) that is conventionally employed as a traditional and therapeutic adjuvant in Asian countries (Chonpathompikunlert et al. 2010). Due to its essential role in cooking and health, black pepper is often termed the "King of Spices." Studies indicate that piperine possesses diverse health-promoting properties, including antioxidant, anti-inflammatory, antimicrobial, and antiulcer functions (Saravanan et al. 2019; Wattanathorn et al. 2008; Murunikkara et al. 2012). Furthermore, piperine enhances the bioavailability of both curcumin and resveratrol when used in combination, such as in piperine–curcumin and piperine–resveratrol mixtures (Johnson et al. 2011; Singh et al. 2012). Recent research in pharmacology indicates that piperine demonstrates significant anti-Alzheimer properties, especially in enhancing cognitive abilities (Wattanathorn et al. 2008; Chonpathompikunlert et al. 2010). Grapevine (*Vitis vinifera*) is the source of resveratrol, a factor that improves mitochondrial function through SIRT1 pathway activation. It also helps maintain cellular energy balance and controls autophagy. Parthenolide is a potential candidate for pharmacotherapy against Alzheimer's disease according to how it prevents oxidative stress induced by amyloid protein through activating AMPK signaling mechanisms in neurons (Singh et al. 2015; Vikal et al. 2024). Toothed Clubmoss (*Huperzia serrata*) contains Huperzine A, an acetylcholinesterase inhibitor. This provides neuroprotection by blocking the excessive activation of glutamate receptor-induced excitotoxicity, thus lessening damage to neuronal membranes which is in turn at risk from oxidative damage through cellulolytic enzymes. Its non-glycinate side chain allows it to pass through cell walls and into cells where it exerts its actions after being hydrolyzed inside the cell (Friedli and Inestrosa 2021; Dadlani et al. 2023; Cheng et al. 2024). The Maidenhair tree (*Ginkgo biloba*) is rich in flavonoids and terpene lactones. It increases cerebral blood flow, relieves oxidative stress, and improves synaptic plasticity. Moreover, it has the capacity to lower tau hyperphosphorylation. Black Pepper contains piperine, which increases the bioavailability of many compounds as well as depressing neuroinflammation alone. On dopamine and serotonin levels, it has a distinct neuroprotective (Birks and Grimley Evans 2002; Anastassakis 2022). Green Tea (Camellia sinensis) is high in such catechins as EGCG, which guards neurons against excitotoxicity directly. It both optimizes mitochondrial function and reduces oxidative stress for those living cells. Hesperidin (*Citrus Fruits*) scavenges ROS, inhibits inflammation pathways, and increases neurotrophic factors like BDNF to support cell survival and blood flow on the neuron level. Snowdrop Plants (*Galanthus* spp.) produce galantamine, it enhances signal strength between neurons, guard cells against apoptotic alarm sound by neurons themselves constituted to go off, and so it may 1 day offer treatment for neurodegenerative diseases from which these functions are suffering or have given rise. Blueberries and Cranberries (*Vaccinium* spp.) are full of anthocyanins and flavonoids. These berries help the nerves by being both antioxidant and anti-inflammatory and promoting new neuronal birth as well. *M. oleifera* is an important plant widely utilized in developing nations to enhance nutrition and strengthen food security. It is distributed in Nigeria, the West Indies, the Americas, and various regions of Asia, including Nepal, India, China, and the Philippines. The French naturalist Jean-Baptiste Lamarck first described this plant in 1785, and it is classified in the Moringaceae family. It is increasingly attracting attention in research due to its significant nutritional benefits and ethnopharmacological applications for both preventive and therapeutic treatment of health issues. Moreover, several pharmacological compounds have been derived from *M. oleifera* included in this list are epigallocatechin gallate, quercetin, and kaempferol that reported to exhibit significant neuroprotective effects against a range of neuroinflammatory conditions, indicating that *M. oleifera* may serve as a remarkable natural herb for neuroprotection (Islam et al. 2021; Ghimire et al. 2021; Azlan et al. 2023; Pareek et al. 2023; Chuene Victor Mashamaite et al. 2024; Worku and Tolossa 2024). Table 19.1 summarizes some traditional Indian medicinal plants having neuroprotective property and their target enzymes/proteins.

4 Molecular Mechanisms of Neuroprotection

Medicinal plants and their phytochemicals offer a multifaceted approach to neuroprotection through the modulation of molecular pathways that are involved in oxidative stress, inflammation, dysfunction of mitochondria, excitotoxicity, and aggregation of proteins. These mechanisms are crucial for the maintenance of neuronal integrity, function, and survival. Here, we explore the molecular underpinnings of neuroprotection conferred by phytochemicals, highlighting their role in preserving neuronal health.

4.1 Antioxidant Mechanisms

Oxidative stress occurs when the production of reactive oxygen species (ROS) surpasses the capacity of the cellular antioxidant defense system. This imbalance results in damage to critical cellular components, including lipids, proteins, and DNA, ultimately causing neuronal dysfunction and cell death. Phytochemicals help counteract oxidative stress by scavenging ROS, activating antioxidant pathways, and safeguarding mitochondrial function. Curcumin stimulates the nuclear factor erythroid 2-related factor 2 (Nrf2) pathway, which plays a crucial role in regulating cellular antioxidant defenses. When activated, Nrf2 moves into the nucleus and attaches to antioxidant response elements (ARE) within the promoters of genes responsible for producing antioxidant enzymes like glutathione peroxidase (GPx), superoxide dismutase (SOD), and heme oxygenase-1 (HO-1). This activation enhances the defense of the cellular system against oxidative damage. In Alzheimer's disease models, curcumin significantly reduced ROS levels, preserved the membrane potential of mitochondria, and prevented the neuronal apoptosis (Anand et al. 2007; Pei et al. 2024). Resveratrol, a polyphenol, enhances mitochondrial function

Table 19.1 Summary of some Indian medicinal plants with neuroprotective action and their inhibitor enzymes/proteins

No.	Plant name	Active compound	Neuroprotective properties	Inhibitor protein/enzyme	Reference
Flavonoid-containing plants					
1	*Ginkgo biloba*	Flavonoids	Enhances cognitive function and memory	MAO (Monoamine Oxidase)	Ahlemeyer and Krieglstein (2003)
2	*Bacopa monnieri*	Bacopasides	Improves memory and reduces anxiety	Acetylcholinesterase	Stough et al. (2001)
3	*Rosmarinus officinalis*	Rosmarinic acid	Antioxidant and anti-inflammatory	Acetylcholinesterase	Kosmopoulou et al. (2024)
4	*Salvia officinalis*	Carnosic acid	Protects neurons from oxidative stress	COX-1, COX-2	Pavić et al. (2019)
5	*Hypericum perforatum*	Hypericin	Antidepressant effects, mood enhancement	MAO (Monoamine Oxidase)	Herraiz and Guillén (2018)
6	*Centella asiatica*	Asiaticoside	Enhances nerve growth factor (NGF)	Phosphodiesterase	Kumar et al. (2009)
7	*Morus alba*	Morin	Neuroprotective effects against neuroinflammation	MAO (Monoamine Oxidase)	Rebai et al. (2017)
8	*Scutellaria baicalensis*	Baicalin	Anti-inflammatory effects	MAO (Monoamine Oxidase)	Liu et al. (2024)
9	*Citrus reticulata*	Hesperidin	Enhances cognitive function	MAO (Monoamine Oxidase)	Li et al. (2023)
10	*Ocimum sanctum*	Eugenol	Antioxidant and anti-inflammatory	COX-2	Said and Rabo (2017)
11	*Allium cepa*	Quercetin	Antioxidant and neuroprotective properties	Acetylcholinesterase	Chakraborty et al. (2022)
12	*Fragaria* × *ananassa*	Fisetin	Protects neurons from oxidative stress	MAO (Monoamine Oxidase)	Navarro-Hortal et al. (2022)
13	*Citrus* × *paradisi*	Naringenin	Exhibits antioxidant and anti-inflammatory properties	COX-2	Cai et al. (2023)
14	*Salvia miltiorrhiza*	Tanshinone IIA	Neuroprotective and anti-inflammatory	COX-2	Fu et al. (2021)
15	*Passiflora incarnata*	Harman	Sedative and neuroprotective effects	MAO (Monoamine Oxidase)	Ingale et al. (2014)
16	*Matricaria chamomilla*	Apigenin	Antioxidant and neuroprotective properties	Acetylcholinesterase	Siddique et al. (2022)

17	*Echinacea purpurea*	Chicoric acid	Protects against neuroinflammation	NO synthase	Qu et al. (2022)
18	*Ziziphus jujuba*	Jujuboside	Anxiolytic and neuroprotective effects	GABA transaminase	Wang et al. (2024)
19	*Thymus vulgaris*	Thymol	Antioxidant and anti-inflammatory	COX-2	Saleem et al. (2022)
20	*Vaccinium myrtillus*	Anthocyanins	Protects neurons from oxidative stress	MAO (Monoamine Oxidase)	Bayazid et al. (2021)
Non-flavonoid plants					
1	*Curcuma longa*	Curcumin	Anti-inflammatory and antioxidant effects.	COX-2	Aggarwal et al. (2007)
2	*Panax ginseng*	Ginsenosides	Enhances cognitive function and protects against neurotoxicity.	Acetylcholinesterase	Reay et al. (2005), Feng et al. (2022)
3	*Withania somnifera*	Withanolides	Exhibits neuroprotective and anti-stress effects.	MAO (Monoamine Oxidase)	Choudhary et al. (2015)
4	*Eleutherococcus senticosus*	Eleutherosides	Enhances resilience to stress and cognitive function.	MAO (Monoamine Oxidase)	Jung et al. (2014)
5	*Zingiber officinale*	Gingerol	Protects neurons from oxidative stress and inflammation.	COX-2	Arcusa et al. (2022), Ayustaningwarno et al. (2024)
6	*Silybum marianum*	Silymarin	Protects liver and brain cells from oxidative damage.	Lipoxygenase	Raza et al. (2011)
7	*Dioscorea opposita*	Diosgenin	Exhibits neuroprotective and anti-inflammatory properties.	Acetylcholinesterase	Som et al. (2022)
8	*Morinda citrifolia*	Morindin	Protects neurons from oxidative stress and inflammation.	MAO (Monoamine Oxidase)	Narasimhan et al. (2019)
9	*Nigella sativa*	Thymoquinone	Exhibits antioxidant and neuroprotective effects.	COX-2	Isaev et al. (2020)
10	*Moringa oleifera*	Moringin	Protects neurons from oxidative stress and inflammation.	MAO (Monoamine Oxidase)	Alqahtani and Albasher (2021)
11	*Allium sativum*	Allicin	Exhibits antioxidant properties and protects against neurodegeneration.	ACE (Angiotensin-Converting Enzyme)	Nadeem et al. (2021)
12	*Carya ovata*	Juglone	Exhibits neuroprotective and anti-inflammatory effects.	Acetylcholinesterase	Ahmad and Suzuki (2019)
13	*Eucalyptus globulus*	Eucalyptol	Exhibits anti-inflammatory and neuroprotective properties.	COX-2	Arooj et al. (2023)

(continued)

Table 19.1 (continued)

No.	Plant name	Active compound	Neuroprotective properties	Inhibitor protein/enzyme	Reference
14	*Rubus idaeus*	Ellagic acid	Protects against oxidative stress and neuroinflammation.	Acetylcholinesterase	Gupta et al. (2021), Alfei and Zuccari (2025)
15	*Taraxacum officinale*	Taraxasterol	Exhibits neuroprotective effects and reduces inflammation.	MAO (Monoamine Oxidase)	Huang et al. (2018), Jiao et al. (2022)
16	*Fucus vesiculosus*	Fucoidan	Protects against neurodegeneration.	MAO (Monoamine Oxidase)	Xing et al. (2023)
17	*Avena sativa*	Avenanthramides	Exhibits neuroprotective effects and reduces anxiety.	COX-2	Wankhede et al. (2023)
18	*Spirulina platensis*	Phycocyanin	Exhibits antioxidant properties and protects neurons.	Acetylcholinesterase	Trotta et al. (2022)
19	*Ginkgo biloba*	Ginkgo flavones	Enhances cerebral blood flow and antioxidant activity.	MAO (Monoamine Oxidase)	Mashayekh et al. (2011)
20	*Coptis chinensis*	Berberine	Exhibits anti-inflammatory and neuroprotective effects.	MAO (Monoamine Oxidase)	Phogat et al. (2024)

and reduces ROS production through the activation of the SIRT1-Nrf2 axis. SIRT1 deacetylates Nrf2, facilitating its activation and nuclear translocation. Furthermore, SIRT1 enhances mitochondrial biogenesis by activating the peroxisome proliferator-activated receptor-gamma coactivator 1-alpha (PGC-1α). Resveratrol's dual action of enhancing antioxidant defenses and supporting mitochondrial health was validated in Parkinson's disease models, where it preserved dopaminergic neurons and reduce the oxidative stress (Moussa et al. 2017; Wu et al. 2024).

4.2 Anti-inflammatory Pathways

Neuroinflammation, driven by the activation of microglia and astrocytes, amplifies neuronal damage by releasing pro-inflammatory cytokines and reactive nitrogen species. Phytochemicals suppress these inflammatory responses by targeting key signaling pathways. Quercetin, a flavonoid, inhibits the nuclear factor kappa-light-chain-enhancer of activated B cells (NF-κB) pathway, a key regulator of inflammatory responses. In pathological conditions, NF-κB becomes activated and migrates to the nucleus, where it stimulates the expression of pro-inflammatory cytokines, including tumor necrosis factor-alpha (TNF-α), interleukin-6 (IL-6), and cyclooxygenase-2 (COX-2). Quercetin inhibits this activation, reducing microglial and astrocytic activation and thereby limiting neuroinflammation and neuronal damage (Costa et al. 2016). Curcumin demonstrates anti-inflammatory properties by suppressing the signaling pathway of Toll-like receptor 4 (TLR4). TLR4, activated by damage-associated molecular patterns (DAMPs), triggers downstream inflammatory cascades via MyD88-dependent and TRIF-dependent pathways. Curcumin prevents TLR4 activation, reducing the release of cytokines and chemokines and mitigating inflammation-induced neuronal apoptosis in experimental neurodegeneration models (Baum et al. 2008; Cox et al. 2014).

4.3 Neurotransmitter Modulation

The dysregulation of neurotransmitter systems is a hallmark of many neurodegenerative conditions. Phytochemicals modulate neurotransmitter levels and receptor activity, restoring balance in neural circuits. Epigallocatechin gallate (EGCG), a catechin derived from green tea, regulates glutamate signaling by preventing the overactivation of N-methyl-D-aspartate (NMDA) receptors. Excessive glutamate activity can cause calcium overload, oxidative stress, and excitotoxicity. EGCG minimizes calcium influx through NMDA receptors, thereby protecting neurons from excitotoxic cell death. In ischemic stroke models, EGCG preserved neuronal viability and improved functional outcomes by modulating glutamate signaling (Xu et al. 1995; Weinmann et al. 2010). Huperzine A, an alkaloid extracted from *Huperzia serrata*, functions as a reversible inhibitor of acetylcholinesterase. By inhibiting the breakdown of acetylcholine, it enhances cholinergic transmission, improving cognitive function and synaptic plasticity. Clinical studies have demonstrated its efficacy in improving memory in Alzheimer's patients, with underlying mechanisms involving increased acetylcholine levels and enhanced neuronal signaling (Xu et al. 1995; Zangara 2003).

4.4 Cell Signaling Pathways

Phytochemicals modulate essential cell signaling pathways that govern neuronal survival, synaptic plasticity, and apoptosis. Key pathways include AMP-activated protein kinase (AMPK), SIRT1, and tropomyosin receptor kinase B (TrkB). Resveratrol, activates AMP-activated protein kinase (AMPK) an energy sensor responsible for maintaining cellular energy balance,thereby enhancing mitochondrial function, autophagy, and cellular resistance to stress. Additionally, it also activates sirtuin 1 (SIRT1), which deacetylates transcription factors such as peroxisome proliferator-activated receptor gamma coactivator 1-alpha (PGC-1α) and forkhead box O (FOXO), promoting mitochondrial biogenesis and reducing apoptosis. In Parkinson's disease models, resveratrol-mediated AMPK and SIRT1 activation improved mitochondrial bioenergetics and reduced alpha-synuclein aggregation, protecting neuronal integrity (Turner et al. 2015). 7,8-Dihydroxyflavone, a selective agonist of tropomyosin receptor kinase B (TrkB), replicates the actions of brain-derived neurotrophic factor (BDNF). Activation of TrkB triggers downstream signaling cascades, such as the phosphoinositide 3-kinase (PI3K)-Akt and mitogen-activated protein kinase (MAPK) pathways, which support neuronal survival, synaptic plasticity, and neurogenesis. Preclinical research has shown that this flavonoid improves cognitive function and provides protection against neurodegenerative damage (Jang et al. 2010). Phytochemicals also contribute to neuroprotection through mechanisms such as modulation of proteostasis, mitochondrial protection, and epigenetic regulation. Curcumin prevents amyloid-beta aggregation by stabilizing non-toxic oligomers, reducing amyloid plaque burden in Alzheimer's disease models. It also enhances autophagic clearance of aggregated proteins, promoting proteostasis. Similarly, resveratrol facilitates the autophagic degradation of alpha-synuclein, reducing its neurotoxic effects and supporting neuronal health. Hesperidin, a flavanone glycoside, enhances mitochondrial function by maintaining mitochondrial membrane potential and decreasing reactive oxygen species (ROS) production. Hesperidin supports mitochondrial energy metabolism, thereby preventing apoptosis and preserving neuronal viability in models of aging and neurodegeneration (Vauzour et al. 2015). The molecular mechanisms of neuroprotection by medicinal plant phytochemicals includes a wide spectrum of activities, including antioxidation, anti-inflammation, neurotransmitter modulation, and the regulation of cell signaling pathways. These mechanisms work in concert to mitigate neuronal damage, enhance synaptic plasticity, and support neuronal survival. Ongoing investigation of these pathways will aid in designing targeted phytochemical therapies for preventing and treating neurodegenerative disorders.

5 Neuroprotective Effects on Specific Conditions

Neuroprotection involves strategies to maintain neuronal health during injury, disease, and environmental stress. The central nervous system (CNS) is highly prone to damage, emphasizing the importance of neuroprotection in addressing the growing prevalence of neurological disorders such as stroke, Alzheimer's, and Parkinson's disease. These conditions significantly impact individuals and healthcare systems. Neuroprotective strategies include pharmacological approaches (e.g., antioxidants, anti-inflammatory drugs, and neurotrophic factors) and non-pharmacological methods (e.g., lifestyle changes and rehabilitation). Understanding these mechanisms can improve patient outcomes. Stroke is a leading contributor to both morbidity and mortality. Neuroprotective effects focus on preserving neuron integrity post-ischemia. Key mechanisms of ischemic damage include excitotoxicity, oxidative stress, inflammation, and apoptosis. Pharmacological agents like NMDA receptor antagonists and antioxidants target these pathways, while non-pharmacological approaches like therapeutic hypothermia support recovery. Current research aims to optimize treatment through personalized medicine and adjunct therapies, enhancing outcomes for stroke patients. Neurodegenerative diseases, including Alzheimer's and Parkinson's, pose significant healthcare challenges. Approaches to Alzheimer's disease target slowing its progression through anti-inflammatory drugs and antioxidants, whereas Parkinson's disease research highlights the role of antioxidants and neurotrophic factors. Novel therapies, including gene and stem cell therapies, are under investigation. Lifestyle factors, including diet and physical activity, contribute significantly to neuroprotection. Ongoing research offers potential for enhancing interventions and improving the quality of life for those affected. TBI (traumatic brain injury) is the result of neuronal damage and cognitive deficits. Neuroprotective strategies include pharmacological agents targeting excitotoxicity and oxidative stress, as well as neurotrophic factors to promote recovery. A comprehensive approach that combines pharmacological and rehabilitative measures is essential for optimizing recovery outcomes. Future neuroprotection research will benefit from multidisciplinary collaboration, personalized medicine, and lifestyle modifications. Advanced technologies like neuroimaging and AI will enhance understanding and monitoring of neuroprotective strategies. Continued innovation holds the potential to transform treatments for neurological conditions, improving recovery, and quality of life.

6 Blood–Brain Barrier and Bioavailability

The blood–brain barrier (BBB) is a selectively permeable structure that safeguards the central nervous system (CNS) from harmful substances while preserving homeostasis. The BBB is made up of tightly packed endothelial cells, pericytes, and astrocytic end-feet, functioning as both a physical and metabolic barrier. It carefully regulates what enters the central nervous system, allowing only specific molecules like glucose and essential amino acids to pass through while blocking harmful substances and pathogens (Abbott et al. 2010; Daneman and Prat 2015). This protective role of the BBB creates major challenges for drug delivery, as it blocks nearly 98% of small-molecule drugs and almost all large molecules from entering the central nervous system (Pardridge 2012). Additionally, efflux transporters (such as P-glycoprotein) actively pump out many compounds, further limiting their CNS penetration (Schinkel et al. 1994; Löscher and Potschka 2005). Despite these obstacles, phytochemicals from medicinal plants, including flavonoids, alkaloids, and terpenoids, have shown considerable promise in neuroprotection owing to their antioxidative, anti-inflammatory, and anti-apoptotic effects (Maiti and Dunbar 2018; Saravanan et al. 2022; Calderaro et al. 2022). However, their therapeutic application is often hindered by poor solubility, rapid metabolism, and inadequate permeability across the BBB. To overcome these limitations, researchers have explored innovative delivery systems and chemical modifications to enhance their bioavailability and CNS efficacy (Tiwari 2020).

6.1 Nanoparticle-Based Delivery Systems

Nanotechnology provides a groundbreaking solution for the delivery of neuroprotective phytochemicals by encapsulating these compounds in nanoparticles that can traverse the BBB. Solid lipid nanoparticles (SLNs) have proven effective in delivering curcumin, a powerful polyphenol known for its anti-inflammatory and neuroprotective benefits. Curcumin, characterized by poor solubility and rapid metabolism, was encapsulated within SLNs to improve its bioavailability and stability. These nanoparticles were further functionalized with low-density lipoprotein (LDL) receptor ligands, exploiting receptor-mediated endocytosis to enhance BBB transport. In Alzheimer's disease models, curcumin-loaded SLNs reduced amyloid plaques and improved synaptic plasticity, demonstrating therapeutic efficacy. Mechanistic studies demonstrated that the nanoparticles were effectively absorbed by BBB endothelial cells, transported across the barrier, and released curcumin into the brain parenchyma (Salehi et al. 2020; Pei et al. 2024). Similarly, lactoferrin-functionalized PLGA nanoparticles were developed to deliver quercetin, a flavonoid with antioxidative and anti-inflammatory properties. Lactoferrin, a natural ligand for specific receptors on BBB endothelial cells, facilitated receptor-mediated transcytosis of quercetin-loaded nanoparticles. In ischemic stroke models, these nanoparticles successfully crossed the BBB, reduced oxidative stress, and preserved neuronal viability.

6.2 Chemical Modifications to Improve Lipophilicity

Chemical modifications have been employed to enhance the lipophilicity of phytochemicals, thereby improving their passive diffusion across the BBB. Resveratrol, a hydrophilic polyphenol, was esterified with fatty acids to increase its lipid solubility. These lipid-soluble derivatives showed increased brain concentrations, reduced dopaminergic neuronal loss, and enhanced motor function in Parkinson's disease models. Notably, the modifications retained resveratrol's antioxidative and anti-inflammatory properties, positioning it as a promising option for treating neurodegenerative diseases (Wu et al. 2024). This approach highlights how chemical alterations can balance the compound's pharmacokinetics while maintaining its therapeutic efficacy.

6.3 Leveraging Endogenous Transport Pathways

Endogenous transport mechanisms, like receptor-mediated transcytosis, offer an alternative approach for transporting phytochemicals across the BBB. Transferrin receptors, which are highly expressed on BBB endothelial cells, serve as effective targets for nanoparticle delivery systems. Baicalin, a flavonoid with neuroprotective properties, was encapsulated in transferrin-conjugated nanoparticles, enabling efficient CNS delivery. In Alzheimer's disease models, these nanoparticles reduced neuroinflammation, enhanced synaptic plasticity, and improved neuronal survival. Mechanistic studies confirmed that the transferrin-baicalin complex utilized receptor-mediated endocytosis, bypassing efflux transporters and directly delivering the therapeutic payload to the CNS (Yan et al. 2013).

6.4 Intranasal Delivery: A Noninvasive Route

Intranasal delivery circumvents the BBB by leveraging the olfactory and trigeminal nerve pathways to achieve direct access to the central nervous system (CNS). This method avoids first-pass metabolism and systemic degradation, offering a rapid and targeted delivery system. Chitosan-coated curcumin nanoparticles administered intranasally demonstrated significant therapeutic benefits in traumatic brain injury models. These nanoparticles adhered to the nasal mucosa and were transported to the brain through axonal pathways. Preclinical findings reported reduced neuroinflammation and oxidative stress, along with improved neural repair and cognitive recovery, showcasing the potential of intranasal delivery for clinical application (Pandya et al. 2024; Kisku et al. 2024).

6.5 Enhancing Bioavailability

Bioavailability is a critical determinant of the therapeutic efficacy of phytochemicals. Factors such as poor solubility, rapid metabolism, and limited BBB permeability often restrict the systemic availability of these compounds. For instance, curcumin has very low bioavailability, primarily because of its rapid glucuronidation and limited solubility. To address this, nanoparticle encapsulation and co-administration with absorption enhancers like piperine have been explored. Piperine, a bioactive alkaloid, inhibits the enzymes responsible for curcumin metabolism, enhancing its bioavailability by over 2000% (Shoba et al. 1998). Similarly, lipid-soluble derivatives of resveratrol have shown improved CNS penetration and higher therapeutic efficacy in neurodegenerative disease models (Katila et al. 2022). The transport of neuroprotective phytochemicals across the BBB and their subsequent bioavailability are major challenges in developing effective therapies for neurological disorders. Advances in nanotechnology, chemical modifications, and alternative delivery routes such as intranasal administration have shown substantial promise in addressing these challenges. Mechanistic insights from preclinical studies highlight the potential of these approaches to enhance the stability, solubility, and CNS targeting of phytochemicals. Continued innovation and interdisciplinary research are essential to fully harness the therapeutic potential of the medicinal plant-derived compounds for neuroprotection.

7 Clinical Studies and Evidence

Medicinal plants and their phytochemicals have acquired quite considerable attention for their neuroprotective properties, particularly in managing the neurodegenerative diseases such as Alzheimer's disease (AD) and Parkinson's disease (PD). These diseases are characterized by the progressive neuronal degeneration caused by oxidative stress, neuroinflammation, and mitochondrial dysfunction. Clinical trials are pivotal in validating the efficacy of these compounds, elucidating their pharmacological targets, and assessing their therapeutic viability.

7.1 Flavonoids: Antioxidant and Anti-inflammatory Agents

7.1.1 Quercetin

Quercetin, a flavonoid widely present in onions, apples, and tea, has been thoroughly researched for its neuroprotective effects. The mechanisms of action of quercetin involve inhibiting amyloid-beta aggregation, reducing neuroinflammation, and enhancing neurogenesis, which together contribute to slowing the progression of neurodegenerative diseases. A randomized controlled trial showed that quercetin supplementation significantly lowered oxidative stress markers and enhanced cognitive function in patients with mild cognitive impairment (MCI) (Dehghani et al. 2020). Despite these encouraging results, quercetin's clinical use is greatly hindered by its low bioavailability, stemming from rapid metabolism and poor solubility. Recent research has further underscored quercetin's potential. A comprehensive review by Grewal et al. (2021) highlighted quercetin's neuroprotective roles in conditions like Alzheimer's disease, Parkinson's disease, Huntington's disease, multiple sclerosis, and amyotrophic lateral sclerosis. The review focused on quercetin's antioxidant and anti-inflammatory properties, particularly its ability to influence critical pathways such as nuclear factor erythroid 2-related factor 2 (Nrf2) and nuclear factor-kappa B (NF-κB), reinforcing its promise as a neurotherapeutic candidate. A study by Nguyen et al. (2023) explored quercetin and its metabolites' neurotherapeutic effects in cognitive impairment and Parkinson's disease. This study identified 28 quercetin metabolites with significant gastrointestinal absorption and blood–brain barrier permeability. Both quercetin and its metabolites exhibited robust anti-inflammatory and antioxidant activities, providing protective effects in preclinical models. However, the study emphasized the necessity for more extensive clinical trials to validate these neuroprotective effects and assess their practical applicability. To address quercetin's bioavailability issues, recent research has explored advanced delivery methods, including nano-encapsulation and structural modifications. Nano-encapsulation enhances quercetin's absorption and stability, allowing sustained release and improved therapeutic efficacy. Structural modifications, such as prodrug approaches, aim to optimize its pharmacokinetics and BBB permeability, further improving its potential for clinical applications. Despite these advancements, challenges remain in translating quercetin's promising preclinical findings into effective therapeutic interventions for humans. Ongoing and future large-scale clinical trials will be critical in validating its neuroprotective effects, optimizing dosing regimens, and assessing long-term safety.

7.1.2 Hesperidin

Hesperidin, a flavonoid primarily found in citrus fruits, has gained recognition for its neuroprotective qualities. Its antioxidant and anti-inflammatory actions play a vital role in combating oxidative stress, alleviating neuroinflammation, and supporting neuronal health. A systematic review demonstrated that hesperidin supplementation improved cognitive performance and reduced systemic inflammation in elderly participants, providing early evidence of its therapeutic potential (Han et al. 2025). Recent advancements have further elucidated hesperidin's neuroprotective effects and the impact on cognitive performance and neurobehavioral functions. The analysis demonstrated that hesperidin notably enhanced memory, learning, sensorimotor function, and locomotion in cognitively impaired animal models. These benefits were linked to its capacity to boost antioxidant defenses, regulate neurotransmitter levels, and influence neuroinflammatory pathways. The study emphasized hesperidin's potential in managing neurodegenerative diseases such as Alzheimer's disease and Parkinson's disease (Olasehinde et al. 2024). Furthermore, a 2023 review emphasized hesperidin's diverse role in addressing various brain disorders, including epilepsy, multiple sclerosis, depression, and neuropathic pain. The review elaborated on hesperidin's capacity to regulate neural growth factors, suppress apoptotic pathways, and reduce the production of pro-inflammatory cytokines such as TNF-α and IL-6. Together, these mechanisms support neuroprotection and help prevent neuronal apoptosis in models of oxidative stress and neuroinflammation (Hajialyani et al. 2019; Bansal et al. 2023). Despite these promising findings, hesperidin's clinical application is hindered by its low bioavailability, primarily due to its hydrophilic nature and rapid metabolism. Recent research has focused on developing advanced delivery systems, such as nanoformulations and phospholipid complexes, to enhance their absorption and therapeutic efficacy. These approaches have shown promise in improving hesperidin's bioavailability and enabling sustained release, thus enhancing its neuroprotective potential in vivo. Further large-scale, randomized clinical trials are necessary to validate hesperidin's neuroprotective effects in human populations. Future studies should focus on determining optimal dosing regimens, assessing long-term safety, and evaluating efficacy across various neurodegenerative conditions. Addressing these challenges could establish hesperidin as a promising therapeutic option for managing cognitive decline and other neurodegenerative disorders.

7.1.3 Kaempferol

Kaempferol, a flavonoid found abundantly in broccoli, kale, and spinach, has shown promising neuroprotective effects in preclinical and early clinical research, though its clinical application remains under investigation. Its mechanisms include reducing oxidative stress by activating the nuclear factor erythroid 2-related factor 2 (Nrf2) pathway and suppressing neuroinflammation by inhibiting nuclear factor-kappa B (NF-κB) signaling. These pathways collectively mitigate neuronal damage and promote cognitive function (Nezhad et al. 2024; Herrera et al. 2025). The trial revealed a significant reduction in inflammatory biomarkers, including tumor necrosis factor-alpha (TNF-α) and interleukin-6 (IL-6), coupled with an enhancement in mitochondrial function, suggesting kaempferol's role in combating the early stages of neurodegenerative diseases. However, the study also highlighted the compound's low bioavailability as a limitation, necessitating advancements in delivery methods. Preclinical studies further underline the potential of kaempferol in neuroprotection. Li et al. (2019) and Zhao et al. (2024) explored kaempferol's effects in models of cerebral ischemia/reperfusion injury. The study reported that kaempferol reduced blood–brain barrier (BBB) dysfunction, attenuated neuroinflammation, and improved neurological outcomes by suppressing oxidative stress and enhancing antioxidant defenses. Similarly, research by Ghosh et al. (2024) demonstrated that kaempferol-loaded nanoparticles significantly enhanced BBB permeability and bioavailability, leading to superior neuroprotective outcomes in animal models of Alzheimer's disease. These formulations improved cognitive and synaptic plasticity, further emphasizing the need for innovative delivery systems. Despite these advancements, no kaempferol-based drug has yet passed late-stage clinical trials or received regulatory approval for neurodegenerative conditions. The primary challenges remain its rapid systemic clearance and limited bioavailability. Current research is focused on developing nanoparticle-based delivery systems and conducting larger, more comprehensive clinical trials to validate its efficacy and safety in human populations.

7.1.4 7,8-Dihydroxyflavone (7,8-DHF)

7,8-Dihydroxyflavone (7,8-DHF), a synthetic flavonoid derivative, has attracted interest for its neuroprotective potential due to its role as a potent agonist of the tropomyosin receptor kinase B (TrkB), mimicking the effects of brain-derived neurotrophic factor (BDNF). TrkB receptor activation supports neuronal survival, synaptic plasticity, and cognitive function, which are essential for combating neurodegenerative diseases. Preclinical studies in animal models of Alzheimer's and Parkinson's diseases have demonstrated that 7,8-DHF crosses the blood–brain barrier, activates TrkB signaling, and provides significant neuroprotection by improving cognitive performance and reducing neuronal loss (Jang et al. 2010). Although 7,8-DHF shows promising neuroprotective effects, recent studies raise doubts about its specificity for TrkB. Some evidence suggests that its metabolites, rather than direct TrkB activation alone, may contribute to neuroprotection. These findings underline the need for further mechanistic validation before advancing to clinical trials. At higher concentrations, 7,8-DHF exhibited neurotoxic effects in some neuronal cultures, highlighting the importance of optimization of the dosage and further evaluation of its safety profile (Liu et al. 2014). Despite these controversies, 7,8-DHF continues to show promise in animal studies. Recent studies have shown that it enhanced motor function, alleviated neuroinflammation, and protected dopaminergic neurons from loss in a mouse model of Parkinson's disease. Furthermore, its capacity to enhance synaptic plasticity and memory performance in Alzheimer's models supports its potential for treating cognitive decline (Tapas et al. 2024). However, as of 2025, no clinical trials have evaluated 7,8-DHF's safety or efficacy in humans. The compound's pharmacokinetics, bioavailability, and potential off-target effects remain key areas for future research.

7.2 Non-Flavonoids: Targeting Neurotransmission and Beyond

7.2.1 Curcumin

Curcumin, a bioactive compound derived from turmeric (*Curcuma longa*), continues to garner significant attention for its neuroprotective properties. Its diverse mechanisms of action, including the activation of nuclear factor erythroid 2-related factor 2 (Nrf2) to boost internal antioxidant defenses and the suppression of nuclear factor-kappa B (NF-κB) to mitigate inflammation, position it as a promising candidate for addressing neurodegenerative diseases. A pivotal clinical trial demonstrated that curcumin supplementation significantly reduced amyloid-beta plaques and improved the behavioral symptoms in Alzheimer's patients (Baum et al. 2008). However, curcumin's therapeutic application is restrained by its poor bioavailability due to rapid metabolism, limited solubility, and poor systemic absorption. Advanced formulations, such as liposomal curcumin and curcumin nanoparticles, are developed to overcome these limitations, which significantly enhance the bioavailability and clinical efficacy (Anand et al. 2007). Recent advancements in 2024 have further underscored curcumin's potential in neurological and other chronic diseases. Curcumin has been shown to possess strong anti-inflammatory properties and the ability to regulate oxidative stress, offering potential therapeutic benefits for neurodegenerative diseases such as Alzheimer's and Parkinson's, as reported by Wei et al. (2024) in their systematic analysis. The review highlighted preclinical studies where curcumin reduced neuroinflammation and oxidative damage while enhancing mitochondrial function, thereby preventing neuronal death. Additionally, a novel randomized, double-blind study in 2024 evaluated a bioavailable curcumin formulation, CurQfen®, in Alzheimer's patients. Participants receiving 400 mg of CurQfen®

twice daily for 6 months exhibited significant improvements in their cognitive and the locomotive functions, as assessed by the Mini-Mental State Examination (MMSE) and the Geriatric Locomotive Function Scale (GLFS). This study also demonstrated reductions in amyloid-beta (Aβ42) levels, tau proteins, and the inflammatory biomarkers such as TNF-α and IL-6, providing robust evidence for curcumin's ability to modulate disease progression (Das et al. 2023). Moreover, a 2024 study in the European Journal of Clinical Pharmacology explored curcumin's applications beyond neurodegenerative diseases. It evaluated its role as a complementary treatment in oncology and other systemic diseases, noting curcumin's potential to alleviate treatment-related side effects through its anti-inflammatory and antioxidant properties. The study emphasized the heterogeneity of results and the necessity for high-quality clinical trials to validate its broad therapeutic effects (Gutsche et al. 2025). Despite these advancements, curcumin's translation into routine clinical use remains a challenge. While advanced formulations like CurQfen® and nanoparticle-based systems have improved bioavailability, the large-scale clinical trials are crucial to standardizing dosing regimens and establishing long-term safety profiles. Continued exploration into curcumin derivatives and hybrid compounds may further optimize its pharmacokinetics and therapeutic efficacy.

7.2.2 Resveratrol

Resveratrol is found in grapes, red wine, and berries, which has demonstrated significant neuroprotective potential. Its neuroprotective mechanisms are primarily attributed to the activation of sirtuin-1 (SIRT1), a protein that is involved in regulating mitochondrial bioenergetics and reducing neuroinflammation. A phase II clinical trial demonstrated that resveratrol supplementation in patients with mild cognitive impairment (MCI) delayed cognitive decline and improved mitochondrial function (Turner et al. 2015; Moussa et al. 2017). However, clinical application is hindered by its short half-life, low bioavailability, and rapid systemic clearance, necessitating advanced formulations to optimize its therapeutic potential. Recent studies have further underscored resveratrol's role in neuroprotection. A systematic review by Singh et al. (2019) highlighted its ability to mitigate oxidative stress, diminish amyloid-beta aggregation, and regulate the inflammatory pathways in preclinical Alzheimer's models. The review emphasized resveratrol's potential in improving cognitive outcomes, although variability in bioavailability among formulations was noted as a limiting factor. Additionally, a meta-analysis by Zeraattalab-Motlagh et al. (2021) investigated the resveratrol's impact on metabolic conditions that are associated with neurodegeneration, such as type 2 diabetes and metabolic syndrome. The findings revealed notable improvements in glycemic control and lipid profiles, suggesting that resveratrol's metabolic benefits may indirectly contribute to its neuroprotective effects. Furthermore, the study noted enhanced neurovascular health, supporting its potential for conditions like vascular dementia. A 2024 review examined novel delivery methods to enhance resveratrol's bioavailability. Nano-encapsulation and liposomal formulations were highlighted as promising strategies to increase systemic absorption and enable sustained release, thereby improving its therapeutic efficacy in neurodegenerative diseases (Salla et al. 2024).

These approaches also aim to address resveratrol's rapid metabolism and enhance its blood–brain barrier permeability, a critical factor in treating neurological disorders. Despite these advancements, challenges remain in translating resveratrol's preclinical success into clinical practice. Extensive, randomized clinical trials are essential to determine its effectiveness in human populations, refine dosing protocols, and evaluate its long-term safety. Ongoing exploration of advanced formulations and combination treatments could further solidify resveratrol's potential as a neuroprotective agent.

7.2.3 Huperzine A

Huperzine A, an alkaloid extracted from *Huperzia serrata*, has been widely researched for its potential therapeutic applications in neurodegenerative disorders, especially Alzheimer's disease. Its primary mechanism involves reversible inhibition of acetylcholinesterase (AChE), leading to enhanced cholinergic neurotransmission by preventing the breakdown of acetylcholine. This mechanism underpins its efficacy in boosting cognitive function, memory, and day-to-day activities in patients with mild-to-moderate AD (Xu et al. 1995). Furthermore, Huperzine A demonstrates neuroprotective properties against glutamate-induced excitotoxicity, a critical factor in neuronal death in neurodegenerative conditions (Zangara 2003). Recent research has provided further insights into Huperzine A's pharmacokinetics, bioavailability, and clinical utility. A study employed an advanced ultra-performance liquid chromatography–tandem mass spectrometry (UPLC–MS/MS) technique to measure Huperzine A concentrations in rat plasma. The findings revealed that Huperzine A possesses a long half-life and high oral bioavailability, characteristics that are favorable for sustained therapeutic effects (Zhang and Wang 2024). These results support its potency for a long-acting therapeutic agent for AD and related conditions. In addition, a clinical trial initiated in 2024 aimed to evaluate the pharmacokinetics and pharmacodynamics of Huperzine A injections in healthy Chinese participants. This trial focuses on optimizing dosing regimens and understanding Huperzine A's metabolic profile, which is critical for its application in broader clinical settings (Wanbangde Pharmaceutical Group Co. 2024).

Huperzine A has also been studied in combination therapies to explore synergistic effects with other neuroprotective agents. Preclinical investigations suggest that combining Huperzine A with antioxidants or anti-inflammatory compounds may enhance its efficacy in mitigating oxidative stress and neuroinflammation, key contributors to neurodegenerative processes. Despite its promising pharmacological profile, the translation of Huperzine A into widespread clinical use faces challenges. Variability in individual responses, potential side effects such as gastrointestinal discomfort, and the need for standardized formulations require further investigation. Large-scale, randomized controlled trials across diverse populations are essential to validate its safety and efficacy comprehensively. Additionally, advancements in drug delivery systems, such as sustained-release formulations or nanoparticles, could enhance its bioavailability and therapeutic outcomes.

7.2.4 Galantamine

Galantamine, an alkaloid derived from *Galanthus* species (snowdrop plants), remains a well-established therapeutic agent for Alzheimer's disease (AD). Its dual mechanistic action involves the reversible acetylcholinesterase (AChE) inhibition, enhancing the cholinergic neurotransmission, and the allosteric modulation of nicotinic acetylcholine receptors, which promotes neuroplasticity and synaptic repair. Clinical trials have consistently demonstrated its ability to improve cognitive performance, the memory, and daily functioning in patients with mild-to-moderate AD, providing significant symptomatic relief (Dengiz and Kershaw 2004; Baakman et al. 2022). Its well-characterized pharmacokinetics and safety profile have further reinforced its clinical utility. Recent advancements have expanded the scope of galantamine's therapeutic applications. In 2024, the U.S. Food and Drug Administration approved Zunveyl (benzgalantamine), a prodrug of galantamine (reference ID: 5419976). This formulation was designed to improve bioavailability and minimize common side effects such as gastrointestinal discomfort, thereby enhancing patient compliance and therapeutic outcomes (Timmins 2024). Prodrug formulations such as Zunveyl represent a significant step in optimizing galantamine for long-term use in AD treatment. Research has also focused on exploring galantamine's potential in combination therapies. A study by Stavrakov et al. (2020) introduced a novel galantamine-curcumin hybrid compound. This hybrid exhibited enhanced inhibitory activity against butyrylcholinesterase, an enzyme implicated in AD progression, and demonstrated superior neuroprotective effects in preclinical models, including reduced oxidative stress and inflammation. Additionally, a review by Frost et al. (2024) investigated the combination of galantamine and memantine in the management of Parkinson's disease. This approach modulated kynurenic acid levels, a potential biomarker for neurodegeneration, and showed promise in addressing both cognitive and motor impairments associated with the disease. Beyond its approved indications, novel derivatives of galantamine are being developed to expand its therapeutic potential. Recent work by Lazarova et al. (2021) evaluated four newly synthesized galantamine derivatives in a scopolamine-induced dementia model in mice. These derivatives exhibited enhanced efficacy in reversing cognitive deficits, reducing neuroinflammation, and improving biochemical markers, supporting their potential for future clinical exploration. Additionally, advanced delivery methods such as transdermal patches and sustained-release formulations are being developed to improve bioavailability and patient adherence. While galantamine continues to be a cornerstone in AD treatment, challenges persist in broadening its clinical applications. Individual variations in therapeutic response, potential adverse effects, and the need for personalized treatment strategies underscore the necessity of further large-scale, randomized controlled trials. Moreover, exploring neuroprotective mechanisms beyond AChE inhibition, such as its impact on neuroinflammation and oxidative stress, remains critical for expanding its therapeutic repertoire.

7.2.5 Piperine

Piperine is an alkaloid from black pepper (*Piper nigrum*), that is renowned for enhancing the bioavailability of various phytochemicals, notably curcumin. Clinical studies have demonstrated that co-administration of piperine with curcumin significantly increases curcumin's systemic availability, thereby amplifying its neuroprotective effects (Shoba et al. 1998). Moreover, piperine exhibits intrinsic anti-inflammatory properties by inhibiting pro-inflammatory cytokines, suggesting potential direct neuroprotective benefits (Tripathi et al. 2022). Recent clinical research has further explored the piperine's therapeutic potential. A randomized clinical trial evaluated the efficacy of oral curcumin combined with the piperine as an adjuvant therapy in COVID-19 patients. The study reported that this combination led to early symptomatic recovery, reduced deterioration, and improved clinical outcomes, indicating piperine's role in enhancing curcumin's therapeutic effects (Pawar et al. 2021). Another clinical trial assessed the impact of curcumin and piperine co-supplementation on the patients undergoing coronary artery bypass grafting (CABG). The results showed a potential reduction in the inflammatory markers, such as C-reactive protein (CRP), and an increase in the total antioxidant capacity (TAC), suggesting that piperine may enhance the curcumin's anti-inflammatory and antioxidant properties in clinical settings (Tehrani et al. 2024). Furthermore, a 2024 study investigated the effects of curcumin and piperine supplementation on oxidative stress and inflammatory markers in patients with inflammatory bowel disease (IBD). The findings indicated improvements in redox balance and a reduction in disease activity, highlighting piperine's potential to augment curcumin's therapeutic efficacy in inflammatory conditions (Martins et al. 2024). Despite these promising findings, piperine's role as a standalone therapeutic agent remains underexplored. Most clinical studies have focused on its bio enhancing properties when combined with other compounds. Future research should aim to elucidate piperine's independent pharmacological effects and determine its efficacy and safety profiles in various clinical contexts.

7.2.6 *Ginkgo biloba* (EGb 761)

Ginkgo biloba extract, particularly the standardized form EGb 761, continues to be a focus of research for its potency of neuroprotective benefits in neurodegenerative diseases. EGb 761 comprises 24% flavonoid glycosides and 6% terpene lactones, which contribute to its neuroprotective properties through multiple mechanisms, including enhancement of cerebral blood flow, reduction of oxidative stress, and anti-inflammatory effects. Early clinical trials demonstrated its efficacy in improving the cognitive function, neuropsychiatric symptoms, and daily activities in dementia patients, though results have been inconsistent, emphasizing the need for standardized dosing and formulations (Weinmann et al. 2010). A 2023 pilot randomized controlled trial explored the efficacy of EGb 761 in improving cognitive function in the patients following mild-to-moderate ischemic stroke. Over a 24-week period, participants receiving 240 mg per day of EGb 761 showed crucial improvements in the cognitive performance that is assessed using the Mini-Mental State Examination (MMSE) and Montreal Cognitive Assessment (MoCA). This study accentuated the potential role of EGb 761 in post-stroke cognitive recovery, linking its benefits to its antioxidant and vascular-enhancing properties (Cui et al. 2023). The findings validated the effectiveness of EGb 761 in reducing cognitive deficits, demonstrating improvements in memory, attention, processing speed, and executive function. Additionally, neuropsychiatric symptoms such as anxiety and depression were significantly reduced in studies that evaluated these outcomes. However, the review also noted variability in outcomes across trials, attributed to differences in dosage, duration, and patient demographics (Hort et al. 2023). A double-blind, placebo-controlled research that was conducted in 2024 evaluated EGb 761 in the patients with AD over a 6-month period. Participants receiving 240 mg per day demonstrated the slower progression of cognitive decline that is compared to those on placebo, as measured by the Alzheimer's Disease Assessment Scale-Cognitive Subscale (ADAS-Cog). Biomarker analyses revealed reductions in inflammatory cytokines such as IL-6 and TNF-α, which further supported its anti-inflammatory role in neuroprotection (Guilherme et al. 2024). There are challenges remaining in optimizing the use of EGb 761. Variability in patient response, particularly in older populations, and inconsistent results across studies underscore the need for large-scale, multicenter trials with standardized protocols. Ongoing research is focused on identifying patient subgroups that are most likely to benefit from EGb 761 and elucidating its precise mechanisms of action, particularly its effects on oxidative stress and neuroinflammation. Advances in formulation technologies, such as sustained-release preparations, are also being explored to improve bioavailability and patient compliance. Moreover, combining EGb 761 with other neuroprotective agents may offer synergistic effects, which could be valuable in managing complex neurodegenerative diseases.

Despite significant progress, several challenges persist in integrating phytochemicals into mainstream clinical practice. Variability in extract preparation, limited bioavailability, and inconsistent therapeutic outcomes across different populations underscores the need for more rigorous and standardized research approaches. Addressing these issues requires large-scale, multicenter clinical trials with standardized formulations and advanced delivery systems to improve efficacy and accessibility. Clinical evidence supports the neuroprotective potential of phytochemicals, highlighting their ability to target key pathological pathways, including oxidative stress reduction, anti-inflammatory responses, and cholinergic modulation. These mechanisms position phytochemicals as potential candidates for addressing neurodegenerative diseases. However, to fully harness their therapeutic potential, ongoing research must focus on overcoming current limitations. Innovations in drug delivery, such as nano formulations and sustained-release systems, coupled with a deeper understanding of pharmacokinetics and pharmacodynamics, will be critical. Ongoing technological innovations and translational research will be crucial in connecting preclinical discoveries with routine clinical practice, facilitating the incorporation of these natural compounds into effective and accessible therapies for neurodegenerative diseases.

8 Future Perspectives

The field of neuroprotection through medicinal plants and their phytochemicals holds immense potential, with opportunities that address the pressing challenges posed by the neurodegenerative diseases. Despite significant progress, the translation of phytochemicals into effective clinical therapies requires overcoming key obstacles, such as bioavailability, targeted delivery, and safety concerns. Future research directions aim to integrate emerging technologies and multidisciplinary approaches to utilize the full therapeutic potential of these compounds.

8.1 Enhancing Bioavailability and CNS Targeting

The limited bioavailability of phytochemicals remains a critical hurdle in their clinical application. Future efforts should focus on innovative delivery systems, such as biomimetic nanoparticles and hybrid nanocarriers, to enhance stability, solubility, and BBB penetration. Stimuli-responsive nanocarriers, designed to release their payload under specific pathological conditions such as oxidative stress or inflammation, have shown great potential for enhancing targeted delivery. For instance, recent research has highlighted the effectiveness of pH-responsive nanoparticles in delivering curcumin in Alzheimer's models, demonstrating their ability to reduce off-target effects (Pei et al. 2024).

8.2 Leveraging Multi-Omics and Systems Biology

Understanding the intricate molecular mechanisms underlying phytochemical actions requires advanced systems biology and multi-omics approaches that are genomics, proteomics, and metabolomics, which can provide comprehensive insights into the cellular pathways modulated by phytochemicals, enabling the identification of novel targets for neuroprotection. For instance, metabolomic studies have identified unique phytochemical-derived metabolites capable of modulating neuroinflammatory pathways, underscoring the utility of multi-omics in therapeutic discovery (Maiti and Dunbar 2018; Calderaro et al. 2022).

8.3 Exploring Novel Phytochemicals and Medicinal Plants

The vast chemical diversity of plant secondary metabolites offers untapped potential for discovering novel neuroprotective compounds. Advanced screening techniques, such as high-throughput metabolomics and bioinformatics-guided identification, can expedite the discovery of bioactive phytochemicals. Additionally, traditional medicinal knowledge remains an invaluable resource for uncovering lesser-known plants with therapeutic potential. Recent exploration of endemic species has revealed novel flavonoids with potent antioxidative properties, suggesting that continued efforts in ethnobotany could yield groundbreaking discoveries.

8.4 Integrating Precision Medicine

Precision medicine offers a promising avenue for individualizing phytochemical therapies based on genetic, epigenetic, and metabolic variations. Pharmacogenomic profiling could identify biomarkers that predict individual responses to specific phytochemicals, enabling tailored treatments for neurodegenerative diseases. Personalized formulations, optimized dosages, and targeted delivery systems could further enhance therapeutic outcomes while minimizing adverse effects (Pardridge 2012; Tiwari 2020).

8.5 Combining Phytochemicals with Conventional Therapies

Phytochemicals could complement conventional treatments for neurodegenerative diseases by targeting complementary pathways or mitigating drug resistance. For instance, combining phytochemicals with existing anti-inflammatory drugs could amplify their efficacy and reduce required dosages, minimizing side effects. Preclinical studies have already exhibited the synergistic effects of quercetin with standard Alzheimer's therapies, paving the way for future combination treatments.

8.6 Advancing Clinical Translation

The transition from preclinical research to clinical applications presents a significant challenge. Rigorous clinical trials are essential to evaluate the efficacy, safety, and pharmacokinetics of phytochemical-based therapies. Collaborative efforts involving academic institutions, pharmaceutical companies, and regulatory agencies can streamline this process, ensuring that effective therapies reach patients. Recent clinical trials focusing on curcumin formulations have demonstrated the potential of phytochemicals to serve as viable therapeutic agents when supported by robust clinical evidence (Hegde et al. 2023).

8.7 Sustainability and Ethical Considerations

The rising demand for medicinal plants necessitates sustainable sourcing practices to prevent overharvesting and ecological harm. Biotechnological approaches, such as plant tissue culture and synthetic biology, offer sustainable alternatives for producing bioactive compounds. Additionally, integrating fair trade and community-based conservation practices can ensure ethical sourcing while preserving biodiversity and traditional knowledge (Daneman and Prat 2015).

The future of neuroprotection by medicinal plants and their phytochemicals lies at the intersection of innovation and multidisciplinary collaboration. By addressing the challenges of bioavailability, safety, and clinical translation, and by integrating cutting-edge technologies such as nanotechnology, AI, and precision medicine, researchers can exploit the full therapeutic potential of all these compounds. Investing in sustainable practices and conducting comprehensive clinical studies are essential for the successful development of phytochemical-based therapies, enabling better treatment options for neurodegenerative diseases.

9 Summary and Conclusion

This chapter underscores the potential of medicinal plants and their phytochemicals in providing neuroprotection through diverse molecular mechanisms. While significant progress has been made, further research is essential to translate these findings into effective clinical interventions for neurodegenerative diseases. Medicinal plants have demonstrated remarkable potential in providing neuroprotective effects across various neurological conditions, offering hope for both prevention and treatment strategies. The natural compounds discussed in this chapter work through multiple mechanisms to protect and enhance neurological function, with each plant offering unique therapeutic properties that can be targeted to specific conditions. In Alzheimer's disease, plants like turmeric, containing curcumin, have shown particularly promising results. Curcumin's ability to cross the blood–brain barrier allows it to directly target multiple pathological processes involved in Alzheimer's progression. It effectively reduces amyloid-beta aggregation, decreases tau hyperphosphorylation, and modulates microglial activation, providing comprehensive neuroprotection. Working alongside curcumin, Ginkgo biloba extract enhances cerebral blood flow and maintains synaptic plasticity, making it a valuable complement in Alzheimer's treatment strategies. Parkinson's disease benefits from a different set of plant compounds, with green tea catechins playing a crucial role. These compounds demonstrate specific protection of dopaminergic neurons and help regulate alpha-synuclein aggregation, addressing core pathological features of the disease. *Bacopa monnieri* supplements by preserving dopamine levels and maintaining mitochondrial integrity, providing a multifaceted treatment approach for Parkinson's patients. In the context of stroke and ischemic injury, Ginkgo biloba and Chinese Sage (Salvia miltiorrhiza) demonstrate particular efficacy. Their ability to improve cerebral circulation and prevent platelet aggregation makes them valuable in both prevention and recovery. These plants also provide substantial antioxidant protection, crucial in the aftermath of ischemic events. Epilepsy management benefits from plants like *Bacopa monnieri* and Passionflower, which work primarily through GABA modulation and neural membrane stabilization. These plants help reduce seizure severity while providing additional benefits such as anxiety reduction and improved sleep quality, addressing multiple aspects of the condition. The treatment of anxiety and depression has found valuable allies in plants like Ashwagandha and Saffron. Ashwagandha's ability to reduce cortisol levels and enhance GABA function, combined with Saffron's modulation of serotonin and enhancement of BDNF expression, provides a comprehensive approach to mental health management. Clinical integration of these plant-based treatments requires careful consideration of several factors. Standardization of preparations, optimal dosing schedules, and potential drug interactions must be carefully managed. Patient-specific factors such as age, comorbidities, and current medications play crucial roles in determining appropriate treatment strategies. Regular monitoring of liver function, blood coagulation, and cognitive effects ensures safe and effective treatment outcomes. Future research in this field continues to evolve, focusing on several key areas. Bioavailability enhancement through novel delivery systems and combination therapies represents a major focus of current research. Additionally, deeper investigation into molecular pathways and potential synergistic effects between different plant compounds promises to unlock new therapeutic possibilities. Clinical validation through larger trials and extended duration studies remains crucial for establishing these treatments in mainstream medical practice. The integration of medicinal plants into neurological treatment protocols represents a promising frontier in healthcare. Their multiple mechanisms of action, relatively low toxicity, and long history of traditional use provide a strong foundation for their continued development as therapeutic agents. As research continues to unveil new properties and applications of these plants, their role in treating neurological conditions is likely to expand, offering hope for patients seeking effective, natural treatment options. Understanding the specific mechanisms and applications of these plants allows for more targeted and effective treatment strategies. However, it is crucial to remember that these natural compounds should be used under proper medical supervision, with careful consideration of individual patient factors and potential interactions with conventional medications. The future of neuroprotective plant medicine lies in the careful balance between traditional knowledge and modern scientific validation, working together to provide optimal patient outcomes.

References

Abbott NJ, Patabendige AAK, Dolman DEM, Yusof SR, Begley DJ (2010) Structure and function of the blood–brain barrier. Neurobiol Dis 37:13–25. https://doi.org/10.1016/j.nbd.2009.07.030

Aggarwal BB, Sundaram C, Malani N, Ichikawa H (2007) Curcumin: the Indian solid gold. In: Aggarwal BB, Surh Y-J, Shishodia S (eds) The molecular targets and therapeutic uses of curcumin in health and disease. Springer US, Boston, pp 1–75

Agrawal AD (2011) Pharmacological activities of flavonoids: a review. Int J Pharm Sci Nanotechnol 4:1394–1398. https://doi.org/10.37285/ijpsn.2011.4.2.3

Aguiar S, Borowski T (2013) Neuropharmacological review of the nootropic herb Bacopa monnieri. Rejuvenation Res 16:313–326. https://doi.org/10.1089/rej.2013.1431

Ahlemeyer B, Krieglstein J (2003) Neuroprotective effects of Ginkgo biloba extract. Cell Mol Life Sci CMLS 60:1779–1792. https://doi.org/10.1007/s00018-003-3080-1

Ahmad T, Suzuki YJ (2019) Juglone in oxidative stress and cell signaling. Antioxidants 8:91. https://doi.org/10.3390/antiox8040091

Alfei S, Zuccari G (2025) Ellagic acid: a green multi-target weapon that reduces oxidative stress and inflammation to prevent and improve the condition of Alzheimer's disease. Int J Mol Sci 26:844. https://doi.org/10.3390/ijms26020844

Alqahtani WS, Albasher G (2021) *Moringa oleifera* Lam. extract rescues lead-induced oxidative stress, inflammation, and apoptosis in the rat cerebral cortex. J Food Biochem 45. https://doi.org/10.1111/jfbc.13579

Anand P, Kunnumakkara AB, Newman RA, Aggarwal BB (2007) Bioavailability of curcumin: problems and promises. Mol Pharm 4:807–818. https://doi.org/10.1021/mp700113r

Anastassakis K (2022) Ginkgo Biloba (Maidenhair Tree). In: Androgenetic Alopecia from A to Z. Springer International Publishing, Cham, pp 459–463

Arcusa R, Villaño D, Marhuenda J, Cano M, Cerdà B, Zafrilla P (2022) Potential role of ginger (Zingiber officinale Roscoe) in the prevention of neurodegenerative diseases. Front Nutr 9:809621. https://doi.org/10.3389/fnut.2022.809621

Arooj B, Asghar S, Saleem M, Khalid SH, Asif M, Chohan T, Khan IU, Zubair HM, Yaseen HS (2023) Anti-inflammatory mechanisms of eucalyptol rich Eucalyptus globulus essential oil alone and in combination with flurbiprofen. Inflammopharmacology 31:1849–1862. https://doi.org/10.1007/s10787-023-01237-6

Ayaz M, Mosa OF, Nawaz A, Hamdoon AAE, Elkhalifa MEM, Sadiq A, Ullah F, Ahmed A, Kabra A, Khan H, Murthy HCA (2024) Neuroprotective potentials of Lead phytochemicals against Alzheimer's disease with focus on oxidative stress-mediated signaling pathways: pharmacokinetic challenges, target specificity, clinical trials and future perspectives. Phytomedicine 124:155272. https://doi.org/10.1016/j.phymed.2023.155272

Ayustaningwarno F, Anjani G, Ayu AM, Fogliano V (2024) A critical review of ginger's (Zingiber officinale) antioxidant, anti-inflammatory, and immunomodulatory activities. Front Nutr 11:1364836. https://doi.org/10.3389/fnut.2024.1364836

Azlan UK, Khairul Annuar NA, Mediani A, Aizat WM, Damanhuri HA, Tong X, Yanagisawa D, Tooyama I, Wan Ngah WZ, Jantan I, Hamezah HS (2023) An insight into the neuroprotective and anti-neuroinflammatory effects and mechanisms of Moringa oleifera. Front Pharmacol 13. https://doi.org/10.3389/fphar.2022.1035220

Baakman AC, Gavan C, Doeselaar L, Kam M, Broekhuizen K, Bajenaru O, Camps L, Swart EL, Kalisvaart K, Schoonenboom N, Lemstra E, Scheltens P, Cohen A, Gerven J, Groeneveld GJ (2022) Acute response to cholinergic challenge predicts long-term response to galantamine treatment in patients with Alzheimer's disease. Br J Clin Pharmacol. https://doi.org/10.1111/bcp.15206

Balakrishnan R, Azam S, Cho D-Y, Su-Kim I, Choi D-K (2021) Natural phytochemicals as novel therapeutic strategies to prevent and treat Parkinson's disease: current knowledge and future perspectives. Oxidative Med Cell Longev 2021:1–32. https://doi.org/10.1155/2021/6680935

Bansal K, Singh V, Singh S, Mishra S (2023) Neuroprotective potential of hesperidin as therapeutic agent in the treatment of brain disorders: preclinical evidence-based review. Curr Mol Med 23. https://doi.org/10.2174/1566524023666230320144722

Baptista FI, Henriques AG, Silva AMS, Wiltfang J, da Cruz e Silva OAB (2014) Flavonoids as therapeutic compounds targeting key proteins involved in Alzheimer's disease. ACS Chem Neurosci 5:83–92. https://doi.org/10.1021/cn400213r

Baum L, Lam CWK, Cheung SK-K, Kwok T, Lui V, Tsoh J, Lam L, Leung V, Hui E, Ng C, Woo J, Chiu HFK, Goggins WB, Zee BC-Y, Cheng KF, Fong CYS, Wong A, Mok H, Chow MSS, Ho PC, Ip SP, Ho CS, Yu XW, Lai CYL, Chan M-H, Szeto S, Chan IHS, Mok V (2008) Six-month randomized, placebo-controlled, double-blind, pilot clinical trial of curcumin in patients with Alzheimer disease. J Clin Psychopharmacol 28:110–113. https://doi.org/10.1097/jcp.0b013e318160862c

Bayazid AB, Chun EM, Al Mijan M, Park SH, Moon S-K, Lim BO (2021) Anthocyanins profiling of bilberry (*Vaccinium myrtillus* L.) extract that elucidates antioxidant and anti-inflammatory effects. Food Agric Immunol 32:713–726. https://doi.org/10.1080/09540105.2021.1986471

Behl T, Makkar R, Sehgal A, Singh S, Sharma N, Zengin G, Bungau S, Andronie-Cioara FL, Munteanu MA, Brisc MC, Uivarosan D, Brisc C (2021) Current trends in neurodegenera-

tion: cross talks between oxidative stress, cell death, and inflammation. Int J Mol Sci 22:7432. https://doi.org/10.3390/ijms22147432
Bellavite P (2023) Neuroprotective potentials of flavonoids: experimental studies and mechanisms of action. Antioxidants 12:280. https://doi.org/10.3390/antiox12020280
Bermejo-Pareja F, del Ser T (2024) Controversial past, splendid present, unpredictable future: a brief review of Alzheimer disease history. J Clin Med 13:536–536. https://doi.org/10.3390/jcm13020536
Bilawal A, Ishfaq M, Gantumur M-A, Qayum A, Shi R, Fazilani SA, Anwar A, Jiang Z, Hou J (2021) A review of the bioactive ingredients of berries and their applications in curing diseases. Food Biosci 44:101407. https://doi.org/10.1016/j.fbio.2021.101407
Birks J, Grimley Evans J (2002) Ginkgo Biloba for cognitive impairment and dementia. Cochrane Database Syst Rev. https://doi.org/10.1002/14651858.cd003120
Bloem BR, Okun MS, Klein C (2021) Parkinson's disease. Lancet 397:2284–2303. https://doi.org/10.1016/S0140-6736(21)00218-X
Boxer AL, Sperling R (2023) Accelerating Alzheimer's therapeutic development: the past and future of clinical trials. Cell 186:4757–4772. https://doi.org/10.1016/j.cell.2023.09.023
Breijyeh Z, Karaman R (2020) Comprehensive review on Alzheimer's disease: causes and treatment. Molecules 25:5789. https://doi.org/10.3390/molecules25245789
Cai J, Wen H, Zhou H, Zhang D, Lan D, Liu S, Li C, Dai X, Song T, Xy W, He Y, He Z, Jinquan T, Zhang J (2023) Naringenin: a flavanone with anti-inflammatory and anti-infective properties. Biomed Pharmacother 164:114990–114990. https://doi.org/10.1016/j.biopha.2023.114990
Calderaro A, Patanè GT, Tellone E, Barreca D, Ficarra S, Misiti F, Laganà G (2022) The neuroprotective potentiality of flavonoids on alzheimer's disease. Int J Mol Sci 23:14835. https://doi.org/10.3390/ijms232314835
Câmara JS, Perestrelo R, Ferreira R, Berenguer CV, Jorge CPC (2024) Plant-derived terpenoids: a plethora of bioactive compounds with several health functions and industrial applications—a comprehensive overview. Molecules 29:3861–3861. https://doi.org/10.3390/molecules29163861
Chakraborty AJ, Uddin TM, Matin Zidan BMR, Mitra S, Das R, Nainu F, Dhama K, Roy A, Hossain MJ, Khusro A, Emran TB (2022) Allium cepa: a treasure of bioactive phytochemicals with prospective health benefits. Evid Based Complement Alternat Med 2022:1–27. https://doi.org/10.1155/2022/4586318
Chen Z-R, Huang J-B, Yang S-L, Hong F-F (2022) Role of cholinergic signaling in Alzheimer's disease. Molecules 27:1816. https://doi.org/10.3390/molecules27061816
Cheng B, Song L, Chen F (2024) Huperzine alkaloids: forty years of total syntheses. Nat Prod Rep 41:59–84. https://doi.org/10.1039/d3np00029j
Chonpathompikunlert P, Wattanathorn J, Muchimapura S (2010) Piperine, the main alkaloid of Thai black pepper, protects against neurodegeneration and cognitive impairment in animal model of cognitive deficit like condition of Alzheimer's disease. Food Chem Toxicol 48:798–802. https://doi.org/10.1016/j.fct.2009.12.009
Choudhary B, Shetty A, Langade D (2015) Efficacy of Ashwagandha (Withania somnifera [L.] Dunal) in improving cardiorespiratory endurance in healthy athletic adults. AYU Int Q J Res Ayurveda 36:63. https://doi.org/10.4103/0974-8520.169002
Costa LG, Garrick JM, Roquè PJ, Pellacani C (2016) Mechanisms of neuroprotection by quercetin: counteracting oxidative stress and more. Oxidative Med Cell Longev 2016:1–10. https://doi.org/10.1155/2016/2986796
Cox KH, Pipingas A, Scholey AB (2014) Investigation of the effects of solid lipid curcumin on cognition and mood in a healthy older population. J Psychopharmacol (Oxf) 29:642–651. https://doi.org/10.1177/0269881114552744
Cui M, You T, Zhao Y, Liu R, Guan Y, Liu J-R, Liu X, Wang X, Dong Q (2023) Ginkgo biloba extract EGb 761® improves cognition and overall condition after ischemic stroke: results from a pilot randomized trial. Front Pharmacol 14. https://doi.org/10.3389/fphar.2023.1147860

Dadlani VG, Pawar HA, Tripathi PK (2023) Huperzine-based derivatives: design, synthesis, and Anti-Alzheimer activity. In: Sharma A, Modi GP (eds) Natural product-based synthetic drug molecules in Alzheimer's disease. Springer Nature Singapore, Singapore, pp 213–244

Daneman R, Prat A (2015) The Blood–Brain Barrier. Cold Spring Harb Perspect Biol 7:a020412. https://doi.org/10.1101/cshperspect.a020412

Daniele S, Giacomelli C, Martini C (2018) Brain ageing and neurodegenerative disease: the role of cellular waste management. Biochem Pharmacol 158:207–216. https://doi.org/10.1016/j.bcp.2018.10.030

Das SS, Gopal PM, Thomas JV, Mohan MC, Thomas SC, Maliakel BP, Krishnakumar IM, Baby (2023) Influence of CurQfen®-curcumin on cognitive impairment: a randomized, double-blinded, placebo-controlled, 3-arm, 3-sequence comparative study. Front Dement 2. https://doi.org/10.3389/frdem.2023.1222708

Dauer W, Przedborski S (2003) Parkinson's disease: mechanisms and models. Neuron 39:889–909. https://doi.org/10.1016/s0896-6273(03)00568-3

Davie CA (2008) A review of Parkinson's disease. Br Med Bull 86:109–127. https://doi.org/10.1093/bmb/ldn013

Dehghani F, Sezavar Seyedi Jandaghi SH, Janani L, Sarebanhassanabadi M, Emamat H, Vafa M (2020) Effects of quercetin supplementation on inflammatory factors and quality of life in post-myocardial infarction patients: a double blind, placebo-controlled, randomized clinical trial. Phytother Res 35:2085–2098. https://doi.org/10.1002/ptr.6955

Dengiz AN, Kershaw P (2004) The clinical efficacy and safety of galantamine in the treatment of Alzheimer's disease. CNS Spectr 9:377–392. https://doi.org/10.1017/s1092852900009378

Dhalaria R, Verma R, Kumar D, Puri S, Tapwal A, Kumar V, Nepovimova E, Kuca K (2020) Bioactive compounds of edible fruits with their anti-aging properties: a comprehensive review to prolong human life. Antioxidants 9:1123. https://doi.org/10.3390/antiox9111123

Dugger BN, Dickson DW (2017) Pathology of neurodegenerative diseases. Cold Spring Harb Perspect Biol 9:a028035. https://doi.org/10.1101/cshperspect.a028035

Elufioye TO, Berida TI, Habtemariam S (2017) Plants-derived neuroprotective agents: cutting the cycle of cell death through multiple mechanisms. Evid Based Complement Alternat Med 2017:3574012. https://doi.org/10.1155/2017/3574012

Esatbeyoglu T, Huebbe P, Ernst IMA, Chin D, Wagner AE, Rimbach G (2012) Curcumin-from molecule to biological function. Angew Chem Int Ed 51:5308–5332. https://doi.org/10.1002/anie.201107724

Feng H, Xue M, Deng H, Cheng S, Hu Y, Zhou C (2022) Ginsenoside and its therapeutic potential for cognitive impairment. Biomolecules 12:1310. https://doi.org/10.3390/biom12091310

Ferreres F, Taveira M, Pereira DM, Valentão P, Andrade PB (2010) Tomato (Lycopersicon esculentum) seeds: new flavonols and cytotoxic effect. J Agric Food Chem 58:2854–2861. https://doi.org/10.1021/jf904015f

Finley JW, Gao S (2017) A perspective on *Crocus sativus* L. (Saffron) constituent Crocin: a potent water-soluble antioxidant and potential therapy for Alzheimer's disease. J Agric Food Chem 65:1005–1020. https://doi.org/10.1021/acs.jafc.6b04398

Fišar Z (2022) Linking the amyloid, tau, and mitochondrial hypotheses of Alzheimer's disease and identifying promising drug targets. Biomolecules 12:1676. https://doi.org/10.3390/biom12111676

Friedli MJ, Inestrosa NC (2021) Huperzine a and its neuroprotective molecular signaling in Alzheimer's disease. Molecules 26:6531. https://doi.org/10.3390/molecules26216531

Frost ED, Shi SX, Byroju VV, Rissardo JP, Donlon J, Vigilante N, Murray BP, Walker IM, McGarry A, Ferraro TN, Hanafy KA, Echeverria V, Mitrev L, Kling MA, Krishnaiah B, Lovejoy DB, Rahman S, Stone TW, Koola MM (2024) Galantamine-Memantine combination in the treatment of Parkinson's disease dementia. Brain Sci 14:1163–1163. https://doi.org/10.3390/brainsci14121163

Fu K, Feng C, Shao L, Mei L, Cao R (2021) Tanshinone IIA exhibits anti-inflammatory and anti-oxidative effects in LPS-stimulated bovine endometrial epithelial cells by activating the Nrf2 signaling pathway. Res Vet Sci 136:220–226. https://doi.org/10.1016/j.rvsc.2021.03.004

Gajender U, Mazumder A, Sharma A, Azad MAK (2023) A comprehensive review of the pharmacological importance of dietary flavonoids as hepatoprotective agents. Evid Based Complement Alternat Med 2023:e4139117. https://doi.org/10.1155/2023/4139117

Ghimire S, Subedi L, Acharya N, Gaire BP (2021) Moringa oleifera: a tree of life as a promising medicinal plant for neurodegenerative diseases. J Agric Food Chem 69:14358–14371. https://doi.org/10.1021/acs.jafc.1c04581

Ghosh S, Singha PS, Ghosh D (2023) Neuroprotective compounds from three common medicinal plants of West Bengal, India: a mini review. Explor Neurosci 2:307–317. https://doi.org/10.37349/en.2023.00030

Ghosh A, Khanam N, Nath D (2024) Solid lipid nanoparticle: a potent vehicle of the kaempferol for brain delivery through the blood-brain barrier in the focal cerebral ischemic rat. Chem Biol Interact 397:111084. https://doi.org/10.1016/j.cbi.2024.111084

Golde TE (2022) Disease-modifying therapies for Alzheimer's disease: more questions than answers. Neurotherapeutics 19. https://doi.org/10.1007/s13311-022-01201-2

Grewal AK, Singh TG, Sharma D, Sharma V, Singh M, Rahman MH, Abdel-Daim MM (2021) Mechanistic insights and perspectives involved in neuroprotective action of quercetin. Biomedicine and Pharmacotherapy, 140:111729

Grotewold E (2006) The science of flavonoids. Springer Nature

Guilherme ML, Osman N, Lamas CB, Laurindo LF, Pomini KT, Guissoni LM, Pereira E, Direito R, Tanaka M, Barbalho SM (2024) Ginkgo biloba: a leaf of hope in the fight against Alzheimer's dementia: clinical trial systematic review. Antioxidants 13:651–651. https://doi.org/10.3390/antiox13060651

Gunasekaran T, Haile T, Nigusse T, Dhanaraju MD (2014) Nanotechnology: an effective tool for enhancing bioavailability and bioactivity of phytomedicine. Asian Pac J Trop Biomed 4:S1–S7. https://doi.org/10.12980/APJTB.4.2014C980

Gupta A, Singh AK, Kumar R, Jamieson S, Pandey AK, Bishayee A (2021) Neuroprotective potential of ellagic acid: a critical review. Adv Nutr 12:1211–1238. https://doi.org/10.1093/advances/nmab007

Gutsche LC, Dörfler J, Hübner J (2025) Curcumin as a complementary treatment in oncological therapy: a systematic review. Eur J Clin Pharmacol 81:1–33. https://doi.org/10.1007/s00228-024-03764-9

Hajialyani M, Hosein Farzaei M, Echeverría J, Nabavi SM, Uriarte E, Sobarzo-Sánchez E (2019) Hesperidin as a neuroprotective agent: a review of animal and clinical evidence. Molecules 24:648. https://doi.org/10.3390/molecules24030648

Han X, Zhang Y, Zhang L, Zhuang Y, Wang Y (2025) Efficacy and molecular mechanisms of hesperidin in mitigating Alzheimer's disease: a systematic review. Eur J Med Chem 283:117144. https://doi.org/10.1016/j.ejmech.2024.117144

Hegde M, Girisa S, BharathwajChetty B, Vishwa R, Kunnumakkara AB (2023) Curcumin formulations for better bioavailability: what we learned from clinical trials thus far? ACS Omega 8:10713–10746. https://doi.org/10.1021/acsomega.2c07326

Heo HJ, Lee CY (2005) Strawberry and its anthocyanins reduce oxidative stress-induced apoptosis in PC12 cells. J Agric Food Chem 53:1984–1989. https://doi.org/10.1021/jf048616l

Herraiz T, Guillén H (2018) Monoamine oxidase-a inhibition and associated antioxidant activity in plant extracts with potential antidepressant actions. Biomed Res Int 2018:1–10. https://doi.org/10.1155/2018/4810394

Herrera TES, Tello IPS, Mustafa MA, Jamil NY, Alaraj M, Atiyah Altameem KK, Alasheqi MQ, Hamoody A-HM, Alkhafaji AT, Shakir MN, Alshahrani MY, Alawadi A (2025) Kaempferol: unveiling its anti-inflammatory properties for therapeutic innovation. Cytokine 186:156846. https://doi.org/10.1016/j.cyto.2024.156846

Hippius H, Neundörfer G (2003) The discovery of Alzheimer's disease. Dialogues Clin Neurosci 5:101–108. https://doi.org/10.31887/dcns.2003.5.1

Hort J, Duning T, Hoerr R (2023) Ginkgo biloba extract EGb 761 in the treatment of patients with mild neurocognitive impairment: a systematic review. Neuropsychiatr Dis Treat 19:647–660. https://doi.org/10.2147/ndt.s401231

Huang S, Meng N, Liu Z, Guo L, Dong L, Li B, Ye Q (2018) Neuroprotective effects of Taraxacum officinale Wigg. Extract on glutamate-induced oxidative stress in HT22 cells via HO-1/Nrf2 pathways. Nutrients 10:926. https://doi.org/10.3390/nu10070926

Hussain G, Rasul A, Anwar H, Aziz N, Razzaq A, Wei W, Ali M, Li J, Li X (2018) Role of plant derived alkaloids and their mechanism in neurodegenerative disorders. Int J Biol Sci 14:341–357. https://doi.org/10.7150/ijbs.23247

Indira Priyadarsini K (2013) Chemical and structural features influencing the biological activity of curcumin. Curr Pharm Des 19:2093–2100. https://doi.org/10.2174/138161213805289228

Ingale SP, Kasture SB (2014) Antioxidant and antiparkinsonian activity of Passiflora incarnata leaves. Oriental Pharmacy and Experimental Medicine 14(3):231–236

Isaev NK, Chetverikov NS, Stelmashook EV, Genrikhs EE, Khaspekov LG, Illarioshkin SN (2020) Thymoquinone as a potential neuroprotector in acute and chronic forms of cerebral pathology. Biochem Mosc 85:167–176. https://doi.org/10.1134/S0006297920020042

Islam Z, Islam SMR, Hossen F, Mahtab-ul-Islam K, Hasan MR, Karim R (2021) Moringa oleifera is a prominent source of nutrients with potential health benefits. Int J Food Sci 2021:6627265. https://doi.org/10.1155/2021/6627265

Jafari A, Abbastabar M, Alaghi A, Heshmati J, Crowe FL, Sepidarkish M (2024) Curcumin on human health: a comprehensive systematic review and meta-analysis of 103 randomized controlled trials. Phytother Res 38. https://doi.org/10.1002/ptr.8340

Jang S-W, Liu X, Yepes M, Shepherd KR, Miller GW, Liu Y, Wilson WD, Xiao G, Blanchi B, Sun YE, Ye K (2010) A selective TrkB agonist with potent neurotrophic activities by 7,8-dihydroxyflavone. Proc Natl Acad Sci USA 107:2687–2692. https://doi.org/10.1073/pnas.0913572107

Jiao F, Tan Z, Yu Z, Zhou B, Meng L, Shi X (2022) The phytochemical and pharmacological profile of taraxasterol. Front Pharmacol 13:927365. https://doi.org/10.3389/fphar.2022.927365

Jimenez-Sanchez M, Licitra F, Underwood BR, Rubinsztein DC (2016) Huntington's disease: mechanisms of pathogenesis and therapeutic strategies. Cold Spring Harb Perspect Med 7:a024240. https://doi.org/10.1101/cshperspect.a024240

Johnson JJ, Nihal M, Siddiqui IA, Scarlett CO, Bailey HH, Mukhtar H, Ahmad N (2011) Enhancing the bioavailability of resveratrol by combining it with piperine. Mol Nutr Food Res 55:1169–1176. https://doi.org/10.1002/mnfr.201100117

Joseph SV, Edirisinghe I, Burton-Freeman BM (2014) Berries: anti-inflammatory effects in humans. J Agric Food Chem 62:3886–3903. https://doi.org/10.1021/jf4044056

Juárez Olguín H, Calderón Guzmán D, Hernández García E, Barragán Mejía G (2015) The role of dopamine and its dysfunction as a consequence of oxidative stress. Oxidative Med Cell Longev 2016:1–13. https://doi.org/10.1155/2016/9730467

Jung CH, Ahn J, Heo SH, Ha T-Y (2014) Eleutheroside E, an active compound from Eleutherococcus senticosus, regulates adipogenesis in 3T3-L1 cells. Food Sci Biotechnol 23:889–893. https://doi.org/10.1007/s10068-014-0119-z

Jurcau A (2022) Molecular pathophysiological mechanisms in Huntington's disease. Biomedicine 10:1432. https://doi.org/10.3390/biomedicines10061432

Kamatham PT, Shukla R, Khatri DK, Vora LK (2024) Pathogenesis, diagnostics, and therapeutics for Alzheimer's disease: breaking the memory barrier. Ageing Res Rev 101:102481–102481. https://doi.org/10.1016/j.arr.2024.102481

Katila N, Duwa R, Bhurtel S, Khanal S, Maharjan S, Jeong JH, Yook S (2022) Enhancement of blood–brain barrier penetration and the neuroprotective effect of resveratrol. J of Controlled Release 346:1–19

Kesidou E, Theotokis P, Damianidou O, Boziki M, Konstantinidou N, Taloumtzis C, Sintila S-A, Grigoriadis P, Evangelopoulos ME, Bakirtzis C, Simeonidou C (2023) CNS ageing in health and neurodegenerative disorders. J Clin Med 12:2255. https://doi.org/10.3390/jcm12062255

Khan H, Marya AS, Kamal MA, Patel S (2018) Flavonoids as acetylcholinesterase inhibitors: current therapeutic standing and future prospects. Biomed Pharmacother 101:860–870. https://doi.org/10.1016/j.biopha.2018.03.007

Khazdair MR, Anaeigoudari A, Hashemzehi M, Mohebbati R (2019) Neuroprotective potency of some spice herbs, a literature review. J Tradit Complement Med 9:98–105. https://doi.org/10.1016/j.jtcme.2018.01.002

Kim M, Kim S-H, Yang W (2014) Mechanisms of action of phytochemicals from medicinal herbs in the treatment of Alzheimer's disease. Planta Med 80:1249–1258. https://doi.org/10.1055/s-0034-1383038

Kisku A, Nishad A, Agrawal S, Paliwal R, Datusalia AK, Gupta G, Singh SK, Dua K, Sulakhiya K (2024) Recent developments in intranasal drug delivery of nanomedicines for the treatment of neuropsychiatric disorders. Front Med 11:1463976. https://doi.org/10.3389/fmed.2024.1463976

Kosmopoulou D, Lafara M-P, Adamantidi T, Ofrydopoulou A, Grabrucker AM, Tsoupras A (2024) Neuroprotective benefits of Rosmarinus officinalis and its bioactives against Alzheimer's and Parkinson's diseases. Appl Sci 14:6417. https://doi.org/10.3390/app14156417

Kumar GP, Khanum F (2012) Neuroprotective potential of phytochemicals. Pharmacogn Rev 6:81. https://doi.org/10.4103/0973-7847.99898

Kumar S, Pandey AK (2013) Chemistry and biological activities of flavonoids: an overview. Sci World J 2013:1–16. https://doi.org/10.1155/2013/162750

Kumar A, Dogra S, Prakash A (2009) Neuroprotective effects of *Centella asiatica* against intracerebroventricular colchicine-induced cognitive impairment and oxidative stress. Int J Alzheimers Dis 2009:1–8. https://doi.org/10.4061/2009/972178

Kumar RR, Vijayakumar R, Kumar RS, Veerappan C, Jagannathan S, Gandhi PR (2010) Isolation purification and analysis of free radical scavenging activity of flavonids and alkaloids from Indigofera longeracemosa Linn. J Med Sci Res 1:63-68

Kushwah S, Maurya NS, Kushwaha S, Scotti L, Chawade A, Mani A (2023) Herbal therapeutics for Alzheimer's disease: ancient Indian medicine system from the modern viewpoint. Curr Neuropharmacol 21:764–776. https://doi.org/10.2174/1570159x21666230216094353

Lazarova M, Tancheva L, Alexandrova A, Tsvetanova E, Georgieva A, Stefanova M, Tsekova D, Vezenkov L, Kalfin R, Uzunova D, Petkova-Kirova P (2021) Effects of new galantamine derivatives in a scopolamine model of dementia in mice. J Alzheimers Dis 84:671–690. https://doi.org/10.3233/JAD-215165

Lee TH, Hong S-M, Yoon DH, Kim SY, Kim CS, Lee KR (2022) Procyanidins and phlobatannins from the twigs of *Rosa multiflora* and their neuroprotective activity. J Nat Prod 85:917–926. https://doi.org/10.1021/acs.jnatprod.1c01033

Li WH, Cheng X, Yang YL, Liu M, Zhang SS, Wang YH, Du GH (2019) Kaempferol attenuates neuroinflammation and blood brain barrier dysfunction to improve neurological deficits in cerebralischemia/reperfusion rats. Brain Research, 1722:146361

Li X, Huang W, Tan R, Xu C, Chen X, Li S, Liu Y, Qiu H, Cao H, Cheng Q (2023) The benefits of hesperidin in central nervous system disorders, based on the neuroprotective effect. Biomed Pharmacother 159:114222. https://doi.org/10.1016/j.biopha.2023.114222

Liu X, Obianyo O, Chan CB, Huang J, Xue S, Yang JJ, Zeng F, Goodman M, Ye K (2014) Biochemical and biophysical investigation of the brain-derived neurotrophic factor mimetic 7,8-Dihydroxyflavone in the binding and activation of the TrkB receptor. J Biol Chem 289:27571–27584. https://doi.org/10.1074/jbc.m114.562561

Liu ZSJ, Truong TTT, Bortolasci CC, Spolding B, Panizzutti B, Swinton C, Kim JH, Hernández D, Kidnapillai S, Gray L, Berk M, Dean OM, Walder K (2024) The potential of baicalin to enhance neuroprotection and mitochondrial function in a human neuronal cell model. Mol Psychiatry 29:2487–2495. https://doi.org/10.1038/s41380-024-02525-5

Livingston G, Huntley J, Sommerlad A, Ames D, Ballard C, Banerjee S, Brayne C, Burns A, Cohen-Mansfield J, Cooper C, Costafreda SG, Dias A, Fox N, Gitlin LN, Howard R, Kales HC, Kivimäki M, Larson EB, Ogunniyi A, Orgeta V, Ritchie K, Rockwood K, Sampson EL, Samus Q, Schneider LS, Selbæk G, Teri L, Mukadam N (2020) Dementia prevention, intervention, and care: 2020 report of the lancet commission. Lancet 396:413–446. https://doi.org/10.1016/S0140-6736(20)30367-6

Lobine D, Sadeer N, Jugreet S, Suroowan S, Keenoo BS, Imran M, Venugopala KN, Ibrahim FM, Zengin G, Mahomoodally MF (2021) Potential of medicinal plants as neuroprotective and therapeutic properties against amyloid-β-related toxicity, and glutamate-induced excitotoxicity in human neural cells. Curr Neuropharmacol 19:1416–1441. https://doi.org/10.2174/1570159X19666210412095251

Löscher W, Potschka H (2005) Blood-brain barrier active efflux transporters: ATP-binding cassette gene family. NeuroRx 2:86–98. https://doi.org/10.1602/neurorx.2.1.86

Madhu SU, Hariprasad MG, Nagalakshmi NC, Pardhe HA (2020) A systematic review on mechanism of neurodegeneration. IP Indian J Neurosci 6:6–13. https://doi.org/10.18231/j.ijn.2020.002

Maiti P, Dunbar G (2018) Use of curcumin, a natural polyphenol for targeting molecular pathways in treating age-related neurodegenerative diseases. Int J Mol Sci 19:1637. https://doi.org/10.3390/ijms19061637

Malík M, Tlustoš P (2023) Nootropic herbs, shrubs, and trees as potential cognitive enhancers. Plants 12:1364. https://doi.org/10.3390/plants12061364

Manolova Y, Deneva V, Antonov L, Drakalska E, Momekova D, Lambov N (2014) The effect of the water on the curcumin tautomerism: a quantitative approach. Spectrochim Acta A Mol Biomol Spectrosc 132:815–820. https://doi.org/10.1016/j.saa.2014.05.096

Martins P, Roberto O, Lívia F, Junior JO, Gomes K, Rodrigues I, Cerqueira IT, Bueno NB, Oliveira M, Moura FA (2024) Effect of curcumin plus piperine on redox imbalance, fecal calprotectin and cytokine levels in inflammatory bowel disease patients: a randomized, double-blind, placebo-controlled clinical trial. Pharmaceuticals 17:849–849. https://doi.org/10.3390/ph17070849

Mashamaite CV, Ramatsitsi MN, Manyevere A (2024) Moringa oleifera Lam.: a versatile climate-smart plant for nutritional security and therapeutic usage in semi-arid regions. J Agric Food Res 16:101217–101217. https://doi.org/10.1016/j.jafr.2024.101217

Mashayekh A, Pham DL, Yousem DM, Dizon M, Barker PB, Lin DDM (2011) Effects of Ginkgo biloba on cerebral blood flow assessed by quantitative MR perfusion imaging: a pilot study. Neuroradiology 53:185–191. https://doi.org/10.1007/s00234-010-0790-6

Mishra S, Palanivelu K (2008) The effect of curcumin (turmeric) onAlzheimer's disease: an overview. Ann Indian Acad Neurol 11:13. https://doi.org/10.4103/0972-2327.40220

Moise G, Jîjie A-R, Moacă E-A, Predescu I-A, Dehelean CA, Hegheş A, Vlad DC, Popescu R, Vlad CS (2024) Plants' impact on the human brain—exploring the neuroprotective and neurotoxic potential of plants. Pharmaceuticals 17:1339. https://doi.org/10.3390/ph17101339

Moreira J, Machado M, Dias-Teixeira M, Ferraz R, Delerue-Matos C, Grosso C (2023) The neuroprotective effect of traditional Chinese medicinal plants—a critical review. Acta Pharm Sin B 13:3208–3237. https://doi.org/10.1016/j.apsb.2023.06.009

Moussa C, Hebron M, Huang X, Ahn J, Rissman RA, Aisen PS, Turner RS (2017) Resveratrol regulates neuro-inflammation and induces adaptive immunity in Alzheimer's disease. J Neuroinflammation 14. https://doi.org/10.1186/s12974-016-0779-0

Murunikkara V, Pragasam SJ, Kodandaraman G, Sabina EP, Rasool M (2012) Anti-inflammatory effect of piperine in adjuvant-induced arthritic rats—a biochemical approach. Inflammation 35:1348–1356. https://doi.org/10.1007/s10753-012-9448-3

Nadeem MS, Kazmi I, Ullah I, Muhammad K, Anwar F (2021) Allicin, an antioxidant and neuroprotective agent, ameliorates cognitive impairment. Antioxidants 11:87. https://doi.org/10.3390/antiox11010087

Nagori K, Nakhate KT, Yadav K, Ajazuddin PM (2023) Unlocking the therapeutic potential of medicinal plants for Alzheimer's disease: preclinical to clinical trial insights. Future Pharmacol 3:877–907. https://doi.org/10.3390/futurepharmacol3040053

Naoi M, Shamoto-Nagai M, Maruyama W (2019) Neuroprotection of multifunctional phytochemicals as novel therapeutic strategy for neurodegenerative disorders: antiapoptotic and antiamyloidogenic activities by modulation of cellular signal pathways. Future Neurol 14:FNL9. https://doi.org/10.2217/fnl-2018-0028

Narasimhan KKS, Jayakumar D, Velusamy P, Srinivasan A, Mohan T, Ravi DB, Uthamaraman S, Sathyamoorthy YK, Rajasekaran NS, Periandavan K (2019) *Morinda citrifolia* and its active principle Scopoletin mitigate protein aggregation and neuronal apoptosis through augmenting the DJ-1/Nrf2/ARE signaling pathway. Oxidative Med Cell Longev 2019:1–13. https://doi.org/10.1155/2019/2761041

Nasb M, Tao W, Chen N (2024) Alzheimer's disease puzzle: delving into pathogenesis hypotheses. Aging Dis 15. https://doi.org/10.14336/ad.2023.0608

Navarro-Hortal MD, Romero-Márquez JM, Esteban-Muñoz A, Sánchez-González C, Rivas-García L, Llopis J, Cianciosi D, Giampieri F, Sumalla-Cano S, Battino M, Quiles JL (2022) Strawberry (Fragaria × ananassa cv. Romina) methanolic extract attenuates Alzheimer's beta amyloid production and oxidative stress by SKN-1/NRF and DAF-16/FOXO mediated mechanisms in C. elegans. Food Chem 372:131272. https://doi.org/10.1016/j.foodchem.2021.131272

Nezhad M, Rasoulizadeh Z, Shabgah AG, Vakili-Ghartavol R, Sargazi G, Navashenaq JG (2024) Exploring the mechanisms of kaempferol in neuroprotection: implications for neurological disorders. Cell Biochem Funct 42:e3964. https://doi.org/10.1002/cbf.3964

Nguyen Duc H (2023) Neurotherapeutic effects of quercetin and its metabolite compounds on cognitive impairment and Parkinson's disease: an in silico study. European J of Drug Metabolism andPharmacokinetics, 48(2):151-169

Nitrini R (2023) The past, present and future of Alzheimer's disease – part 1: the past. Arq Neuropsiquiatr 81:1070–1076. https://doi.org/10.1055/s-0043-1777722

Nunes YC, Mendes NM, Pereira E, Chehadi AC, Lamas CB, Haber S, Bueno S, Araújo AC, Strozze VC, Catharin RC, Laurindo LF, Tanaka M, Barbalho SM, José M (2024) Curcumin: a golden approach to healthy aging: a systematic review of the evidence. Nutrients 16:2721–2721. https://doi.org/10.3390/nu16162721

Olasehinde TA, Ekundayo TC, Ijabadeniyi OA, Olaniran AO (2024) The impact of hesperidin on cognitive deficit and neurobehavioural disorders: a systematic review and meta-analysis of preclinical individual studies. Curr Behav Neurosci Rep 11. https://doi.org/10.1007/s40473-024-00284-9

Pandey V, Chakrabarty D (2024) Phytochemical importance of medicinal plants as potential sources against neurodegenerative diseases. In: Kumar L, Bharadvaja N, Singh R, Anand R (eds) Medicinal and aromatic plants. Springer Nature Switzerland, Cham, pp 111–124

Pandey A, Kumar Gupt R, Srivastava R (2011) Curcumin-the yellow magic. Asian J Appl Sci 4:343–354. https://doi.org/10.3923/ajaps.2011.343.354

Pandya JD, Musyaju S, Modi HR, Okada-Rising SL, Bailey ZS, Scultetus AH, Shear DA (2024) Intranasal delivery of mitochondria targeted neuroprotective compounds for traumatic brain injury: screening based on pharmacological and physiological properties. J Transl Med 22:167. https://doi.org/10.1186/s12967-024-04908-2

Pannu A, Sharma PC, Thakur VK, Goyal RK (2021) Emerging role of flavonoids as the treatment of depression. Biomolecules 11:1825. https://doi.org/10.3390/biom11121825

Papagiouvannis G, Theodosis-Nobelos P, Rekka EA (2024) A review on therapeutic strategies against Parkinson's disease: current trends and future perspectives. Mini-Rev Med Chem 24. https://doi.org/10.2174/0113895575303788240606054620

Pardridge WM (2012) Drug transport across the blood–brain barrier. J Cereb Blood Flow Metab 32:1959–1972. https://doi.org/10.1038/jcbfm.2012.126

Pareek A, Pant M, Gupta MM, Kashania P, Ratan Y, Jain V, Pareek A, Chuturgoon AA (2023) Moringa oleifera: an updated comprehensive review of its pharmacological activities, ethnomedicinal, phytopharmaceutical formulation, clinical, phytochemical, and toxicological aspects. Int J Mol Sci 24:2098. https://doi.org/10.3390/ijms24032098

Patel SS, Raghuwanshi R, Masood M, Acharya A, Jain SK (2018) Medicinal plants with acetylcholinesterase inhibitory activity. Rev Neurosci 29:491–529. https://doi.org/10.1515/revneuro-2017-0054

Pavić V, Jakovljević M, Molnar M, Jokić S (2019) Extraction of carnosic acid and carnosol from sage (Salvia officinalis L.) leaves by supercritical fluid extraction and their antioxidant and antibacterial activity. Plants 8:16. https://doi.org/10.3390/plants8010016

Pawar KS, Mastud RN, Pawar SK, Pawar SS, Bhoite RR, Bhoite RR, Kulkarni MV, Deshpande AR (2021) Oral curcumin with piperine as adjuvant therapy for the treatment of COVID-19: a randomized clinical trial. Front Pharmacol 12. https://doi.org/10.3389/fphar.2021.669362

Pei J, Palanisamy CP, Natarajan PM, Umapathy VR, Roy JR, Srinivasan GP, Panagal M, Jayaraman S (2024) Curcumin-loaded polymeric nanomaterials as a novel therapeutic strategy for Alzheimer's disease: a comprehensive review. Ageing Res Rev 99:102393–102393. https://doi.org/10.1016/j.arr.2024.102393

Perry E, Howes M-JR (2010) Medicinal plants and dementia therapy: herbal hopes for brain aging? CNS Neurosci Ther 17:683–698. https://doi.org/10.1111/j.1755-5949.2010.00202.x

Phogat A, Singh J, Malik V (2024) Pharmacological effects of Berberine – a Chinese medicine, against xenobiotics toxicity. Pharmacol Res - Mod Chin Med 13:100507. https://doi.org/10.1016/j.prmcm.2024.100507

Pilipović K, Jurišić Grubešić R, Dolenec P, Kučić N, Juretić L, Mršić-Pelčić J (2023) Plant-based antioxidants for prevention and treatment of neurodegenerative diseases: phytotherapeutic potential of Laurus nobilis, Aronia melanocarpa, and Celastrol. Antioxidants 12:746. https://doi.org/10.3390/antiox12030746

Qiao A, Wang Y, Zhang W, He X (2015) Neuroprotection of brain-targeted bioactive dietary Artoindonesianin O (AIO) from mulberry on rat neurons as a novel intervention for Alzheimer's disease. J Agric Food Chem 63:3687–3693. https://doi.org/10.1021/acs.jafc.5b00396

Qu Y, Shen Y, Teng L, Huang Y, Yang Y, Jian X, Fan S, Wu P, Fu Q (2022) Chicoric acid attenuates tumor necrosis factor-α-induced inflammation and apoptosis via the Nrf2/HO-1, PI3K/AKT and NF-κB signaling pathways in C28/I2 cells and ameliorates the progression of osteoarthritis in a rat model. Int Immunopharmacol 111:109129. https://doi.org/10.1016/j.intimp.2022.109129

Rajendran V, Ponnuraj K (2024) High-throughput virtual screening and molecular dynamics simulation reveals NPC170742 a novel chalconoid compound as a potential inhibitor of D-glycero-D-manno-heptose-1,7-bisphosphate 7-phosphatase in *Helicobacter pylori*. J Biomol Struct Dyn 42:10911–10921. https://doi.org/10.1080/07391102.2023.2259483

Ramasamy S, Rajendran V, Rangaraj R, Chinnayan V, Palanisamy R, Prasannan D (2012) Effect of Elaeocarpus sphaericus in freund's complete adjuvant induced rheumatoid arthritis in albino rats. Indo Glob Res J Pharm Sci 2:378–382

Rana L, Harwansh RK, Deshmukh R (2025) Recent updates on phytopharmaceuticals-based novel phytosomal systems and their clinical trial status: a translational perspective. Crit Rev Ther Drug Carr Syst 42:1–54. https://doi.org/10.1615/CritRevTherDrugCarrierSyst.v42.i1.10

Raza SS, Khan MM, Ashafaq M, Ahmad A, Khuwaja G, Khan A, Siddiqui MS, Safhi MM, Islam F (2011) Silymarin protects neurons from oxidative stress associated damages in focal cerebral ischemia: a behavioral, biochemical and immunohistological study in Wistar rats. J Neurol Sci 309:45–54. https://doi.org/10.1016/j.jns.2011.07.035

Reay JL, Kennedy DO, Scholey AB (2005) Single doses of Panax ginseng (G115) reduce blood glucose levels and improve cognitive performance during sustained mental activity. J Psychopharmacol (Oxf) 19:357–365. https://doi.org/10.1177/0269881105053286

Rebai O, Belkhir M, Boujelben A, Fattouch S, Amri M (2017) Morus alba leaf extract mediates neuroprotection against glyphosate-induced toxicity and biochemical alterations in the brain. Environ Sci Pollut Res 24:9605–9613. https://doi.org/10.1007/s11356-017-8584-6

Rehman MU, Wali AF, Ahmad A, Shakeel S, Rasool S, Ali R, Rashid SM, Madkhali H, Ganaie MA, Khan R (2019) Neuroprotective strategies for neurological disorders by natural products: an update. Curr Neuropharmacol 17:247–267. https://doi.org/10.2174/1570159X16666180911124605

Ross CA, Tabrizi SJ (2011) Huntington's disease: from molecular pathogenesis to clinical treatment. Lancet Neurol 10:83–98. https://doi.org/10.1016/s1474-4422(10)70245-3

Saha P, Ahmad F (2024) Neuroprotective, anti-inflammatory and antifibrillogenic offerings by Emodin against Alzheimer's dementia: a systematic review. ACS Omega 9:7296. https://doi.org/10.1021/acsomega.3c07178

Said MM, Rabo MMA (2017) Neuroprotective effects of eugenol against aluminiuminduced toxicity in the rat brain. Arch Ind Hyg Toxicol 68:27–37. https://doi.org/10.1515/aiht-2017-68-2878

Saleem A, Afzal M, Naveed M, Makhdoom SI, Mazhar M, Aziz T, Khan AA, Kamal Z, Shahzad M, Alharbi M, Alshammari A (2022) HPLC, FTIR and GC-MS analyses of Thymus vulgaris phytochemicals executing in vitro and in vivo biological activities and effects on COX-1, COX-2 and gastric cancer genes computationally. Molecules 27:8512. https://doi.org/10.3390/molecules27238512

Salehi B, Calina D, Docea AO, Koirala N, Aryal S, Lombardo D, Pasqua L, Taheri Y, Marina Salgado Castillo C, Martorell M, Martins N, Iriti M, Suleria HAR, Sharifi-Rad J (2020) Curcumin's nanomedicine formulations for therapeutic application in neurological diseases. J Clin Med 9:430. https://doi.org/10.3390/jcm9020430

Salla M, Karaki N, El Kaderi B, Ayoub AJ, Younes S, Abou MN, Baksh S, El Khatib S (2024) Enhancing the bioavailability of resveratrol: combine it, derivatize it, or encapsulate it? Pharmaceutics 16:569–569. https://doi.org/10.3390/pharmaceutics16040569

Saravanan K, Kalaiarasi C, Kumaradhas P (2017) Understanding the conformational flexibility and electrostatic properties of curcumin in the active site of rhAChE via molecular docking, molecular dynamics, and charge density analysis. J Biomol Struct Dyn 35:3627–3647. https://doi.org/10.1080/07391102.2016.1264891

Saravanan K, Sivanandam M, Hunday G, Pavan MS, Kumaradhas P (2019) Exploring the different environments effect of piperine via combined crystallographic, QM/MM and molecular dynamics simulation study. J Mol Graph Model 92:280–295. https://doi.org/10.1016/j.jmgm.2019.07.019

Saravanan K, Hunday G, Kumaradhas P (2020) Binding and stability of indirubin-3-monoxime in the GSK3β enzyme: a molecular dynamics simulation and binding free energy study. J Biomol Struct Dyn 38:957–974. https://doi.org/10.1080/07391102.2019.1591301

Saravanan K, Karthikeyan S, Sugarthi S, Stephen AD (2021) Binding studies of known molecules with acetylcholinesterase and bovine serum albumin: a comparative view. Spectrochim Acta A Mol Biomol Spectrosc 259:119856. https://doi.org/10.1016/j.saa.2021.119856

Saravanan K, Sugarthi S, Suganya S, Kumaradhas P (2022) Probing the intermolecular interactions, binding affinity, charge density distribution and dynamics of silibinin in dual targets AChE and BACE1: QTAIM and molecular dynamics perspective. J Biomol Struct Dyn 40:12880–12894. https://doi.org/10.1080/07391102.2021.1977699

Schinkel AH, Smit JJM, Van Tellingen O, Beijnen JH, Wagenaar E, Van Deemter L, Mol CAAM, Van Der Valk MA, Robanus-Maandag EC, Te Riele HPJ, Berns AJM, Borst P (1994) Disruption of the mouse mdr1a P-glycoprotein gene leads to a deficiency in the blood-brain barrier and to increased sensitivity to drugs. Cell 77:491–502. https://doi.org/10.1016/0092-8674(94)90212-7

Shoba G, Joy D, Joseph T, Majeed M, Rajendran R, Srinivas P (1998) Influence of piperine on the pharmacokinetics of curcumin in animals and human volunteers. Planta Med 64:353–356. https://doi.org/10.1055/s-2006-957450

Siddique YH, Rahul AG, Afzal M, Varshney H, Gaur K, Subhan I, Mantasha I, Shahid M (2022) Beneficial effects of apigenin on the transgenic Drosophila model of Alzheimer's disease. Chem Biol Interact 366:110120. https://doi.org/10.1016/j.cbi.2022.110120

Singh DV, Godbole MM, Misra K (2012) A plausible explanation for enhanced bioavailability of P-gp substrates in presence of piperine: simulation for next generation of P-gp inhibitors. J Mol Model 19:227–238. https://doi.org/10.1007/s00894-012-1535-8

Singh CK, Liu X, Ahmad N (2015) Resveratrol, in its natural combination in whole grape, for health promotion and disease management. Ann N Y Acad Sci 1348:150–160. https://doi.org/10.1111/nyas.12798

Singh AP, Singh R, Verma SS, Rai V, Kaschula CH, Maiti P, Gupta SC (2019) Health benefits of resveratrol: evidence from clinical studies. Med Res Rev 39:1851–1891. https://doi.org/10.1002/med.21565

Som S, Antony J, Sp D, Ponnusankar S (2022) Neuroprotective role of Diosgenin, a NGF stimulator, against aβ (1–42) induced neurotoxicity in animal model of Alzheimer's disease. Metab Brain Dis 37:359–372. https://doi.org/10.1007/s11011-021-00880-8

Stavrakov G, Philipova I, Lukarski A, Atanasova M, Zheleva D, Zhivkova ZD, Ivanov S, Atanasova T, Konstantinov S, Doytchinova I (2020) Galantamine-curcumin hybrids as dual-

site binding acetylcholinesterase inhibitors. Molecules 25:3341. https://doi.org/10.3390/molecules25153341
Steinmetz JD et al (2024) Global, regional, and national burden of disorders affecting the nervous system, 1990–2021: a systematic analysis for the Global Burden of Disease Study 2021. Lancet Neurol 23. https://doi.org/10.1016/s1474-4422(24)00038-3
Stough C, Lloyd J, Clarke J, Downey L, Hutchison C, Rodgers T, Nathan P (2001) The chronic effects of an extract of Bacopa monniera (Brahmi) on cognitive function in healthy human subjects. Psychopharmacology 156:481–484. https://doi.org/10.1007/s002130100815
Subash S, Essa M, Al-Adawi S, Memon M, Manivasagam T, Akbar M (2014) Neuroprotective effects of berry fruits on neurodegenerative diseases. Neural Regen Res 9:1557. https://doi.org/10.4103/1673-5374.139483
Tapas KM, Joseph B, Bosung S, Kaspar K, Harry D, Manik G, Ajay K, Ishani G, Shreya M, Alonso H, Malcolm L, Marc JS, Robert CG, Volodymyr G, Istvan M (2024) Neuroprotective effect of 7,8-dihydroxyflavone in a mouse model of HIV-Associated Neurocognitive Disorder (HAND). J Neurosci Neurol Disord 8:090–105. https://doi.org/10.29328/journal.jnnd.1001101
Tehrani SD, Hosseini A, Shahzamani M, Heidari Z, Askari G, Majeed M, Sahebkar A, Bagherniya M (2024) Evaluation of the effectiveness of curcumin and piperine co-supplementation on inflammatory factors, cardiac biomarkers, atrial fibrillation, and clinical outcomes after coronary artery bypass graft surgery. Clin Nutr ESPEN 62:57–65. https://doi.org/10.1016/j.clnesp.2024.05.003
Timmins P (2024) Industry update: the latest developments in the field of therapeutic delivery, July 2024. Ther Deliv:1–10. https://doi.org/10.1080/20415990.2024.2414732
Tiwari A (2020) Piperine: a comprehensive review of methods of isolation, purification, and biological properties. Med Drug Discov 7:100027. https://doi.org/10.1016/j.medidd.2020.100027
Tripathi AK, Ray AK, Mishra SK (2022) Molecular and pharmacological aspects of piperine as a potential molecule for disease prevention and management: evidence from clinical trials. Beni-Suef Univ J Basic Appl Sci 11:16. https://doi.org/10.1186/s43088-022-00196-1
Trotta T, Porro C, Cianciulli A, Panaro MA (2022) Beneficial effects of spirulina consumption on brain health. Nutrients 14:676. https://doi.org/10.3390/nu14030676
Turner RS, Thomas RG, Craft S, van Dyck CH, Mintzer J, Reynolds BA, Brewer JB, Rissman RA, Raman R, Aisen PS (2015) A randomized, double-blind, placebo-controlled trial of resveratrol for Alzheimer disease. Neurology 85:1383–1391. https://doi.org/10.1212/wnl.0000000000002035
Uddin MS, Kabir MT, Niaz K, Jeandet P, Clément C, Mathew B, Rauf A, Rengasamy KRR, Sobarzo-Sánchez E, Ashraf GM, Aleya L (2020) Molecular insight into the therapeutic promise of flavonoids against Alzheimer's disease. Molecules 25:1267. https://doi.org/10.3390/molecules25061267
Ullah A, Munir S, Badshah SL, Khan N, Ghani L, Poulson BG, Emwas A-H, Jaremko M (2020) Important flavonoids and their role as a therapeutic agent. Molecules 25:5243. https://doi.org/10.3390/molecules25225243
Vauzour D, Martinsen A, Layé S (2015) Neuroinflammatory processes in cognitive disorders: is there a role for flavonoids and n-3 polyunsaturated fatty acids in counteracting their detrimental effects? Neurochem Int 89:63–74. https://doi.org/10.1016/j.neuint.2015.08.004
Velmurugan B, Rathinasamy B, Lohanathan B, Thiyagarajan V, Weng C-F (2018) Neuroprotective role of phytochemicals. Molecules 23:2485. https://doi.org/10.3390/molecules23102485
Vikal A, Maurya R, Bhowmik S, Khare S, Raikwar S, Patel P, Das Kurmi B (2024) Resveratrol: a comprehensive review of its multifaceted health benefits, mechanisms of action, and potential therapeutic applications in chronic disease. Pharmacol Res - Nat Prod 3:100047. https://doi.org/10.1016/j.prenap.2024.100047
Walker FO (2007) Huntington's disease. Lancet 369:218–228. https://doi.org/10.1016/s0140-6736(07)60111-1
Wang J, Song Y, Gao M, Bai X, Chen Z (2016) Neuroprotective effect of several phytochemicals and its potential application in the prevention of neurodegenerative diseases. Geriatrics 1:29. https://doi.org/10.3390/geriatrics1040029

Wang J, Shi C, Jiao W, Wang X, Sang Y (2024) Study on the neuroprotective and sedative effects of the combination of jujuboside B and spinosin from Ziziphi Spinosae Semen. J Funct Foods 119:106274. https://doi.org/10.1016/j.jff.2024.106274

Wankhede NL, Kale MB, Bawankule AK, Aglawe MM, Taksande BG, Trivedi RV, Umekar MJ, Jamadagni A, Walse P, Koppula S, Kopalli SR (2023) Overview on the polyphenol Avenanthramide in oats (Avena sativa Linn.) as regulators of PI3K signaling in the management of neurodegenerative diseases. Nutrients 15:3751. https://doi.org/10.3390/nu15173751

Wattanathorn J, Chonpathompikunlert P, Muchimapura S, Priprem A, Tankamnerdthai O (2008) Piperine, the potential functional food for mood and cognitive disorders. Food Chem Toxicol 46:3106–3110. https://doi.org/10.1016/j.fct.2008.06.014

Wei Y, Li H, Li Y, Zeng Y, Quan T, Leng Y, Chang E, Bai Y, Bian Y, Hou Y (2024) Advances of curcumin in nervous system diseases: the effect of regulating oxidative stress and clinical studies. Front Pharmacol 15:1496661. https://doi.org/10.3389/fphar.2024.1496661

Weinmann S, Roll S, Schwarzbach C, Vauth C, Willich SN (2010) Effects of Ginkgo biloba in dementia: systematic review and meta-analysis. BMC Geriatr 10. https://doi.org/10.1186/1471-2318-10-14

Worku B, Tolossa N (2024) A review on the neuroprotective effect of *Moringa oleifera*. Oxidative Med Cell Longev 2024. https://doi.org/10.1155/2024/7694516

Wu J, Tian Z, Wang B, Liu J, Bi R, Zhan N, Song D, He C, Zhao W (2024) Exploring resveratrol against Alzheimer's disease and Parkinson's disease through integrating network pharmacology, bioinformatics, and experimental validation strategy in vitro. Heliyon 10:e37908. https://doi.org/10.1016/j.heliyon.2024.e37908

Xing M, Li G, Liu Y, Yang L, Zhang Y, Zhang Y, Ding J, Lu M, Yu G, Hu G (2023) Fucoidan from Fucus vesiculosus prevents the loss of dopaminergic neurons by alleviating mitochondrial dysfunction through targeting ATP5F1a. Carbohydr Polym 303:120470. https://doi.org/10.1016/j.carbpol.2022.120470

Xu SS, Gao ZX, Weng Z, Du ZM, Xu WA, Yang JS, Zhang ML, Tong ZH, Fang YS, Chai XS (1995) Efficacy of tablet huperzine-A on memory, cognition, and behavior in Alzheimer's disease. Zhongguo Yao Li Xue Bao 16:391–395

Xu SL, Bi CWC, Choi RCY, Zhu KY, Miernisha A, Dong TTX, Tsim KWK (2013) Flavonoids induce the synthesis and secretion of neurotrophic factors in cultured rat astrocytes: a signaling response mediated by estrogen receptor. Evid Based Complement Alternat Med 2013:1–10. https://doi.org/10.1155/2013/127075

Yan F, Wang Y, He S, Ku S, Gu W, Ye L (2013) Transferrin-conjugated, fluorescein-loaded magnetic nanoparticles for targeted delivery across the blood–brain barrier. J Mater Sci Mater Med 24:2371–2379. https://doi.org/10.1007/s10856-013-4993-3

Yang Y, Yuan Y, Zhang G, Wang H, Chen Y-C, Liu Y, Tarolli CG, Crepeau D, Bukartyk J, Junna MR, Videnovic A, Ellis TD, Lipford MC, Dorsey R, Katabi D (2022) Artificial intelligence-enabled detection and assessment of Parkinson's disease using nocturnal breathing signals. Nat Med 28:1–9. https://doi.org/10.1038/s41591-022-01932-x

Yang Y, Yu L, Zhu T, Xu S, He J, Mao N, Liu Z, Wang D (2023) Neuroprotective effects of Lycium barbarum polysaccharide on light-induced oxidative stress and mitochondrial damage via the Nrf2/HO-1 pathway in mouse hippocampal neurons. Int J Biol Macromol 251:126315–126315. https://doi.org/10.1016/j.ijbiomac.2023.126315

Zangara A (2003) The psychopharmacology of huperzine A: an alkaloid with cognitive enhancing and neuroprotective properties of interest in the treatment of Alzheimer's disease. Pharmacol Biochem Behav 75:675–686. https://doi.org/10.1016/s0091-3057(03)00111-4

Zeraattalab-Motlagh S, Jayedi A, Shab-Bidar S (2021) The effects of resveratrol supplementation in patients with type 2 diabetes, metabolic syndrome, and nonalcoholic fatty liver disease: an umbrella review of meta-analyses of randomized controlled trials. Am J Clin Nutr. https://doi.org/10.1093/ajcn/nqab250

Zhang K, Wang H (2024) Development and validation of a highly sensitive UPLC–MS/MS method for the determination of Huperzine A in rat plasma. Biomed Chromatogr. https://doi.org/10.1002/bmc.6011

Zhang J, Zhang Y, Wang J, Xia Y, Zhang J, Chen L (2024) Recent advances in Alzheimer's disease: mechanisms, clinical trials and new drug development strategies. Signal Transduct Target Ther 9. https://doi.org/10.1038/s41392-024-01911-3

Zhao L, Luo T, Zhang H, Fan X, Zhang Q, Chen H (2024) Kaempferol enhances intestinal repair and inhibits the hyperproliferation of aging intestinal stem cells in Drosophila. Frontiers in Cell and DevelopmentalBiology, 12:1491740

Zhu J, Cui Y, Zhang J, Yan R, Su D, Zhao D, Wang A, Feng T (2024) Temporal trends in the prevalence of Parkinson's disease from 1980 to 2023: a systematic review and meta-analysis. Lancet Healthy Longev 5:e464–e479. https://doi.org/10.1016/s2666-7568(24)00094-1

Prospects and Challenges in Developing Neuroprotective Phytomedicines

20

Lenin Kumar Bompalli

Abstract

The escalating incidence of neurodegenerative diseases, including Alzheimer's, Parkinson's, and Huntington's, poses a significant global health challenge, prompting an urgent need for effective therapeutic interventions. Neuroprotective phytomedicines, derived from various medicinal plants, represent a promising avenue for the development of novel treatments. These phytomedicines are rich in bioactive compounds that may confer neuroprotection through multiple mechanisms, including the modulation of oxidative stress, anti-inflammatory pathways, and the inhibition of neuronal apoptosis. Recent studies have identified several plant-derived compounds, such as curcumin, resveratrol, and ginkgo biloba extract, as having potential neuroprotective properties. For instance, curcumin has demonstrated the ability to reduce amyloid-beta aggregation in Alzheimer's models, while resveratrol has shown promise in enhancing mitochondrial function and neurogenesis. Furthermore, *Ginkgo biloba* has been reported to improve cognitive function by increasing cerebral blood flow and exhibiting antioxidant effects. Despite the therapeutic potential, the development of neuroprotective phytomedicines faces several challenges. Variability in the phytochemical composition of plant extracts can lead to inconsistent therapeutic effects, complicating standardization efforts. Moreover, several bioactive compounds exhibit low bioavailability, which limits their efficacy in clinical settings. Regulatory hurdles and the necessity for rigorous clinical evidence also present significant barriers to the commercialization of these phytomedicines. To overcome these challenges, innovative strategies such as advanced extraction techniques, nanotechnology-based formulations, and targeted delivery systems are essential. Additionally, comprehensive clinical trials that investigate the pharma-

L. K. Bompalli (✉)
Dr. B. R. Ambedkar University, Etcherla, Srikakulam, Andhra Pradesh, India

K. Kuca et al. (eds.), *Medicinal Plants and Their Bioactives in Human Diseases*,
https://doi.org/10.1007/978-3-032-01356-9_20

cokinetics, safety, and efficacy of these compounds are crucial for substantiating their therapeutic claims. Addressing these multifaceted challenges will be pivotal in unlocking the full therapeutic potential of neuroprotective phytomedicines, paving the way for effective treatments for neurodegenerative diseases.

Keywords

Neurodegenerative diseases · Alzheimer's, Parkinson's · *Ginkgo biloba* · Bioactive compounds · Phytomedicines · Therapeutic effects

1 Introduction

Central nervous system (CNS) disorders affect both the brain and spinal cord, encompassing conditions such as neurodegenerative diseases, tumors, infections, and cerebrovascular illnesses. Due to their progressive nature and limited treatment options, these disorders represent a significant global health challenge. Their underlying pathology involves a complex interaction of several factors and structure, including the blood–brain barrier (BBB). Although the BBB maintains the stability of the CNS, it also blocks more than 98% of potential therapeutic compounds, creating a formidable obstacle to effective treatment. Developing effective therapies for CNS disorders requires overcoming two major challenges: crossing BBB and addressing the underlying complex pathological mechanisms (Liu et al. 2025). Neurons are essential for the proper functioning of the human brain, as they play a vital role in communication. While most neurons are generated in the brain, they are distributed throughout the body. The increasing incidence of neurodegenerative disorders, including Alzheimer's disease, Parkinson's disease, Huntington's disease, and multiple sclerosis, constitutes a major public health problem worldwide. There is growing interest in neuroprotective phytomedicines derived from medicinal plants, which have been used for their therapeutic benefits in various cultures for centuries. Cognitive dysfunction is a significant health challenge in the twenty-first century, with many neuropsychiatric and neurodegenerative disorders, including schizophrenia, depression, Alzheimer's disease, cerebrovascular disorders, seizure disorders, traumatic brain injury, and Parkinson's disease, often associated with severe cognitive dysfunction. Over time, various neurotransmitters and signaling molecules have been recognized as potential therapeutic targets. Traditional and novel compounds have been explored in relation to these targets. Phytochemicals derived from medicinal plants are essential for maintaining the chemical balance of the brain by modulating the activity of receptors associated with key inhibitory neurotransmitters. Resveratrol, a natural polyphenol found in wine, red grapes, berries, chocolate, and peanuts, is attracting increasing interest for its potential neuroprotective effects (Dhingra et al. 2021). Polyphenols are abundant in fruits, vegetables, tea, and wine. They possess potent antioxidant and anti-inflammatory properties, making them effective in reducing oxidative stress and neuroinflammation, which are key contributors to neurodegeneration (Dos Santos et al. 2024). A variety of chronic

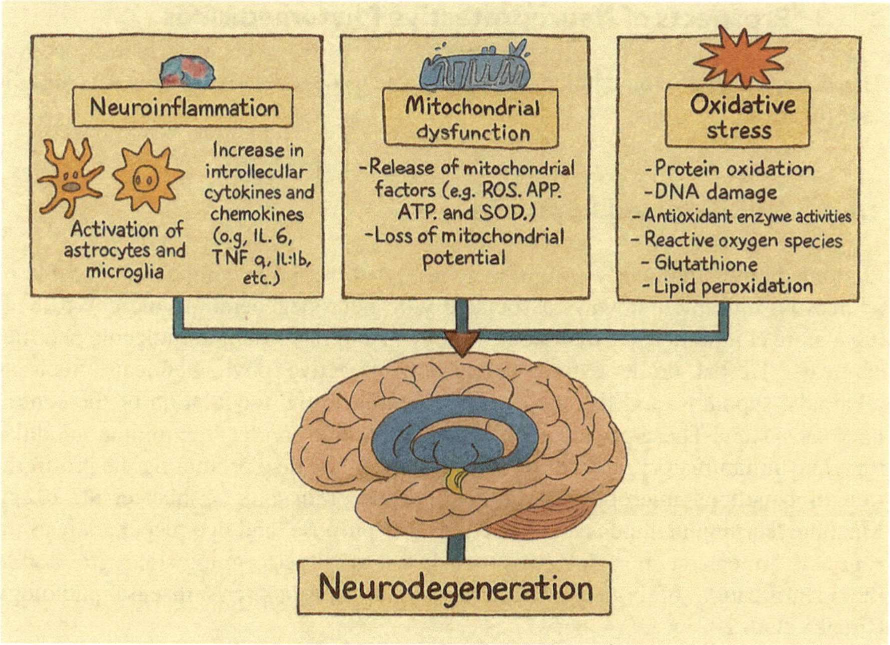

Fig. 20.1 Different biological processes for the development of pathogenesis of neurodegenerative diseases including oxidative stress, mitochondrial dysfunction, and neuroinflammation

progressive central nervous system disorders triggered by deterioration. Furthermore, oxidative stress, neuroinflammation, dysfunction in mitochondria, dysfunctional protein misfolding and agglomeration, and other biological processes have been linked to neurodegeneration (Fig. 20.1). These biological pathways have been implicated in the development of NDs and their pathogenesis (Rahman et al. 2021).

Alzheimer's disease primarily causes memory loss, cognitive decline, and behavioral changes, making it the most common form of dementia. Parkinson's disease leads to motor issues like tremors, stiffness, and slow movement. Huntington's disease is inherited and causes involuntary movements, psychiatric symptoms, and cognitive decline (Houldsworth 2024). Neurological disorders (NDs) are the second leading cause of death and the primary cause of disability worldwide. The global population is aging, leading to a growing social and health burden. As neurological diseases become more prevalent, their progressive nature makes them incurable. Current treatments only help alleviate symptoms or slow disease progression. NDs are caused by a wide variety of mechanisms, including programmed cell death, aggregated proteins, oxidative stress, excitotoxicity, and aging (Singh et al. 2025). There are 850,000 people in the United Kingdom living with dementia, at a cost of over £26 billion per year. This is predicted to double by 2040 (Franca et al. 2023). This chapter reviews the potential of these phytomedicines for neuroprotection, highlighting both their prospects and the challenges encountered in their development.

2 Prospects of Neuroprotective Phytomedicines

The potential of phytomedicines in treating neurodegenerative diseases is significant for several reasons:

2.1 Multitargeted Approach

Phytomedicines frequently comprise a variety of bioactive compounds capable of addressing multiple pathways associated with neurodegeneration, thereby providing a more comprehensive treatment strategy in contrast to single-molecule pharmaceuticals. Herbal crude extracts and their bioactive phytochemicals, such as alkaloids, saponins, and flavonoids, are essential in the modulation of the central nervous system. These phytoconstituents exhibit neuroprotective, immunomodulatory, anti-inflammatory, and antioxidant effects, while also promoting the proliferation of beneficial microbes that affect the gut–brain axis (Ashish et al. 2021). Machine learning methods can be used for this purpose, and in a recent analysis on routinely collected cohort data, this was shown to outperform logistic regression in the identification of both Alzheimer's and non-Alzheimer's disease pathology (Franca et al. 2023).

A multitargeted strategy utilizing phytomedicines for neurodegenerative conditions involves the application of plant-derived compounds that concurrently engage multiple biological pathways involved in the development of these diseases. Neurodegenerative disorders, including Alzheimer's disease (AD), Parkinson's disease (PD), Huntington's disease (HD), and amyotrophic lateral sclerosis (ALS), are intricate in nature, with their progression influenced by a range of factors such as oxidative stress, inflammation, protein misfolding, and neuronal degeneration. Phytomedicines, which are plant-based remedies or compounds, have been explored for their ability to intervene in multiple of these mechanisms simultaneously, making them promising candidates for treating or managing neurodegenerative diseases.

2.2 Common Side Effects of Phytomedicines for Neurodegenerative Diseases

Numerous phytochemicals demonstrate reduced toxicity levels in comparison to traditional pharmaceuticals, rendering them appealing options for the long-term treatment of chronic neurodegenerative disorders (Alam and Mubeen 2023). *Rosmarinus officinalis*, commonly known as rosemary, has a long history of use in traditional medicine. Research has indicated that *R. officinalis* and its secondary metabolites exhibit neuroprotective effects through various molecular mechanisms, including cholinesterase inhibition, modulation of the dopaminergic system, and regulation of oxidative and inflammatory pathways (Oresanya and Orhan 2024). Although phytomedicines are sourced from natural plants, they are not entirely free from potential adverse effects. While they may offer therapeutic advantages for neurodegenerative conditions, their safety profiles can differ based on various factors, including dosage, duration of administration, individual patient circumstances, and the presence of other medical conditions or medications.

3 Challenges in Development

As the global use of herbal medicinal products continues to grow and many more new products are introduced into the market, public health issues and concerns surrounding their safety are also increasingly recognized. Although some herbal medicines have promising potential and are widely used, many of them remain untested, and their use is also not monitored. This makes knowledge of their potential adverse effects limited and identification of the safest and most effective therapies as well as the promotion of their rational use, more difficult. It is important to reiterate the staggering rate at which interest and use of herbal medicines are expanding. Over the past decade, the use of herbal medicines represents approximately 40% of all healthcare services delivered in China, while the percentage of the population that has used herbal medicines at least once in Australia, Canada, USA, Belgium, and France is estimated at 48%, 70%, 42%, 38%, and 75%, respectively. In spite of the positive perception of patients on the use of herbal medicines and alleged satisfaction with therapeutic outcomes, coupled with their disappointment with conventional allopathic or orthodox medicines in terms of effectiveness and/or safety, the problem of the safety of herbal remedies continues to remain a major issue of concern.

4 Bioavailability Issues

Bioavailability is defined as the fraction of an active compound that successfully enters the bloodstream and arrives at its intended target to produce a therapeutic effect. Consequently, even if phytomedicine is rich in an active compound, its effectiveness may be constrained by poor bioavailability. The bioactive composition of phytomedicines can vary significantly based on factors such as plant species, growing conditions, and extraction methods. This variability can affect the consistency and efficacy of the final product (Bhowmik et al. 2023). Numerous bioactive compounds exhibit low bioavailability because of inadequate solubility and swift metabolic processes. This challenge significantly impairs their therapeutic potential, thereby highlighting the need for innovative delivery systems (Khan and Zafar 2023).

5 Regulatory Hurdles

The regulatory framework governing phytomedicines is often intricate, necessitating comprehensive clinical trials to demonstrate both safety and efficacy. Nootropics, or brain supplements, are specifically designed to enhance memory and cognitive function. This category of nutraceuticals has cultivated a substantial global market, estimated at approximately $200 billion. In contrast to conventional pharmaceuticals that assert specific therapeutic benefits, dietary supplements do not carry explicit claims and are not subject to FDA approval. While these nutraceuticals typically contain one or two key ingredients, such as vitamins and fatty acids, there is a scarcity of clinical data regarding their overall physiological effects and potential interactions with other medications. It has been observed that brain supplements may provide advantages for individuals with healthy conditions rather than for those who are healthy. Consequently, greater scrutiny regarding their application is warranted. Dietary herbs and herbal supplements have garnered significant interest as neuro-nutraceuticals due to their balanced composition of various compounds beneficial for brain health, including flavonoids, saponins, omega-3 fatty acids, vitamin B6, vitamin C, and essential micronutrients like magnesium, iron, and zinc. Furthermore, herbs are regarded as safe and offer a holistic approach to enhancing brain health, in contrast to single-target pharmacological drugs that frequently result in adverse side effects (Ashish et al. 2021).

6 Limited Clinical Evidence

While preclinical investigations frequently highlight the potential benefits of phytomedicines, there remains a pressing need for more comprehensive clinical trials to substantiate these results and endorse their therapeutic application in humans. Despite a long-standing tradition of use and encouraging preclinical findings, phytomedicines encounter considerable obstacles due to insufficient clinical evidence, particularly concerning neurodegenerative disorders such as Alzheimer's, Parkinson's, and Huntington's disease. Although laboratory and animal research often yields optimistic outcomes, there is a notable absence of large-scale, rigorously designed human clinical trials that conclusively establish the therapeutic effectiveness of numerous plant-based treatments.

7 Formulation Techniques and Clinical Trials

Innovative extraction methods, such as supercritical fluid extraction and ultrasound-assisted extraction, can enhance the yield and purity of bioactive compounds. Additionally, formulation strategies such as nanoparticle delivery systems may improve bioavailability (Bhowmik et al. 2023). Neurodegenerative disorders affect millions of people worldwide. Although age is the single most contributing risk factor to the development of all neurodegenerative disorders, recent findings reveal that a combination of an individual's genetic makeup and environmental factors can equally contribute to increasing the risk for neurodegenerative disorders (Richard et al. 2022). Conducting well-designed clinical trials that focus on specific phytomedicines can provide the necessary evidence to support their use in clinical settings. Collaboration between researchers, healthcare providers, and regulatory agencies will be essential (Shah et al. 2023).

8 Standardization and Quality Control

The development of a good agricultural and collection Practices (GACP) subcommittee to facilitate the availability of high-quality herbal medicines on the market by providing training and support to small producers and farmers was one of the proposals. Incentives for cultivators of herbal phytoconstituents might be offered to encourage GACP implementation. These services include offering technical and logistical assistance in identifying optimal agricultural production areas, delivering seeds and seedlings, selecting fertilizers and pesticides, and providing advice on harvesting and primary processing gear. The establishment of standardized protocols for the preparation and quality control of phytomedicines is essential for ensuring uniformity in their composition and effectiveness (Khan and Zafar 2023). The standardization of herbal formulations plays a crucial role in assessing the quality of these drugs based on their chemical composition, as well as physical, chemical, phytochemical, and in vitro parameters. Conducting quality assessments of herbal formulations is vital for their acceptance within modern medical systems. The World Health Organization (WHO) has highlighted the necessity of regulating medicinal plant products through contemporary methodologies in several resolutions. The WHO advocates traditional and herbal remedies in natural health care programs, citing their affordability, safety, and reliability. Furthermore, the WHO underscores the significance of both qualitative and quantitative methods for sample characterization, biomarker quantification, and fingerprint profiling (Prajapati et al. 2023). The standardization of herbal medicines involves defining a set of standards or inherent characteristics, consistent parameters, and specific qualitative and quantitative values that ensure quality, efficacy, safety, and reproducibility.

9 Conclusion

Neuroprotective phytomedicines present significant potential in combating neurodegenerative diseases, attributed to their diverse mechanisms of action and favorable safety profiles. Nonetheless, their development faces considerable obstacles, including variability in composition, issues related to bioavailability, and regulatory challenges. By tackling these issues through innovative research and collaborative efforts, the therapeutic capabilities of phytomedicines can be fully realized, offering renewed hope for individuals affected by neurodegenerative disorders. The multitargeted strategy of phytomedicines in treating these diseases represents a promising pathway for creating safer and more effective therapies. By influencing various disease pathways—such as oxidative stress, inflammation, protein aggregation, and neurotransmitter dysfunction—phytochemicals may help alleviate symptoms and slow the progression of these conditions. However, additional clinical trials are essential to validate their efficacy and safety in human subjects, as most of the existing research has been conducted in vitro or using animal models. While phytomedicines exhibit potential as multitargeted therapeutic agents for neurodegenerative diseases, their side effects must be carefully considered. These side effects can range from mild gastrointestinal discomfort to more serious concerns, such as liver toxicity and interactions with other medications. It is imperative for patients to seek guidance from healthcare professionals prior to using these remedies, particularly if they are taking other medications or have existing health conditions. Phytomedicines have a long-standing history in the treatment of neurodegenerative diseases, with varying levels of cultural acceptance across different regions. However, challenges such as standardization, scientific validation, regulatory oversight, and the risk of side effects impede their wider adoption and development. Addressing these challenges through thorough scientific inquiry, quality assurance, and increased public awareness will be vital in unlocking the full therapeutic potential of phytomedicines for neurodegenerative conditions. Variability in composition and bioavailability issues are significant hurdles in the development and use of phytomedicines for neurodegenerative diseases. These challenges can impact the consistency, safety, and effectiveness of plant-based treatments. However, ongoing research and advancements in formulation techniques, along with a focus on standardization and quality control, hold promise in improving the reliability and therapeutic potential of phytomedicines. Addressing these issues will be essential for their wider acceptance and integration into mainstream medicine. Phytomedicines represent a valuable and underutilized resource in the treatment of neurodegenerative diseases, with their ability to target multiple disease pathways simultaneously. While there is significant potential, it is crucial to address the challenges related to standardization, bioavailability, clinical evidence, and safety. By leveraging modern technologies, improving clinical research standards, and integrating these treatments within a broader, personalized healthcare framework, phytomedicines could play an increasingly important role in managing neurodegenerative diseases. However, their widespread use will depend on the development of a solid evidence base and improved regulatory oversight to ensure their efficacy and safety.

References

Alam MS, Mubeen S (2023) Medicinal plants and their bioactive compounds in the management of Alzheimer's disease: a review. Neurochem Int 156:105402

Ashish K, Arpita K, Sukant G, Sunil CK, Renu W (2021) Experimental evidence and mechanism of action of some popular neuro-nutraceutical herbs. Neurochem Int 149:105124

Bhowmik D, Kumar KS, Choudhury R (2023) Medicinal plants: an overview of their potential in the management of chronic diseases. Phytother Res 37(10):4567–4582

Dhingra AK, Rathi V, Chopra B (2021) Resveratrol. In: Naturally occurring chemicals against Alzheimer's disease. Academic Press, Cambridge, MA, pp 33–47

Dos Santos MG, Da Luz DD, De Miranda FD, De Aguiar RF, Siebel AM, Arbo BD et al (2024) Resveratrol and neuroinflammation: total-scale analysis of the scientific literature. Nutraceuticals 4(2):165–180

Franca D, John G, Zoe K, Ivan K, Paul MM, Neil PO, Laura MP, Viola P, James BR, Stephen WS, Henrik Z (2023) Neurodegenerative disease of the Brain: a Survey of interdisplinary approaches. J R Soc Interface 20:20220406

Houldsworth A (2024) Role of oxidative stress in neurodegenerative disorders: a review of reactive oxygen species and prevention by antioxidants. Brain Commun 6(1):fcad356

Khan MI, Zafar A (2023) Phytochemicals from medicinal plants: A potential approach for the treatment of neurodegenerative diseases. Front Pharmacol 14:1068754

Liu J, Wang T, Dong J, Lu Y (2025) The blood–brain barriers: novel nanocarriers for central nervous system diseases. J Nanobiotechnol 23. https://doi.org/10.1186/s12951-025-03247-8

Oresanya IO, Orhan IE (2024) Deciphering neuroprotective effect of rosmarinus officinalis L. (Syn. Salvia rosmarinus Spenn.) through preclinical and clinical studies. Curr Drug Targets 25:330–352

Prajapati M, Girase PS, Rajput GP (2023) Quality control and standardization of herbal drugs. Int J Res Publ Rev 4(8):2800–2803

Rahman MH, Bajgai J, Fadriquela A, Sharma S, Trinh TT, Akter R, Jeong YJ, Goh SH, Kim CS, Lee KJ (2021) Therapeutic potential of natural products in treating neurodegenerative disorders and their future prospects and challenges. Molecules 26:5327

Richard NLL, Chaulagain B, Trivedi R, Gothwal A, Layek B, Singh J (2022) A review of the common neurodegenerative disorders. Current therapeutic approaches and the potential role of nanotherapeutics. Inter J Mol Sci 23(3):1851

Shah RA, Khan MA, Ahmad F (2023) The role of herbal medicine in neurodegenerative disorders: A review. J Ethnopharmacol 295:115432

Singh AA, Katiyar S, Song M (2025) Phytochemicals Targeting BDNF Signaling for Treating Neurological Disorders. Brain Sci 15:252

GPSR Compliance

The European Union's (EU) General Product Safety Regulation (GPSR) is a set of rules that requires consumer products to be safe and our obligations to ensure this.

If you have any concerns about our products, you can contact us on ProductSafety@springernature.com

In case Publisher is established outside the EU, the EU authorized representative is:

Springer Nature Customer Service Center GmbH
Europaplatz 3
69115 Heidelberg, Germany

Batch number: 09019519

Printed by Printforce, the Netherlands